T0295909

Project Management Next Generation: The Pillars for Organizational Excellence

Harold Kerzner, PhD, MS, MBA
Al Zeitoun, PhD, PgMP
Ricardo Viana Vargas, PhD, PMP

This book is printed on acid-free paper.

Published by John Wiley & Sons, Inc., Hoboken, New Jersey
Published simultaneously in Canada

For general information about our other products and services, please contact our Customer Care Department within the United States at (800) 762-2974, outside the United States at (317) 572-3993 or fax (317) 572-4002.

Wiley publishes in a variety of print and electronic formats and by print-on-demand. Some material included with standard print versions of this book may not be included in e-books or in print-on-demand. If this book refers to media such as a CD or DVD that is not included in the version you purchased, you may download this material at http://booksupport.wiley.com. For more information about Wiley products, visit www.wiley.com.

Library of Congress Cataloging-in-Publication Data

Names: Kerzner, Harold, author. | Zeitoun, Al, author. | Vargas, Ricardo Viana, author.
Title: Project management next generation : the pillars for organizational excellence / Harold Kerzner, PhD, MS, MBA, Al Zeitoun, Phd, PgMP, Ricardo Viana Vargas, PMP.
Description: First edition. | Hoboken, New Jersey : Wiley, 2022. | Includes index.
Identifiers: LCCN 2022017519 (print) | LCCN 2022017520 (ebook) | ISBN 9781119832270 (cloth ; alk. paper) | ISBN 9781119832294 (adobe pdf) | ISBN 9781119832287 (epub)
Subjects: LCSH: Project management. | Organizational change.
Classification: LCC HD69.P75 K4937 2022 (print) | LCC HD69.P75 (ebook) | DDC 658.4/04—dc23/eng/20220411
LC record available at https://lccn.loc.gov/2022017519
LC ebook record available at https://lccn.loc.gov/2022017520

Cover Design: Wiley
Cover Image: © Trademarks are the property of their respective owners

SKY10034487_051822

To
our wives – Jo Ellyn, Nicola, and Zelia –
and our kids – Jason B, Lindsey, Andrea, Jacalyn, Jason K,
Adam, Zeyad, Sarah, Ana, and Gabi –
for being our continued source of inspiration

Contents

Preface

Executives in both the public and private sectors worldwide are beginning to recognize the full benefits that effective project management can bring to their organizations and are willing to make the necessary changes – especially changes in culture. As such, the landscape for project management is changing, and 10 pillars of project management have emerged. These critical 10 pillars are expected to drive project management for the next decade and to significantly improve the performance of organizations that take the time to understand and implement them.

This book builds on a strong and balanced combination of research and multiple global practical experiences across programs, projects, and transformation initiatives. A critical element of the book are the vast contributions of global companies and organizations throughout the book's chapters. These shed light on the strategic changes and project management capabilities improvements that future organizations and academia will focus on.

This book will enable executives, professionals, and students of project management to be better prepared for the jobs of the future as the project economy takes more of a center stage for organizations aspiring to sustain their excellence.

The book is broken down as follows:

- **Chapter 1:** Discusses how project management has matured to a true strategic delivery capability that sets the foundation for organizational excellence and high performance.
- **Chapter 2:** Discusses why the humanitarian and social missions provide an excellent example for understanding the true impact project management builds worldwide.
- **Chapter 3:** Discusses the importance of creating an innovative culture and introduces a model to help integrate the many ingredients necessary to drive transformative change.
- **Chapter 4:** Discusses how digitization is a critical component of delivering projects outcomes and showcases a framework for unifying the digital disruptions with the human element of change.
- **Chapter 5:** Discusses how project management skills have been evolving and what the future focus of that role would be requiring of project manages.
- **Chapter 6:** Discusses the new forms of leadership necessary for the next decade of organizational excellence and addresses the tight linkages between culture, leadership, and the associated project management implications.
- **Chapter 7:** Discusses how the shift to a project way of working is going to dominate the future of work and highlights critical examples of the cultural shifts required for the future mega and complex global projects.
- **Chapter 8:** Discusses the changing dynamics of adaptive frameworks and project life cycles and the need for creating a fine balance between alignment and autonomy in organizations seeking sustainable excellence.

- **Chapter 9:** Discusses how project management offices will continue to evolve and become a must-have strategic governing body in an agile and volatile world.
- **Chapter 10:** Discusses how project management will benefit from the shifting nature of metrics into value and strategic based ones and illustrates how to consider establishing a sustainable metrics management program.

We are indebted to all the professionals and companies that were forthcoming in sharing key information about their Excellence in Action practices and their views of the future of project management. The contributing companies list includes:

- Airbus
- Ambev
- ASGC
- Astra Zeneca
- Bosch
- Cisco
- Dubai Customs
- Dundas Data Visualization
- Eli Lilly
- Fabio Doehler
- Farm Credit
- GEA
- GM
- Hospital Albert Einstein
- IBM
- IdeaScale
- IIL
- Medtronic
- Merck KGaA
- Philips
- PMO Global Alliance
- Progressive Insurance
- Project Management United
- Repsol
- ServiceNow
- Siemens
- SITA
- Solvo360
- Sunrise upc
- Wuttke & Team, a Gita brand

With our sincere gratitude to our followers and readers,

Harold Kerzner,
Al Zeitoun,
Ricardo Viana Vargas

1

Pillar #1: Strategic Delivery Capability

1.0 Setting the Stage

The profession of project management has been changing and fast. This is a different world than the one that has driven the early days of the birth of this discipline. When the authors first came up with the idea behind this book, their driver was to think about and illustrate the next iteration of this profession, which could take us through the remainder of the decade. The more closely we reviewed this ambition and worked with many of the companies around the globe to pulse their views of the future of work, the more we realized that project management is truly at an inflection point and that it is finally ready to be tackled from a wider ecosystem that encompasses the culture, business value, and the sharper focus on co-creating solutions with customers and other stakeholders. The 10 pillars we selected as the foundation for this book cover this wider ecosystem. They allow us to look at where projects are strategically poised to create a distinct way of working into the future.

As we start tackling the 10 pillars, we are reminded that predicting the next generation of project management is a complex topic. Strategy is hard and to find the patterns that connect the pillars to how organizations excel in working in the future is not an exact science anymore. The chaos that the world encounters that combines a multitude of attributes of uncertainty is immense. It is finally resting upon the project managers of the future to take on the leading role of change making that we had been predicting over the past years. The system-wide mindset that these future leaders will bring has reached the right moment of being highly valued. It is our hope that this book and its pillars serve as a critical guidepost that the organizations of the future would follow in driving their focus, investing in the right skills, in recreating how their work is done, and deciding on what data truly matters.

With the increasing vast demands for complex infrastructure programs, green energy, and the number of organizations that are committing to ambitious goals on their journey to climate neutrality, and to possibly achieving net zero emissions by 2050 while they hit some intermediate targets by 2030, the role of strategic projects that are effectively delivered will only multiply. Based on industry trends, the authors' experiences, and the multiple studies that still indicate the large gap between envisioned goals and the executed outcomes, this book is focusing on the excellence practices that will enhance the opportunities for this world to see what project management principles could help us achieve.

Before getting into some of the background and barriers behind the first pillar that sets the tone for the set of critical shifts we see into the next generation of project management, namely project management being a true strategic competency, let us start with one of the *Excellence in Action* sections that support the move in that direction. Throughout the book, we will demonstrate the next generation trending we see around the selected guiding 10 pillars with examples of world class organizations that have managed to show great evidence of excellence in their operations, nicely coupled with the maturing of the principles of project management.

In this first one, a dedicated professional to driving and changing the project management profession has gone through a clear maturity path in her leading strategic work with the Mayo Clinic. As a certified Portfolio Management Professional by the Project Management Institute (PMI), she has tackled the critical transformation objectives required to get project management to a strategic competency via orchestrating a set of interconnected change initiatives as part of her portfolio responsibility.

Excellence in Action: Strategy Management Services, Mayo Clinic[1]

One thing that is certain in all aspects of our lives is that "things will change." Regardless of how much planning, preparation, and project management we do; change will continue to occur in both expected and often very unexpected ways. While this is evident in our professional and personal lives, it's also very relevant to the project management profession as well.

Past, Present, and Future

Many of us who have been involved in project management for many years, or many decades as in my case, may remember the days when project management wasn't really "a thing" and it definitely was not a career to be aspired to. People typically got assigned to "get something done" and would use whatever means they could to get people together, figure it out, and use "project heroics" to accomplish it. These early days of basic project management techniques had several successes, but more than likely ended in failures with long project timelines, the need for additional resources, and a variety of unanticipated surprises.

Fortunately, some really smart people figured that there had to be better ways of doing projects and created a series of methodologies, standards, and tools and templates. This began the era of formalized project management when organizations, such as the Project Management Institute (PMI), became known for sharing their knowledge, providing education, certifying practitioners, and establishing the project management profession.

This was particularly important as projects were becoming larger, more complex, and more costly. Project management gained worldwide recognition and adoption as the rate of project success increased and organizations realized the necessity and value of project management. During this time, hundreds of thousands of people became certified Project Management Offices (PMOs) were formed, and the term *project management* became part of standard business terminology.

---

1 Material in this section was provided by Terri Knudson, *Senior Director, Strategy Management Services, Mayo Clinic.* She has a diverse background having served in leadership roles in Finance, Operations, Audit, Strategic Planning, Business Management, and Project Portfolio Management for one of the largest non-profits, integrated healthcare organization in the world. *The information and opinions shared in this article are those of the author based on her expertise and should not be attributed to the Mayo Clinic.*

As the years passed by and things continue to change there is a need for project management to adapt again as we enter "The Next Generation." This iteration has an increase in technology, business agility, and accelerated deliverables that is requiring project management approaches to pivot and adapt to meet the new needs and challenges for our organizations more quickly than ever before. As Dr. Kerzner has noted, "More project managers are expected to manage strategic projects rather than just traditional or operational projects." With this occurring over the past few years, it was time to introduce this next generation within my organization and move forward in establishing "Strategy Management Services" (SMS) (Figure 1-1).

After establishing and leading the Enterprise Portfolio Management (EPMO) at my organization for 10 years, it was time to move onto the next level, and fortunately I had the opportunity to pursue my SMS vision by taking a new position within a large, innovative department in my organization. In this role, I've been empowered to create the next generation of project management with an established PMO team that was needing a new direction. While it took a few months to formulate, reorganize, and introduce the new approaches; after only two years we are now a well-established high-performing team, have strong business partnerships, and are fully recognized as SMS.

Spectrum of Strategy Management Services

The vision of SMS began with the desire to move beyond the historical perspectives related to project management and a PMO by fully recognizing the critical importance of focusing on Strategic Execution and Value Delivery. While many of the services provided by SMS are based on

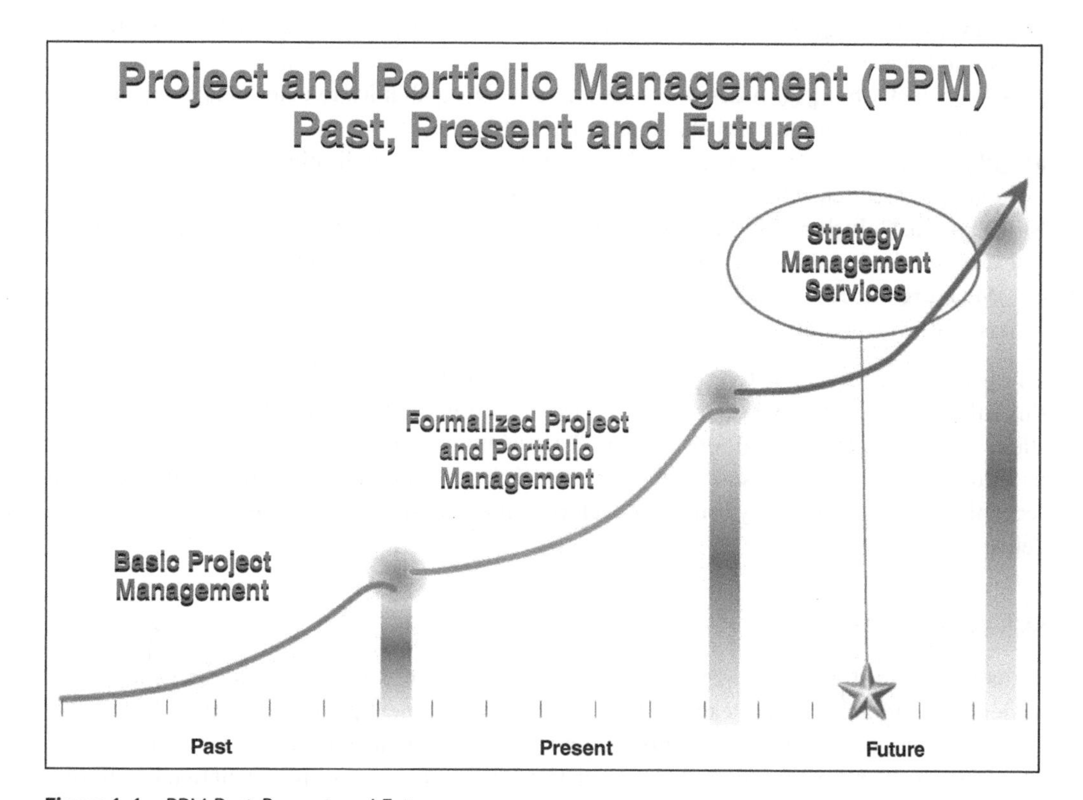

Figure 1-1 PPM Past, Present, and Future

Figure 1-2 Strategy Management Services

the foundation provided by project management, business analysis, change management, and other standards, SMS recognizes that the ultimate goal is to deliver the value directed by the strategic vision of the organization.

Although there are dozens of books on strategic planning and execution, very few actually bring all the pieces together. Most of us fully realize that all the best strategic planning in the world gets you nowhere if you can't execute it successfully. The SMS "full spectrum of strategic services" is focused on using a logical approach, recognizing the interrelationships across the roles, and beginning with strategy and ending with successful results. This model allows all project stakeholders to recognize their specific roles and clearly understand how all phases need to work together to get the desired outcomes.

The stages of the SMS model are defined as follows (Figure 1-2):

01 – Develop Strategy. Partner with business areas to develop approaches to deliver the strategic goals and objectives.

02 – Architect Changes. Assess business and system capabilities to determine competencies, weaknesses, and gaps to highlight changes needed.

03 – Plan Initiatives. Create plans to address business needs by indicating project goals, leaders, resources, costs, schedule, metrics, and other attributes.

04 – Execute Solutions. Successfully deliver business solutions through a multi-disciplinary team comprised of skilled resources committed to achieving business needs.

05 – Measure Success. Provide tangible and visible measures of progress toward meeting business goals and objectives.

Project management has typically focused heavily on the model's stages 3 and 4. Providing this full spectrum of services emphasizes the need for the SMS team to be engaged in all stages – both "upstream" and "downstream." Developing abilities in these other stages occurs naturally over time, as SMS staff become viewed as business partners in strategic delivery and recognized for the skills they bring to development of the strategy, architecting changes, and delivering results.

SMS Evolution

Transitioning to strategic management services requires executive-level support, experienced leadership, and an engaged team willing to adapt to this next generation. Just like the last generation and our transition into project and portfolio management, SMS is a journey with a various steps and phases that evolve over time. To be successful, it is beneficial to have a

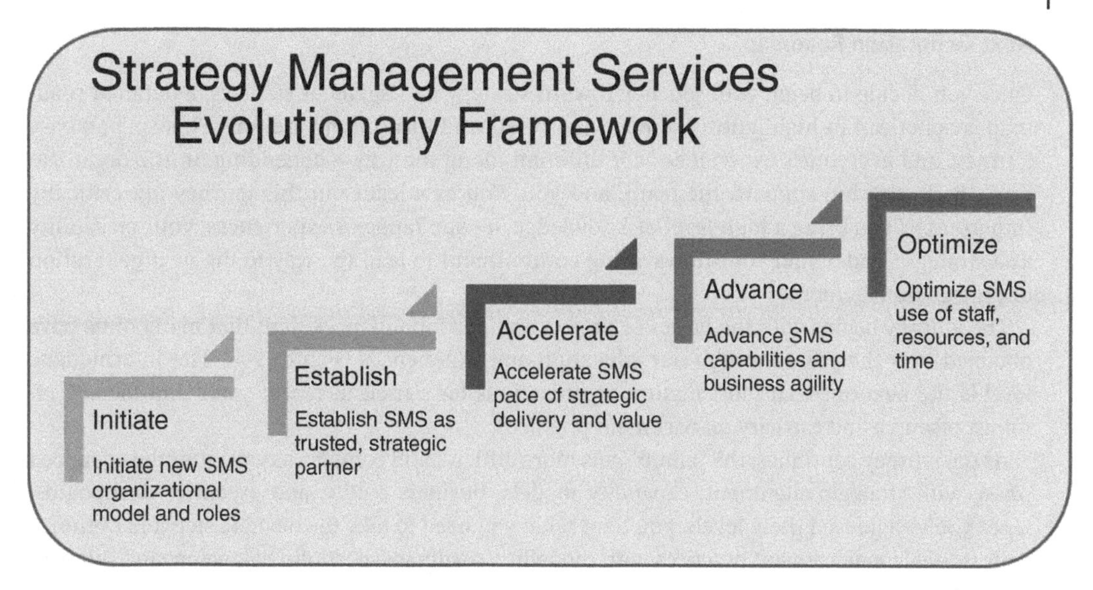

Figure 1-3 Services Evolutionary Framework

plan that outlines an intentional path and the development stages along the way (Figure 1-3). The following are the five stages created by our SMS leadership team based on a typical maturity model:

1) **Initiate.** The transition begins with the "Initiate" phase by introducing the new strategy management concepts to leaders, the team, and stakeholders. This includes educating the staff on new approaches, along with infusing some of the basics of PPM to emphasize its ongoing importance. Another major component of this phase is the need for organizational alignment to the strategy to ensure clear business relationships with each of the strategic pillars.
2) **Establish.** Once organizationally aligned, the next phase is for the leaders and members of each team to create strong and trusting relationships with their strategic business partners. This provides full engagement and involvement with the business leaders and teams as an active and accountable member that is fully available to assist, support, and respond to business needs.
3) **Accelerate.** With a solid business partnership and a high-performing team in place, you are ready for the level stage by building the ability of your team to respond and adapt quickly to changing business needs. This includes being involved in strategic ideation discussions, providing business analysis, assessing options for decision-making, and planning and execution.
4) **Advance.** Continual improvement and advancement is an integral part of any high-performing team and is the next phase of evolution as the SMS staff continues to develop new skills, build business capabilities, utilize new tools and technologies, and create innovative solutions across the SMS spectrum.
5) **Optimize.** The pinnacle of success is reaching the level of optimization where the SMS team are fully engaged in all aspects of strategic management, are valued members of the business team, and are consistently successful in delivering on strategic and business needs.

Next-Generation Roadmap

Once you decide to begin your journey toward strategy management services, a detailed roadmap is beneficial in highlighting some of the twists and turns along the way. This isn't an easy journey, and everyone's experiences are different along the way – depending on the organization, the leadership support, the team, and you. You as a leader in this journey are critically important as you bring a high level of knowledge, a wide range of experiences, your credibility as a strategic leader, and your unwavering commitment to lead the way to the next generation of project management.

The journey begins with the basics of project and portfolio management that many of us have obtained over the years through our education and experiences (Figure 1-4). The intermediate level is the next one that most mature organizations have used in recent years and focuses on robust resource and capacity management practices.

As the journey continues, the "climb" gets more difficult and complex as you enter the advanced phase with strategic alignment, capability models, business agility, and executive dashboards. Once you've achieved these levels, you have what you need to take the next big step and optimize your strategic management practices with modeling, roadmaps, portfolio balancing, and full benefit realization.

The ultimate "prize" for achieving this final level is one that is shared by your organization, by your business stakeholders, and by your SMS team. This is the strategic value from the results you and your team have delivered, the lessons learned and relationships built along the way, and the satisfaction that you have accomplished the full spectrum of strategic services. Best wishes on your journey!

Let us now start introducing the journey that this project management revolutionary move toward a strategic competency has taken and then we will follow with a few more diverse and concrete examples of excellence stories that contribute to the practices across the 10 pillars.

1.1 Background

Project managers are now managing strategic as well as operational or traditional projects. Executive management appears to recognize and appreciate the contributions that the project managers are making to the growth of the business. This has increased the recognition that effective project management practices can bring to a company.

The benefits of effective project management have been known for some time. Some of the benefits include:

- Performing more work in less time and with fewer resources
- An increase in profitability and customer satisfaction
- An increase in organizational effectiveness and efficiency
- An increase in competitiveness
- Improvements in quality
- Better control of scope changes and scope creep
- Application to all business units and all types of projects
- Better approach to problem-solving and decision-making
- Decision-making in the best interest of the company rather than its business units
- Delivering solutions to customers rather than just products and deliverables

Strategic Execution and Value Delivery
Degree of Difficulty
High
Low
Optimized – Strategic Management Services
Strategic Planning and Modeling
Strategy Roadmaps
Portfolio Prioritization and Balancing
Benefit Realization
Advanced – Alignment, Architecture, and Agility
Strategic-Alignment
Business Architecture and Capability Models
Agility and Adaptability
Metric Management and Dashboard
Intermediate – Resource and Capacity Management
Time Tracking
Allocation/Capacity Management
Financial Management
Demand Management
Basic – Project and Portfolio Management
Project Portfolio Management
Business Requirements and Scope Management
Schedule Management
Issue/Risk/Change Management

Figure 1-4 Next-Gen Roadmap

As more organizations recognize the benefits, project managers are now seen as managing part of a business rather than just projects. The challenge facing companies is in determining the best way that the company can scale and expand the required diverse project management skills.

1.2 Line-of-Sight

Trust in asking project managers to manage strategic projects has resulted in the establishment of a line-of-sight from project teams to senior management such that the teams are kept informed about strategic business objectives to ensure that strategic projects are aligned correctly. Line-of-sight not only creates the correct decision-making mindset for the workers but also provides the workers with more knowledge about the organization, thus reducing the chance for ineffective behavior. Line-of-sight can also make it easier to develop the proper risk management mindset. The notion that "information is power" is disappearing in the project management landscape as strategic information is widely shared.

1.3 Sustainable Competitive Advantage

Companies that thrive on competitive bidding for a large portion of their revenue stream, such as with project-driven organizations, promote their delivery system as well as the outcomes or deliverables. Companies are recognizing that excellence in project management can lead to a sustainable competitive advantage. Achieving excellence is not that difficult, but maintaining and continuously improving for excellence is a challenge. Maintaining excellence is a never-ending journey.

All too often, organizations that have reached some degree of excellence in project management become complacent, and then they realize too late that they have lost their competitive advantage. This occurs when organizations fail to recognize the importance of continuous improvement to maintain the competitive advantage.

Figure 1-5 illustrates the risk and why there is a need for continuous improvement. As companies begin to mature in project management and reach some degree of excellence, they achieve a competitive advantage. The competitive advantage might very well be the single most important strategic objective of the firm if it chooses to exploit this advantage.

Unfortunately, competitive advantage is usually short-lived. The competition does not sit by, idly watching you exploit your competitive advantage. As the competition begins to counterattack, you

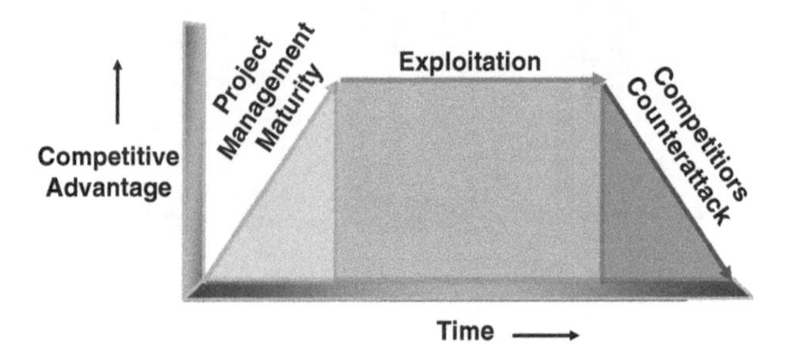

Figure 1-5 Risk with Maintaining the Competitive Advantage

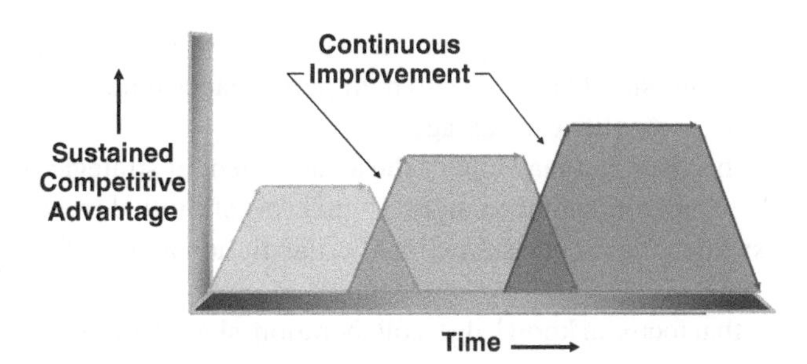

Figure 1-6 The Need for Continuous Improvement

may lose a large portion, if not all, of your competitive advantage. To remain effective and competitive, the organization must recognize the need for continuous improvement in project management, as shown in Figure 1-6. Continuous improvement allows a firm to maintain its competitive advantage even when the competitors counterattack. It is through continuous improvement efforts that the competitive advantage becomes sustainable.

1.4 High-Performance Teams

When discussing continuous project management improvement efforts, companies usually look at enhancements to the processes, tools, and techniques. What is often bypassed is the need for high-performance teams that include enhancements to people skills. Not all companies need high-performance teams, and the definition of **high performance** can change within each company. But by providing proper education and training with an emphasis on people skills and leadership, high-performance teams can contribute significantly to a firm's sustainable competitive advantage and frequently outperform similar teams.

There does not exist a clearly defined list of characteristics of high-performance teams. Academia and researchers focus on specific industries and types of projects. Warrick (2014) identified the following characteristics of high-performance teams:

- Effective leadership
- Team members that are competent, compatible, and committed to the success of the team
- Strong commitment by the leader and the team members to excellence
- Clearly understood mission and goals that team members are committed to achieving
- Clearly understood member roles and responsibilities
- Effective planning procedures
- Effective follow-up procedures and control
- Flexibility to respond quickly to needed change

Other characteristics that are often identified include effective collaboration, high levels of trust among team members, and rapid response to environmental changes.

1.5 High-Performance Organizations

What companies are now realizing is that the strategic focus should be on creating a high-performance organization (HPO) rather than just high-performance teams. The rationalization is as follows: If we believe that we are managing our entire business by projects, then the organization rather than individual teams should be driven to high-performance outcomes. This will then lead to a potential sustainable competitive advantage.

HPOs can eliminate barriers that may hinder the achievement of strategic goals by responding and adjusting quickly to environmental changes. High levels of mutual trust and clear and open communications exist, allowing for cross-functional collaboration and the flattening of organizational hierarchies.

HPOs have cultures that focus on knowledge, collaboration, shared visions, and the right worker skillsets. Workers are empowered, asked for their opinions, and willingly accept responsibility and accountability. They are provided with the necessary information to meet customer and stakeholder needs to ensure business success.

There's no universally accepted definition of a high-performance organization. Akdemir et al. (2010) identified 26 characteristics of HPOs, as shown in Column 1 in Table 1-1. Column 2 shows typical project management practices that help support the high-performance organization characteristics. Column 3 identifies project management pillars discussed in this book that will also support the HPO characteristics.

Table 1-1 How project management supports the HPO

HPO Characteristic	Project Management Contribution to an HPO	Pillar #
Well-understood vision and values	Executives provide a clear vision and goals to project teams by establishing a line-of-sight	1
Proper use of discipline	A nonthreatening environment where teams can speak their mind without punishment and acceptance that some projects may fail due to the risks	3
Clear set of specific goals	Line-of-sight provides goals and strategic objectives to project teams	1
Strong communication	Effective social leadership and a cooperative culture promote clear and open communications	3
Trust and confidence	Effective social leadership and a cooperative culture is promoted; empowerment leads to high levels of mutual trust between team members	6
Fun	Effective servant and social leadership allow the team members to see the fruits of their efforts and make the work challenging and enjoyable	6
Decision-making at the lowest level	Effective social leadership and empowerment allows team members, even at the lowest levels, to participate in decisions	5
Effective training	The company is committed to life-long project management education	2
Performance feedback	Company believes in capturing lessons learned and best practices from performance feedback	10
Customer focus	Fosters high levels of customer and stakeholder collaboration and interaction	
Measurement techniques	Continuous exploration of new tangible and intangible metrics for performance measurement	10

Table 1.1 (Continued)

HPO Characteristic	Project Management Contribution to an HPO	Pillar #
Strategic change management	Clearly understands and supports the need for change management when necessary	
Encourage innovation	Encourages team members to use brainstorming techniques, design thinking and creative problem solving	4
Team-based effort	Effective social leadership practices reenforce the team concept and establishes high levels of collaboration and cooperation; roles and responsibilities are clearly defined and understood	8
Participative leadership	Social leadership techniques encourage team members to participate in idea generation and decision-making; emphasis is on participative rather than individual leadership	8
Incentives and recognition	Team members are recognized and rewarded for contributions to a successful outcome	3
Recruiting and hiring best talent	Recognizing the contribution of project teams to the company's success helps create a pipeline for attracting talented project management personnel	9
Work-life balance	Effective social leadership encourages an understanding of the work-life balance	9
Managing diversity	The globalization of project management fosters a better understanding of how to manage and control diversity on project teams	2
Motivation	Effective social leadership focuses on motivation through empowerment	6
Compensation and performance appraisal	360° performance appraisal techniques are used, and rewards are provided either through compensation or recognition	6
Knowledge management	Company maintains a knowledge management system	4
Meaningful job	Social leadership and empowerment encourage recognition of a meaningful job	3
Effective succession planning	Project team members are usually qualified to fill more than just one critical position on a project team	7
Effective planning	The organization maintains several frameworks that support effective planning practices including ways to eliminate barriers in achieving strategic goals	8
Maintain ethical standards	Effective social leadership is based on high ethical standards	6

1.6 Strategic Competency

Column 2 in Table 1-1 clearly indicates that project management practices have a significant impact in helping companies become a high-performance organization. As shown in Figure 1-7, project management knowledge and expertise are now a critical competency in companies.

Many companies today conduct a study every year or two to identify the four or five strategic career paths in the company that must be cultivated so that the growth of the firm is sustainable. Project management makes the short list of these four or five career path slots. As such, project

Figure 1-7 Project Management Becomes a Strategic Competency

management is now treated as a "strategic competency," rather just another career path position for the workers.

Part of this is evident by looking at to whom project managers now report project status and make presentations. Historically, PMs conducted briefings for the project sponsors and occasionally senior management. Now, with the responsibility to manage strategic projects that may impact the future of the firm, project managers may be conducting briefings for all senior management, and even the board of directors.

By understanding and identifying the barriers prior to implementation, the chances of successful implementation will improve as with the recognition as a strategic competency.

1.7 Background to Barriers[2]

The Project Management Institute (PMI) recently celebrated its fiftieth anniversary. Even though most of the core concepts of project management have been recognized and used successfully for decades, there is still resistance in the form of barriers that can prevent successful implementation of all or specific components of project management. As new techniques begin being used in the project management environment, such as the impact of digitalization, artificial intelligence, the internet of things (IOT), big data, blockchain, and disruptive project management practices, new barriers are expected to appear. An understanding of the barriers can help us prevent or diminish their impact.

Up until about 10 years ago, there appeared to be limited published research on the identification and impact of these barriers. Part of the problem was that the literature at that time seemed to focus heavily on successes rather than failures because nobody wants to admit to having

2 The remainder of this section has been adapter from Barriers to Implementing Project Management by Harold Kerzner and J. LeRoy Ward; white paper, 2020. J. LeRoy Ward, PMP, PgMP, PfMP, CSM, CSPO is the President of Ward Associates and the past Executive Vice President for Enterprise Solutions at IIL.

made a mistake. Today, we recognize that we may discover more opportunities for continuous improvement efforts from failures and mistakes rather than from best practices and lessons learned.

In an early paper by Kerzner and Zeitoun (2008), the authors focused heavily on the barriers that existed primarily in emerging markets. The authors stated:

> Growth in computer technology and virtual teams has made the world smaller. Developed nations are flocking to emerging market nations to get access to the abundance of highly qualified and relatively inexpensive human capital who want to participate in virtual project management teams.
>
> A multi-national virtual project management team, however, may come with headaches. Because of the growth of project management worldwide, many executives openly provide lip service to its acceptance, yet behind the scenes, they erect meaningful barriers to prevent it from working properly. This creates significant hardships for those portions of the virtual team that must rely upon their team members in emerging market nations for support.
>
> Barriers to effective project management implementation exist worldwide, but in emerging market nations, the barriers are more apparent. To be aware of the possible barriers and their impact on project management implementation allows us proactively to begin to surmount them.

Today, many of the barriers that previously appeared in primarily emerging markets are now quite apparent in developed nations and within areas of companies that may have been using project management for decades. Barriers are no longer restricted just to specific countries or nations. Some barriers may be industry-specific, appear in certain functional disciplines of a company, or occur because of the personal whims of some managers and executives. Barriers can appear anywhere and at any time.

Some industries appear to be more prone to project management implementation barrier than others. The barriers in the IT industry have been discussed in the literature by Johansen & Gillard (2005), Khan et al. (2011), Khan & Keung (2016), Marly Monteiro de Carvalho (2014), Niazi et al. (2010), Polak & Wójcik (2015), and Terlizzi et al. (2016). Research and development barriers have been addressed by Sakellariou et al. (2014), Santos et al. (2012), and Sommer et al. (2014). Recently, there has also been research in public sector barriers, as described by Blixt & Kirytopoulos (2017).

Another industry commonly discussed is the construction industry, as indicated by Arnold & Javernick-Will (2013), Hwang and Tan (2012), Loushine et al. (2006), Moore & Dainty (2001), and Senesi et al. (2015). Some authors focus on barriers in specific countries. As examples, Wenzhe et al. (2007) looked at the Chinese construction industry whereas Hwang et al. (2014) researched small construction projects in Singapore. Magnier-Watanabe & Benton examined barriers facing Japanese engineers.

There has also been research in barriers that can affect certain ***PMBOK® Guide*** Areas of Knowledge or specific project management processes, tools, and techniques. Kutsch & Hall (2009 & 2010) and Paterson & Andrews (1995) looked at the barriers that impact risk management. Ambekar & Hudnurkar (2017) focused on the use of Six Sigma. Ali & Kidd (2014) examined configuration management activities, and Hwang et al. (2017) investigated barriers affecting sustainability efforts.

Figure 1-8 Categories of Barriers

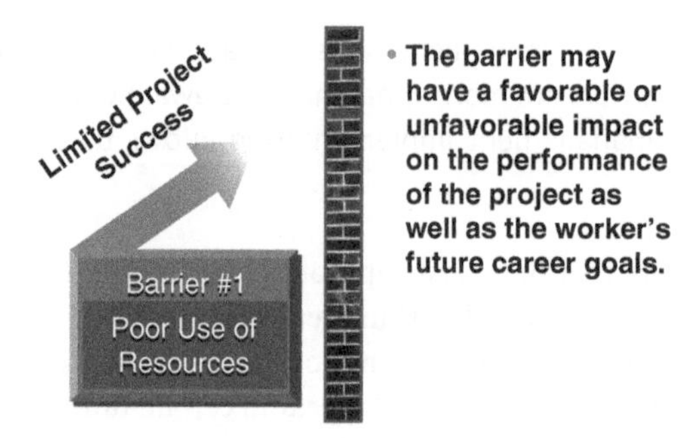

Figure 1-9 Human Resources Management Barrier

There are numerous possibilities for classifying the different barriers affecting project management. This paper briefly discusses some of the more prominent barriers in the categories identified in Figure 1-8.

The authors solicited feedback concerning the barriers from almost all the PMI chapters outside of North America. The authors are indebted to those PMI chapters that took the time and effort to respond to our requests.

Human Resources Barriers

Whenever we change or introduce new management processes, whether it is for project management, Agile, Scrum, Six Sigma, or other practices, we must consider the impact on the wage and salary administration program. Workers expect to be recognized, or even rewarded, for good performance. Unfortunately, we often introduce new programs without considering employee performance review implications until some damage occurs and barriers form to prevent the new processes from being implemented correctly.

Sometimes human resources barriers are created that can cause a conflict between what appears to be in the best interest of the project and the best interest of the worker. It is not uncommon for the project team to fail to realize the impact of the barrier or even that it exists until the project is over. In most cases, as shown in Figure 1-9, the result may be limited project success, or possibly even failure, based on when the barrier is recognized.

Situation 1 (Co-Location Barrier)

A project manager working for a government agency was placed in charge of a two-year project and wanted a co-located team. The PM was fearful that, if the team members were to remain in their functional areas, the functional managers might use the workers frequently on other projects, thus impacting the schedule of his project. During project staffing, the PM also demanded the best resources, knowing full well that many of the workers would be overqualified for the tasks and therefore underutilized. Although the demand for the best resources benefited this project manager's assignment, other projects that required workers with these specific skills were short-handed and struggled. The project manager's decision may have benefited the project but not necessarily the entire company.

The project manager found a vacant floor in a government building and all the workers were relocated to this location on a full-time basis, even though the assignment did not necessarily mandate full-time workers. However, even though the workers were removed from their functional organizations, their functional managers were still responsible for their performance reviews.

At the end of the two years, the project was deemed as a success. However, many of the workers were quite upset because:

- Most of the workers were given mediocre performance reviews during the two-year period because their functional managers were unaware of their performance.
- When given the choice of who deserved a promotion during the two-year period, the functional managers promoted those workers first who resided in the functional area and benefited from a multitude of functional area projects.
- Some of the workers discovered that their functional managers filled their vacated position with other workers and that these now-displaced team members had to find positions elsewhere, and possibly lose some seniority.

In Situation 1, which occurred in a developed nation, the project was a success and the project manager received a promotion. Unfortunately, the workers did not see any benefits to their career goals by working on this project and stated that they had no desire to work for this PM again. The organization had to rethink the benefits of using a co-located team approach. Barriers can exist anywhere.

Situation 2 (Prolonged Employment Barrier)

A government-run utility in an emerging market nation embarked on a three-year project to build a new power generation plant. To minimize the cost of the project and support the local economy, the decision was made to use local and country-wide labor rather than hiring more expensive contractors external to their country. This would have the additional benefit of providing employment for many local workers.

The workers were happy with the opportunity for employment but were fearful of what might happen when the project came to an end. To guarantee long-term employment, and possibly retirement benefits, the workers began slowing down the project and began making mistakes, to the point where the schedule was now stretched out to 10 years. Project management implementation was blamed as the barrier for the schedule delays.

Situation 3 (Building an Empire Barrier)

This situation has some of the characteristics of Situation 2. In some countries, your salary, power, and authority are based on the size of the empire you control. In such a case, hiring three below-average workers to do the same work as two average workers is better for empire building. Additionally, even though finding adequate human resources may be difficult, sometimes companies simply do not put forth a good search effort; friends and family members may be hired first, regardless of their qualifications. In this situation, the project's schedule is usually elongated so that the empire that is built will last as long as possible.

Situation 4 (Overtime Barrier)

Overtime is usually needed when pressure is placed on the team to maintain a schedule. However, in some cultures, overtime is used as a reward system to give workers the opportunity to earn additional income. This can occur even if the overtime is not necessary.

Some countries put restrictions on overtime and may require that the government must authorize the overtime especially if it is paid overtime. This occurs when the country is fearful that the overtime, if prolonged, may create a new class of citizens. There is also the danger that mistakes may be made intentionally on a great many projects to justify the use of extended overtime.

Situation 5 (Career Path Barrier)

A government agency discovered that as they began outsourcing more work to national and non-national contractors, their ability to evaluate project management performance was becoming difficult because each contractor would report status differently. Some contractors appeared to be performing at a higher level of project management than others by using *PMBOK® Guide* processes, but the government was unable to compare contractors' performance for awarding future contracts. The government then encouraged all contractors to use project management according to the *PMBOK® Guide* and highly recommended that the individuals managing their contracts become certified as PMP®s.

While the government recognized the value in promoting project management professionalism in the contractors' organizations, the government could not see the benefits of professionalism in project management in the public sector. Public sector project managers were treated more so as project monitors than project managers. The Office of Personnel Management in the government was unable to write job descriptions for government project managers because their duties, authority, responsibility, and decision-making capability did not fit in the standard "mold" use for other job descriptions. As such, an assignment as a government project manager was viewed as a non-promotable position that could impact your career.

Situation 6 (Certification Barrier)

An emerging market nation recognized the benefits of project management implementation and encouraged both government contractors as well as government agencies to support training programs that lead to individuals becoming PMP®s. Training programs were put in place by private firms as well as universities. In most cases, the company or government agency paid for the training of the workers.

Once the individuals became PMP®s, they began asking their company for a salary increase. The company argued that the cost that the company invested in their education should be viewed as a near-term salary increase and that other financial benefits would occur in the future. The workers were unhappy with the response and had expected immediate financial benefits when becoming certified. As such, the employees found employment in other companies and other countries that would provide them with salary increases. The companies stopped paying for project management training, no longer supported certification efforts, and in many cases discontinued the encouragement to follow the *PMBOK® Guide*.

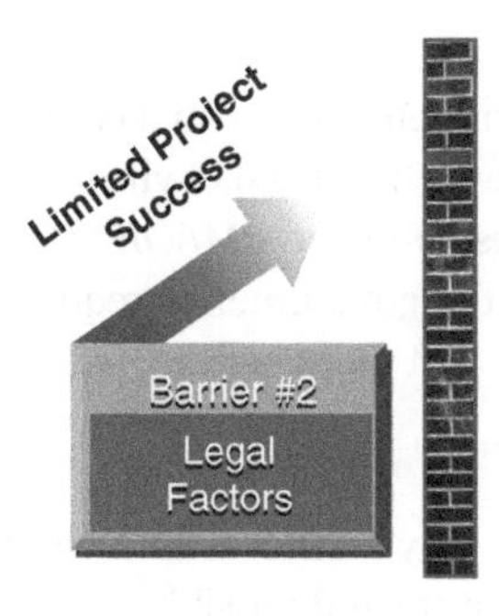

Figure 1-10 Legal Barriers

Situation 7 (Educational Barrier)

A company that recognized the need for project management implementation sent their workers to training programs. The employees returned to work with the expectation that they would be able to implement the tools and techniques they learned in the training. When the employees discovered that their company was reluctant to implement many of the tools and techniques they had learned, the employees sought employment elsewhere.

Legal Barriers

Some countries establish laws that provide limitations on how much, if any, of the financial resources that the country possesses can leave the country. This creates the barriers shown in Figure 1-10. The country can put limitations on procurement activities that leave the country. There can also be laws on paid overtime for workers.

Companies that wish to do business within these countries must abide by these laws even if the laws seem improper. An example might be the worker's right to hold a job even if the worker's performance is substandard. Some laws may even foster corruption possibilities by making it clear that bribes and "gifts" may be appropriate under certain circumstances during competitive bidding activities.

Situation 8 (Procurement Barrier)

The government in an emerging market nation wanted to limit the procurement of goods and services from outside the country. During competitive bidding, companies were asked to prepare for the government's approval a list of qualified vendors from within the country. Project managers were pressured to use contractors from within the country even if external contractors provided higher quality goods and services. To make matters worse, additional pressure was imposed to select contractors in cities that had the greatest unemployment rates regardless of the capabilities of the vendors.

Situation 9 (Unemployment Barrier)

A project manager was pressured into awarding a procurement contract to a vendor in a city that had significant unemployment. As the project progressed, the project manager realized she could accelerate the schedule by using overtime. Unfortunately, government permission was required for authorization of overtime pay. The project manager soon discovered that not only would the government not authorize the overtime, but it was also reluctant to allow the project to finish early and with fewer resources, which could increase the unemployment level and poverty in the community.

Situation 10 (Rigid Policies Barrier)

A company was awarded a contract for a government agency in a country that utilized rigid official public procurement processes, rules, and laws. The legal environment created a great deal of inflexibility and many of the traditional processes in the *PMBOK® Guide*, such as with change management activities, were not in line with the government's requirements. To adhere to the inflexibility, the project's budget was increased, and the schedule had to be elongated.

Situation 11 (Restrictions Barrier)

Some countries have government policies that may restrict who a contractor will be allowed to work with, both in the government agency's country and possibly in other countries due to political concerns or competitive factors. Some policies can also dictate who the prime contractor must hire as subcontractors even if the subcontractor's interest in and importance of the project are different than that of the prime contractor.

Even though a customer or government agency follows the processes and principles within the *PMBOK® Guide*, there can still exist laws or policies related to the criteria for decision-making, the time to make the decision and the parties involved in making the decision. Every government agency can have their own interpretation for the acceptance/rejection criteria for project deliverables, the evaluation of quality and decisions related to required permits. Some of the differences that can exist between countries include:

- The definition of the level of customer requirements, technical specifications and quality required since the differences are observed when the client and the contractor are from different countries.
- Timing and requirements for the process of obtaining permits and construction licenses, in accordance with the applicable regulations and laws of the country.
- Interpretation or agreement of requirements or conditions of acceptance of the project and its deliverables.
- Involvement of areas of the organization when negotiating with the client, in terms of scope changes in the project.
- Definition of norms, laws, international treaties, and rules that must be complied with.
- Process of management, administration, and negotiation of leases of machinery and equipment with third parties, when dealing with other countries.
- Be at the forefront in terms of performance and productivity of machinery, equipment, and software existing in the market in other countries and manage to bring them to your project.
- Definition and adequate management of the change control system, through authorized change requests, analysis of impact on risks, time, and project resources.
- Generation of the PMO's value in the organization, which monitors the governance of projects and establishes a common framework of methodologies, processes, policies, and information systems.

Project Sponsorship Barriers

We assign sponsors or governance committees to projects to provide project teams with a line of sight to senior management for strategic information, assistance for decisions that cannot be made entirely by the project team, coordination of large stakeholder groups and resolution of problems that can be resolved more effectively by the sponsors. It is not uncommon for individuals to be asked to serve as sponsors without understanding project management, or their role and

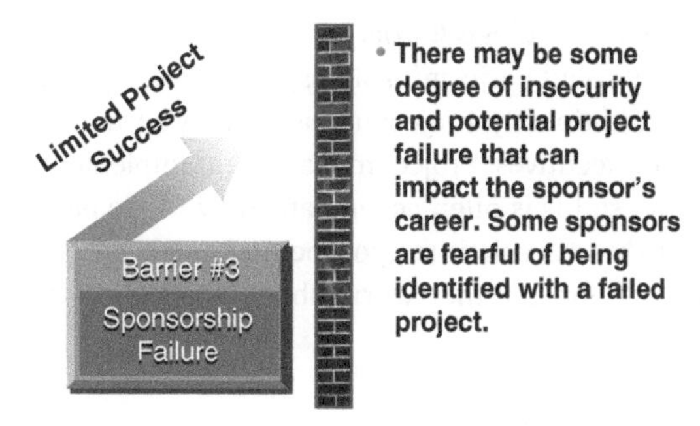

Figure 1-11 Sponsorship Barriers

responsibility as a sponsor, and the fact that project governance is not the same as functional governance. There is also the risk that the individual may abuse the position. In any event, barriers as seen in Figure 1-11 can be created that prevent the implementation of effective project management.

Situation 12 (Centralization of Authority Barrier)[3]

Many countries maintain a culture in which very few people have the authority to make decisions. Decision-making serves as a source of vast power in both privately held companies and governmental organizations. Project management advocates decentralization of authority and decision-making. In such countries, the executive level will never surrender its authority and decision-making power to project managers. Project managers may then function as puppets and are unable to effectively manage the projects.

Situation 13 (Lack of Executive Sponsorship Barrier)

Project sponsorship may exist somewhere in the company but often not at the executive level, for two reasons: First, executives may recognize that they have no knowledge useful to the project. Therefore, they could make blunders that would become visible to the people that put them in power. Second, and possibly most important, acting as a sponsor on a project that fails could end the executive's career, politically. Therefore, sponsorship, if it exists at all, is often at a low level in the organizational hierarchy, a level at which workers are expendable. The result is that sponsors cannot or will not make decisions or help project managers in times of trouble. We then have invisible sponsorship.

Situation 14 (Organizational Hierarchy Barrier)

In traditional project management practices, we tend to believe that problems are resolved at the project sponsor level. But in nations in which organizational hierarchy is sacred, following the chain of command can elongate the project management process to a point at which schedules become irrelevant. Also, the infrastructure to support project management may exist only to filter bad news from the executive levels and to justify the existence of functional managers. Some decisions and news may go as high as government ministers. Simply stated, project managers may not know where and when a decision will be made and cannot be sure where project information will end up. There may exist excessive bureaucracy that is not visible at the project management level.

3 Situations 12–16 and 19–22 have been adapted from Kerzner and Zeitoun (2008).

Situation 15 (Insecurity at Executive Levels Barrier)
Executives may feel insecure about performing as a sponsor because their positions are the result of political appointments. Additionally, project managers may be viewed as the stars of the future and, as such, are a threat to executives. Project management implementation could force the loss of an executive's status, and status is often accompanied by fringe benefits and other privileges. Before executives consider throwing their support behind a new approach, such as project management implementation or a project, they worry about its effect on their power, authority, and chance for advancement.

Situation 16 (Social Obligations Barrier)
In emerging markets, social obligations due to religious beliefs and politics are a way for executives to maintain alliances with those people who put them in power. As such, project managers may not be allowed to interact socially with certain groups that may possess critical information. Unlike traditional project management practices where PMs may have the right to communication with everyone, it may not be possible without going through the project sponsor.

Situation 17 (Lack of Education Barrier)
Not all sponsors understand project management or have a desire to attend courses on project management. Being unsure about their role as a sponsor, they focus mainly on how to deliver the results faster and at a lower cost regardless of the quality, risks, or the best way to achieve the results. Sometimes, sponsors will make promises to customers for rework or additional testing at no cost to the customer. This creates havoc for project teams.

Situation 18 (Project Charter Barrier)
Without at least a cursory knowledge of project management, sponsors are reluctant to prepare and sign a project charter for fear that the costs and schedule are not well estimated. This then forces some team members and even stakeholders to begin to perform some of the tasks without prior authorization.

Cost of Implementation Barriers

Most organizations today understand the benefits that can be forthcoming from effective project management implementation but are unsure about the costs associated with obtaining the strategic benefits. As such, there may be apprehension or even fear in a commitment and the creation of a barrier as shown in Figure 1-12 will occur.

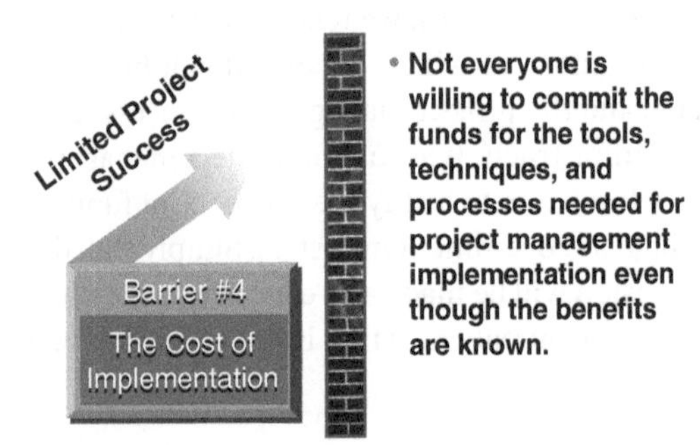

Figure 1-12 Implementation Barriers

The fear is that readers of the 756-page 6th edition *PMBOK® Guide* will believe that all the processes, inputs, outputs, tools, and techniques must be implemented for the strategic benefits to be recognized. This is not the case. There are extensions to the *PMBOK® Guide* where it can be downsized for specific needs. This could be the reason why the 18-page *SCRUM Guide* has become popular, or the shift to principles focus in the 7th edition *PMBOK® Guide* became useful.

Situation 19 (The Cost of Implementation Barrier)
The costs associated with the implementation of project management include purchasing hardware and software, creating a project management methodology, and developing project performance reporting techniques. Costs such as these require a significant financial expenditure that the company may not be able to afford. They also require tying up significant resources in implementation for an extended time. With limited resources and the better human resources – which are required for implementation – removed from ongoing work, companies may avoid project management even when they recognize its benefits.

Situation 20 (Risk of Failure Barrier)
Even if a company is willing to invest the time and money for project management implementation, there is a significant risk that implementation will fail. And even if implementation is successful but a project begins to fail, for any number of reasons, blame may be placed on faulty implementation. Executives who must then explain the time and money expended for no real results may find that their position in the hierarchy is now insecure. Therefore, some executives refuse to either accept or visibly support project management.

Situation 21 (Costs of Training Barrier)
Project management implementation is difficult without training the workers. This includes project managers, team members, and sponsors. The need for training can create additional problems. First, how much money must be allocated for training? Second, who will provide the training and what are the credentials of the trainers? Third, can people be released from their current project work to attend training classes? It is time-consuming and expensive to train people in project management. Adding together the cost of implementation and the cost of training may frighten executives away from accepting project management.

Situation 22 (Need for Sophistication Barrier)
Project management requires sophistication, not only in the technology but also in the ability of people to work together. In emerging market countries, employees may not have been trained in teamwork and may not be rewarded for their teamwork contributions. Communication skills may be weak, including report writing. People may see teamwork as a medium through which others are able to recognize their lack of competencies and mistakes.

Culture Barriers

Sustainable strategic project management success requires a cooperative culture where team members work together and make decisions in the best interest of the firm. This is most often accomplished without considering authority, power, or pay grade. Cooperative cultures often determine the type of organizational structure to be used for projects. As an example, the matrix structure seems to work best with cooperative cultures. Yet even within cooperative cultures, barriers, as shown in Figure 1-13, can be erected if the employees feel threatened or have hidden agendas.

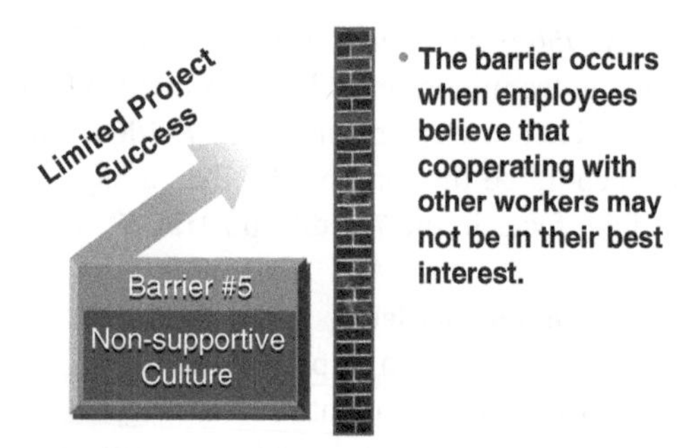

Figure 1-13 Cultural Barriers

Situation 23 (Planning Barrier)
If an organization lacks standards or commitment for project management, then the planning process may struggle with poor estimates for work effort, duration, and cost. The planning barrier can occur if the organization fails to support the use of a methodology. The result can be ambiguous scope and requirements. Poor planning usually translates into plans that change too often, unrealistic milestones, and a lack of faith at all levels of management that project management can succeed. If there are risks that the plans may fail, workers may create barriers and find excuses as to why they cannot participate in planning activities.

Situation 24 (Paperwork Nightmare Barrier)
When looking at the number of activities in the ***PMBOK® Guide***, there is apprehension that an excessive amount of unproductive time may be spent completing all the paperwork required. A barrier can then be created with a multitude of reasons why certain reports and paperwork is not necessary. The barrier can be significant if any organization believes in the philosophy that "what is not on paper has not been said."

Situation 25 (Project Completion Barrier)
As projects begin to wind down, workers begin to worry about their next assignment. If they are unsure about their next assignment, they may elongate the project closure process. They may also leave their current project before the work is finished to guarantee employment elsewhere. This may create a hardship for the remaining team members.

Sometimes, there exists a lack of dedication to project closure. Employees are often afraid to be attached to the project at closure when lessons learned and best practices are captured. Lessons learned and best practices can be based on what was done well and what was done poorly. Employees may not want anything in writing that indicates that best practices were discovered through their mistakes.

Situation 26 (Management Reserve Barrier)
Cooperative cultures tend to find ways to protect themselves and their colleagues from some risks, rework, poor estimating, and other such situations. The management reserve is one such way. However, a barrier can be created when the customer believes that the management reserve is created solely for the benefit of the contractor rather than the customer as well.

Situation 27 (PMP® Barrier)
Not all organizations encourage their people to become certified as project management professionals. One of the benefits of certification is that it becomes easier to understand each person's

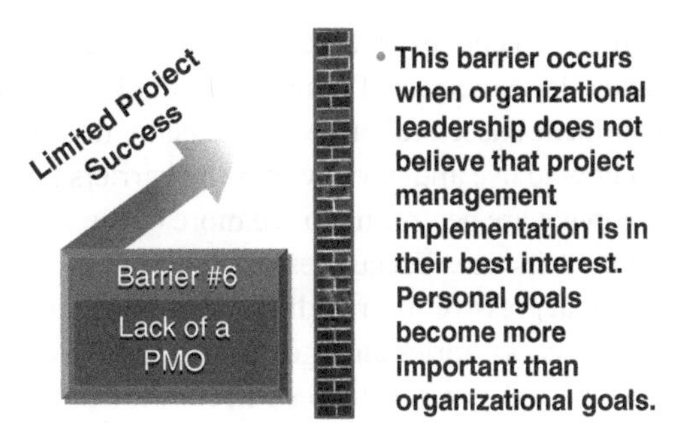

Figure 1-14 The PMO Barrier

role and responsibility, thus providing a good foundation for a cooperative culture. In noncooperative cultures, the role of the project manager becomes that of a firefighter.

Project Management Office (PMO) Barriers

Even though there are several types of PMOs, the existence of a PMO usually implies that there exists an organization dedication to continuous improvements in project management. Unfortunately, executives may view the PMO as a threat if the executives are happy with the status quo and their position in the firm, and do not want to see any changes take place. This barrier is shown in Figure 1-14.

PMOs can be used not only to promote effective project management practices, but also to capture and share best practices and lessons learned. They can also assist senior management in the monitoring of enterprise environmental factors that can impact decision-making. This is important in countries that struggle with hyperinflation and must control the country's scarce resources rapidly.

Situation 28 ("Information Is Power" Barrier)

When managers believe that information is power, they create barriers that may prevent the formation of organizations that gather and disseminate information. Sometimes, several functional PMOs are created to centralize functional information and control its release. Infighting can occur in functional units as to who controls each PMO if the information is seen as a source of power.

Situation 29 (PMO Financing Barrier)

All PMOs require human and nonhuman resources to be effective. Resources require capital expenditures. Managers that believe information is power will always create barriers that justify not financing a PMO. Barriers can be established by either staffing the PMO with nonqualified individuals or limiting the tools provided to the PMO for monitoring and controlling projects.

Conclusion

All countries, including emerging market nations, have an abundance of talent that has yet to be fully harvested. Barriers can appear in any organization for a multitude of reasons as identified in the situations presented here. Virtual project management teams as well as a recognition of the strategic benefits of project management may be the starting point for the full implementation of project management.

As project management grows, executives will recognize and accept the benefits of project management and see their business base increase. Partnerships and joint ventures may become more prevalent. The barriers that impede successful project management implementation will still exist, but we will begin to excel in how to live and work within the barriers and constraints imposed.

Corporate executives worldwide are beginning to see more of the value of project management and have taken steps to expand its use. Executives in some of the rapidly developing nations appear today to be much more aggressive in providing the support needed for breaking through many of the barriers. As more success stories emerge, we will see the various economies strengthening and becoming more connected, and executives will start to implement project management more fully.

In the next section, as we review the Medtronic project management journey, we find that the company has approached the growth journey of implementing project management practices holistically. The prioritization of this journey was tackled globally with a high degree of focus on enabling the success of spreading project management across all key business areas. Careful strategic thought was exercised to build the executive sponsorship needed and creating the right educating and community.

1.8 Excellence in Action: Medtronic[4]

I have seen many changes in project management over my last 35 years. From starting in the project management profession and not feeling confident in knowing what I was doing, to advancing to a formal project management professional membership and professional project management certification(s), I have seen the industry significantly evolve. Today, there are many new aspects and training opportunities for project management in soft skills, online learning, Agile, hybrid and many, many more.

Our organization has benefited from some grass roots efforts in project management, called *Project Management Rigor for Innovation,* in which we are continuously working to improve the overall footprint of project management. PMRI is focused on project management, which has been identified by technical leaders as a core competency for our global research, development, innovation, and cross functional community involved in product and services development and documentation to demonstrate best-in-class innovation. We also have another grassroots effort at our organization called *Program Management Leader Community* PMLC. This grass root effort is building a community of practice that serves to provide opportunities to our program managers to enhance their program management leadership skills, accelerate career development and expand their work.

Both efforts focus on training, networking / communication, and career development. They are voluntary and optional initiatives and focus on building an enterprise-wide community of project and program leaders. It is for individuals in project and program management roles leading initiatives that span operating units, regions and corporate. The focus is for new and existing project and program managers who are aspiring to expand their related skills and ultimately, into leadership roles. It is also a community for curating best practices. These efforts align with our mission (tenet 5) to "recognize the personal worth of all employees by providing an employment

4 Material in this section has been provided by Michael O'Connor, Ph.D., MS, MBA, PMP®, PgMP®, IPMA-B, CPD, CSM, CSPO, DASM, Director Strategy and Project Management, Medtronic, 710 Medtronic Parkway, Minneapolis, MN 55432, 763-526-0203. ©2021 by Medtronic. All rights reserved. Reproduced with permission.

framework that allows personal satisfaction in work accomplishes, security, advancement opportunity, and means to share in the company's success."

These programs are not new, mandatory, corporate led, or a formal training program to teach basics. They are also not a governing body to mandate tools or standards or certifications. Instead, the overall goal of these programs is to build future project / program managers based on the specific needs and requirements of leadership and project managers in operating units, regions and corporate.

The PMI Pulse of the profession has been a good resource for our team to review, help us guide and best determine a course of action. We read the PMI Pulse every year and utilize the information to update our project management PMRI strategy. We find the information and research very useful, and it ultimately helps us plan for the future of project management at our organization.

It is critical that we prove to senior leadership how important the roles of project / program management are to the organization. Recently, our organization has undergone a major restructuring plan that has put us in a position to be more agile, while also playing small and playing big.

These changes have moved us into the new future of our organization and has shifted our focus to needing to add Agile to our project management training and definitions. The question now is how do we roll out Agile with a few different options? There are the Scrum Alliance, Scrum Org., and the Project Management Institute and Disciplined Agile (DA). I have been trained as a Scrum Master and Scrum product owner through the Scrum Alliance, and as a PMI DA Scrum Master. Both have their pros and cons but overall, I think we will lean toward the PMI DA process as it is a toolbox of methodologies like Scrum, Lean, and Kanban. I also believe that Agile provides a more up to date and consistent method to create dashboards that provide necessary updates to executives. I need to be able to work with one methodology, to make sure a large organization like ours has a preferred method and keeps consistent vernacular and definitions. This is, of course, just one very small area to increase focus, awareness, and training.

At my organization we have undertaken a very large structural change to be more flexible, adapt to change quickly and keep things simple. The Agile philosophy will work very well for how we have made changes in our organization. I see this as an opportunity to change the way we work from, at least Waterfall to a hybrid, and in some cases, more Agile methodologies. I believe the PMI DA Agile methodology will give us the necessary tools to adapt to projects and situations as they evolve. The utilization of metrics will also need to increase and provide measures of success for projects. This is not always easy, but you won't know if you do not measure. Of course, all of this will take people, leadership, and a commitment to training to create a more productive organization.

I would also like to incorporate this type of Agile and hybrid training to add to our project managers toolbox, while also creating a more consistent set of training and techniques, making it possible for project managers in research and development to move to clinical, marketing regulatory, quality, and other areas of the organization. We have currently three large buckets in our organization for careers, including: People Management, Technical Leadership, and Project Management. This is very important to every stage of careers in an organization as it sets itself up for success. Project Management has been an area in which we created something called the *Project Management Knowledge Pathways*. This currently includes areas for career development, as well as opportunities to engage, and collaborate, leverage expertise, recognize and reward, discover external knowledge and innovate. This has proven to be a solid collection of subjects that any person in the organization can explore and learn. I am also planning to explore AI and project management knowledge pathways and better understand how they can be defined, communicated, and change the culture of our organization.

We are a global organization, and we have many cultures around the world. It is important for us to create and implement these grass roots efforts by ensuring we are inclusive. Currently, I am happy to say that we have representation from the United States, Europe, India, and China.

In order for true innovation to be successful, it's critical that we look around the globe, and within our organization, to create true innovation at record speed. Since we are in a regulated industry, we rely on these regulations to drive our work. I am also working toward trying to simplify this process, making it easier to navigate but regardless, we still must meet standards. Again, this is where having Agile in the toolbox would be extremely beneficial and to help change the focus of the project when needed to many iterations. These sprints will be more valuable than the Waterfall methodology. Even a hybrid approach will become very useful soon as I plan to offer more training on both Agile and hybrid methods.

Currently the grassroots Project Management Rigor for innovation offers internal and external training, roundtables, office hours, articles, knowledge pathways, project management community and transferrable skills across the organization. Every one of these items create stronger career pathways, exposure to senior leaders, coaching and mentorship and best in class resources and training.

I believe in our organization, with its large size and global footprint, that we need a project management toolbox, including many different items. This will depend on the education, experience, and passion that each person needs to bring to the job. In my opinion, every person needs to be treated differently, and they should have the ability to pick and choose the best items they need to perform the work. We are entering a new phase of the organization led by a new CEO, who would like to play small and play big. To do this, we will need serve our customers in a more flexible and Agile focus.

Our organization is on a journey to create more efficiency and productivity, while also making work easier. This is not an easy task, and it will take time to develop and create success. It is my journey and passion to create improved tools, training, and communication in project management. This will require change that may not be popular at first. However, to be successful, we must create new ways of working and think "BIG" to stay competitive as a market leader. We need to remain focused on delivering a strong innovation strategy and financial performance as an organization. Project management can create the change that is needed to drive this continued success. Overall, project management, and all the tools associated with it, can empower the organization to meet its mission of alleviating pain, restore health, and extend life.

1.9 Strategically Improving

Projects are becoming clear strategic change and transformation vehicles. The old world of project management used language like final and finite mindset. In many instances we were stuck in following specific rules and structures. Even today many of the world organizations are very short term in their strategic planning and the associated running the business practices. As we shift to the focus on continuously improving, we want to see this improvement guided by a difference enhanced language, around adaptive, collaborative, context driven, and inclusive. The new generations demand it, the world needs it, and the possibilities for the discipline applied right are immense.

The big *Ah-ha!* in transforming originations happens when personal ownership of a changed way of continuously improving the way of working is widespread. As we demonstrate the practices to achieving organizational excellence, many of the organizations contributing material to this book discuss just one or two pillars. Some companies have implemented many of the pillars. It is our hope that these contributions provide good starting points for experimentation and improvement that you can consider in the context of your own organization.

Innovation is one form of a strategic project. Repsol contribution, which appears in Section 1.10, is an example of how strategic innovation can introduce companies to many of the pillars of project management. The journey the organization has taken shows how their achieved new way of working was an integrated destination that combined the critical elements of vision, the right-fitting organization, the enabled digital workplace, and the adaptive methodologies and mindset.

1.10 Innovation in Action: Repsol[5]

Repsol is a global and integrated company in the energy sector. We operate across 37 countries with a team comprising over 25,000 people who work on building a sustainable future. Our vision is to be a global, energy company based on innovation, efficiency, and respect creating value sustainably to promote progress in society.

Innovation is an important leverage in our vision and is also one of our company values, together with transparency, responsibility, results orientation, and collaboration. At Repsol, we believe that the key to our competitiveness and development resides in our ability to generate new ideas and put them into practice in a spirit of cooperation and continuous collective learning. It has been a long and arduous journey to get innovation as part of our DNA and there are still many challenges to be overcome.

The following is how we are building the culture and capabilities for being the global and integrated energy company that we want to be and keep sustainably in the market.

First Phase: Acquiring Knowledge (2011–2012)

Until 2011, innovation at Repsol was focused on R&D activity. In 2011, the Innovation Program was launched to response to the Strategic Plan (2011–2014) about the quality control and knowledge management functions. The Program was sponsored by Upper Management that decided to incorporate innovation as one of the corporate values since 2012. The main objective was to embed the innovation in our culture and day-to-day activities.

In addition, an organizational structure was created to support and encourage the Innovation Program. The Corporate Innovation Unit, Business Innovation Units and the Innovation Committee were the organizational units in this structure. An innovation network was created including entrepreneurs, innovation teams and facilitators.

Second Phase: Strengthening the Organization: Building the Capabilities and Measuring the Progress (2013–2015)

In 2013, the focus was to build the capacity for the organization. The first edition of the Facilitators Training Program was held to support continuous improvement processes with lean-based methodologies.

Besides that, another pilot within the Chemical Business started. This pilot is a successful case in our company, and it has become the Lean Transformation program for this business, other businesses, and corporate areas where it has been deployed according to their needs.

The aim was to promote a culture of innovation in 2014. The Strategic Innovation Reflection (SIR) was held with the participation of all the innovation units generating a company model to

5

add more value from Innovation. Meanwhile the first edition of the IN Awards was launched with the participation of +5.000 employees and +500 initiatives. In 2014, the Innovation network continued growing until around 75 innovation and improvement teams.

In 2015, some KPIs were defined and put in place to measure the impact of innovation. A global corporate program called "Go" started with the aim to generate innovative proposals in order to improve the EBITDA. In addition, business units deployed the innovation model through specific roadmaps to leverage their strategic plans.

Figure 1-15 shows actions put in place to support and encourage innovation from the Innovation Units in Corporation and Business areas.

Third Phase: Leverages for Transformation: Innovation in Management and New Ways of Working (2016–2018)

- **Management principles:**

  Since 2016, an evolution of the operating model (processes, structure, policies and management criteria, work dynamic and decision-making, knowledge management, etc.) is accomplished by each business unit in accordance with the new strategic challenges. The Innovation Programs accompanying the new Strategic Update include the identification and prioritization of initiatives by the Business Units committees.

The alignment and mobilization of the organization to attain a shared purpose and vision are essential in this transformation. Besides this, communication is a key element to ensure the final goals.

> New models of people and organization management: Agile philosophy is the engine of the change.

Some global initiatives foster a more flexible and efficient corporate environment, taking advantage of the new technologies:

- Develop and implement policies and models to ensure that the culture of innovation is promoted.
- Propose new ways of working to be implemented in the units depending on their own and global needs.
- Promote continuous improvement projects in order to optimize processes (Lean).
- Define and propose the cultural change necessary to achieve a flexible and collaborative organization.
- Develop leaders to ensure behaviors associated with innovation and entrepreneurship.
- Recognize entrepreneurs and innovation teams who develop initiatives.
- Generate and develop capacity in the organization, both in project teams and units.
- Conduct surveillance, validation, and divulgation of new effective and replicable approaches with high potential impact in order to respond to the problems and opportunities of the businesses.

The result, as shown in Figure 1-16, is a cultural transformation based on innovation and new ways of working that have provided Repsol some important leverages of this transformation toward becoming a lean company.

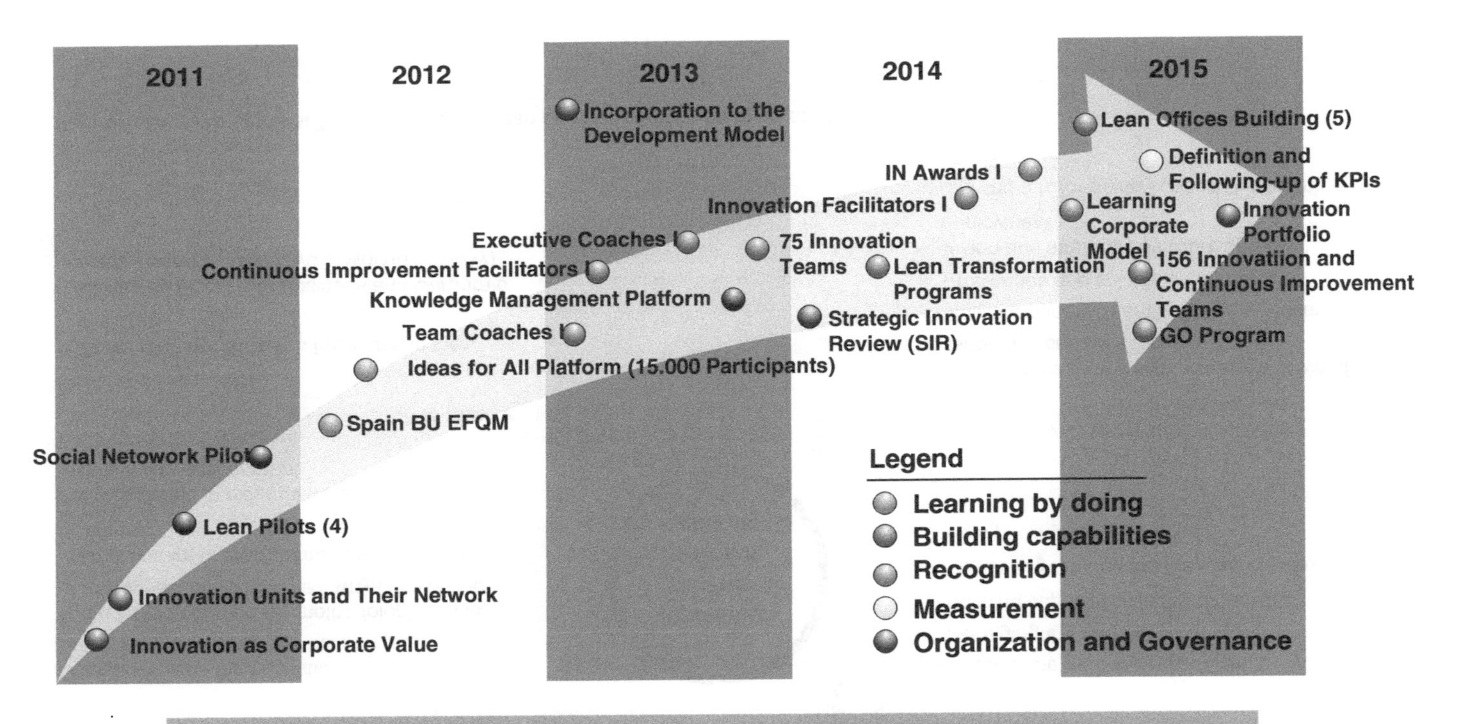

Figure 1-15 Details of the Actions Accomplished

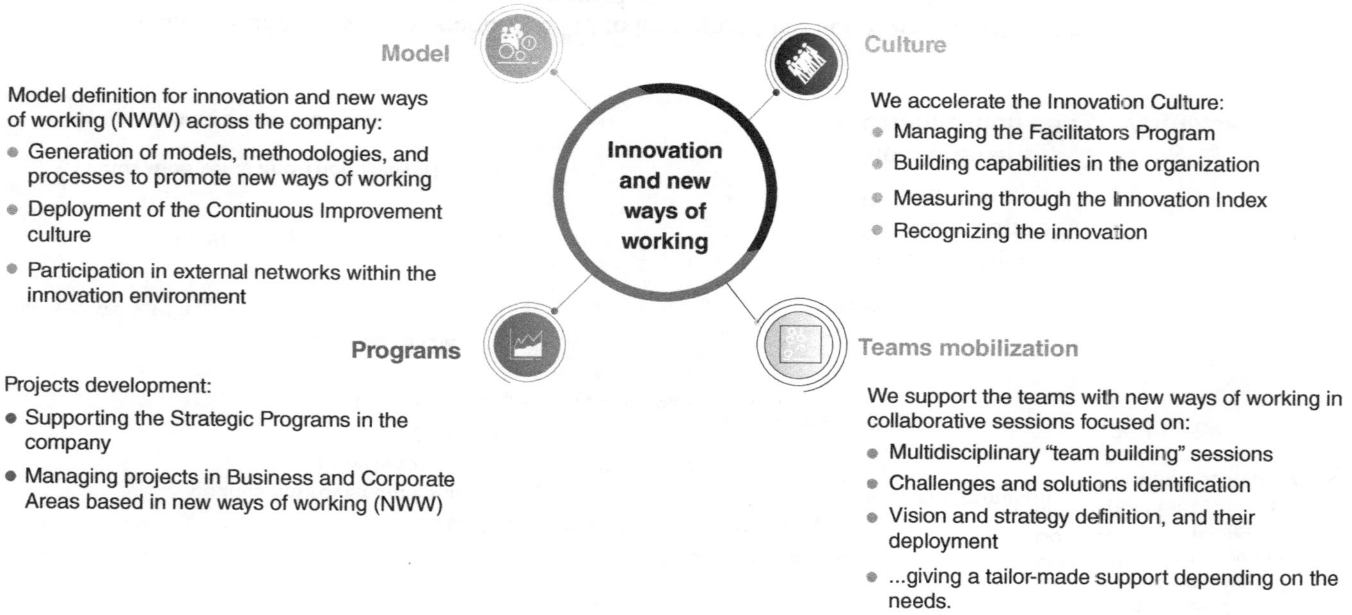

Model

Model definition for innovation and new ways of working (NWW) across the company:

- Generation of models, methodologies, and processes to promote new ways of working
- Deployment of the Continuous Improvement culture
- Participation in external networks within the innovation environment

Culture

We accelerate the Innovation Culture:

- Managing the Facilitators Program
- Building capabilities in the organization
- Measuring through the Innovation Index
- Recognizing the innovation

Innovation and new ways of working

Programs

Projects development:

- Supporting the Strategic Programs in the company
- Managing projects in Business and Corporate Areas based in new ways of working (NWW)

Teams mobilization

We support the teams with new ways of working in collaborative sessions focused on:

- Multidisciplinary "team building" sessions
- Challenges and solutions identification
- Vision and strategy definition, and their deployment
- ...giving a tailor-made support depending on the needs.

Figure 1-16 The Cultural Transformation: A Lean Company with New Ways of Working

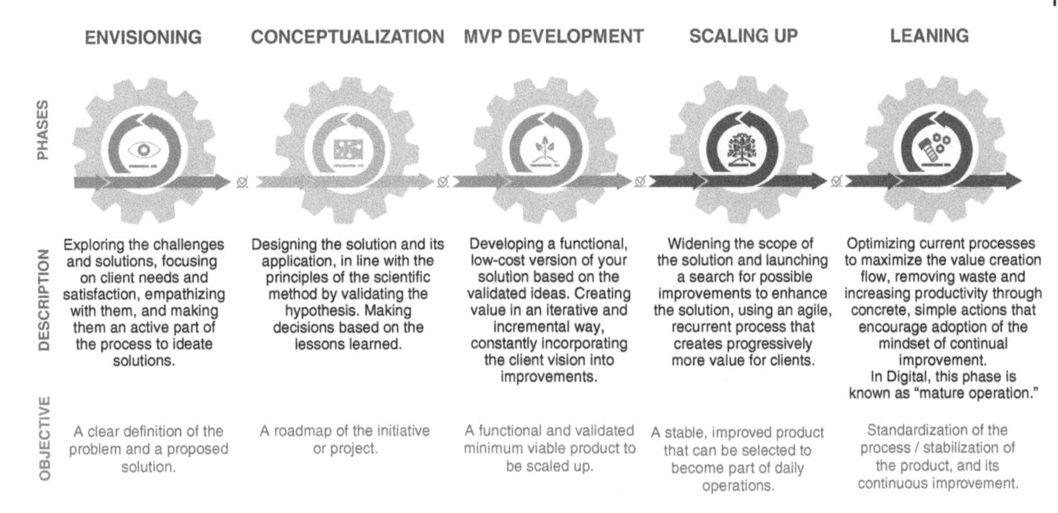

ENVISIONING
CONCEPTUALIZATION
MVP DEVELOPMENT
SCALING UP
LEANING
PHASES
DESCRIPTION
OBJECTIVE
Exploring the challenges and solutions, focusing on client needs and satisfaction, empathizing with them, and making them an active part of the process to ideate solutions.
Designing the solution and its application, in line with the principles of the scientific method by validating the hypothesis. Making decisions based on the lessons learned.
Developing a functional, low-cost version of your solution based on the validated ideas. Creating value in an iterative and incremental way, constantly incorporating the client vision into improvements.
Widening the scope of the solution and launching a search for possible improvements to enhance the solution, using an agile, recurrent process that creates progressively more value for clients.
Optimizing current processes to maximize the value creation flow, removing waste and increasing productivity through concrete, simple actions that encourage adoption of the mindset of continual improvement.
In Digital, this phase is known as "mature operation."
A clear definition of the problem and a proposed solution.
A roadmap of the initiative or project.
A functional and validated minimum viable product to be scaled up.
A stable, improved product that can be selected to become part of daily operations.
Standardization of the process / stabilization of the product, and its continuous improvement.

Figure 1-17 Innovation Process Model

Fourth Phase: New Ways of Working as a Transformation Driver (2018–2020)

The previous phase showed that in Repsol was considering both an Agile and a Lean approach. The result was that we created the Agile&Lean Model to deliver value across the entire chain of value as it is shown in Figure 1-17. Repsol Agile&Lean is the management model for a portfolio of initiatives and projects from conception to scaling up and subsequent optimization and continuous improvement. In addition, the acceleration program project generated clear value through the results achieved and actually the main key learnings were adopted widely across the company.

These outcomes enabled Repsol to go a step further to define what a New Ways of Working concept meant in the organization. As far as the organization was facing an environment whose constant are uncertainty, complexity and technological disruption, the whole organization needed and needs a new way of organizing, working, and relating.

New Ways of Working was defined as a set of tools and actions geared to improve and increase collaboration, agility and productivity and it is deployed through five elements shown in Figure 1-18, which are implemented within Operating Model Transformation Processes.

Transformation processes are characterized by a change in the operating model driven by the leadership team and activating five elements. They are enabled through an organizational change that articulates the leader's vision to give clarity to the organization, laying the foundations to generate new dynamics and usually it is split into three different but not isolated phases: To Start, To Adapt, To Adopt.

- **To Start:** The organization knows the purpose of the transformation, understands its need, and is committed to adapting and embracing the transformation.
- **To Adapt:** Define and design the path of transformation based on the needs and maturity of the organization. It involves identifying challenges and generating solutions.
- **To Adopt:** Make the transformation a reality by consolidating the new ways of working. It may involve adjusting the changes that we have adapted in the previous phase, with a vision of continuous improvement.

During the implementation of a transformation process, critical roles (Figure 1-19) are put in practice in the organization, beginning by focusing on the client itself and considering the leader has the accountability on transformation in all its phases. This fact has brought out leadership as

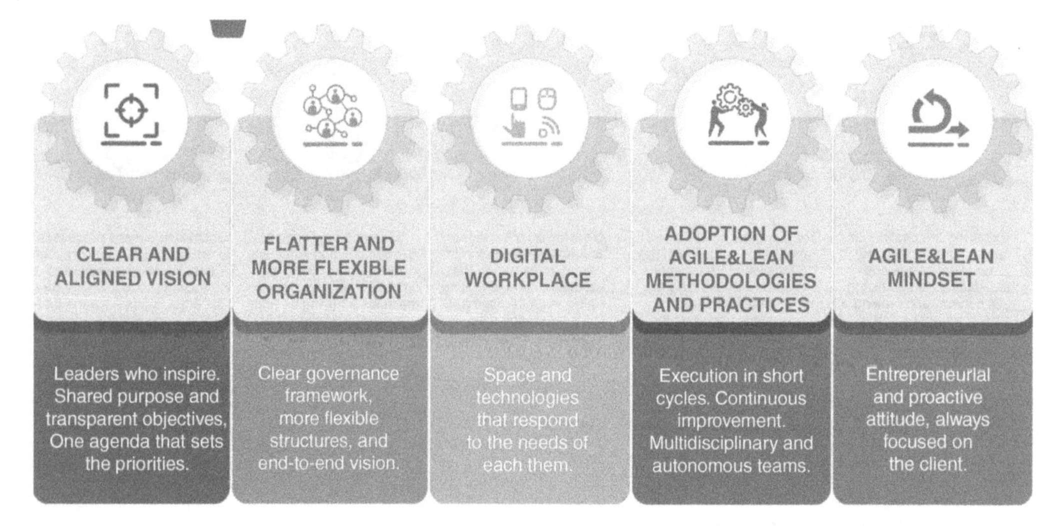

Figure 1-18 Repsol's New Ways of Working Elements

another enabler of the transformation processes and, due to this, leaders are being trained on how to inspire tomorrow and transform today.

In the past several years, Repsol has launched important initiatives: the Lean rollout in Chemical and Refining, the Agile rollout in the Digital department, transformation of departments such as Technology and Corporate Ventures, People and Organization, Auditing, Control and Risks, Finance, and Corporate Strategy. The decision to transform the organization is a strategic decision and requires hard work from people, investment, and time.

The experience at Repsol shows the potential of improvement that these New Ways of Working have brought about in where they were applied (Figure 1-20):

- In the culture survey, significant improvements in the lean index (an average six-point positive difference) and in the percentage of most effective employees (average 14 points positive difference), when comparing divisions that underwent transformation and those that did not.
- Additionally, at the organizational level, the hierarchy has been reduced, increasing the average number of reports to leaders by 75 percent increasing the Span of Control (SoC).

Fifth Phase: Stepping up the Transition (2021–2025)

At the end of 2020, Repsol launched a new Strategic Plan aiming to transform the whole company to achieve the net zero emissions ambition by 2050. This Plan is considering Agile & Lean New Ways of Working all across the value chain driven by a new operating model run by:

- A strategic talent management, by enhancing workforce planning, reskilling, and upskilling to face digitalization, new business and decarbonization; by boosting data driven culture; with a new and adapted professional development framework with diversity and inclusion.
- An organizational agility, by simplifying the organization; by promoting flexibility, productivity, and work-life balance.

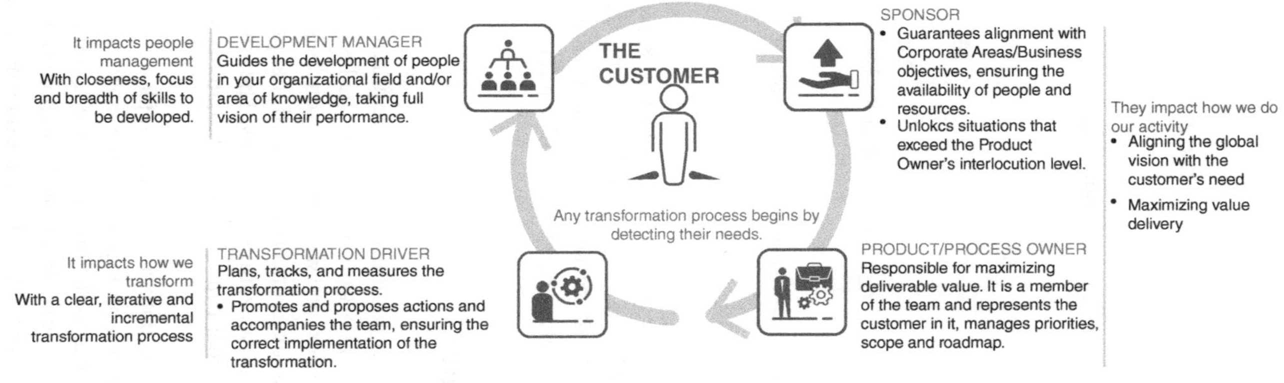

THE LEADER
It has the "accountability" on transformation in all its phases: start, adapt, and adopt.
It impacts people management
With closeness, focus and breadth of skills to be developed.
DEVELOPMENT MANAGER
Guides the development of people in your organizational field and/or area of knowledge, taking full vision of their performance.
THE CUSTOMER
Any transformation process begins by detecting their needs.
SPONSOR
• Guarantees alignment with Corporate Areas/Business objectives, ensuring the availability of people and resources.
• Unlokcs situations that exceed the Product Owner's interlocution level.
They impact how we do our activity
• Aligning the global vision with the customer's need
• Maximizing value delivery
It impacts how we transform
With a clear, iterative and incremental transformation process
TRANSFORMATION DRIVER
Plans, tracks, and measures the transformation process.
• Promotes and proposes actions and accompanies the team, ensuring the correct implementation of the transformation.
PRODUCT/PROCESS OWNER
Responsible for maximizing deliverable value. It is a member of the team and represents the customer in it, manages priorities, scope and roadmap.

Figure 1-19 Main New Ways of Working Roles in Transformation Processes

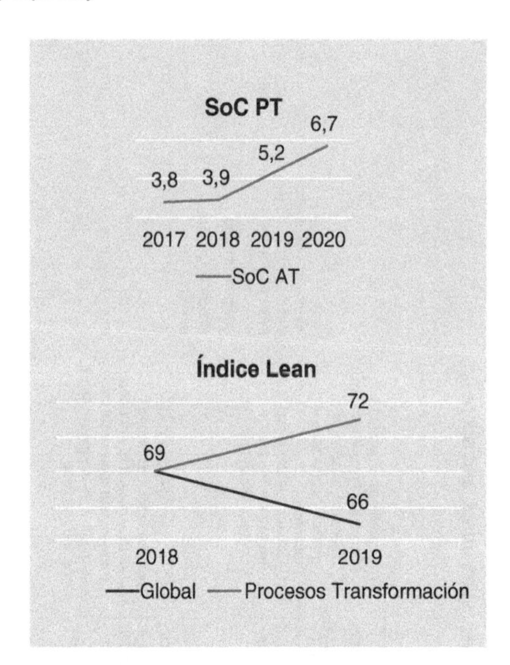

Figure 1-20 Improvements in Span of Control (SoC) and Lean Index

Repsol aims to become a more competitive company, a Repsol that develops new decarbonized businesses, a Repsol that with its actions and investments contributes to a world that still has fewer CO_2 emissions. Repsol will accelerate its transformation for the future they want, with customers at the center of the strategy, moving toward a decarbonized world.

1.11 Strategic Agility

The next three sections cover a multitude of the required dimensions of strategic agility. The multitude of Agile changes in organizations over the last decade indicate that the remainder of this decade and into the next one will witness a continued stretching of the ways of working to accommodate the demands of uncertainty, continual pressures of customers, and the need to inject urgency into our organizational cultures. The customer centricity dimension continues to grow as one of the root causes of the adaptability shift and the solutions co-creation expectations.

Merck KGaA highlights the impact of global complex events on how we adapt and build our resilience. Then, the Cisco section focuses on the customer and employee experience coupled with new forms of governance as the organization goes through a disciplined maturity journey. Following Cisco, the ServiceNow contribution drives home how strategic agility resulted in exploring new roles for the future PMOs. In addition, the dimension of product management is added to the mix to confirm the continual development of products and services that dominate the organizations of the future.

These three sections highlight the important role of the holistic and integrated progressing on a strategic agility path.

1.12 Excellence in Action: Merck Kgaa[6]

The world is evolving quickly. This is not new, but what we have seen during the pandemic in 2020 is a complete disruption of what was known previously. It is just a first taste of what could happen in the coming decade. At the beginning of this pandemic, some companies made the decision to reallocate budget and resources in a short period of time to fight the virus: mask manufacturing, vaccine research, tests manufacturing etc. The outcome was uncertain, pandemic duration also. Thus, making a decision without a lot of data is risky. This is where the project and portfolio management functions could demonstrate its added value: which project to be put on deck in order to free resources and budget? What is the level of risk taken to do this? Is it worth it? What are the expected adjusted values for the new projects coming in the pipeline?

All markets have been impacted and not only pharmaceutical industry and companies like AstraZeneca, Pfizer, Moderna, culture, fashion, automotive, etc. In all areas, decisions have been made to re-align the project portfolios and to stay competitive and to adapt to the new environment. In my point of view, it is just a first sign, and this will become a standard in the upcoming decade: how to move/change fast? If you are not ready to do so, you will be out of the game quickly. The fact that you have a good set of metrics to run your portfolio does not tell you that these metrics will be still relevant in two or three years from now. Let's be prepared to change.

The world is changing quickly, and we had a clear overview of this during the COVID-19 pandemic. In a short period of time, the world faced an unprecedent situation and new solutions accompanied by ways of doing have emerged.

Project management is a key actor of this change. Let's have a look back on what happened. An unknown virus started to infect the population in several countries at the same time and it spread very quickly. All countries, industries, and services were impacted. To fight this pandemic, solutions have been developed How to control the pandemic? How to test the people? How to protect the person not infected yet? How to develop in a short period of time a vaccine? How to ensure the delivery of key food, water etc.? So many questions and all of these included a project management role.

As a project and portfolio manager, we have been heavily involved to deal with this new situation and, on my point of view, it is only a starting point. The world won't be as it was in all dimensions. There was a **BEFORE** and there will be an **AFTER** pandemic.

When I say that project management has been heavily involved to deal with the pandemic, I can provide several examples: in the pharmaceutical industry, resources have been shifted to the research of a vaccine and testing kits which means that portfolio prioritization has been used widely as we cannot drop off everything from the table just by snapping our fingers. In addition, logistic topics were at the heart of the manufacturing activities and is still the case today. Here also, project managers are key.

I could find many examples, but the key message here is that project management importance is growing and so is its impact. The other point is that our methods have evolved and will continue to evolve again and again in the future. We cannot manage our projects today as we did 5 or 10 years ago. I would like to share with you some areas, where we will be challenged in order to be prepared for the future.

---

6 Material in this section was provided by Alexandre Fara, Head of the Project Management Office.

- Project management activities are moving more and more toward strategy. I don't want to say that strategy is going to be defined by a PMO organization as the key axis are and will be defined by the business, but our role is evolving to a strategic component. Portfolio is here to execute the strategy and a clear link between the activities done and the strategy should be developed and available. It is not enough to say "all of our projects are executing the strategy" because at a certain point of time you will need to make crucial decision as we have seen with COVID-19 pandemic.

  Operational projects will be continued to be managed by the PMO but more and more we see strategic initiatives also requesting a Professional Project Management leader. This will grow for sure in the coming years. In order to be ready to do so, strategy knowledge, customers and market segment needs should be known by our project managers. We are not requesting them to manage a scope, budget, and timing but we will ask them to have additional knowledge and understanding of the business and its environment. The future project manager will be a real business challenger.

  Project managers will be hired for a specific mission for a certain period of time and need to be flexible enough in order to move quickly to another company/business for another mission. This means **ADAPTABILTY** and fast learning curve on business environment. The PM will need to have much broader view than now. Focusing not only on the project but also on the surrounding world to be able to anticipate risk and issues. I am sure you have heard about data lake and all information we can have from there. Project manager skill is to be able to deal with a lot of information and to retain only what is key for the project he/she is managing. In a more and more competitive world,

  ROI is key for a project. Even if it is aligned with your strategy, the project prioritization on business impact and value is growing more than ever before. Access to the information is easier with data lake tools and reports available. The more info you have as a PMO, the more you need digest them and to exploit them to provide a clear message to your sponsors – PMO will evolve in this area too: finances, business acumen and understanding of the business will be key to manage a project to a success
- Managing remote teams is starting to be the basis today. We are using so many tools now to manage our meetings and our projects remotely. Our stakeholders, partners, sponsors are used or starting to be used to that. This was not the case some years ago when you had your Core Team around the table to discuss the upcoming activities or to make an update to the top management to make them aware about your portfolio projects' situation. Today, this kind of activities is requiring new skills for a project manager: no possibility to see the body language of the participant, no possibility to use your personal body language to convey the right message. This is not easy to deal with, but this will be the future. How do you manage a remote team with cultural differences and in a different environment? In a traditional PMO with processes well established and using a Waterfall approach, governance in place is not a success factor: human relationship is key! You can have all documents and governance in place, but if you don't develop your human skills you will fail.

  One of the key components for a project manager role is the human relationship. Managing people without having a hierarchical link is the most critical skills we need to have. This is coming more and more critical with virtual and remote team members across the globe. Dealing with other people in other location without having the possibility to have regular face to face meeting is a key challenge which needs to be handled in a successful way. If you don't have this, you cannot achieve success without difficulty. How do you engage your team remotely? How do you keep them aware of the situation in an efficient way?

A lot of digital tools are available, and we need to pick the right ones to be successful. The key success factors won't be only to select the right tool but the usage of them. As an example, when you are conducting a project presentation, especially face to face and in one room, it is pretty easy to capture the attention of the audience: tone of your voice, movement on the stage, movement in the room in order to be closer to some people to answer question etc. With a remote presentation you will need to change your presentation methods to ensure everybody is engaged.

This topic has taken more and more importance during the last year and will be a key differentiator for a project manager in the coming decade. During the pandemic, we have tried several ways to manage this situation and I can tell you that one solution is not applicable to all situation and all teams. Each of them has its own dynamic and the role of the project manager is to propose and to define the right setup with the Core Team. This can be virtual coffee once a week/month, virtual game, or just a dedicated channel to discuss and share our thoughts on a regular basis for instance. There is no magic solution, and this is the reason why the role of project management is key! We are here also to define and implement the right setup with our team members. This is part of our job and critical.

We see some skills and competencies emerging from our daily job and work. Our sponsors are requiring more and more empowerment, leadership behavior, engagement and not only from the project manager but from the entire team and the PM is the trigger.

- When we are talking about remote team members, we need to have in mind also our stakeholders and sponsors. It was not easy before the pandemic, but it will be more complex in the future. Mapping the stakeholders, embark them in the project, keep them aware of the situation is key and will be more challenging with remote situations. A good project can be a complete disaster if the stakeholders are not well-managed and the project manager's role on this is key and where he/she can demonstrate the PMO added value to a project.

  Do not expect to have one stakeholders' map at the beginning on your project which will be the same at the end. As said before, everything is evolving fast and your job is to ensure to be up to date every day: who are my stakeholders, where are they located? What are their names? And the most important: what do they want to know and when?

- A question which is regularly asked by some stakeholders or partners is: how could we measure the PMO impact on our projects? Unfortunately, there is no key performance indicator for this question. Every project is different and so it is not because you don't have a project manager leading a project that it will fail. You never know what would have happened if you had one! When we started to build our PMO group, we had few project managers, but we built our vision, we made several proposals to our leaders to have better visibility on what was going on, and we built a communication path to them. Several years afterward, the group had grown, and project managers are now requested for each project and for me, this is the PMO impact measure.
- We are entering a digital world more than ever: communication, data management. Everybody is aware of this situation. The question is: are we ready? If we are talking about project management tools and practices, we are at the beginning. Until now, Waterfall process was the main rule we followed, and we see Agile and hybrid methods raising up.

  On my point of view, the role of a project manager is not to apply a specific process, but to choose the right one and on top of that to be able to create, and manage several processes mixed together. The future project manager will be creative. Adaptability is key to ensure the most efficient way to develop what is requested by the business: this can be a new organization, new product, new infrastructure. The project manager will come with the toolbox, and it will be up to him/her to choose the right ones to achieve the goal. There won't be a single standard, but there will be several possible tools and the PMO will be the reference to choose the right one, the most adapted to the situation.

As the world is moving fast, process should be flexible. In any company, we need to define a clear process to ensure a harmonization of all project methodologies but also a comprehensive and accurate reporting system and portfolio management. Even if the processes are set up, the PMO should ensure the adaptability of the process to the current project situation and environment. Project managers are not here to apply the process but to adapt the process to the situation.

Do not expect the business to tell you what process to use in the future; they rely on you to tell them. This is where PMO as an added value: tracking activities base on a defined process vs adapting the process to the required activities to ensure success. The process is a guideline and a check list to be sure not to forget anything. ADAPTABITY will be key for a SUCCESS.

As you can see, the future project manager will have additional and improved skills and competencies. He/she should be more creative, adaptable, manage remote team and stakeholders, have a huge business knowledge. If we are not able to go that way, we won't be part of the game anymore. The business needs us to be successful. Let's be part of this new world!

1.13 Excellence in Action: Cisco[7]

Cisco Customer Experience – Project Management Role Evolution – Transformation in Action

Cisco Customer Experience (CX)

The industry is in the most demanding competitive environment in history. These tumultuous times are driving dramatic changes. Changes that have already transformed how people around the world live, work, play and learn.

But one thing has not changed at Cisco amidst these demanding times. That's Cisco dedication to customers, who are and always have been at the heart of what Cisco does. The result: bold and continuous customer-focused innovation that in recent years has revolutionized the Cisco Customer Experience (CX). Today it's a companywide mindset.

It's a commitment driving the industry's most trusted Customer Experience (CX) with our extraordinary people and technologies. Cisco enables this through the *Collaborative Intelligence* approach which consists of three customer-centric experiences: *CX Lifecycle Success*, *CX Cloud,* and the *Cisco Success Portfolio*.

CX ensures that our customers receive the best possible experience at every interaction at every single Cisco touchpoint. How is this accomplished? By being customer obsessed. Obsessed about each customer's business; obsessed about each customer's challenges; and obsessed about the problems each and every one of our customers are facing.

At Cisco, CX is a deliberate and holistic company-wide effort. Our teams work in coordination with all parts of the enterprise -- CX, Engineering, Marketing, People and Communities.

Cisco CX is the result of a genuine transformation. One that involves new roles, new platforms and technologies, new processes, and new portfolios. It's an accomplishment that to date has never been replicated at this scale, in a business and portfolio as complex as Cisco's.

---

7 Material in this section has been provided by Julian L. Morris, BSc (Hons), MBA, PMP, SA, Global Project Management Community Leader, CX, Cisco.

Just as our employees deliver exceptional customer experiences through professional and support services, the Cisco CX journey delivers additional value in digital capabilities, data, portals, APIs and more. Each CX component from our people to our technology is critical in delivering the best comprehensive customer experience at scale.

As the industry transforms, the way customers define value is also changing. They expect value from every step of the life cycle. In many cases, customer experience even surpasses price and product as a key brand differentiator. At Cisco, CX is at the center of ground-breaking innovations.

The Cisco customer life cycle defines the stages of the customer journey through key touch points. If the customer is delighted at one touchpoint, then then transition to and through the next is seamless. Woven together these experiences strengthen customer confidence and are fundamental in building close, meaningful, and enduring relationships across our global ecosystem.

CX – Global Role Communities and Employee Experience

The concept of Role Communities was implemented as part of Cisco's CX transformation. As already stated, industry transformation now occurs at breakneck speed. As a result, Cisco is evolving at an even faster pace and continues to lead the way forward in this journey.

We've learned that continuously elevating the value Cisco brings to our customers requires a laser-focus on driving best-in-class Customer Experience. We've also learned that delivering professional growth opportunities for our employees is equally important.

Cisco's internal "Role Communities" were launched to instill a strong sense of "*belonging*" and a "*commitment*" to role professionalism along with consistency and quality through active and global peer communities. Such affinity is critical to a great culture because truly *great Customer Experience is born from a great Employee Experience.*

Employee Experience is fundamental in building amazing customer experiences, you need to first start with the moments that matter to the employee. Role Communities is a key part of enhancing employee experience by making engagement and culture a priority within the organization. It's a manifestation of people and culture coming together to drive norms and cultural shift. It's a deliberate approach in creating the most fulfilling empathy-based employee experience by focusing on the moments that matter. Taking the persona focus to an individual level, as an enriching experience is personal. Every role knows the value of the work they are doing, which creates a culture of empowerment, value, and innovation. Cisco had the biggest endorsement in 2020, by being named **#1** on the "World's Best Workplaces" list by *Great Places to Work and Fortune Magazine* for the *second year* in a row!

CX Role Communities represent the horizontal lens through which the CX journey is viewed via each core role's unique perspective. These peer-based perspectives also serve as the change agent for those roles.

Each Role Community represents a single point of contact that owns the role definitions, processes, assets, and evolution for the overarching core CX roles. The vision for role communities is to *build and grow the most highly engaged global community of customer-obsessed professionals that share, innovate and scale learning & best practices to deliver extraordinary customer experience together.* All based on three deeply held values of *Customer Obsession, Trusted Expertise and being Extraordinary Together.*

Role Communities are a key mechanism in the acceleration of organizational transformation. The impact of communities is broad and deep addressing strategic core role evolution,

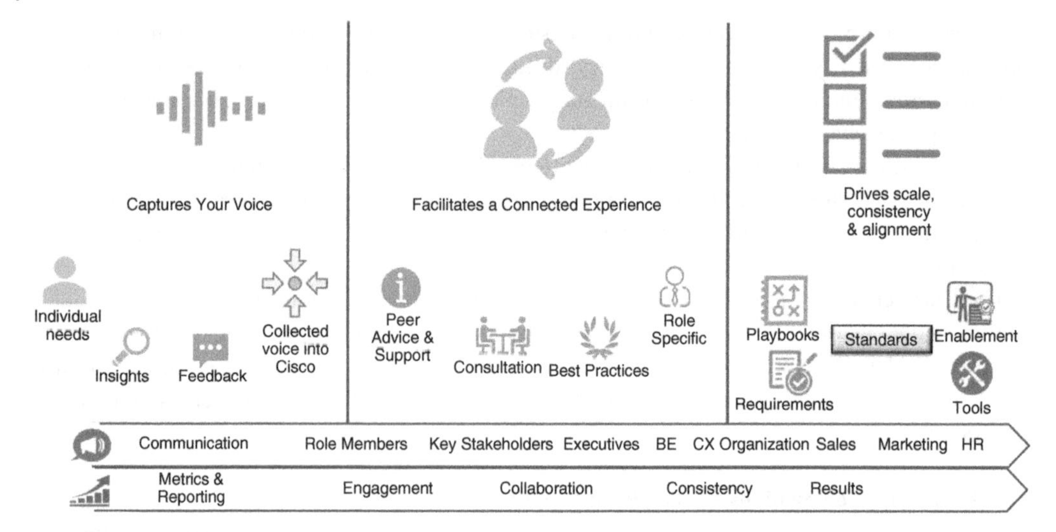

Figure 1-21 What Is Role Community?

functional excellence, and individual performance by making knowledge transparent and easy to access.

Each unique Role Community captures the voice of individual contributors and facilitates a connected experience across the globe, ensuring consistency and alignment to business strategy and objectives.

In summary, Role Communities drive employee experience via role engagement, role-consistency, change-management, and transformation.

Figure 1-21 illustrates the general concept of what a Role Community does.

Global Project Management Role Communities

As illustrated in Figure 1-22, Project Management is as one of the 12 core CX customer facing roles. The Cisco CX global Project Management team comprises over *one thousand* Project and Program managers delivering over *ten thousand* active projects and programs with multiple billions of dollars in revenue.

There are three main customer facing project management delivery roles: *program manager, project manager*, and *project specialist*. Program managers successfully manage and lead programs to deliver customer business outcomes for complex, transactional, and multicomponent interdependent programs. Project managers successfully manage and lead projects to deliver business outcomes for our customers. Both roles also provide leadership to the program team and delivery teams to deliver programs and projects on time, within budget and with high quality, driving customer satisfaction. Project specialists support project and program managers through the delivery life cycle.

One of the key areas in implementing the Project Management role community was to create the right governance structures to ensure key constituents within the global Project Management community were represented. This was achieved by the creation of the *Global Project Management Council*, which is made up of the three main regional geographies and underlying business segments plus the CX Centers and key corporate functions. This was to ensure we were able to manage global coordination, scalability, consistency, and transformation. This model accelerated the evolution of the Project Management role in line with CX transformation.

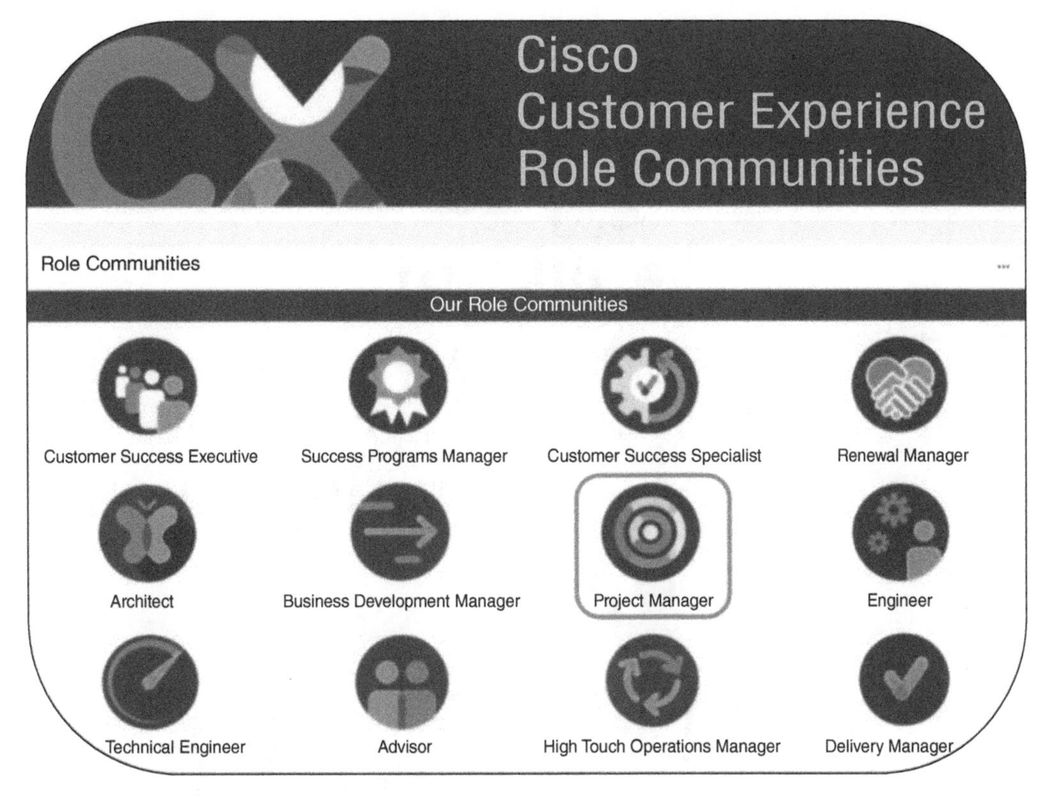

Figure 1-22 Cisco CX Core Customer Facing Roles

Global Project Management Council

Figure 1-23 shows the global and regional structure of the Project Management Councils. The Global Council is made up of 12 seats representing the key parts of the Project Management population. The global and regional council flow information in both directions. To anchor the role of the Global Council, a comprehensive Constitution was introduced to bring clarity to the role of the council.

The Council Constitution

A clear vision statement was created: To Build a BEST-IN-CLASS Project Management community that delivers the HIGHEST LEVEL of VALUE throughout the CUSTOMER JOURNEY.

Alongside the vision a mission statement was defined:

- To identify and respond to employee engagement, adoption, tools, method, and skills needs of the Global Project Management Community.
- To engage with employees, Customers, and other stakeholders to identify current and future skills needs, plan & ensure global execution accordingly.
- To align PM Community talent growth and delivery capabilities to future business/customer demands and stimulate innovation.

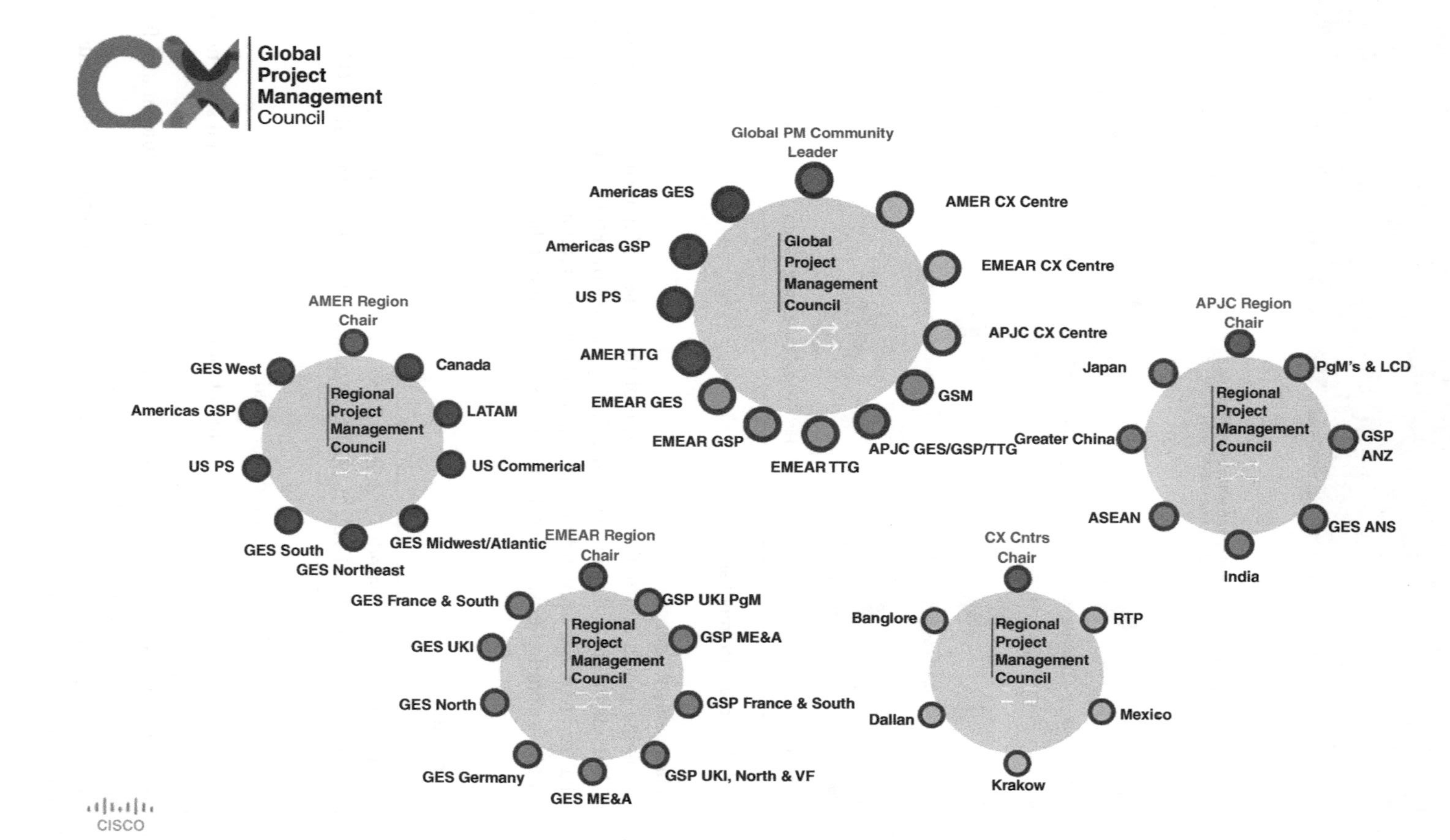

Figure 1-23 Global Project Management Council Structure

- To enable an inclusive community response to Global Project Management and wider Customer Experience Policy and initiatives.
- Provide a central body to drive global consistency with regional nuance to advance the project management capabilities in alignment to the CX strategy.

Plus, the Project Management Council members are required to commit to the following principles:

- **Objectivity:** Members should act for the widest benefit of others.
- **Accountability**: Every Member sits on the PM Council on behalf of the stakeholders they represent, and it is expected that every Member ensures that reasonable arrangements are in place to report back on their work.
- **Openness:** Members should be as open as possible about all their actions as Members of the PM Council. It should be ensured that reasons are given for decisions taken by the PM Council.
- **Confidentiality:** Every Member should respect confidentiality, where that is relevant, and ensure that confidential material is protected and that it is not used without permission from the PM Council.
- **Promoting equality and respect for others:** Every Member should undertake their responsibilities with due regard to the need to promote equal opportunity for all and demonstrate respect and consideration toward others.
- **Non-bias:** Toward decisions made on global standards, regardless of personal views on the topic.

Defining the core principles of a PM Council Member is critical to the success of the Council. Members are expected to work on the basis of mutual support, shared values and a culture of joint working and collaboration. This behavior and maturity of approach is at the heart of everything the PM Council represents.

Global Project Management Community – Governance Model

Figure 1-24 shows the complete Project Management Community governance model. It was created to directly align the *Executive Sponsor* for the Project Management role within CX with the *Global Project Management Council* (as shown in Figure 1-23) and ultimately the global *Project Management Community*. We use the concept of *Create, Approve & Adopt* to implement initiatives across the community.

Global Project Management Community – Transform for Success

To begin the Project Management transformation journey, we first needed to understand our starting point. To establish this, we undertook several assessments to inform our position.

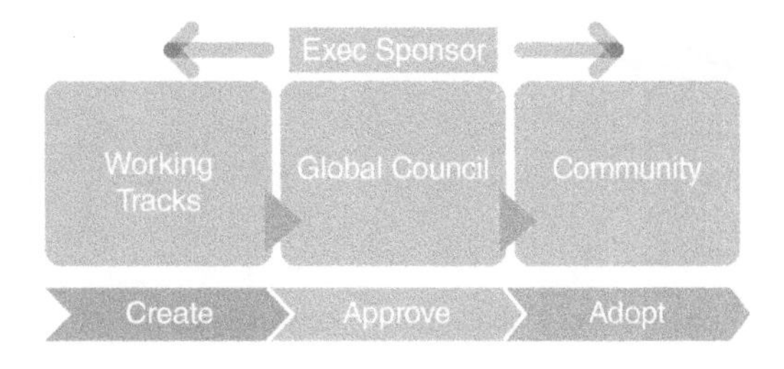

Figure 1-24 Global Project Management Community – Governance Model

The *Leadership Transform for Success survey* was made up of four distinct areas:

- What has "Changed" in 1 year in your PM area/community?
- What PM "Maturity" score would you give your area?
- What are the "Barriers" to Success?
- What is the one thing you would "Change"?

Additionally, a maturity assessment of the PM capabilities using the five levels of the *Capability Maturity Model Integrated* (CMMI) was completed. The Project Management Community individual contributor feedback was captured via direct verbatim comments using the Community Platform and the official quarterly role community survey results. All these date points were consolidated and fed into the key analysis input stage.

Identifying the Focus Areas

Taking the *Transform for success* analysis (shown in Figure 1-25) and using the *Global Project Management governance model* (shown in Figure 1-24) to work on the key analysis to identify and determine a clear set of *Focus Areas* (Figure 1-26) to address on behalf of the Global Project Management community in line with the wider CX transformation.

These focus areas and the wider narrative were taken back to the global Project Management community and tested to ensure that everyone was aligned. Community and leadership feedback at this stage was greatly encouraged. Before moving to the next stage, we captured all ideas until we had a comprehensive list for each of the four focus areas.

Global Project Management – Transformation Map

For the next stage we organized a face-to-face envisioning workshop spread over three days onsite at our Cisco office in Raleigh, North Carolina. In attendance were all representatives from the Global Project Management Council (Figure 1-23), plus selected Project Management subject matter experts (SMEs) and professional facilitators.

Coming out of the intensive envisioning workshop was a comprehensive *Transformation Map* (Figure 1-27) defining all the key initiatives for each focus areas, prioritized into a time horizon

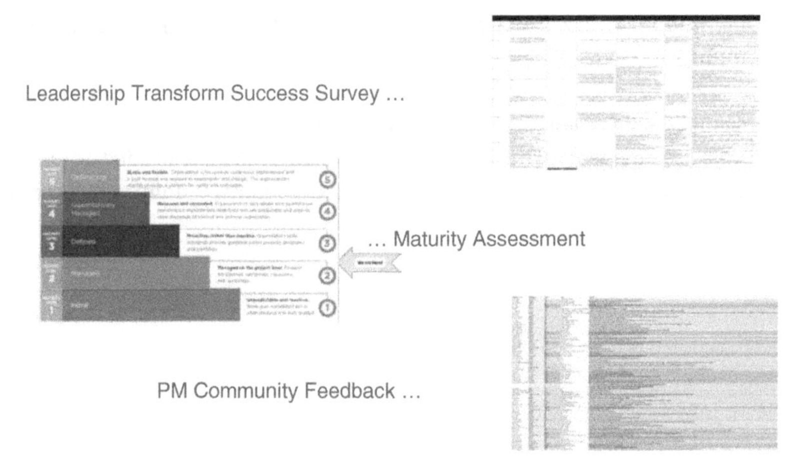

Figure 1-25 Global Project Management – Key Analysis

Community
Governance
Capability & Skills
Platform & Method
Great Customer Experience is born from a great Employee Experience
We strive to develop a strong sense of belonging and a commitment to the PM Profession, consistency and quality through active community participation
Build the right capabilities, equipped with the skills, tools, and standard approaches to deliver functional excellence.

Figure 1-26 Global Project Management Community – Focus Areas

and sequenced. Under each initiative were several deliverable actions that were recorded in a Transformation action log as we moved to execution phase.

As you can see from Figure 1-27, the overriding evolution path for the Global Project Management Community was from "*Deliverable Coordinators*" with a focus on *cost to serve* to ultimately "*Value Orchestrators*" with a drive to *time to value*. Everything in between spread over the various focus areas and time horizons made this evolution possible.

Conclusion

Implementing Role Communities as part of the Cisco Customer Experience transformation was a critical step in creating a powerful employee experience. Role Communities allowed Cisco CX to pivot the core roles into the CX transformation journey. In particular, the Project Management Community allowed the role to evolve at a faster rate by focusing the role evolution on moving from "*Deliverable Coordinators*" with a focus on *cost to serve* to ultimately "*Value Orchestrators*" with a drive to *time to value*. It also allowed us to successfully embed industry-standard Project Management competencies and culture at scale while keeping project managers tightly integrated with the delivery teams providing proximity to customers.

The Project Management Role Community gives each project professional the ability to use their voice to provide direct feedback and play their part in the role journey at every step of the way. It also creates a strong sense of *belonging* and is a primary focal point for all things Project Management. And because it is accessible to all employees, the platform provides a focal point to access resources, advice, and expertise to establish project management as a strategic competency.

This approach supports a clear narrative for mutual understanding and allows all members to participate in the journey. Achieving the right blend of individual feedback, alignment with the business objectives, and converting that into a clear narrative backed up with a set of initiatives that can be measured is a very powerful thing.

"Coming together is a beginning; keeping together is progress; working together is *success*."

– *Henry Ford*

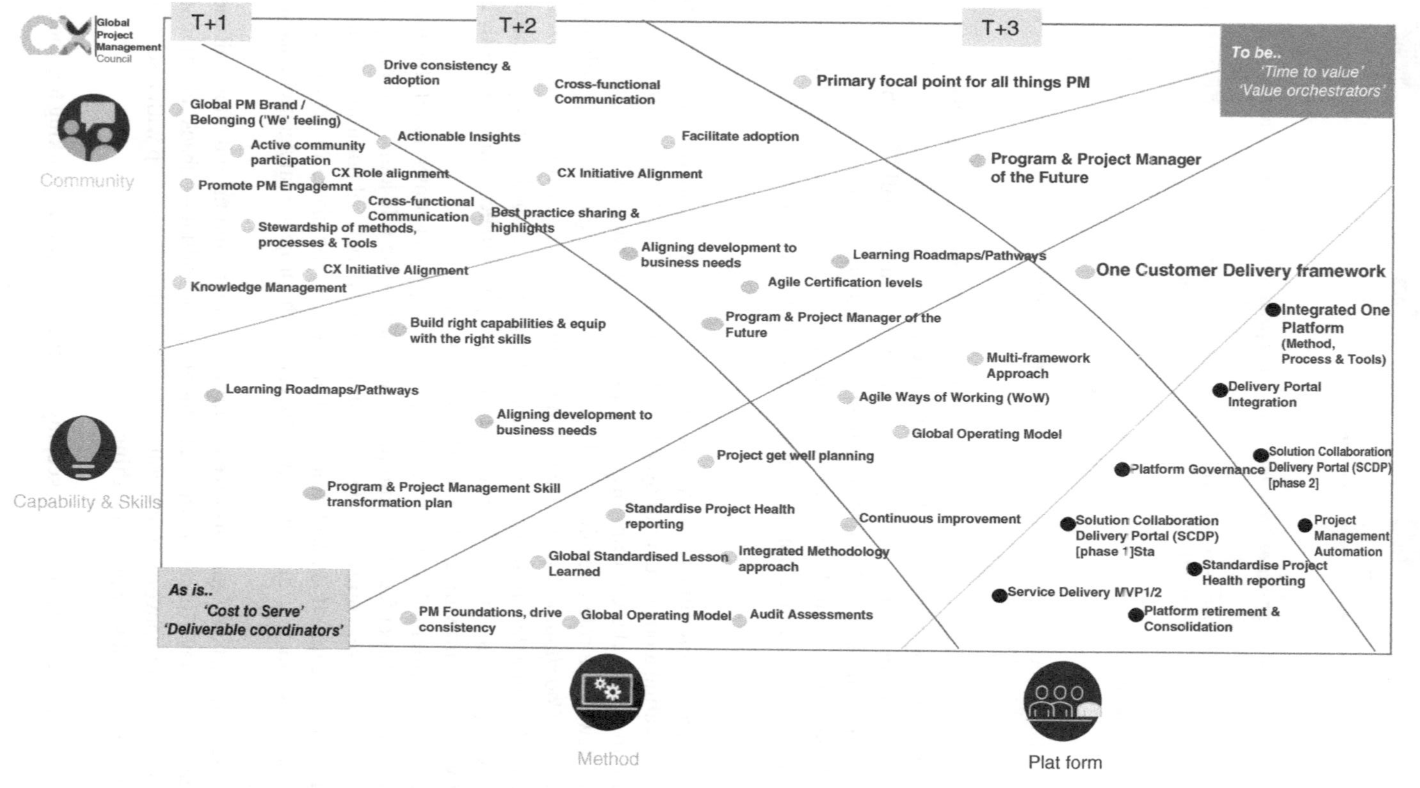

Figure 1-27 Global Project Management – Transformation Map

1.14 Excellence in Action: Servicenow[8]

Uncertainty has always been a characteristic of the modern business environment but was made dominant by the COVID-19 pandemic. For companies adapting to the pandemic, the learning curve to take on new challenges – an 'all-remote' workforce, changing revenue streams, and an uncertain economic outlook – was steep as organizations attempted to pivot people, processes, strategies, and technology architectures. As a result, less nimble companies struggled to adapt.

This ability to pivot is the heart of *strategic agility*. Successful companies weather disruption by not only maintaining business continuity, but by quickly adapting to new business challenges and opportunities.

At the start of the pandemic, a global survey conducted by ESI Thought Lab and ServiceNow found that just 10 percent of large enterprises described themselves as *agile*. By the end of 2020, that 10 percent had almost tripled to 28 percent.

Yet even that percentage seems modest, given the increasing velocity of change in the business environment. Companies are scrambling to adapt to massive shifts in customer demand and employee needs. Disruption on both of these fronts has put a new premium on an organization's ability to respond quickly to changing market conditions.

Corporate digital transformation is key to digital transformation.

Modern companies now realize that to achieve strategic agility, they must digitize manual processes and workflows. This need for digital transformation is driving massive corporate investment in an array of technologies, including automation and machine learning.

According to IDC, companies are expected to invest $6.8 trillion globally between 2020 and 2023 in digital technology initiatives (Manek 2020).

By focusing on digital transformation efforts, the modern C-suite can achieve its goal of strategic agility in response to business uncertainty.

The Shift for the Project Management Office (PMO)

The modern PMO requires new capability and expertise to advise business leaders on where and how to quickly reallocate resources in response to shifts in business demand and risk. The emphasis is on outcome realization across business lines; the PMO must be positioned to advise and manage the nimble reallocation decisions needed for strategic agility.

Project management has traditionally focused on establishing certainty through risk reduction in areas such as time, scope, and cost management. But controlling risk becomes moot if market demand shifts significantly. Often the traditional approach can deliver a project on time and on budget but not deliver the product or service that's genuinely needed.

In the post-pandemic world, PMOs have to shift their mindset from risk reduction to strategic agility. In practice, this means a shift in focus to guiding realization of business outcomes. To do this, PMOs have to expand their range of expertise from project management to program and product management.

The challenge: develop the strategic muscle to oversee multiple projects, products, and programs to ensure business outcomes and value realization.

---

8 Authors: Simon Grice (Senior Director, Innovation), Doug Page (Senior Manager, Product Management), Rani Pangam (Senior Director, IT Project Management), Tony Pantaleo (Director, Product Success) Acknowledgement: Used with permission from ServiceNow, Inc.

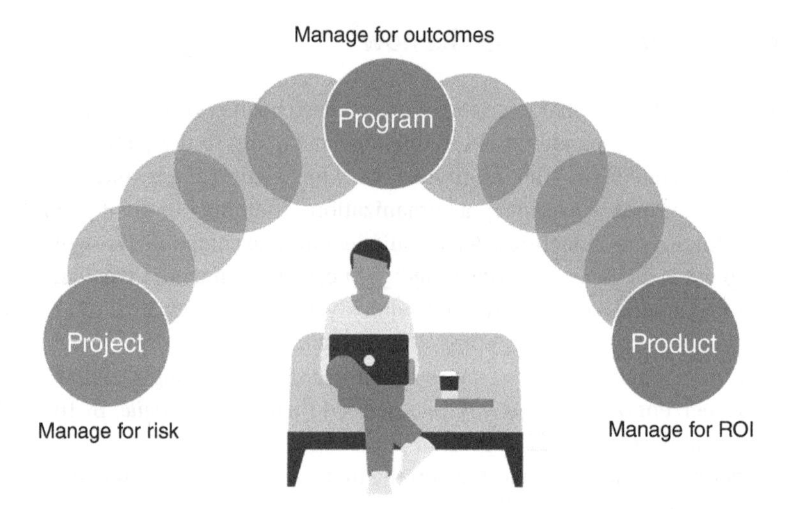

Figure 1-28 Moving the PMO Focus from Risk Reduction to Strategic Agility

Changing the Definition of Project Management

To develop these new competencies, a new definition of project management is emerging (Figure 1-28).

Today's fast-changing world requires major changes in project manager skills and mindsets. Agility, team collaboration, and change management are replacing traditional methods of tactical control and strict governance.

Project management has been typically defined as a set of activities designed to produce a specific deliverable tied to a product or service. The focus has been on governance: managing in terms of time, scope, and cost.

The new definition adds *program* and *product* management to the PMO's approach.

Product management focuses on the development and continuous delivery of a product or service. In this case, success is measured not by effective governance but by user adoption and experience; feedback is then used to drive ongoing improvement. Successful product management practices enable companies to navigate through change to deliver and sustain a competitive product or service.

Program management is the ability to encompass a set of projects and/or products that deliver outcomes linked to a company's strategic goals; these programs become part of a portfolio. An IT portfolio, for example, is a collection of programs and projects that share productivity and operational efficiency goals, which are then linked to corporate objectives.

Program management represents a new operational approach that moves from tactical risk management at the project level to enterprise management of value and outcome realization. This orientation is important in business environments where the value of a project isn't always obvious or results in unintended consequences. This situation is likely to become more common as the proportion of enterprise value derived from the customer and employee experience increases.

A shift toward program management (Figure 1-29) positions the modern PMO to:

- Influence stakeholders around how value and measurable outcomes are defined across enterprise capabilities

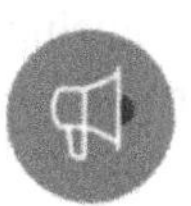
Influence stakeholders around how value and measurable outcomes are defined across enterprise capabilities
Drive return on investments in crucial company strategies and business priorities
Measure and communicate progress against various project and product outcomes to business executives

Figure 1-29 Modern PMO Responsibilities

- Drive return on investments in crucial company strategies and business priorities
- Measure and communicate progress against various project and product outcomes to business executives

A modern PMO should be able to quantify outcomes, perform against business priorities, drive a return on investment, and communicate progress to executive leadership.

Drawing upon its experience across projects, products, and programs, the modern PMO develops understanding and insight into the levers that impact customer and employee needs. With this understanding, the PMO can provide consistent reporting and advice to business leaders on how outcomes will be delivered and achieved – and where pivots are needed – to support company goals and priorities.

A good recap of many of the changing practices in the practices of project management is highlighted in the Farm Credit example in the next section.

1.15 Excellence in Action: Farm Credit Mid-America[9]

The Farm Credit System is comprised of over 70 independent financial institutions, referred to as Associations, with a unified goal of securing the future of rural communities and agriculture. Our member-owned cooperative structure ensures our business decisions remained focused on the customers we support. "Our loans and related financial services support farmers and ranchers, farmer-owned cooperatives and other agribusinesses, rural homebuyers and companies exporting U.S. ag products around the world" (Farm Credit 2021).

Farm Credit Mid-America, one of the largest Associations in the Farm Credit System, serves the credit needs of farmers and rural residents across Indiana, Ohio, Kentucky, and Tennessee. We provide a wide range of financial services including real-estate loans for land purchases, operating loans designed to meet the feed, seed, and fuel needs for daily operations, business loans,

9 Material in this section has been provided by Dr. Chuck Millhollan and Daro Mott.

equipment financing and crop insurance products. We have approximately 80 retail offices spread throughout rural communities in our four-state area of operations.

In 2012, Farm Credit Mid-America was 19.7 billion earning asset Association (Farm Credit Mid-America 2021 Annual Report 2012). Led by the desire to serve their member-owners, Farm Credit Mid-America stood up their Operations team in early 2012 with the objective of mapping their business processes and identifying opportunities to not only gain efficiencies, but also enhance their customer experiences through an in-depth understanding of their core business processes and how they delivered value to their members. The Operations team, staffed with continuous improvement professionals with backgrounds in Six Sigma and Lean Manufacturing, conducted extensive Voice of the Business workshops to elicit not only a deep understanding of the current state business processes, but also to identify and prioritize needs of our team members related to process stability and enhancement.

The Operations team identified hundreds of process improvement opportunities, and Farm Credit Mid-America senior leadership identified the need for structured approach for not only prioritizing, selecting, and implementing the projects, but also making decisions based on strategic alignment and value. In 2014, Farm Credit Mid-America established the Process Improvement and Execution team with the responsibilities for managing the Association's strategic project portfolio, including the allocation of resources and prioritization of projects to meet organizational strategic imperatives. Due to realized successes founded in the application of portfolio management best practices, the Operational Process Excellence team was established in 2017 and expanded the Process Improvement & Execution team's scope to include providing leadership for the Association's strategic planning processes in collaboration with the executive committee and board of directors, leading business owners for key business processes, provide real-time business process support, and the collection and management of business process-related knowledge management.

As of April 2021, "Farm Credit Mid-America is backed by the strength of more than 80,000 customers and $28.8 billion in earning assets, both owned and managed" (Farm Credit Mid-America 2021). The following outlines Farm Credit Mid-America's project management and continuous improvement journey from 2012 to 2021, and how we designed our structured approach to strategic planning, strategic project portfolio management, business process ownership, and continuous improvement in a financial services environment based on a combination of practitioner experience and academic research.

Tri-Focal Lens for Project Success

As stated by Kerzner (2021a), "Many executives understand the benefits of project management, but are often at a loss on how to achieve them (p.1)." Project management has matured significantly in the past several decades; however, there remain conflicting findings on which factors result in successful projects. Published research and conceptual papers reflect inconsistencies in definitions about what successful project management is all about and what skills are needed by project managers.

An extensive literature review conducted by Millhollan and Kaarst-Brown (2016) identified the potential conflict in goals and measurement of success from three different perspectives: the process of project management, the project manager, and the project outcomes. Each of these perspectives shifts the focus on not only measurement of "success," but also the project leadership skills necessary to maximize value delivery in an organization (Figure 1-30). In practical application, each perspective of success is defined by various stakeholders at various points in time (Millhollan 2015).

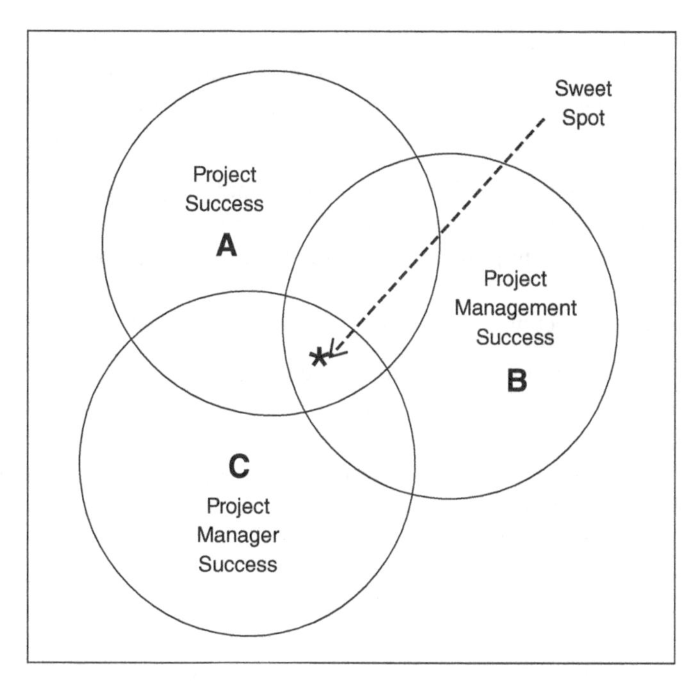

Figure 1-30 Tri-focal Project Success Lens

Paradoxes in Literature on Successful Projects

The challenge of project success comes with balancing differing expectations and perceptions (Judgev & Muller 2005; Shenhar et al. 2001). Perceived success is defined by not only meeting the technical requirements and providing a product, service, or result as defined in the project objectives, but also by achieving high levels of satisfaction from the stakeholder groups (Baker et al. 1988). Analysis of the literature supports that the challenges related to these different sets of project success factors and different categories or groupings are threefold (Millhollan 2015):

1) Some of the factors that contribute to project success are realized during a project, such as meeting project related constraints like budget and schedules and creating new products or services.
2) Other project success factors might not be realized until long after project completion, such as customer satisfaction or commercial success.
3) The factors influencing project success measurements and perceptions are often in conflict. For example, meeting a budgetary or schedule constraint can have a negative impact on satisfying technical or functional requirements.

These points also highlight that a successful project is a function of metrics not usually considered in the literature on project success: effective outcomes associated with stakeholder analysis, decision-making, negotiation, conflict resolution, change management, and politics of change.

Paradoxes in Literature on Project Management Success

Consistent across the literature are arguments that project management has not provided the expected benefits it is common for stakeholders to place blame on project management when projects fail. For this perception to be true, the failure would need to be rooted in the ineffective application of the project management methodology, failure to effectively plan, or a lack of structure related to managing delivery according to agreed-upon constraints and objectives. Millhollan and Kaarst-Brown (2016) concluded that three paradoxes exist in the literature, aligning with Munns and Bjeirmi (1996), Azim et al. (2010), and Lacerda, Ensslin and Ensslin's (2011) views:

1) Effective project management methodology can contribute to project success because it provides a structured approach and standard tools or procedures.
2) Effective project management provides a structured approach, but does not ensure success of the project.
3) Absence of effective project management methodology contributes to project failure.

What is noticeably absent from the literature on successful project management is the decision-making associated with the selection of tools and techniques. There are skillsets that we say are important and can tie back to various project success metrics, however, there are also gaps in how project management literature and reputable project management standards focus on these. The key paradox in the project management (process) success, there is an implied dependence on skills outside project management methodology.

Paradoxes in Literature on Project Manager Success

There is agreement in the literature that project manager competencies are an essential ingredient for project success, that a project's success or failure is influenced by who manages that project, and that project managers must leverage their soft skills to effectively manage often complex, multi-disciplinary teams (Patanakul 2011; Pinto and Slevin 1988). Given that perceptions of success are heavily dependent upon project outcomes and how the project management tools and techniques are leveraged to assist in producing expected outcomes, it becomes apparent that a project manager's ability to elicit, understand, and manage stakeholder expectations throughout a project life cycle, and often even into the product life cycle, is paramount for project success. The project manager's opportunity to influence perceptions about project success rests in their ability to understand what stakeholders value, manage the real-world factors that influence how the project delivers value, and ensure the reality delivered and expectations are aligned (Millhollan 2008).

Based on the gaps presented in the previous sections and the author's experiences, it appears that application and decision-making around the techniques (hard skills) of project management are the critical skills of the successful project manager. This begins with the following (Millhollan 2015):

1) Skills associated with interpersonal interactions to elicit and understand stakeholder expectations related to a specific project so that one can use this information to identify and prioritize factors that will influence their perceptions of success.
2) Ensuring aligned expectations between different stakeholder groups through communication, negotiation, and conflict resolution skills, as these expectations could not only be in conflict, but also evolve over time as the project progresses from early planning through execution and delivery.
3) Decision-making and negotiation skills to develop strategies to manage not only the project, but also stakeholder expectations about agreed upon end-state goals.

If research indicates a specific set of skills, or range of skills, is necessary to be an effective project manager, why do the professional standards not provide descriptions that are more detailed or provide guidance for procuring and developing these skills? If we know that successful project

Table 1-2 Summary of key paradoxes in literature (Adapted from Millhollan and Kaarst-Brown 2016)

Category of Literature	Paradoxes
Project Success	• Some of the factors that contribute to project success are realized during a project, such as meeting project related constraints like budget and schedules and creating new products or services. • Other project success factors might not be realized until long after project completion, such as customer satisfaction or commercial success. • The factors influencing project success measurements and perceptions are often in conflict. For example, meeting a budgetary or schedule constraint can have a negative impact on satisfying technical or functional requirements.
Project Management Success	• Effective project management methodology can contribute to project success because it provides a structured approach and standard tools or procedures. • Effective project management provides a structured approach, but does not ensure success of the project. • Absence of effective project management methodology contributes to project failure.
Project Manager Success	• Skills associated with interpersonal interactions to elicit and understand stakeholder expectations related to a specific project so that one can use this information to identify and prioritize factors that will influence their perceptions of success. • Ensuring aligned expectations between different stakeholder groups through communication, negotiation, and conflict resolution skills, as these expectations could not only be in conflict, but also evolve over time as the project progresses from early planning through execution and delivery. • Decision-making and negotiation skills to develop strategies to manage not only the project, but also stakeholder expectations about agreed upon end-state goals.

managers need a broader range of skills, why are we not including these skills in basic or advanced project management curriculum? Is it because it is too hard, or because we consider these dispositional skills rather than skills that can be taught? These are important questions to ask. Table 1-2 provides a summary of the key paradoxes related to the tri-focal success lens.

Kerzner (2021b) emphasizes that "project success does not fit into any predefined criteria. There can be a different definition of success in each industry and even in the same industry, the meaning of success can change from project to project and in each life cycle phase (p.1)." Understand this changing definition of success phenomenon as a practitioner, and drawing upon the tri-focal lens, we designed the Operational Process Excellence team and our methodology to allow for a shift in focus on success to the intersection, or sweet spot, of each perspective.

Farm Credit Mid-America's Value-Driven Strategic Portfolio Management

Strategic Imperatives

One of the foundational concepts that we focused on early in our journey was creating a culture of focusing on realized benefits from strategic projects. Establishing a top-down ethos for benefit realization began with the executive committee and the board of directors. Using our strategic imperatives (see Table 1-3), we dedicate time during our annual strategic planning kick-off in July discussing strategic projects and their value proposition through the eyes of our board of directors and executive leaders.

Table 1-3 Farm Credit Mid-America strategic imperatives

Provide an Exceptional Customer Experience	Add value beyond customer expectations so that customers desire to repeat the experience and tell others about it.
Grow Constructively	It is important that the portfolio continues to grow and effectively serve the marketplace in a manner that benefits customers now and in the future.
Maintain Credit Quality and Administration	Doing the right things in the right ways allows us to work with customers as they experience challenges. Make quality new loans, provide solid loan administration, and responsively service distressed loans.
Maintain Sustainable Financial Operations	Ongoing diligence will help provide financial strength and position the Association for future success despite economic challenges and changes in regulatory and government policy.
Live our Purpose and Values	Our purpose and values are reflected in Our Compass (see Figure 1-31) and everything we do. We work together, care for one another, and coach and grow talent to better serve our customers.

Figure 1-31 Farm Credit Mid-America, Our Compass

Table 1-4 Farm Credit Mid-America value propositions

Increased Productivity	Doing more with the same number of resources. E.g., Increasing the number of loans, a credit analyst can decide. This implies a backlog of loans ready for work (available input).
Cost Reduction	Removing unwarranted expenses (operating or capital) through gained efficiencies. E.g., Eliminating waste in operational processes.
Decreased Turn Time	Reducing the amount of time it takes to complete tasks in the loan origination process. E.g., The average time it takes to decision loans is reduced through automating obvious decisions (approvals & denials).
Enhanced Customer Experience	Measurable increase in volume, margin, or fees. E.g., Creating an experience that incentivizes members to consolidate debt held by competitors.
Enhanced Employee Engagement	Measure of the emotional connection to our Association, workplace relationships and collaboration, commitment to our purpose, and commitment to the success of our organization. E.g., Reduced turnover or higher engagement surveys.
Long-term Business Sustainability	Holistic perspective with the objective of long-term growth, financial strength, and Association stability to sustain operating capabilities through regulatory and financial challenges. E.g., Focus on long-term business impact vs short-term profit, price, or results.

Our Compass

As referenced in our "Live our Purpose and Values" strategic imperative, we leverage Our Compass as a decision-making lens through which we keep our customers, both internal and external customers, at the heart of our business planning and decision-making.

Creating a Culture of Strategic Benefit Management

Using our strategic imperatives to brainstorm metrics as a group, we defined a set of value propositions to lead a discussion related to each strategic project and the expected benefits and benefit realization timelines. Table 1-4 lists the identified value propositions and their definitions.

Our objective was not to estimate with precision and set expectations for specific benefits; however, we did want our board of directors and executive leaders to engage a discussion about the primary and secondary benefits associated with each of our strategic projects and have the context to explain how each aligns with our Strategic Imperatives and contributions to our long-term strategic goals. We defined primary benefits as the intended outcomes directly linked to the business decision or program's objectives, i.e., benefits accrue from the intended effects of the business decision or program, and these primary benefits are connected to the initiative's perceived success. We defined secondary benefits as additional benefits that might result from realizing primary benefits; however, secondary benefits are not primary objectives. By nature, the secondary benefits are less important or impactful, smaller in scale, or will not be realized in a typical business cycle.

In the next steps, our board of directors, executive leaders, and senior leaders collaborated in an exercise to identify the primary and secondary benefits for each of the strategic projects in the project portfolio. The project Executive Sponsor presented their projects and explained each of the business objectives, and then small groups would discuss the anticipated benefits. Each group would present and defend their findings and, collectively, the team would come to an agreement and build the "Value Dashboard." A sample of the Value Dashboard is provided in Figure 1-32.

Primary Benefit(s) / Secondary Benefit(s)	Operational Efficiency			Increased Revenue	Enhanced Customer Experience	Enhanced Employee Engagement	Long-term Business Sustainability
	Increased Productivity	Cost Reduction	Decreased Time				
Comprehensive Sales Philosophy	Primary			Primary	Primary	Secondary	Secondary
Customer Selection	Primary	Primary	Secondary	Primary			Primary
Credit Structure	Primary	Primary	Secondary		Secondary	Secondary	Secondary
Retail Structure	Primary	Primary	Secondary	Secondary	Primary	Primary	Primary
AgriLytic (Ag Loan Origination)	Primary	Secondary	Primary	Secondary	Primary	Secondary	Primary
Country Home Loan System	Primary	Secondary	Primary	Primary	Primary	Secondary	Primary

Figure 1-32 Strategic Project Value Dashboard

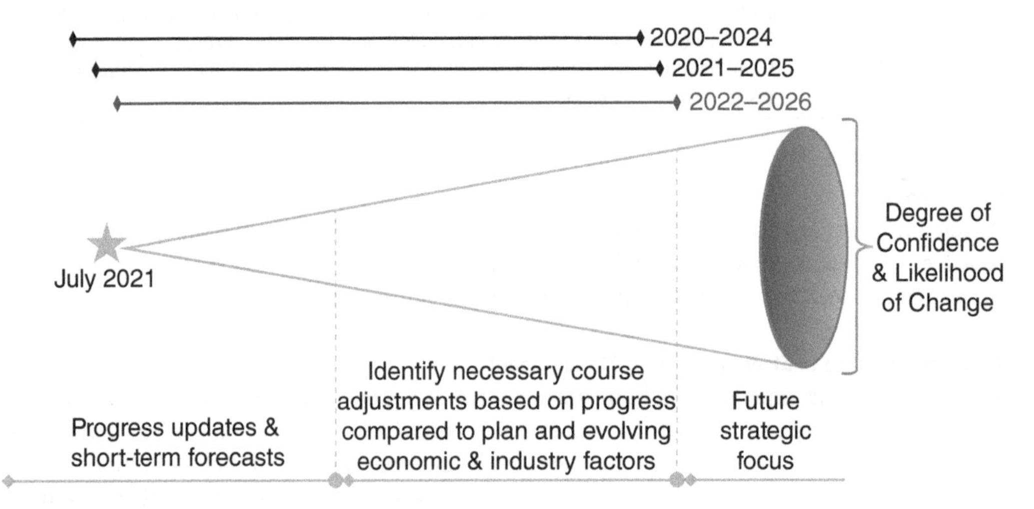

Figure 1-33 Strategic Planning Cone of Uncertainty

This entire process laid the foundation for not only why we select specific strategic projects, but also created an expectation for post-implementation benefit measure.

Strategic Planning Process

Farm Credit Mid-America has a five-year strategic planning cycle that is refreshed each year based on factors that influence our strategic direction and go-to-market strategies. Our annual strategic planning process begins in July of each year with our board of directors in collaboration with the Association's executive and senior leaders. We begin the process, facilitated by the Senior Vice President Operational Process Excellence, with a progress report on the current year business plan, including current status for our strategic projects and short-term operations forecasts for the remainder of the year. Focus then shifts to future-state strategic planning, taking into consideration progress to date on strategic initiatives and an evaluation of our competitive position and marketplace factors that influence strategic direction. We refer to Figure 1-33 as our "Cone of Uncertainty" that illustrates not only our relative degree of confidence in our strategic direction, but also emphasizes the importance of revisiting and refreshing our strategic plan on a regular basis. The outcome is clear guidance on not only our short-term focus, but also our future strategic direction as an Association.

The Senior Leadership Team uses this guidance to develop the 5-year strategic roadmap for each individual business segment and for the entire Association. Facilitated by the Vice President Process Improvement and Execution, the Senior Leadership Team evaluates current state as compared to expected future state, and leverages that analysis to identify the potential strategic projects necessary to realize the Association's strategic goals. Members of the Senior Leadership Team also serve as the Executive Sponsors for the strategic projects. This five-year, all-inclusive view of the Association's body of work is then reviewed and refreshed quarterly based on progress and marketplace influences to provide a view into the strategic projects, and operational work, we can anticipate in the coming months.

Including an initiative on the five-year strategic roadmap does not imply guaranteed approval. As projects approach initiation, the executive sponsors identify the business sponsor and business technology sponsor that will, in partnership with an assigned program manager and enterprise business analyst, prepare a formal Strategic Project Request for review and approval by the senior leadership team considering the current project portfolio and resource capacity. Once approved in concept based on the request merit, business case, and alignment with our strategic plan, the strategic project formally approved and prioritized by the executive committee. Projects are prioritized using whole numbers, and numbers are only used once. For example, there is only one number one strategic priority, one number two strategic priority, etc. This prioritization process then informs our resource allocation model. Resource managers allocate resources to strategic projects based on their prioritization, or based on business value and contribution to our strategic goals.

Operational Process Excellence Team

The Operational Process Excellence team is structured to contribute to not only effective project leadership, but also to define and manage value delivery before, during and long after project completion and transfer to operations. Our long-term objective in evolving the team to the current structure (see Figure 1-34) was to both deliver on the Association's strategic plan and to

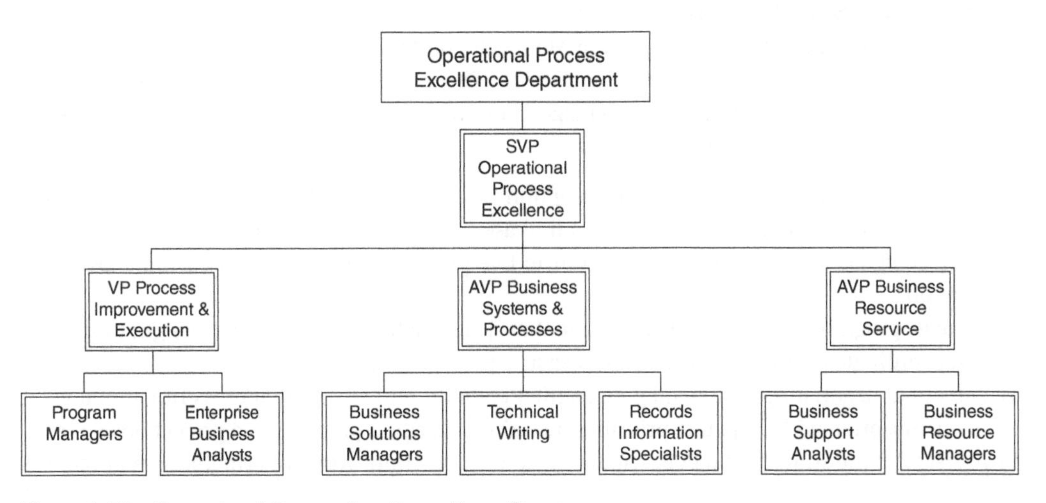

Figure 1-34 Operational Process Excellence Team Structure

proactively manage executing on long-term benefit realization through dedicated business process and product life cycle management.

Process Improvement and Execution Team

The Process Improvement and Execution team is comprised of Program Managers, Six Sigma Black Belts, Lean Practitioners, and Enterprise Business Analysts. The methodological diversity on the team was designed to not only continue with the original Operations Team's focus on continuous process improvement, but also to incorporate a tool set of skills necessary to address to meet organizational needs beyond the standard project management body of knowledge. Many factors influence a project manager's efficacy in applying project management tools and techniques.

As outlined by Danity, Cheng, and Moore (2003), there is a relationship between project success, individual project manager performance, and perceptions about project manager "performance criteria." Given that perceptions play a role in defining project success (Baker, Murphy, & Fisher, 1988), it stands to reason that project managers must possess skills in addition to the project management tools and techniques outlined in a single set of professional standards. If we asked experienced project managers, how many of them would suggest that their depth of knowledge of project management processes, tools and techniques was the key to their success? Likewise, if we went to senior leaders and project sponsors, how many would suggest they needed project managers with a deeper understanding of project management standards (Millhollan 2015)?

"Today, more and more companies believe that they are managing their business as though it is a series of projects. Trust in the abilities of project managers has increased significantly" (Kerzner 2021a, p. 2). With a focus on the future needs related to strategic project management, coupled with experience as practitioners and academic research in this domain, we purposefully built a team with a more comprehensive set of knowledge, skills, and abilities. In addition to the interpersonal skills we seek in team members, covered later in this section, we either hire for, or develop, team members to not only execute of a given scope of work, but also serve the Association as problem solvers, critical thinkers, and trusted advisors to our senior leadership team.

The value that structured business analysis and program management brings to the Association is evidenced in the grade, or salary, ranges for our Process Improvement and Execution team members. Both positions are compensated similar to Association leaders.

Project/Program Managers: The project/program management role with our Association combines knowledge and skills in Program Management, Statistical Process Control (SPC), Lean Six Sigma (LSS), Design for Six Sigma (DFSS), Theory of Constraints (TOC), Information Technology (IT), Agile/Scrum, project management, internal management consulting and group facilitation. Primary responsibilities include:

1) Manage all aspects of projects, from inception through project closure, ensuring that project management methodology is applied to all phases of the project life cycle.
2) Overseeing an individual portfolio of four to five projects in multiple phases and geographic locations.
3) Manage programs with multiple component projects, including providing oversight to support and coordinate project-level activities to ensure program goals are met.
4) Directing, negotiating, and managing conflict resolution to ensure that cross-functional team members, often with disparate priorities, remain focused on program and project objectives.
5) Assist with the strategy and deployment of Continuous Improvement principles and practices throughout the association.

6) Help lead Continuous Improvement Voice of the Business teams in identifying, scoping, and prioritizing opportunities for improvement.
7) Lead CI events and projects to successful completion and realization of the value forecasted in the event or project charter.

Enterprise Business Analysts: To fill a critical need for tracing business requirements from project ideation through to delivery and benefit measurement, we added Enterprise Business Analysts to the Process Improvement and Execution team. Primary responsibilities include:

1) Leads the business analysis approach for strategic projects.
2) Effectively communicates with all stakeholders throughout the life cycle of projects.
3) Elicit, document, and manage requirements and related acceptance criteria.
4) Ask probing questions to understand user requirements with the ability to provide feedback on possible implications of all requests. Accurately and concisely articulate issues, questions and recommendations to business stakeholders, team members, and management.
5) Works with the business owners, program managers, and other stakeholders to manage the expectations for benefits realized after project completion.
6) Collaborate with the business owner and other stakeholders by discussing business and technical impacts of and dependencies related to priority decisions.
7) Identify, establish, map, improve, and measure association business systems, processes, and value streams.

Business Systems and Process Team

Comprised of Business Solution Managers that own key business processes and serve as business owners for projects, Technical Writing that functions as the editor-in-chief for our process and procedure manuals, and Records Information Specialists that provide centralized loan document preparation, scanning, indexing and quality assurance.

Business Solutions Managers: The geographical dispersed structure of our Association, with 80 retail offices in four different states, led the identified need to have a centralized team of business process subject matter experts to serve as the liaison between our user groups and project teams. Primary responsibilities include:

1) Engaging key stakeholders to execute the overall business process vision and deliver solutions that allow the business segments to achieve operational excellence.
2) Representing the business and contributing to the achievement of association goals and objectives through strategic projects, critical run the business initiatives and departmental requests.
3) Enabling the lending process through business ownership of technical systems. Proactively soliciting feedback from a diverse group of business users on processes, new system development, features and existing system functionality for enhancements and continuous improvement.
4) Facilitating the development, documentation, retention, and maintenance of lending and retail office operations process and procedures.
5) Managing and prioritizing backlog; working with the business to manage expectations for completion.
6) Providing timely, efficient support and service to end-users and leaders through consistent, repeatable work processes and interpretation of policy and procedures.

Technical Writing: The Technical Writing function was established to not only create and manage the taxonomy governance process, but also to engage with our project teams to maintain our process and procedure manuals as business processes and technologies evolve. Primary responsibilities include:

1) Manage the organizational process and procedure manuals through the manual change management process.
2) Manage all new/change requests, request consolidation, communicate all manual change requests to the change management team and manuals subject matter experts (SMEs).
3) Conduct manual content needs assessment, knowledge transfer, and approval sessions with the change requester, manual SMEs.
4) Manage the communication process for manuals changes.
5) Manage the manual and intranet taxonomy governance process, which includes supporting taxonomy and the creating relevant metadata for content to improve search and find, ensuring the governance policies and procedures are sustained and communicated, and training the association on the importance of taxonomy, as relevant.
6) Collaborate with core Manual Change Management Team members and manual SMEs and other stakeholders on manual change management process improvement opportunities, document and manage all team actions plans.

Records Information Specialists: Loan documents are one of the primary outputs of our lending processes and systems. The Records Information Specialist team was established to ensure access to, and easy identification of, loan documents throughout the lending and servicing processes. Primary responsibilities include:

1) Index documents by ensuring integrity of documents, electronic images, and/or records metadata within set service level agreements and accuracy guidelines.
2) Review batches indexed by other team members for accuracy and completeness.
3) Provide technical support to other departments on document imaging issues and assist users in requests for information related to indexed documents.
4) Identify, document, and provide input to management related to business processes to improve the quality of records management programs.

Business Resource Service Team

Again, due to the geographically dispersed nature of our business, we identified the need for a centralized team that would support operations and be the first in the line of defense to identify process inefficiencies and system enhancement opportunities.

Business Support Analyst: Business Support Analysts are analogous with a helpdesk for business process. They are subject matter experts on our key business processes and experienced users of our core lending systems. Primary responsibilities include:

1) Provide first point of contact for internal business user support on business systems, policies/procedures, and processes.
2) Engage in active troubleshooting and problem-solving in regard to systems, credit policies/processes, and retail/consumer operations. This requires an in-depth understanding of Association's internal and external business systems and processes, workflows, products, credit philosophy, policies, and procedures.
3) Recommend solutions by asking probing questions, analyzing, and documenting detailed information from business users regarding complex business situations utilizing subject matter experience as well as a network of resources.

4) As a project team lead, research and analyze current and historical concerns based on data captured to evaluate best practice solutions or trends.
5) Establish, maintain, and share documented processes and solutions for workflows, products, credit philosophy, policies, and procedures to serve as a knowledge resource for the team.
6) Provide subject matter knowledge as team lead on project teams and continuous improvement efforts by discussing the identification of trends and business needs and best practices in the area of lending processes and retail/consumer operations and implement proactive risk mitigation practices.

Business Resource Manager: Leveraging the knowledge discovery and creation processes from the Business Support Analysts, Business Resource Managers research, review, and analyze existing documented business knowledge and identify opportunities to design, develop, implement, support, and enhance business process solutions. Primary responsibilities include:

1) Manage all aspects of complex or escalated business process concerns raised up from Business Support Analysts regarding business systems, policies/procedures, and processes; serve as a specialized resource that clearly communicates short-term solutions and evaluates and collaborates to determine long-term solutions.
2) Oversee concerns regarding systems, credit policies/processes, and retail/consumer operations and create, evaluate, and review appropriate entries into knowledge base.
3) Research and advise long-term solutions for complex business situations by asking probing questions, analyzing, utilizing subject matter expertise and a network of resources, and documenting detailed information from business users.
4) Evaluate comprehensive lending and operational procedures, manuals, and forms; and serve as subject matter expert influencing updates. Coordinate and cooperate with Operational Process Excellence department resources, business owners and partners.
5) Proactively share knowledge with team members through documentation and updates to internal key processes in order to maintain balance between effective processes, and safe and sound lending practices.
6) Evaluate and measure trends in business needs and provide regular ongoing feedback to stakeholders regarding best practices in the area of lending processes and ag retail/consumer operations and implement proactive risk mitigation practices. Share and consistently implement best practices with team on credit, retail, and consumer operational matters.

Project Environment

"Over the past two decades, there has been growing empirical research showing that servant leadership enables companies to develop and maintain a competitive advantage. Servant leadership is a philosophy where the main goal of the leader is to serve the team rather than the team working to serve the leader" (Kerzner 2021b, p. 11). This phenomenon, coupled with our academic research and practitioner experiences, led us in designing our project environment and project leadership processes with a focus on servant leadership and trusted advisor philosophies, organizational change management, and adaptive methodologies.

Servant Leadership and Trusted Advisor Philosophies

A trusted advisor is trusted by leaders to solve their biggest challenges (Green & Howe 2012). Becoming a trusted advisor is a coveted and privileged position for anyone who works in the capacity of solving problems on behalf of leaders. The Operational Process Excellence Team recognizes that it must spend the necessary time to cultivate relationships with key leaders within the organization; once mutually beneficial relationships are developed, the Operational Process Excellence team assigns its most talented and skilled personnel to maintain relationships with teams and influential individuals within the organization. Due to our focus on partnering with, and serving, the business, it is common for Operational Process Excellence team members, i.e., program managers, enterprise business analysts, and business solutions managers, to be embedded within the decision-making, planning, and execution processes with business segment leadership teams.

Organizational Change Management

"These new or strategic projects, which required a strategic project leadership model, focused on creating competitive advantages, change management, and thinking outside of the box" (Kerzner 2021b, p. 8). Organizational change management, as related to strategic projects, is facilitated by the Operational Process Excellence Team. Change management is a comprehensive, cyclic, and structured approach for transitioning individuals, groups, and organizations from a current state to a future state with intended business benefits (PMI 2013, p. 7). Strategic projects represent the Association's priorities, and those priorities are articulated as officially authorized changes which will best transition Farm Credit Mid America to its desired future state. We designed our processes with the understanding that effective project selection, planning, and implementation is only the beginning of value delivery. The Operational Process Excellence Team partners with Association team members who are deeply interested and play a role in organizational change management, including organizational development, business technology, corporate communications, and other shared services.

The Operational Process Excellence team leads organizational change management in several ways.

1) The Senior Vice President (SVP), Operational Process Excellence is the chairperson for the Senior Leadership Team, facilitates the business planning process, and partners with Executive Committee members to address critical business challenges. The Senior Leadership Team is a cross-functional team of business leaders and shared service leaders who sponsor all the projects within the strategic project portfolio.
2) The Vice President, Process Improvement and Execution facilities cross-functional change management and communication meetings. The purpose of those meetings is to improve company-wide communication practices and bring consistency to the process of implementing change.
3) The Operational Process Excellence Team, specifically the enterprise business analysts and program managers, work each day to fulfill the business objectives of Farm Credit Mid-America's prioritized strategic project portfolio.
4) The enterprise business analysts, while gathering business requirements from business owners, leaders and contributing subject matter experts, identify potential changes to business processes and the associated knowledge, skills and abilities required for effective project implementation to inform the change management processes.

Hybrid, or Adaptive, Methodologies

"The landscape for project management changes almost every year. Some changes are relatively small or incremental and can be implemented quickly whereas other changes may require significant organizational support before becoming reality" (Kerzner 2021a, p. 1). As we designed the Operational Process Excellence team and our project management processes, we understood the importance of designing processes that avoid unnecessary rigidity, adapted to evolving business needs, and applied the tools and techniques specific to the business problem.

The Operational Process Excellence Team leads projects that require the practical application of Agile, traditional, and hybrid project delivery methods. Applying the concept of methodological diversity, the Operational Process Excellence Team selects the approach that matches the business need. For example, when developing a new loan origination process for smaller agricultural loans that did not require the detailed underwriting processes necessary for large loans, the team selected an Agile methodology. We did not have a legacy system or process that required dual entry, and the business environment allowed for incremental delivery of value-added functionality. Accordingly, the project management team, partnered with business owners, cross-functional subject matter experts, and business technology to define priority functionality to deliver in phases that provided measurable business value in each release. As another example, the project management team purposefully selected a Waterfall approach when replacing our general ledger system due to the data integrity, business process, and reporting risks associated with using two systems concurrently.

Author Biographies

Dr. Chuck Millhollan, Contact Information: chuck.millhollan@gmail.com

Chuck is the Senior Vice President of Operational Process Excellence for Farm Credit Mid-America. He has over 25 years of experience in strategic planning, quality management, and project, program & portfolio management. Past positions include Vice President Process Improvement & Execution for Farm Credit Mid-America, Director of Program Management for Churchill Downs Incorporated and Manager of the Applications Engineering Project Management Office for Humana Inc.

In addition to his work as a practitioner, Chuck develops curriculum and teaches for several accredited universities. He is a member of the International Institute of Business Analysis (IIBA), a senior member of the American Society for Quality (ASQ), and a member of the Project Management Institute (PMI) and Kentuckiana PMI chapter.

Chuck earned a Doctorate of Professional Studies in Information Management from Syracuse University in 2015, a Master of Science in Project Management from the University of Wisconsin in 2003, a Master of Business Administration from the University of Florida in 2000, and a Bachelor of Science in Management in 1993 through Southern Illinois University.

Chuck's certifications include Project Management Professional (PMP), Program Management Professional (PgMP), Agile Certified Practitioner (PMI-ACP), Certified Business Analysis Professional (CBAP), Six Sigma Black Belt, Certified Manager of Quality/Organizational Excellence (CMQ/OE), Certified Software Quality Engineer (CSQE), and Certified Managed Healthcare Professional (MHP).

You can find additional information about Chuck via his online business card at www.millhollan.net.

Daro Mott's personal mission statement is to help family, friends, colleagues, and organizations achieve their strategy. He is a multi-sector leader who applies the core concepts of Good to Great and How to Win Friends and Influence People to his everyday interactions. Daro is passionate about leadership, management, systems, transformation, building cultures of excellence and applying data and evidence-based solutions to achieve strategy. He speaks at more than a dozen venues each year including national and international conferences. He co-authored a book entitled, Coaching (Six Sigma) Green Belts for Sustainable Success, which the American Society for

Quality published in May of 2015. He is the author of numerous articles and is a certified Project Management Professional, Six Sigma Black Belt, Lean Leader, Predictive Index Analyst and Baldrige Examiner. He is a board member for the Association of Internal Management Consultants, the Louisville Zoo, and the Susan Polgar Foundation.

Daro has lectured at the University of Louisville, the University of Indiana Southeast, the University of Phoenix and ATA College. He currently teaches in the MBA program at the University of Louisville. Mr. Mott holds an undergraduate degree from Yale University and two graduate degrees from the University of Louisville.

Daro is currently the Vice President of Process Improvement and Execution at Farm Credit Mid-America, and responsible for the effective implementation and benefit realization of the Association's strategic project portfolio.

As we start shifting to addressing the next nine pillars, we will wrap up the strategic positioning of project management for the future with two contributions. The first comes from PM Conference BRIDGE 2021, as the platform that is creating an avenue for collaboration in supporting the future envisioning of project management practices. The second is a letter to future project managers that is intended to motivate professionals and practitioners in tomorrow's organizations to ready themselves for what's ahead.

You can contact Daro at daromott@gmail.com.

1.16 Excellence in Action: Project Management United[10]

In the past few years, the pandemic situation, climate crisis, equality movements and other events that affect our lives are sending a message to our society that we have to change in order to survive. The progress of technologies and innovation are key leading factors if we think about the speed of vaccination, sustainable environment creation or human protection. However, transformation does not happen overnight and in most of the cases people feel they are competing with time. Looking across all transformations, we find a human and in the multiple cases the human is a project manager.

If we simply define a standard project management as supervising the three main constraints (time, budget, and scope), in most cases it does not help to overcome the challenges we are facing around the globe. With the greater levels of uncertainty, organizations are becoming more and more complex, mandating projects that require change management. In such organizations, project managers are getting constant pressure not only for standard project management but also for operational efficiency and innovation with reduced resources. All of these requirements cannot be fulfilled without understanding the new reality, keeping up with ongoing innovations, and finding ways to branch out from standard project management.

As a senior project manager working in a company that delivers innovative software to advance the world's infrastructure – sustaining both the global economy and environment – I do believe that in order to implement any change successfully, we have to have a clear vision of what future

10 Material in this section has been provided by Edita Kemzuraite, PMP – Co-founder | Head of Content at Project Management United, a senior project manager at Bentley Systems with demonstrated history in complex software development and transformation projects that deliver innovative software to advance the world's infrastructure – sustaining both the global economy and environment. Former PMI Lithuania Chapter President (2020–2021).

project management looks like, what we are doing now and how we as project managers should change if we want to be the New Generation Project Managers.

What Might the Future of Project Management Look Like?

Following the latest trends of project-based organizations, understanding the successful changes and effective project management that happens there, and identifying where standard project management components can be operationalized by adding new technologies and experimenting with the new methods of working is terrific. However, in the beginning these innovations require greater investment, quick learning, and higher levels of risk tolerance to finish without the expected outcomes as technologies might be too complex or teams are not ready for the change.

If we look to the future of project management without the existing constraints and imagine that all technical challenges are in the past, its vision starts with projectized organizations. In general, it means that the project manager has full power in any decision involved to the project. In addition to that, standard project management should be simple. The word "simple" has different meanings in each knowledge area of the traditional project management and these might be potential examples for the simplified standard of project management:

- **Simple planning, no scope creep, and an always updated project plan.** Nowadays, even being Agile or using Kanban boards do not help to be as operationalized as we would love to be – things get chaotic before we even finish one iteration. The common problem that still exists is that when we add something, we forget to remove lower priority tasks from the iteration, and this usually leads to scope creep, overloaded teams and unhappy users who are waiting for quick project delivery. The project managers often have to be the ones who send constant reminders to adjust the backlog, to remove leftovers and focus on priorities. In an ideal way of operating, project managers would not have to be the ones who constantly look at the documents and manually adjust the plans. In the future, project management of these obstacles should be forgotten – machines will be taking the notes and reminding project managers and team members what was discussed, what action items we should be focusing on now and what we plan to do next.

  From my perspective, changes to projects happen with almost every new strategic session or stakeholder meeting. When human beings and technologies work hand in hand on projects, project plans will not be created in the beginning and forgotten. Project plans are going to be nothing but an audit trail – a collection of records and adjustment with ability to analyze the important steps we made and highlight the lessons learnt. We will forget the word "outdated," as "real-time automated updates" will be our new normal. With every step the project team is making, the robots or simulators will be acting as a service for project managers.
- **Eliminate risks easily.** Risk management of every project is different and highly dependent on the industry, sector, or relying parties. In the future vision, even risk management can be simple! Imagine that a project manager does not need to go anywhere while running a construction project: every single detail he needs to analyze, touch or even smell can by simulated with the help of digital twins that would represent the virtual copy of the physical delivery you need to create in virtual reality.
- **Stakeholder management.** According to the latest research, communication and stakeholder management are the common challenges of every initiative or project. In the future, stakeholder management can create convenience using smart technologies. Imagine that based on project requirements, common user personas would be created to help project managers to have reliable RACI models and avoid miscommunications in important project collaborations.

If this future vision becomes attainable, we can imagine that the operational part of projects would be much easier and less or even no human interaction would be required for these tasks. However, after reducing the load of manual operational tasks, future project managers would have more time to be the ones who are setting the tone inside and outside of the project. The complexity of the projects would not go anywhere, we would just be better prepared by predicting the future with reliable techniques that replace the repetitive human work not only for project managers but for all team members. That is why leadership and understanding what role every project plays in the sustainable environment would become critical components of the future project management.

Together with the future vision of project management the role of project managers is leveling up to strategic projects. The new level requires constant collaboration with organizational executives so that organizational success is measured by projects and the impact project results bring for social good. Future project management should seek to improve quality of life by implementing sustainable ideas and ensuring long lasting positive impact.

Major Changes Expected in Project Management in the Upcoming Years

The personal experience and the multiple examples how the world leading companies are breaking the organizational silos inspired me to support this future-oriented thinking in my profession. As a co-founder of an international project management conference BRIDGE 2021 I am a believer of the power of networking. We want to create a platform for next generation project managers to collaborate without the barriers of different PM schools, certification levels or industries. Our projects are unique but the competencies we need to build for the future are all common: organizational ability, communication, speed to deliver, quick learning, technology-oriented management, and adoptability to any uncertain situation. Future Project Management starts NOW. Tomorrow starts NOW.

The good news is that in some industries, the future project management vision is already transforming into reality with the help of advanced technologies. For example, one of the first signs of the application of different project management skills is happening in 2015 ESM Productions (ESM), a local company tasked with event planning for the Pope's visit in Philadelphia, Pennsylvania in the United States. ESM Productions retained AEROmetrex to help engineer the complex logistics for this historic event. Hosting the event required massive planning and coordination with municipal and state agencies, as well as with the US Secret Service. To accommodate the 56,400 temporary structures needed, 33 miles of security barricade perimeter, and the impact of local road closures, while ensuring smooth traffic flow and seamless communication throughout the planning process and the event, ESM needed comprehensive, realistic, and integrated 3D modeling and simulation software.

Reality modelling techniques provided by Bentley Systems redefined the planning and managing of the city infrastructure as well as collaboration with asset and building owners, government agencies, and the public.

The example of the digital city can help the event planners and project managers not only to manage the complexity, but with the help of digital twins it can develop a flood resilience strategy, manage public works projects and support the many things city governance project managers do to provide a high quality of life for the residents (more information about the project: AEROmetrex Designs 3D Map of Philadelphia for 2015 Papal Visit (bentley.com)). As technologies are evolving rapidly, nowadays we have multiple Digital Cities over the world including Dublin's city digital twin as a result of Bentley Systems and Microsoft partnership (https://cities-today.com/bentley-systems-and-microsoft-team-up-on-digital-twins/), Helsinki City and many more.

However, in some industries and smaller countries we should not expect such rapid changes in project management. The technology adoption and usage also require additional skillsets and acceleration in this direction require projects with additional investment. In the next few years we should also expect that the Citizen Development and Citizen Revolution announced by PMI in 2020 (https://www.pmi.org/citizen-developer) would help to digitize simpler solutions easier without big investment in IT departments as low code-no code platforms do not require programming skills.

Looking from another side to achieve the future vision, technological progress is not enough. Project managers are already facing additional risks and constraints that must be managed as a part of the project. I selected a couple of key components that have become critical to consider and that might significantly change the scope of every project:

- **Cyber security.** Digital project management can be time and resource saver. However, it can destroy the whole project or even put multiple organizations in jeopardy if it becomes a victim of cyberattacks. Together with digitalization, we should expect that probability and impact of the security related risks will be increasing, and project managers will need to have security champions or experts in the project teams that would make sure the project data is protected in our digital environment and that endpoints and networks are safe.
- **Respect privacy.** Any digital solution that can be implemented in order to make project management more efficient should be also carefully considered. Artificial intelligence (AI) techniques should be centered around interactions that are helpful and effective but not intrusive. Future organizations and project managers have to make sure that every interaction of AI is simple and creates convenience rather than additional complexity and frustration among team members.
- **Choosing trusted technology providers.** The success of future project management will be highly dependent on the data and selected technologies as a part of organizational strategy. For example, digital dashboards that monitor performance of key project performance indicators without any human interaction will be hard to trust for a while. Organizations must choose reliable vendors and technology providers as it will be critical for project managers to get insights and plan next steps based on the trusted information. Reliable data is a critical success factor.

In order to accelerate future project management, we should expect changes in the strategic level of the organizations. The top-level managers, decision makers responsible for investment to the technologies inside organizations should not forget the changing role of project managers. Their role is to empower project managers to lead the projectized strategies by providing the necessary resources and training in order to continuously improve and learn about the future vision.

Unlike the typical functional organization, project managers in a fully projectized organization are not limited to tasks such as decision-making and resource utilization. This change would require standard project managers to adopt new ways of operating. With the additional power to utilize resources, future project managers must ensure they are surrounded by proficient technology experts and other leaders in order to overcome all the technical challenges and realize the future vision of digital project management. With this accountability, we should expect future project managers working closely with human resource managers in order to build stronger teams with the required skillset for each function.

With the acceleration of projectized organizations, we should expect the volume of projects for a single project manager to increase as the operational parts of project management would not be manual anymore. Project managers must adopt the right methods to projects that would highlight the importance of the different project management life cycles and ability to adopt the right one in the unique situations. It would require quick learning and rapid change management in order to keep project management operationalized. In such cases, the future project manager would have enough capacity to jump from one project to another in a significantly short amount of time. However, it is critical to understand how it all fits into the big picture.

Otherwise, if you see a project result as some independent delivery, that "does not fit" anywhere else, it is easy to burn out and lose the passion for projects. Organizations together with project managers should work on the projectized strategies that are resilient and long lasting to avoid those risks.

As we can imagine, change management would be critical in moving to future project management methodologies. Implementing all the necessary innovations will require embracing the change in the whole organization and its culture. There is no way that a project manager can make the difference alone – everyone needs to be playing a part. In transformations like these, it is always helpful to imagine a colony of penguins that are trying to find a creative solution in a life changing situation described in a book "Our Iceberg Is Melting" by John Kotter. In every project you can face situations when a team member or important stakeholder will be complaining: "This is not how we do it here." However, looking from the other angle, within the same team you will find the persons who will ask: "How can I help." And this is the beauty of it. If a project manager knows how to approach and unify people with multiple opinions and different personalities to a One Team – it can change the situation significantly. In the future, projects must eliminate the fear of change, help the team to face the future and make related actions achieving common goals. These challenges will require project managers to build power skills such as adoptability and courage.

In addition to change management, it is inevitable for the future project manager to be a person who "always connects the dots." Either it is an internal situation, when you need to solve a conflict between different groups or an external one in which the project manager needs to send an important message from a customer to a team to fulfil the project requirements. In situations like these, I call a project manager a Connector or a Bridge. From my personal experience, I used to manage a couple of projects that were unique by their definition. The joint projects brought technology experts from two different companies together to create something powerful – digital twins. Surprisingly, the biggest difficulty in the projects was not technical – both companies had strong specialists and experts who were doing their job perfectly.

The cultural difference, variety of different personalities and inability to find a common way of working were our main challenges when I was assigned to manage these projects. What a project manager should do in the situations like these? The answer is to become the bridge connecting the different pieces. It is a complicated task, but active listening, finding painful points and changing team's mindset from "Two Different Companies" to "One Team" can make the expected difference!

How Do We Plan for the Future?

The future vision of project management is not utopia or illusion. Gradually, the world is moving in that direction by necessity, and we have multiple good examples even in a country as small as Lithuania. Through investment in strategic initiatives, multiple international companies are well-positioned to deliver powerful insights across risk categories trending in the world.

At least two major financial operations companies are leading separate initiatives toward sustainability. Swedbank Lithuania is a main partner of "Sustainable business" as a platform to tell the stories about the new sustainable ideas and a place for other companies to rethink their strategies. In addition to that, SEB Bank has a Corporate Sustainability policy and since 2009 they have been providing a yearly report illustrating how their business is connecting responsibility for environment, people and community through separate projects and initiatives. And behind all those great initiatives we see humans – project managers/project leaders.

"Environmental, Social and Governance (ESG) and climate change considerations are critical to managing risk and seizing opportunities in today's global capital markets" (https://esg.moodys.io/) – Moody's is announcing a family of business partners who are implementing projects and providing solutions for a greener economy and globally sustainable businesses. Moody is one of the international companies that is playing an important role by helping other organizations to rate them based on different levels of interoperability. Sustainability is one of the criteria in the interoperability rating that help to start changing the entire culture with more sustainable assets.

Overall, these are just a few examples of implementing world-changing solutions in big organizations. These big companies are sending message to the smaller ones that could be illustrated by a Hellen Adams Keller's quote. Hellen Adams Keller was an American author, disability rights advocate, political activist, and lecturer: "Alone we can do so little, together we can do so much." The aforementioned SEB Bank is announcing that more than 50 percent new companies in Lithuania are including sustainable solutions to their business strategies and it is even more popular among startups and small businesses rather in medium ones. A project manager is playing a critical role in every one of those projects as a leader for strategic implementation.

In order to accelerate the future vision, we should start learning from the existing examples and rewarding the projects that are already following the projectized strategy concept. Personally, I get excited watching Bentley Systems *Year in Infrastructure Awards* (https://yii.bentley.com/en/awards) as year after year it highlights the amazing infrastructure projects that are winning in multiple categories. These rewards probably best reveal how the company I work for is committed to sustainability: we are providing the tools and services for industry professionals and organizations to help create or improve quality of life through sustainable project deliveries. Every project nominated to these awards is judged on the positive impact that it will have on the local social, economic, or environmental situation.

In addition to those practical examples, the planning and execution toward the future vision cannot happen without support of education providers and scientific discoveries. The young professionals are the ones who are the proficient users of the technologies, and they might be the ones who can help to accelerate in the digital environment and to find smart solutions for more futuristic project management! In past few years, universities, different PM schools, certification providers and other nonprofit-based associations have started playing a significant role in building the generation project management.

They have the power to spread the word, starting from Young Generation to mature Project Management Professionals who need to think how to adopt their skillset and improve their knowledge toward trending technology and sustainable ideas. And we already have great examples that can help future project managers to change the mindset and build the necessary skillset toward projectized strategies:

- In a Pulse of the Profession 2021 (2021) PMI (Project Management Institute) is announcing a new term for the future-fit organizations called "Gymnastic enterprises" that "are leading the way by empowering their people to master new ways of working, emphasizing the human element, and understanding the central role that organizational culture plays in enabling all of these capabilities."

- IPMA (International Project Management Association) is organizing Global Best Practice Week promoting the practical examples on "How to reach and maintain Resilience."
- Multiple universities and business schools in Lithuania are partnering with PM Schools and international companies to support sustainability and accelerate the sustainable ideas through projects involving students and professionals to the process.

To illustrate that, I could also mention that Bentley Systems was one of the strategic and business partners establishing the Centre of Smart Cities and Infrastructure, which Kaunas University of Technology (KTU) opened in 2019th. The result of this partnership is multiple smart city solutions that are created by students to the local cities. Smart waste collection systems, advanced street lighting, mobile applications informing on the traffic and even more to ensure and accelerate the sustainability.[11] Does it illustrate the future of project management? Yes – because, again, all the proposed solutions require implementation, and we need humans – project managers leading these projects to the result.

It is great to see that the world leading organizations are stepping up – acceleration happens through partnerships and openness to innovative ideas. The two popular project management associations – PMI and IPMA – are also highlighting the importance of the young generation. PMI has organized "Young Professional Awards" and is recognizing our top future leaders. Similarly, IPMA has a "Young Crew" as a key component of their organization, supporting and building the network of young professionals for their growth and maturity.

Personally, I believe that young professionals and mature project managers together can create a powerful diversity as they use different practices and methods of operating. The multiple PMI Chapters are also following these initiatives and encouraging to become stronger by creating the joint initiatives together: starting with "Joint Initiatives of the 3 Italy Chapters to Face the Lockdown" in early 2020 and continuing with PMI UK & PMI Chapter "Exchange Webinars on the Future of Work" and others.

To summarize the multiple examples above, I believe that these best practices are worth following. Multiple organizations are already including the impact for social good in their organizational strategies. However, transforming those strategies to projectized organizational structure is not as easy. We are getting there slowly, but my hope is that world's leading companies will set the tone for multiple industries, and the bright future of next generation project managers will come sooner rather than later.

Conclusions

The vision of future project management has to be in alignment with an organizational strategy of empowering project managers to be in control of projects. According to the future vision, standard project management will be simplified by technologies, but project managers will need to gain better skills for leadership, decision-making and talent management inside their projects. The future project managers should be shifting their focus to a more strategic and pragmatic level.

For the upcoming years, we should expect multiple changes in project management. Spreading the ideas to the world is easier together, but the transformations to future project management

11 based on the article https://en.ktu.edu/news/five-smart-city-solutions-from-ktu-students-for-kaunas/

should start from multiple levels: organizational, social, and personal. PM Conference BRIDGE 2021 as a platform for organization leaders and project, program, portfolio managers to collaborate is supporting the future vision of project management.

Multiple examples show that changes toward the future project management vision are already happening. Even starting with small steps, we can have an impact and spread the important message about sustainability to the world through projects, products, or initiatives. Learning from the success and failure stories in the conferences and other available sources can help to strengthen the future vision and accelerate the change.

1.17 Letter to Future Project Manager

Dear Future Project Manager,

Congratulations! This letter is specifically for you as you consider a career in the diverse field of project management.

This career path you have chosen is both exciting and challenging. The field of project management will allow you to explore different industries, and aid in transformational change. To prepare for this journey, take stock of societal trends, organizational practices, and the ever-changing influence of technology. To be the best at what you do, you must continuously hone your talents and skills through every opportunity, no matter how small or large, to lead and manage a project.

And remember, knowing the science around what to do differs vastly from mastering the art of the profession. Expect to operate in an environment where volatility, uncertainty, complexity, and ambiguity (VUCA) will be a reality.[12] Adaptability and learning will be your lead-in for your most fulfilling years in the profession.

As experienced project managers, we hope the following nuggets of advice will resonate with you in your journey to master the art of project management.

1) **Believe in the leader that you are and never stop learning.** The discipline of project management offers project managers the opportunity to lead with delegated authority for a specified period. To advance in your role, you need to embody the qualities akin to that of an ambassador assigned to a nation for a designated time. It is imperative to settle quickly into the environment in which you will work, and spend time immersing yourself in the nuances of the business to build credibility. Be prepared to familiarize yourself with stakeholders' interests and their motivations. You will need to engage with key decision makers and diplomatically use your influence to advance the work.

   Influence is your authority as you manage those above you, beside you, your team, and external parties. At all times, communicate value and continually make a business case for a project in various presentations and forums. However, that message will need to be tailored to the audience you address. Producing situational reports will be critical to enable strategic business decisioning.

   Taking these actions will prepare you and your stakeholders in the event your project goes in a different direction. Your investment in learning will give you the confidence to handle any

12 **VUCA** stands for volatility, uncertainty, complexity, and ambiguity. It describes the situation of constant, unpredictable change that is now the norm in certain industries and areas of the business world. **VUCA** demands that you avoid traditional, outdated approaches to management and leadership, and day-to-day working. https://www.mindtools.com/pages/article/managing-vuca-world.htm

situation you encounter. It is not enough to be the smartest in your profession – rather, apply your project management leadership skills in the context that best serves the business leaders and customers.

2) **Be a champion for achieving organizational goals as much as you are a champion for your project:** While projects are becoming more dynamic you should expect to manage and deliver outcomes without a specific end date. Projects need to pivot quickly to be able to address planned and unplanned conditions. Stay focused on delivering benefits, and the other project variables – such as scope, schedule, and funding – will follow. Learn what benefits are important to your leaders and customers, then navigate with that end-goal in mind throughout the project. Be aware that some projects may benefit the organization when stopped or canceled versus completing.
3) **Go beyond the basics when measuring success:** One of the most important activities a project manager performs is collaborating with stakeholders at all levels to define and measure project success. When defining what success should look like, go beyond known and expected performance metrics such as tracking progress on budget, scope, and schedule, and define assessment criteria and metrics that will help you and the stakeholders vigorously evaluate if the resources and investments are delivering on the business strategy and outcomes. Your role is to stress upon business leaders the importance of assessing if there still is business value to continue the work based on point in time conditions.
4) **Data rules and analysis reigns.** Most everything we do produces data and leaves a digital footprint. Automation of metrics can transform data into useful information to enable better decision-making. However, metrics alone are not enough to make informed decisions. Rather, metrics, coupled and analyzed with project context, organizational priorities, stakeholder input, and risk tolerance thresholds enable timely decision-making with relevant information.
5) **Elevate the value of project management by embracing diverse methodologies.** With expectations to increase time to market for delivering business and customer value, project managers need to recognize the importance of blending diverse practices from other disciplines into a cohesive project management approach. If these practices prove to add greater value than only relying on traditional project management practices, embrace them, scale them, and incorporate them into your project management approach. Further, do not fret. As automation streamlines routine project management tasks, this will free you to focus on maximizing the business value of your project.

Do not be alarmed that this list is short. These nuggets of advice are not meant to be exhaustive nor prescriptive. They reflect simple but powerful lessons learned from both successful and failed projects. If you choose to adopt these principles as guideposts, you will reap the rewards of a successful and gratifying project management career.

Bios

Michelle Brunn has provided project and program management expertise and leadership for over 25 years for both business and technology services. She has specialized in establishing enterprise level functions.

Grace Navas, PMP, has provided project and program management leadership for over 25 years in private sector industries and has dedicated a good portion of her career to public service. Her project management work has encompassed product delivery, implementation of project management functions, and oversight and compliance practices.

Throughout their careers with the Federal Reserve System, Michelle and Grace have collaborated on work that has elevated and recognized project and program management as a professional discipline.

The views expressed in this section 1.17 are strictly those of the authors and do not represent the position or views of the Federal Reserve System or the United States.

References

Akdemir, Bünyamin, Erdem, Orhan, and Polat, Sedat. (2010). Characteristics of high-performance organizations, *Suleyman Demirel University Journal of Faculty of Economics & Administrative Sciences* 15 (1): 155–174.

Ali, Usman, and Kidd, Callum. (2014, April). Barriers to effective configuration management application in a project context: An empirical investigation. *International Journal of Project Management* 32 (3): 508--518. 11p. DOI: 10.1016/j.ijproman.2013.06.005.

Ambekar, Suhas, and Hudnurkar, Manoj (2017). Factorial structure for Six Sigma project barriers in Indian manufacturing and service industries. *TQM Journal* 29 (5): 744–759. 16p. DOI: 10.1108/TQM-02-2017-0021.

Arnold, Paul, and Javernick-Will, Amy (2013, May). Projectwide Access: Key to effective implementation of construction project management software systems. *Journal of Construction Engineering & Management* 139 (5): 510--518. 9p. DOI: 10.1061/(ASCE)CO.1943-7862.0000596.

Azim, S., Gale, A., Lawlor-Wright, T., Kirkham, R., Khan, A., and Alam, M. (2010). The importance of soft skills in complex projects. *International Journal of Managing Projects in Business* 3 (3): 387–401. doi: http://dx.doi.org/10.1108/17538371011056048.

Baker, B., Murphy, D., and Fisher, D. (1988). Factors affecting project success. *Project Management Handbook*. 2nd Edition. pp. 902–919. New York: Van Nostrand Reinhold.

Blixt, Carley, and Kirytopoulos, Konstantinos. (2017). Challenges and competencies for project management in the Australian public service. *International Journal of Public Sector Management* 30 (3): 286–300. 15p. DOI: 10.1108/IJPSM-08-2016-0132.

Carvalho, Marly Monteiro de. (2014). An investigation of the role of communication in IT projects. *International Journal of Operations & Production Management* 34 (1): 36–64. 29p. DOI: 10.1108/IJOPM-11-2011-0439.

Farm Credit Mid-America 2012 Annual Report. (2012). Retrieved from https://e-farmcredit.com/fcs/media/Assets/Media%20Downloads/2012-Annual-Report.pdf on April 14, 2021.

Farm Credit Mid-America. (2021). Retrieved from https://e-farmcredit.com/about on April 14, 2021.

Farm Credit. (2021). Retrieved from farmcredit.com/overview-and-mission on April 14, 2021.

Green, C.H., and Howe, A.P. (2012). *The Trusted Advisor Fieldbook: A Comprehensive Toolkit for Leading with Trust*. Hoboken, NJ: John Wiley and Sons.

Hwang, Bon-Gang, and Tan, Jac See (2012, September/October). Green building project management: obstacles and solutions for sustainable development. *Sustainable Development* 20 (5): 335–349. 15p. 3 Charts. DOI: 10.1002/sd.492.

Hwang, Bon-Gang, Zhao, Xianbo, and Toh, Li Ping (2014, January). Risk management in small construction projects in Singapore: Status, barriers, and impact. *International Journal of Project Management* 32 (1): 116–124. 9p. DOI: 10.1016/j.ijproman.2013.01.007.

Hwang, Bon-Gang, Zhu, Lei, Tan, and Joanne Siow Hwei (2017, June). Green business park project management: Barriers and solutions for sustainable development. *Journal of Cleaner Production* 153: 209--219. 11p. DOI: 10.1016/j.jclepro.2017.03.210.

Johansen, Jane, and Gillard, Sharlett (2005). Information resources project management communication: personal and environmental barriers. *Journal of Information Science* 31 (2): 91–98. 8p. DOI: 10.1177/0165551505050786.

Jugdev, K., and Muller, R. (2005). A retrospective look at our evolving understanding of project success. *Project Management Journal* 36 (4): 19–31.

Kerzner, H. (2021a). Predictions on the future of project management.

Kerzner, H. (2021b). The growth of project management social leadership.

Kerzner, Harold, and Zeitoun, Al, (2008). *Barriers to Implementing Project Management in Emerging Market Nations*, International Institute for Learning (IIL) white paper; . 2008

Khan, Arif Ali, and Keung, Jacky (2016). Systematic review of success factors and barriers for software process improvement in global software development. *IET Software*. 2016, Vol. 10 Issue (5, p): 125-–135. 11p. DOI: 10.1049/iet-sen.2015.0038.

Khan, Siffat Ullah, Niazi, Mahmood, and Ahmad, Rashid (2011, July). Barriers in the selection of offshore software development outsourcing vendors: An exploratory study using a systematic literature review. *Information & Software Technology* 53 (7): 693-–706. 14p. DOI: 10.1016/j.infsof.2010.08.003.

Kutsch, Elmar, and Hall, Mark (2009, September). The rational choice of not applying project risk management in information technology projects. *Project Management Journal* 40 (3): 72–81. 10p. 4 Charts. DOI: 10.1002/pmj.20112.

Kutsch, Elmar, and Hall, Mark (2010, April). Deliberate ignorance in project risk management. *International Journal of Project Management* 28 (3): 245–255. 11p. DOI: 10.1016/j.ijproman.2009.05.003.

Lacerda, R., Ensslin, L., and Ensslin, S. (2011). A performance measurement view of IT project management. *International Journal of Productivity and Performance Management* 60 (2): 132–151. doi: http://dx.doi.org/10.1108/17410401111101476

Loushine, Todd W., Hoonakker, Peter L.T., Carayon, Pascale, and Smith, Michael J. (2006, November). Quality and safety management in construction. *Total Quality Management & Business Excellence.* 17 (9): 1171–1212. 42p. 1 Diagram, 4 Charts. DOI: 10.1080/14783360600750469.

Magnier-Watanabe, Remy, and Benton, Caroline (2013, April). Knowledge needs, barriers, and enablers for Japanese engineers. *Knowledge & Process Management.* 20 (2): 90–101. 12p. 5 Charts, 1 Graph, 6 Maps. DOI: 10.1002/kpm.1408.

Manek, S. (2020). Digital transformation investments to top $6.8 trillion globally as businesses and governments prepare for the next normal, IDC, December 8, https://cdn.idc.com/getdoc.jsp?containerId=prMETA47037520.

Millhollan and Kaarst-Brown. (2016). Lessons for IT project manager efficacy: A review of the literature associated with project success. *Project Management Journal* 47 (5).

Millhollan, C. (2008). Scope change control: Control your projects or your projects will control you. Paper presented at the PMI North America Global Congress 2008, Denver, CO.

Millhollan, C. (2015). A phenomenological study of factors that influence project manager efficacy: The role of soft skills and hard skills in IT-centric project environments. (Unpublished dissertation). Syracuse, NY: Syracuse University.

Moore, David R., and Dainty, Andrew R. J. (2001, October). Intra-team boundaries as inhibitors of performance improvement in UK design and build projects: a call for change. *Construction Management & Economics* 19 (6): 559–562. 4p. DOI: 10.1080/01446190110055508.

Munns, A. K., and Bjeirmi, B. F. (1996). The role of project management in achieving project success. *International Journal of Project Management* 14 (2): 81.

Niazi, Mahmood, Babar, Muhammad Ali, and Verner, June M. (2010, November). Software process improvement barriers: A cross-cultural comparison. *Information & Software Technology* 52 (11): 1204–1216. 13p. DOI: 10.1016/j.infsof.2010.06.005.

Patanakul, P. (2011). Project manager assignment and its impact on multiple project management effectiveness: An empirical study of an IT organization. *Engineering Management Journal* 23 (4): 14–23.

Paterson, Christopher J., Andrews, Richard N.L. (1995, Spring). Procedural and substantive fairness in risk decisions: Comparative risk assessment procedures. *Policy Studies Journal* 23 (1): 85–95. 11p. DOI: 10.1111/j.1541-0072.1995.tb00508.x.

Pinto, J. K., and Slevin, D. P. (1988). Project success: Definitions and measurement techniques. *Project Management Journal* 19 (1): 67.

Polak, Jarosław, Wójcik, Przemysław (2015, August). Knowledge management in IT outsourcing/offshoring projects. *PM World Journal* 4 (8): 1–10. 10p.

Project Management Institute (2013). Managing Change in Organizations: a Practice Guide. The Project Management Institute (PMI).

Sakellariou, Evy, Karantinou, Kalipso, and Poulis, Konstantinos (2013/2014, Winter). Managing the global front end of NPD: lessons learned from the FMCG industry. *Journal of General Management* 39 (2): 61–81. 21p. DOI: 10.1177/030630701303900204.

Santos, Vitor Ricardo, Soares, António Lucas, and Carvalho, João Álvaro (2012, April–June). Information management barriers in complex research and development projects: an exploratory study on the perceptions of project managers. *Knowledge & Process Management* 19 (2): 69–78. 10p. DOI: 10.1002/kpm.1383.

Senesi, Christopher, Javernick-Will, Amy, and Molenaar, Keith R. (2015, April). Benefits and barriers to applying probabilistic risk analysis on engineering and construction projects. *Engineering Management Journal* 27 (2): 49–57. 9p. 5 Charts. DOI: 10.1080/10429247.2015.1035965.

Shenhar, A. J., Dvir, D., Levy, O., and Maltz, A. C. (2001). Project success: A multidimensional strategic concept. *Long Range Planning* 34 (6): 699–725.

Sommer, Anita Friis, Dukovska-Popovska, Iskra, and Steger-Jensen, Kenn. (2014, August). Barriers towards integrated product development – Challenges from a holistic project management perspective. *International Journal of Project Management.* 32 (6): 970–982. 13p. DOI: 10.1016/j.ijproman.2013.10.013.

Terlizzi, Marco Alexandre, Meirelles, Fernando de Souza, de Moraes, Heverton Roberto and Oliveira Cesar. (2016, April). Barriers to the use of an IT Project Management Methodology in a large financial institution. *International Journal of Project Management* 34 (3): 467–479. 13p. DOI: 10.1016/j.ijproman.2015.12.005.

Warrick, D. (2014). What leaders can learn about teamwork and developing high performance teams from organizational development practitioners, . *OD Practitioner* 46 (3): 68–75.

Wenzhe Tang, Maoshan Qiang, Duffield, Colin F., Young, David M.; ., and Youmei Lu. (2007, December). Risk management in the Chinese construction industry. *Journal of Construction Engineering & Management* 133 (12): 944–956. 13p. 1 Diagram, 8 Charts. DOI: 10.1061/(ASCE)0733-9364(2007)133:12(944).

2

Pillar 2: Applying Project Management in Humanitarian and Social Initiatives

2.0 What Makes Humanitarian Projects Different?

Most projects that organizations implement are motivated by financial results like revenue, profit, and cost reduction. However, projects in humanitarian and social initiatives are driven by a different set of benefits (Figure 2-1).

The deep need for resources, logistical assistance, and support is the key driver for their existence. They may aim to relieve the challenges faced by a country, a region, or an ethnic group (beneficiaries) facing disasters (natural or man-made), consequences of war, conflict, and poverty.

Another critical aspect of these types of projects is their impact on the team's behavior. The natural human inclination to support each other and bring a better life or reduce the suffering of others shapes the motivation of those working on these initiatives.

In the same angle, humanitarian projects also have a strong psychological impact on those involved. Working surrounded by suffering, death, pain, violence, and tragedy brings another level of challenges to the operations.

Most of the time, these projects happen with extremely limited resources, with deadlines that mean the difference between life and death for those affected by the disaster or tragedy.

They require complex coordination of different people: from engineers to health professionals, from procurement personnel to experts in logistics. All of that happens in coordination with a strong need for developing local capacity, which is where a critical concept of *disaster risk reduction* comes into play: "build back better."

2.1 The Impact of Project Management Practices in Humanitarian Projects

One of the biggest challenges that social and humanitarian projects face is the unique dimension to their work: the team's motivation and the desire to support others.

It sounds contradictory that what brings a deep sense of meaning to the project is also the one bringing challenges. However, the sense of purpose and community, associated with incredibly high motivation, can create a sense of "superpowers" that makes any planning or coordinated solution impossible.

Figure 2-1 Humanitarian and Social Projects Are Different

Whether a project is related to humanitarian efforts or not, there is no project that does not benefit from proper planning and coordinated execution: from an emergency response to an earthquake to the reconstruction of a war-torn village or region.

All of them need competent and effective management.

In this chapter, we will discuss three case studies. The first one is based on Ambev, one of the largest beverage companies in the world, and its portfolio of initiatives to support works related to the Covid-19 pandemic. The second one is the construction of emergency hospitals to treat Covid-19 infections developed by one of the leading hospitals in Brazil: Albert Einstein. The third one covers the approach used by the United Nations Office for Project Services to manage more than 1,000 humanitarian and development programs globally.

2.2 Excellence IN Action: Ambev: A Humanitarian Approach to Addressing Challenges During the Covid-19 Pandemia[1]

This is the story of a company with the purpose of **bringing people together for a better world**, built by people who live for a higher cause and are always seeking evolution, both personally and in their environment (Figure 2-2). This is Ambev Brewery.

Ambev is a Brazilian company present in all states of Brazil, **with 100 distribution centers and 32 breweries**. Ambev has a Brazilian CEO, and most of its executives are also Brazilian. So, it should come as no surprise that they recognize their responsibility to contribute to the country's growth and at the same time deeply understand the issues that Brazil has faced during the Covid–19 pandemic. Ambev and its team are always looking for a way to use their resources to mobilize and do something that matters to people, including its 32,000 employees and its entire ecosystem of partners, customers, and clients.

1 Material in this section was provided by Ambev SA and was written by Juliana Fernandes Alves, Beatriz Guitzel Borghi, Rodrigo Moccia, José Finocchio Júnior and Lucas Dato. Juliana F Alves is mother and project director executive in industrial projects for Ambev, building new breweries, innovation, and dreams. Beatriz Borghi is a food engineer, PMO at Ambev, passionate about life and sharing experience. Bringing people together and working for a better world are the purpose of Rodrigo Moccia, Ambev´s corporate affairs director. He strongly believes in the new generation leading the business, the country, and the society for a better future. Jose Finocchio Júnior is a researcher and author of project management methodologies and PMO consultant. Lucas Fachine Dato is a biotechnologist and brewmaster who is currently a Brewery/Plant Manager at AMBEV involved in the execution of projects throughout the pandemic.

Figure 2-2 Bring People Together for a Better World

Based on Ambev's motivation to transform the world, the company has joined forces with different partners to turn ideas into reality and help the most diverse causes in times of crisis or difficulties. The arrival of the coronavirus and the pandemic increased this spirit of collaboration, even though it was already present in Ambev's culture and DNA.

At the beginning of the pandemic, more than protecting its employees with essential initiatives such as installing hand sanitizers stations, manufacturing and distributing face masks, providing an app to monitor contacts, and ensuring social distancing, Ambev's team was looking for **big dreams with purpose**, putting projects in place that could change people's lives. As the saying goes, in their culture: "When the going gets tough, the tough get going."

Working with purpose and thinking about the impact that these actions can have on people's lives brings a special sense of belonging to the team, partners, and consumers.

All the initiatives presented here, in addition to the initiative and dedication of Ambev's team, had countless partnerships with public and private companies, which were willing to contribute with the best they had and the best they could do, just like Ambev.

How Change Happened

The first challenge of the pandemic in Brazil was the high demand for hand sanitizers; the demand was much higher than the supply, which caused a shortage of hand sanitizers. Ambev had the opportunity to use an existing capacity and raw materials to produce and distribute this product throughout the country in 10 days. To centralize initiatives like this, the Squad of Good was created, ensuring quick approval and execution of what needed to be done.

The Squad of Good is an autonomous cross-functional team that started with six people focused on the same drive: "How can we adapt what we have or what can we create in order to solve society's problems?" They worked together on all actions, supporting each other, with a horizontal

Figure 2-3 The Squad of Good: Six People Focused on the Same Drive

command structure. More than donating hand sanitizers or donating resources to hospitals, it was important that the company donated the skills of its people: their expertise, leadership, agility, and so on.

To ensure that the initiatives reached as many people as needed, it was very important to stay connected to the current moment, evaluating the potential impact of each action to make things happen. The Squad was responsible for studying and seeking high-impact actions for each moment. They met daily in the mornings to create, propose, and discuss initiatives that could improve the situation and life in the community.

The team quickly began an important movement toward an agile, holistic, and principle-oriented management. The focus on execution, which was already very strong at Ambev, played an important role for the Squad of Good from the start of the pandemic in early 2020 (Figure 2-3).

At the same pace of execution, the crisis committee handled the approval of projects. This crisis committee promoted a daily meeting at the end of the day with the company's vice-presidents to make decisions and discuss topics related to the pandemic situation. This schedule allowed the Squad of Good to conceive, structure and approve the initiatives in less than a week.

Figure 2-4 Ambev Used Its Knowledge and Historical Legacy to Develop Many Initiatives During the Crisis of Covid-19

All the skills and abilities available in the Ambev team were important to execute the initiatives. It was not just about running a project, but evaluating the employees' potential, and understanding how they could do more and offer great projects, bringing people together as one team.

Challenges to Deal With

A key step in the execution and management of the project included implementing a new culture of experimentation without fear of making mistakes, facing challenges, and understanding the new reality, aligned with the existing innovation culture: more collaborative, autonomous, and less punitive. This culture allowed a multidisciplinary team to work together. These distinct areas working together gave a new vision for the projects, and the proximity of the team increased, which helped the fluidity of the initiatives.

Ambev's actions have always focused on solving a problem that the society was facing at that time. This was true not only for the hand sanitizer issue as previously mentioned but also for the demands for oxygen, coolers for vaccine storage, and much more (Figure 2-4). This has been the company focus of the Squad of Good.

In this movement, important tools and skills were developed across the Ambev team, allowing the quick assessment of projects, starting plans, and moving toward a robust execution. The learnings of the "Squad of Good" were disseminated to other projects and influenced the entire team, resulting in improvements such as:

- Innovating in a consumer-centric way, delivering high-quality products.
- Reducing the lead-time and establishing the goal of getting to market delivering products and projects with excellence, while also making a difference in people's lives.
- Promoting a complete integration of the Business Analysis, Project Management, Portfolio Management, and Program Management areas.

More Than Donating Money, Donating Its Best Capabilities

Given the difficulties faced by the country and its population, many companies mobilized through in donations and initiatives. Ambev knew that all kinds of help would be useful at that moment, and following its principles of culture, the company used its entire set of skills to find the best solutions. And by doing that, the company became much more involved with causes and people.

The first action taken was the transformation of Ambev's production lines and bottles to produce alcohol-based hand sanitizers, as mentioned before. This product was very critical for Brazil at that moment. Ambev had the raw material – ethanol – available, so that was the time to understand how the company could use it, work on it, and turn it into a product that was essential at that moment, in order to donate it to hospitals and the community.

Ethanol from Ambev's breweries was used to produce **1.5 million units of hand sanitizers** in 237 ml containers. For the distribution of the first 500,000 units, public hospitals in the most affected areas of the cities of São Paulo, Rio de Janeiro, and Brasília were prioritized. Later, Ambev announced that it would double that production, distributing another 500 thousand units to serve all the Federative Units in the country. All 1.5 million units have already been delivered.

In the meantime, Ambev manufactured and donated **3 million face shield masks** to healthcare professionals in the country. This amount would be enough to serve every healthcare professional in Brazil, considering that there are almost 3 million professionals officially registered in the Datasus database. The production was made from PET, the same material used to manufacture bottles for Guaraná Antarctica – a traditional beverage in Brazil. All face shields were handed over to the Ministry of Health.

These two pieces of personal protection equipment were also critical for hospitals and people in general at the beginning of the pandemic in Brazil, but the country – and its citizens – needed more.

The Squad of Good kept looking for ideas and initiatives that would contribute to the ecosystem (Figure 2-5). Ambev is an expert in project management and in the construction of breweries, so they had the idea of building a hospital, because there was a shortage of beds in all Brazilian states during the pandemic.

The hospital initiative took five days of discussions, from the conception of the idea to the beginning of the construction of a complete and permanent hospital on the east side of the city of São Paulo.

Figure 2-5 The Importance of a Teamwork.

The hospital met the needs caused by Covid-19 and remained a permanent legacy for this poor and vulnerable region of São Paulo. Thanks to a lot of work and commitment from the team and partners, the hospital was built in record time; after 36 days it was already running and helping the community.

At that time, this was the **fastest hospital development in the history of Brazil**. These beds would not only help fight Covid-19, but also become an asset that would remain for this very important and respected public hospital in São Paulo. The best part of this initiative is that it has multiplied, giving rise to many similar initiatives, with hospitals being built following the same model!

Even bigger projects were envisioned, such as a partnership for vaccine manufacturing in Brazil, and the conversion of part of a brewery into an oxygen plant, aiming to fill oxygen cylinders that would be donated to healthcare units in the countryside of São Paulo.

All of this was only possible with the engagement of many professionals, including engineering, procurement, and legal support. The support from the leaders was also essential for the execution of all these projects. There was freedom to create, suggest, and discuss, with unconditional support at all levels.

One Plus One Does Not Always Equal Two

Building a hospital is a considerable challenge for those who are used to building breweries and producing beer. However, by bringing people with different backgrounds together, it became a reality. People with the same purpose and a bold mindset can do things that have never been done before.

Fear holds people back. The fear of making mistakes ends careers and organizations. To avoid this, the Squad decided to support each other and take risks together, outside their comfort zone, moving away from the usual outcomes and searching for the unimaginable. Everyone at Ambev understood that they could not keep doing the same thing repeatedly and expecting different results. Innovation is about taking risks!

Hospital: Less than 40 Days to Build a New Healthcare Structure

At the beginning of the pandemic and following its principles and culture, Ambev wanted to dream bigger. The Squad of Good conceived the construction of a new hospital, in a region of São Paulo City with 1.3 million people and 0.5 hospital beds per thousand inhabitants (Figure 2-6).

This number gets even more concerning when compared to what is recommended by the Ministry of Health – the ideal ratio is between 2.5 and 3 hospital beds per thousand people.

The results were remarkable:

- 5 days of planning and 36 days of execution.
- The hospital was built using modular construction (offering more agility).
- Opened on April 27, 2020.
- Nowadays, this hospital is a reference for the treatment of Covid-19 in the public healthcare system in the state of São Paulo.
- 100 new hospital beds.
- 2,500 people assisted.
- Permanent legacy for the entire local population.

Figure 2-6 Hospital Construction

Oxygen Plant: Turning a Beer Factory into an Oxygen Producer

In mid-January, the oxygen shortage in hospitals in the city of Manaus reached a critical stage. Manaus is a city with complex logistics, situated in middle of Amazon rainforests. There was a collapse in the healthcare system due to the lack of oxygen, which led to many deaths.

Ambev quickly decided to join its partners and help, sending initially 500 oxygen cylinders to Manaus (Figure 2-7). The oxygen was sent in MD model cylinders, which is one of the most indicated for this type of situation, given that it is easy to handle and transport, offering agility in the service. The cylinders were donated to the State and can be reused.

This initial step mobilized local companies to donate oxygen plants to the region. Afterwards, Ambev invested in its own plant, since the company uses oxygen to carbonate beer and soft drinks. Part of the Colorado brewery in the city of Ribeirão Preto, located in the countryside of the state of São Paulo, was transformed into an oxygen factory. Even with no knowledge about the subject, Ambev gathered its employees and faced this challenge.

- 500 oxygen cylinders donated.
- Conversion of part of a brewery into an oxygen factory to deliver 120 oxygen cylinders per day.
- 40 days to execute the project and get approval to begin production.
- Approximately 500 people assisted in 10 Healthcare Units around the city.

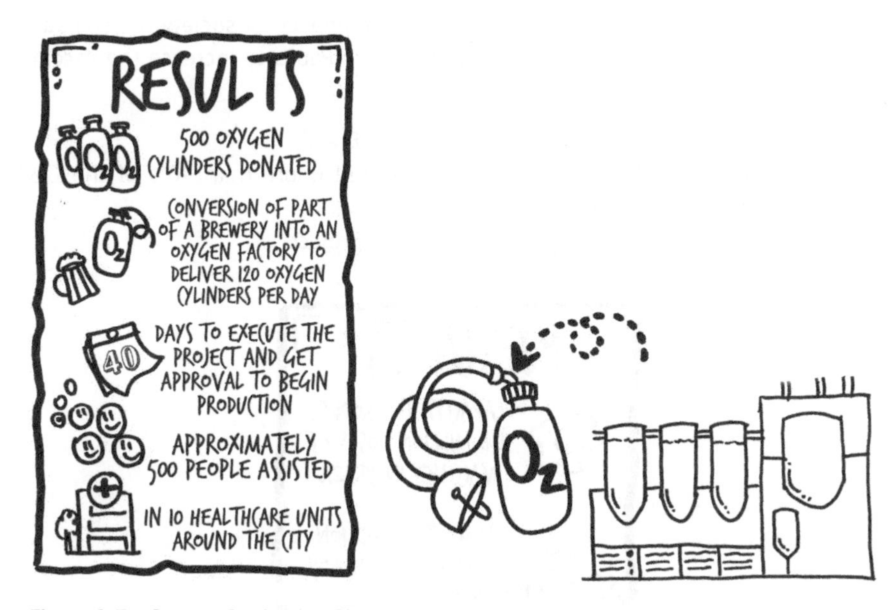

Figure 2-7 Oxygen Production Plant

Vaccine Production Facility: Knowledge, speed, and union to produce vaccines in Brazilian territory

Since the beginning of the pandemic, most Brazilians (and that includes Ambev, which is a 100% Brazilian company) had a dream that the cure for Covid-19 could be found or a vaccine could be produced (Figure 2-8). The crisis brought the possibility for some manufacturers to develop a vaccine to protect humanity, but there was still a huge challenge: local production and distribution.

Ambev understood that the **vaccine would be the greatest weapon against the pandemic**, but Brazil, with continental dimensions and a strong need for scalable production, demanded a very high number of doses. The company's know-how on topics such as infrastructure,

(continued)

(continued)

project management, and agility applied to projects was an extremely valuable resource at that time.

With the purpose of speeding up the vaccine production and allowing Brazil to have more autonomy to produce the inputs locally as well, Ambev led an **alliance of companies to donate resources**, contributing with the necessary infrastructure to set up a vaccine production facility, which was promptly donated.

In this donated facility, its laboratory would be able to produce, in Brazil Covid-19 vaccine, which was created with the dedication of thousands of researchers worldwide.

The speed with which the facility was developed as well as the mobilization of resources to start the production of **100 million doses** as soon as possible was fundamental. Ambev worked on several fronts, including conducting commercial negotiations, interface with suppliers and monitoring the field execution process.

Still within Ambev's purpose of generating a positive impact for society, the company also replicated this initiative at the Butantan Institute in São Paulo, to increase the vaccine production capacity in the national territory.

- Capacity to produce 100 million doses of vaccines in Brazil.
- Almost 40 million doses have already been delivered (more than 40% of the Covid-19 immunization available in the country).
- 11 Ambev professionals exclusively dedicated to the project.

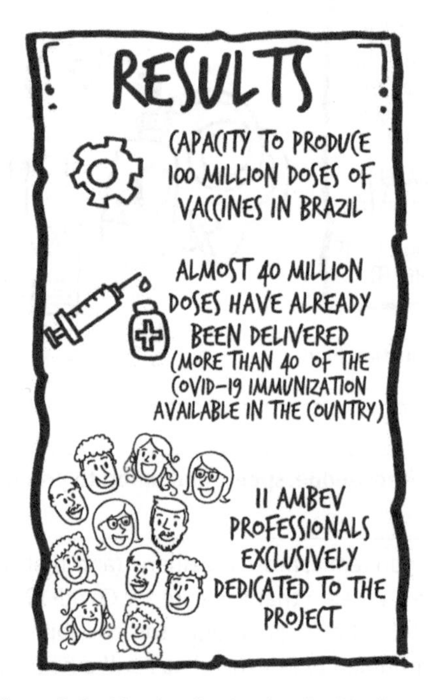

Figure 2-8 Vaccine Production Facility Results

Project Management and Nonlocality

The pandemic brought a new reality to managers, professionals who are normally not far from the execution side of the process. At the same time, engineering teams were also far from their positions and professionals of all areas worked far from the traditional environment they used to live in.

Teams are no longer using the same physical space, but have resorted much more to virtual sharing, which is happening not only in project management but involves all functional organizations. It generates an opportunity to redesign and create collaborative and holistic models.

The Extended View of the Project Value Chain

One of the main lessons learned in the face of the Covid-19 pandemic was that a project starts with an idea, not with the project charter. And, likewise, it doesn't end at the startup, with lessons learned. Nowadays, there are no more borders or silos between operations and projects. To have exponential results, you need to see long-term value. This transformation is so valuable for stakeholders and the entire community involved in the projects.

The traditional model of PM covered one segment of the value chain, extending it from the preparation of the opening letter to the conclusion of the project's final product, collecting customer acceptance. This is just a small part of the value chain (Figure 2-9).

Project management starts before this, when the problem is formulated, brainstormed, when business is being analyzed to understand the company's needs.

Figure 2-9 Project Value Chain

At the beginning, from studies regarding the problem to the product launch; on the other hand, life does not end when a product is launched. On the contrary, it transforms the lives of users by delivering business functionalities. The prolonged use of these characteristics will bring value for the organization or some significant transformation for its stakeholders.

What does it mean to be holistic in project management?

Being holistic means seeing the system as a whole, without individual goals ahead of the strategy, and an integrated project process in all of the working teams.

Being holistic has also another meaning that goes beyond the concept of a multidisciplinary approach: where teams, together and integrated, provide solutions combining their different skills, capabilities, and knowledge.

Let's Go a Little Further

New approaches of project management are based on principles. This change provides conditions for a team to deal with larger contexts, where prescriptive tools were not enough anymore. It is necessary to give a more general orientation, expanding the range to different situations, regardless of the technique or tool.

Some of these principles are perfectly understandable within a context, for example, complexity. When there are multiple parts interacting, you can only understand the big picture by looking at a bottleneck, at the whole systemic behavior, and this is a general orientation that we can also look for tools that promote this, such as, constraint theory, system dynamics, complexity pattern formation, etc.

Ambev believes that stewardship can be used to guarantee a better future for everyone. Stewardship means treating people well, taking sustainable actions for the environment, taking care of the society, taking care of the teams and colleagues, treating shareholders well, honoring sponsors, respecting good practices and commonly accepted ethics.

Ecosystem in Project Management

In the future, the organization of a project will transcend organizational boundaries. The project will be the integration of components provided by an ecosystem, where each part offers a catalog of services with the rules of integration (Figure 2-10).

In the hospital construction project, for example, Ambev had an engineering partner, with experience in modular construction for breweries. Project management will evolve to model through componentization and integration of the ecosystem.

Figure 2-10 Environmental, Social, and Corporate Governance

Quick reactions, the ability to multiply the impact, to inspire companies and people, combined with the freedom to try, discover, and innovate. . . These were some of the many legacies left and learned by Ambev, which certainly brought changes to the lives of its employees, customers and, above all, of the people impacted by any of its initiatives. To conclude, purpose and impact have always guided Ambev to go further, to dream big, and to bring people together for a better world.

Our commitment does not end here. This article was concluded in June 2021, but several initiatives are in the pipeline, and the process continues. . .

2.3 Excellence in Action: Albert Einstein Hospital: Application of Project Management to Address the Covid-19 Health Crisis and Lessons Learned[2]

Introduction

When Hospital Albert Einstein diagnosed the first Covid-19 case in Brazil, the disease had affected approximately 80,000 patients, spreading beyond Asia, Europe, and the Americas. It was then Brazil's turn to be hit by the pandemic. A 62-year-old patient, who had just arrived from Italy, was admitted to the hospital's Emergency Room with typical symptoms of the disease. His diagnosis would then be confirmed through complementary tests the following day.

Some weeks earlier, professionals from the institution discussed aspects related to the pandemic, namely patient care and structural aspects. The need to organize the main actions, mitigate risks, and develop synchronized multidisciplinary plans led the Organization's Management to create a Crisis Committee formed by specialists in infectious and respiratory diseases, in addition to members of the Hospital Infection Service, and by leaders from all areas, whereby action plans would be discussed and designed through a complex model of interactions between patient care, care support, and management.

In order to provide support to the **Crisis Committee**, a **Planning Unit** was implemented, consisting of a **multidisciplinary team with experts** in project management, operational efficiency methodologies, as well as a decentralized medical team with management training, focused on harmonizing different task packages, project integrations, and communication between professionals who, from that moment on, would be dedicated to **handling the crisis**.

Among the essential actions of the planning unit are the following:

- Promotion of the **optimization of essential resources** for the operation
- Management of **several fronts** in progress **simultaneously**
- Facilitation of the performance of care teams, that is, **100% dedication to patient care** and permanent attention to the needs and safety of employees
- Mitigation of failure **in communication**

2 Material in this section was provided by The Sociedade Beneficente Israelita Brasileira Albert Einstein and was written by Juliana Pan e Haggéas da Silveira Fernandes. Juliana Pan is Strategic Planning, Programs and Portfolios Manager. She has worked in several sectors, such as industry, upstream, consulting and healthcare. Haggéas is doctor, Critical Care Medicine specialist. After years treating severe patients, working in intensive care units, he attended a Healthcare Management MBA, and developed a special interest in quality improvement, patient safety, Lean Six Sigma, and project management and agile in healthcare. Today Haggéas is the director of Medical Practices, Quality and Patient Safety in Hospital Albert Einstein.

The purpose of this article is to present the work developed by the **Planning Unit of the Crisis Committee of Hospital Israelita Albert Einstein,**[3] in the scenario of the pandemic caused by the coronavirus, as of February 2020.

Different action fronts were set up to support the organization in its demands, as well as the adoption of safety measures for employees. One of the major challenges in complex projects is to ensure that it is **carried out well** and that the final delivery is consistent with what was proposed, mainly in a **volatile and uncertain scenario**, in which it is necessary to **change uncertainties into risk management**.

Planning Model

The Covid-19 pandemic has caused a humanitarian crisis on a global scale and unprecedented complexity in the modern world. It challenges the fiscal and economic capacity of governments, places a strain on health systems and on the ability of organizations to cope with constant changes and readily adapt to different scenarios imposed by the virus over time.

The pandemic developed in a period of great socioeconomic changes in a world of adaptation accelerated by the digital revolution and by turmoil related to government policies and events capable of creating consequences in the lives of the population of virtually every country on the planet. As a result, uncertainty reached extreme levels.

For many organizations, the most visible effects were related to their **operating and business models**. From the way professionals would work to the speed of information arrival, the hardships required the **acceleration of decisions, adaptation and resilience of the workforce, and leadership**.

The heterogeneity of systems and the scarcity of resources for trained professionals contrasted with the significant increase in the demand for work, posing increasingly greater and unpredictable risks. Projections were made and quickly revised, owing to the change in data, the epidemiological behavior of the disease, and the impact of coping actions. The pandemic actually characterized a **scenario of war in the healthcare sector**.

Fast and assertive planning, with data obtained in an accelerated way would be central to **decision-making**.

Agile models[4], not fully adopted until then, were used as a basis for responding to the demands of the pandemic.

In fact, at the onset of the pandemic, organizations that performed better were adopting agile practices in their operating models. This meant the need for a response structure that would transform information, scientific data, expert guidance, decisions, and resource allocation, into a

3 The Sociedade Beneficente Israelita Brasileira Albert Einstein operates in both private and public healthcare in all stages of healthcare, teaching and education, consulting, research and innovation, and social responsibility. It is headquartered in São Paulo (SP) and develops activities in the city of São Paulo, in the interior of São Paulo, in the states of Rio de Janeiro, Goiás, Minas Gerais, Espírito Santo, Pará and Pernambuco, and in the Federal District. Its service provision structure is formed by 12 private health units in São Paulo and 1 in Sorocaba (SP), 27 units of the Brazilian Universal Healthcare System (SUS), operated through a management contract and agreements with the governments of the cities of city Mogi das Cruzes and São Paulo, six teaching units in São Paulo, one in Rio de Janeiro (RJ) and one in Belo Horizonte (MG).

4 The Agile Manifesto is a declaration of essential values and principles for software development; however, the impacts of the Manifesto and the proposed agile development are undeniable for organizations of different industries. The Agile Manifesto values are: Individuals and interaction between over processes and tools; working software over comprehensive documentation; customer collaboration over contract negotiation, and responding to changes over following a plan.

planning model whose response would be efficient and ensure the coordination necessary to rise to the challenges posed by the situation.

1 The Agile Model Adapted to the Moment of Crisis

The health crisis created the need for the institution to adapt to a model of assertive deliveries, often simultaneously, occurring within limited periods. The organization therefore adhered to the Agile concept, suitable for conditions of high uncertainty, complexity, risk, and characterized by constant change.

In order to help facilitate the decision-making process, the concept of **squads**[5], typical of agile methodologies, was used in the building of **multidisciplinary teams** in order to maintain assertiveness and efficiency.

Squads are multidisciplinary teams with autonomy and time dedicated to actions to handle the crisis, but with **systematic reports** for quick decision-making.

Maintaining the balance of a large organization is not easy.

The operating system is made up of many factors – objective and values, talent management, database, and technology – and each of them can get completely out of hand, becoming static and conservative on the one hand or risky and chaotic on the other.

Balance is the major challenge, and leadership made the difference by creating a **carefully balanced system** that **offered stability and agility**. And that agility, in sum, requires humbleness, which became evident. Humble people recognize the **pointlessness of predicting the unpredictable** and instead create **quick feedback loops** to ensure that initiatives will remain on the right track.

The following are some of the advantages of working with **agile methods** over traditional project management approaches:

- Greater **alignment between teams** for the quick resolution of potential problems and conflicts
- **Risk reduction** and **high-quality results**
- Resource savings through **more assertive delivery**
- **Agility** and **efficiency** in project delivery and in its execution
- **Flexibility** to **propose alternatives** and reach the best solution

2 Initial Planning Steps

The crisis plan considered integrated practices for hospitals and units managed by the organization.

The preparation began by building **a team** to **monitor infections in the world**, and the squad was named the **Intelligence Unit**, formed by people who daily sought information about what was happening in the world. In addition to reading studies, newsletters, and the news, the team also spoke with **international experts**. Later, a core of this team, made up of medical researchers, stood out and began to obtain information from the countless publications that were quickly made available (**Research Unit**). The Intelligence Unit then started to work with updated data on projections, essential for handling the pandemic. Its Epidemiology Unit made calculations that served to structure the hospital early on. Thus, the demands for equipment were projected – from protective masks to mechanical ventilators – as well as the number of professionals.

Based on this information, the **epidemiology team** was able to make projections for the future (Figure 2-11). The calculations helped **structure** the **hospital ahead of time**.

---

5 Squad is the name of the organizational model that separates employees into small multidisciplinary groups with specific goals.

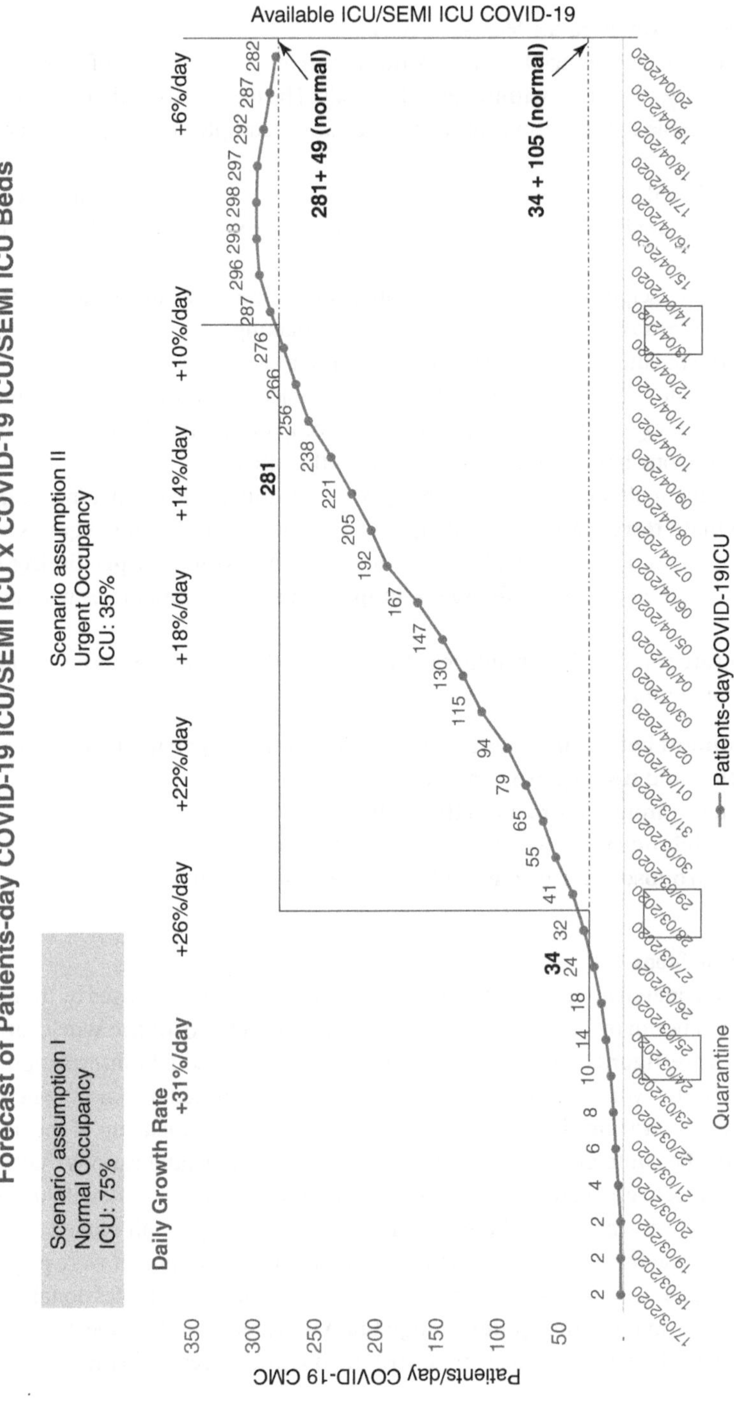

Figure 2-11 Demand Curve

3 Planning Unit

With the projections in hand, the **Planning Unit** was then split up and started to work with the Squads that were being formed.

A workspace was determined, which became known as the **War Room**, away from the healthcare areas, respecting the evidence that arose at the time for the prevention of the progression of the Coronavirus infection, such as social distancing and the use of masks. Digital meeting tools began to be used intensively, which contributed for meetings, such as the Safety Huddle[6] for the update of Healthcare Risk Management, to continue being held, given the current status of the Quarantine. These same tools were used for the Planning Unit and Crisis Committee meetings.

The **Safety Huddle** followed the steps below:

1) Return of items pending completion from the previous day
2) Update of patients diagnosed with Covid-19
3) Update of employees on a leave suspected and/or confirmed to have Covid-19
4) "Watcher" patients (legal/image risk management or other situations worthy of Senior Management support)
5) Problems with the Clinical Staff and/or multidisciplinary team that would pose a risk to patient safety (unsafe or unusual practices, lack of medical assessment, disruptive behavior)
6) Any serious adverse event or event with potential for serious harm worth sharing
7) Problems with infrastructure, electronic systems, equipment/lack of materials or medicines
8) Any new notice or critical situation that could impact other areas and was worth sharing or required Management's support
9) Success case (positive fact of the day, to be celebrated)

Immediately, the lack of critical supplies, such as masks and gloves, started to be observed. Supply chains became fragile in view of the dependence on imported raw materials, with the business spirit often overlapping the necessary empathy for the situation.

Faced with a public health threat, many suppliers have chosen to raise their prices in order to capitalize on the high demand, rather than recognizing the extreme urgency of protecting healthcare professionals who represent the frontline in the fight against the pandemic.

More than ever, managers played a key role, in which their adaptability and resilience made a difference in the performance of the services during this period, ensuring, above all, security and reliability to the front teams, understanding the ecosystem as a whole, and working toward the future horizon as much as possible.

When managing a large project with several stakeholders, systematic follow-up is pivotal, and a **Logbook** was developed in this regard, which focused on all the information and **points of attention**, providing stakeholders with an overview of the activities.

As the activities kept growing at an accelerated pace, the **Planning Unit** would make the **Logbook** available three times a day in order to update all the squads on the progress of each front.

6 The Safety Huddle, also "safety meeting," was proposed by the Institute Healthcare Improvement (IHI). This method raises safety awareness at operational or front-line level and helps the organization develop a safety culture.

In the Logbook it would report the actions, but was not limited to the following:

- **Epidemiological** data
- **Economic impacts** on Brazil and Worldwide
- Projections and production curves
- **Employee Program** – with the purpose of ensuring well-being and the physical and psychological safety of the teams during the crisis, including employee monitoring, through occupational health mechanisms
- Status of operation of the departments, reallocation of resources, hiring, employee leaves, training plan for employees in operation and the hiring of new employees
- **Inventory coverage** (critical items and oxygen), services, and acquisitions (equipment)
- Status of beds in operation, **expansion for the conversion of ICU beds and Surgical Clinic**, new structures and facilities
- Actual status of costs, expenses, revenues, margin, and **capital expenditure control**
- **Evidence-based medicine**, development of clinical research submitted to the National Research Ethics Committee and the Research Ethics Committee on the Use of Human Subjects
- **Donations** to Public Health

Critical issues were mapped in order to **optimize the response plans**:

4 Governance Model

Set up to oversee and manage the Crisis Management Plan, this model was supported by members of the Board, of the Executive Board, and of the Technical Areas, and was led by the **Crisis Committee**.

Governance actions consisted of the following:

- Monitoring the main changes occurring between market agents, mainly health plan companies and regulators, within a context in which changes have been occurring very often.
- Developing and implementing a crisis management communication strategy for all stakeholders.
- Defining through implementing event-detection tools to support the detection of events that can contribute to deepen scenarios and serve as a decision-making reference for operational issues and the execution of the Agile model.

5 Personnel Management Unit

Developed to follow-up on actions integrated in human resources management. From the increase and reallocation of internal staff, through integrated healthcare actions, the unit played a fundamental role in the proper allocation of human resources.

Staff Increase and Reallocation

In order to meet the increased healthcare demand, new professionals were selected and hired in March, a fully online process carried out over a period of 14 days, in which it was possible to reinforce the staff with 1,137 professionals, between hiring for both a limited and an unlimited period, and another 1,081 employees were reallocated internally.

New trainings were formatted as a result of the pandemic, combining practical and theoretical approaches based on continuous training platforms. In the months of March and April 2020, over 5,400 professionals went through training at practical stations, which simulate real service situations, and the online training content received more than 116,700 hits.

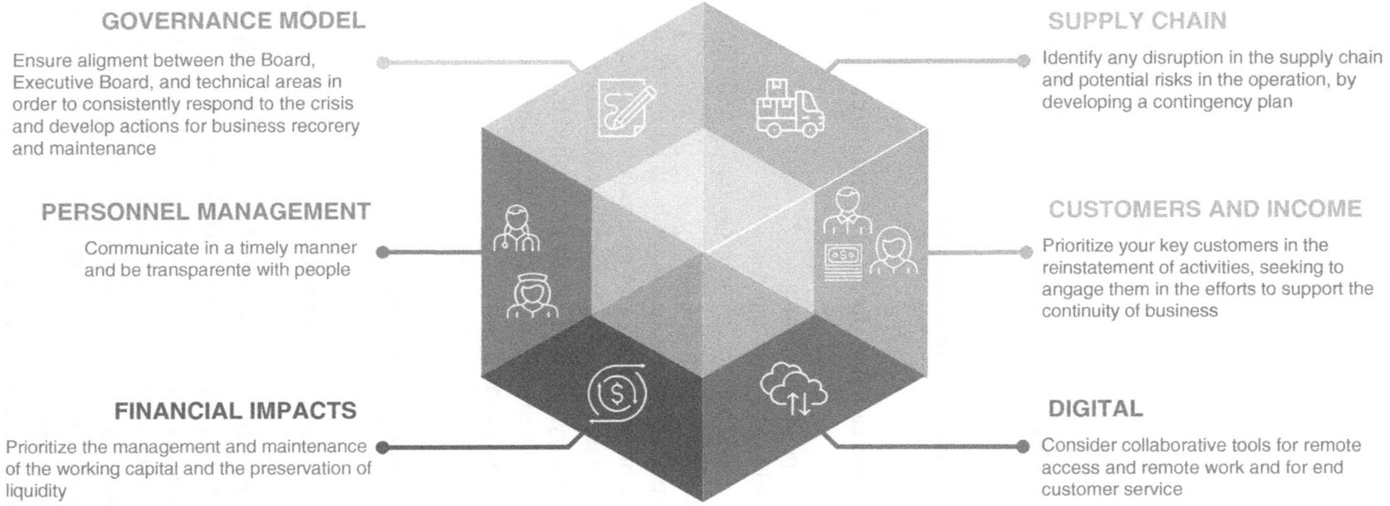

GOVERNANCE MODEL

Ensure aligment between the Board, Executive Board, and technical areas in order to consistently respond to the crisis and develop actions for business recorery and maintenance

PERSONNEL MANAGEMENT

Communicate in a timely manner and be transparente with people

FINANCIAL IMPACTS

Prioritize the management and maintenance of the working capital and the preservation of liquidity

SUPPLY CHAIN

Identify any disruption in the supply chain and potential risks in the operation, by developing a contingency plan

CUSTOMERS AND INCOME

Prioritize your key customers in the reinstatement of activities, seeking to angage them in the efforts to support the continuity of business

DIGITAL

Consider collaborative tools for remote access and remote work and for end customer service

Figure 2-12 Map of Critical Actions

Integrated Employee Support Actions

One of the risks related to the management of the pandemic was frontline professionals that had participated in the care of hospitalized patients with Covid-19 being absent due to contamination. The unit worked to:

- Create mechanisms to guide, monitor 24/7, and assist those in need, among employees (confirmed and suspected cases of Covid-19, as well as other health problems).
- Provide emotional support for dealing with stressful conditions and burnout, which have become common and have an extreme impact on the pandemic.
- Set quarantine rules.
- Stipulate succession and prevention plans for leaders and technical teams.
- Implement and monitor working conditions, providing the required technologies and information.
- Reevaluate licensing, travel, local and global mobility policies.

6 Financial Impacts

The management and maintenance of the working capital and the preservation of liquidity were prioritized, with the:

- Reassessment of the projected cash flow (forecast) for the 100 days ahead and definition of corrective actions.
- Focus on cost control and operating expenses.
- Reassessment of scheduled capital expenditures.
- Negotiation with priority stakeholders in the ecosystem – with those who also had similar issues, and agendas were opened for discussion.
- Understanding of the impacts of contractual obligations for events of force majeure; likewise, the necessary documentation was gathered for renegotiations and contractual claims.
- Negotiation of more flexible supplier terms and additional financing lines.
- Consideration of the temporary benefits recently established by the government in relation to the suspension, postponement, or reduction of tax contributions and the reduction of salaries and suspension of employment contracts.

7 Supply Chain

Identify any disruption in the supply chain and potential risks in the operation, by developing a **contingency plan**:

- A survey on raw materials, packaging, materials, and essential services needed for critical deliveries was conducted.
- Joint actions with strategic suppliers were considered, such as load optimization, integrated production planning, and collaborative capacity management.
- The most financially vulnerable suppliers that could cause greater impact on the chain were mapped, and the possibility of making financial contributions and/or advances were evaluated, in order for them to acquire raw materials, keep services active, and comply with deliveries.
- The logistical planning was revisited, focusing on cost reduction, which could imply the temporary or permanent closing of some services.
- Available assets and resources were optimized, such as capacities, cargo transportation, and relocation of specialists.

- The innovation plan to identify the chance of anticipating the launch of products that would help the organization at a time of crisis was reassessed.

The responsibility for managing inventories, anticipating demands, and establishing a partnership relationship with suppliers was vital to keep the structures supplied with equipment and supplies in order to handle the Covid-19 pandemic.

In negotiations with partners, it was possible to secure the necessary purchases and identify in advance situations in which the regular suppliers would not be sufficient to meet the demand.

Internally, information technology supported the supply activities, with a material planning system running on a digital platform and automated order processing.

8 Customers and Income

Physicians and patients were prioritized in the reinstatement of activities, seeking to engage them in the efforts to support the continuity of the services.

- Loyalty was reinforced, ensuring support to physicians, under special conditions, such as reduced fees and Telemedicine support, due to the significant reduction in outpatient movement and to the reduction in surgical care.
- The consumer journey and experience were assessed, adapting the strategy of going to the hospital and implementing the transition from physical channels and service delivery to digital models and expansion of home care.
- A clear and consistent communication strategy was developed, informing physicians, patients, and employees about the measures adopted by the organization.

9 Digital

Implementation of collaborative tools for remote work and for health and patient care.

- The infrastructure for new traffic and usage patterns was prepared, evaluating security, authentication, and network capacity standards, in addition to considering its partners and suppliers
- 24/7 structure and team dedication in order to ensure technological support for business continuity in view of a crisis requiring remote work
- Potential cyber and system vulnerabilities were timely assessed and addressed, at a time when the exposure and traffic of information in the online environment has expanded even further
- Real-time data and alerts **dashboard** were developed to serve the several executive levels
- Digital patient and employee communication channels were created, which was essential in an institutionalized quarantine scenario, such as Telemedicine.

Results

Between March 2020 and January 2021, 13,590 patients with Covid-19 were treated at Einstein's units in an inpatient or Day Clinic regimen (Figure 2-13). Of these, 74%, or 10,027, were admitted to public units: 6,330 patients in public hospitals managed by Einstein (Hospital Municipal M'Boi Mirim – Dr. Moysés Deutsch, Hospital Municipal Vila Santa Catarina – Dr. Gilson de Cássia Marques de Carvalho, and Pacaembu Field Hospital), and 3,697 in Emergency Care Units (UPA), in inpatient care, some requiring mechanical ventilation while waiting for permanent beds. This volume represented 18 percent of all admissions of patients with Covid-19 in

RESULTS

Between March 2020 and January 2021

6,000 TRAINED EMPLOYEES
caring for patients infected with the disease

694 REALLOCATED EMPLOYEES
For the two public hospitals managed by Einstein and in the Field Hospital

13,6 thousand
HOSPITALIZED PATIENTS
in an inpatient or Day Clinic regimen from March 2020 to January 2021

3,6 thousand
IN THE PRIVATE HEALTHCARE

10 thousand
IN THE UNIVERSAL HEALTHCARE SYSTEM, WHICH REPRESENTS 18% OF THE TOTAL HOSPITALIZED PATIENTS
in the entire public system in the city of São Paulo

DONATION OF
BRL 36.1 million
IN PERSONAL PROTECTIVE EQUIPMENT AND RUBBING ALCOHOL
to five Special Indigenous Health Districts (DSEI), with approximately 600 villages, and over 130 public and philanthropic hospitals in several states, financed with their own funds and non-financial donations received

EINSTEIN FUND IN THE FIGHT AGAINST COVID-19:
BRL 48.5 million
RAISED AND FULLY INVESTED IN THE UNIVERSAL HEALTHCARE SYSTEM
for the expansion and improvement of public units, donation of hygiene kits and food baskets for vulnerable communities, and the donation of personal protective equipment and rubbing alcohol for health institutions in several states

Figure 2-13 Results Highlights

the city of São Paulo in that period –the percentage considers data from the Municipal Health Department, which recorded 34,962 discharges of patients with Covid-19 in the public hospital network.

To meet this demand in the private and public sectors, Einstein increased its facilities of mechanical fans, hired and reallocated professionals, and identified opportunities to optimize the available resources, with the creation of new operating beds. The initiatives, adopted within the scope of private health and in the units of the public system, were possible through an integrated approach at the primary, secondary, and tertiary levels of healthcare.

1 Pacaembu Field Hospital

At the request of the Government of the City of São Paulo, Einstein Hospital took over the Pacaembu Field Hospital operation, set up on an emergency basis to alleviate the demand for hospital beds within the city's public system.

During the three months of its operation, in the most critical period of the pandemic, the hospital provided 200 beds of low and medium complexity and treated 1,515 patients transferred from other public health units, in order to free up those structures for the treatment of more serious cases.

The hospital occupied an area of 6,300 square meters on the lawn of the Pacaembu soccer stadium and was partly equipped by Einstein with its own resources, either relocated from other

Figure 2-14 Pacaembu Field Hospital

units or acquired with funds from donations (Figure 2-14). After the closing of the Field Hospital, Einstein donated all the equipment used at the site to the City Government, with an estimated value of BRL 7 million.

The structure received all the resources of a complete hospital. Electronic medical records, diagnostic imaging equipment (radiography, ultrasound, tomography), pharmacy area, and a cafeteria for professionals are some examples, in addition to Telemedicine, which connected the local team with the central unit for a second opinion.

As visits were not allowed, contact between family members and patients was made on WhatsApp via mobile phones and tablets. The daily medical report was also transmitted to family members over the app.

The entire work was focused on the safety, quality, and humanistic care to patients and to the safety of the 520 professionals who would take turns in three shifts at the site. The teams were made up of professionals who were already working in the organization and of others hired specifically to respond to the pandemic.

As this was first field hospital in the country in the action against Covid-19, there was no standard to be followed, which meant the organization needed to define the operating model. The task was assigned to a multidisciplinary team, made up of professionals from the medical care, logistics, supplies, and pharmacy areas. Later, the definitions for physical structure, operating model, circulation flow, care protocols and protection protocols for health professionals, and the humanistic care for patients and families adopted were able to be applied to similar structures created as emergencies in other regions of the country.

2 Hospital M'Boi Mirim

Hospital Municipal M'Boi Mirim – Dr. Moysés Deutsch, managed by the organization, was one of the main structures of the Universal Healthcare System[7] in São Paulo for the care of the pandemic, with 514 beds (220 in the ICU and 294 in the infirmary) dedicated to Covid-19 patients (Figure 2-15). More than 4,000 patients were treated at the unit throughout the year, an important contribution to avoid the collapse of the city's health system. The unit is managed by Einstein, in association with the Dr. João Amorim Study and Research Center.

Management's speed and sense of urgency were crucial for the results achieved. Also in March, the hospital was the first in the public network to make adaptations to meet the imminent health emergency, with the increase of operational beds, the conversion of infirmary beds into ICU beds, and the reinforcement of respiratory support equipment.

In April, the unit was expanded, funded by the private sector and built in record time.

The hospital was also an example of best practices in human care, and created innovative ways to minimize the distress and loneliness of both patients and families. Televisits were made available in the ICU and, in more serious cases, face-to-face visits. Equipped with safety suits, the closest family members that were not in an at-risk group were able to say goodbye to their loved ones. A multidisciplinary team was assigned to provide support to patients and families.

Resumption Plan

The **Continuity Plan – Resumption** was revised and activated, ensuring that the applications and services would be operational, as well as the continuity of essential functions for patients, in addition to the frequency in the vulnerability assessment of the Einstein Hospital units (public and supplementary).

It took 33 **days** from the start of construction to the delivery of the expansion work of Hospital Municipal M'Boi Mirim. The annex has a total area of 1.4 thousand square meters, distributed across two floors, connected to the hospital, and an operational capacity of 100 beds.

The work involved an investment of **BRL13.5 million** by Ambev, Gerdau, Brasil ao Cubo, and Einstein, and is now a permanent part of the hospital's structure.

Figure 2-15 M'Boi Mirim Hospital

7 The Universal Healthcare System (SUS) is the denomination of the Brazilian public healthcare system created by the Federal Constitution of 1988.

The organization demonstrated that the healthcare environment was safe in the midst of the pandemic.

Governance Model

- Lessons were reflected and **action plans** were mapped with resilience.
- **Crisis Committee** protocols were considered.

Personnel Management

- **Employees' return** to "normal" activities was managed.
- The **strategy for absorbing talents** hired and identified in the pandemic in regular operations was revised.
- **Financial Impacts**
- The **capital expenditure portfolio** and the additional financing for low-cost and long-term debt were reprioritized.
- The normalized **financial reporting** was accelerated.
- **The size** and **the operating model** (e.g. fixed vs. variable resources; people vs. technology) were systematically assessed.

Supply and Operations Chain

- The transparency and visibility into the risks of the entire supply chain were explored.
- Collaboration with customers and suppliers in order to synchronize operations and meet priorities despite constraints.
- Implementation of agile inventory and cash flow management.
- Orders, appointments, and inventory were revised.

Health Plan Companies and Income

- The operational model to strengthen customer engagement was redesigned.
- Advanced technologies available in customer interactions and distribution channels were implemented.
- The cycle of receivables, collection processes, and OKRs were redefined.
- E-commerce and channel strategy and the relationship with B2C and B2B were optimized.
- Control mechanisms were established for risk prevention and responses.
- Contingency plans for financial disruptions were designed.

Digital

- The capacity to intensify flexible and digital working conditions was developed.
- Cyber risk management mechanism and the entire technological infrastructure mechanism were structured.

Through the uniform oversight of the main actions taken by the organization in the period, over the course of months it was possible to adjust the resources to smaller needs during the most critical fluctuations of the pandemic across different regions of Brazil.

Lessons Learned

This will not be the last pandemic: Invest **now** to reimagine health systems.

The Covid-19 pandemic exposed neglected weaknesses in the **world's capabilities for surveillance and response to infectious diseases** – weaknesses that have persisted despite the obvious damage they caused during previous outbreaks.

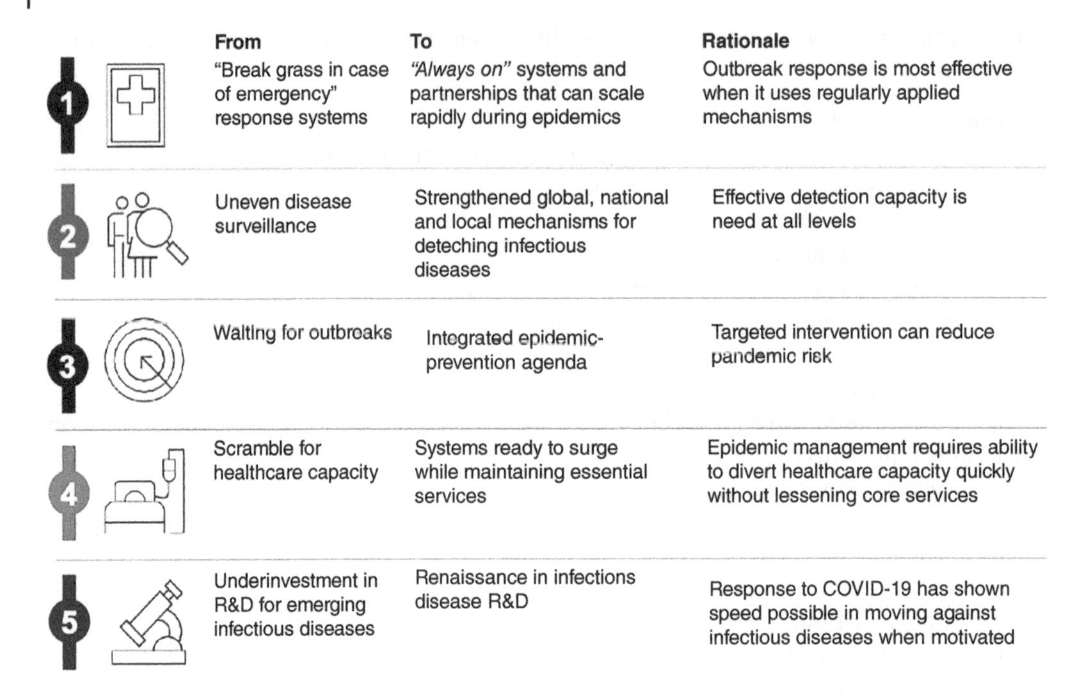

From
To
Rationale
1
"Break grass in case of emergency" response systems
"Always on" systems and partnerships that can scale rapidly during epidemics
Outbreak response is most effective when it uses regularly applied mechanisms
2
Uneven disease surveillance
Strengthened global, national and local mechanisms for deteching infectious diseases
Effective detection capacity is need at all levels
3
Waiting for outbreaks
Integrated epidemic-prevention agenda
Targeted intervention can reduce pandemic risk
4
Scramble for healthcare capacity
Systems ready to surge while maintaining essential services
Epidemic management requires ability to divert healthcare capacity quickly without lessening core services
5
Underinvestment in R&D for emerging infectious diseases
Renaissance in infections disease R&D
Response to COVID-19 has shown speed possible in moving against infectious diseases when motivated

Figure 2-16 Changes in the Health System – Part 1

Five changes to health systems may help reduce the chance of future pandemics (Figures 2-16 and 2-17):

The scenario of uncertainties calls for a clear form of direction from organizations: "We recognize that there is a crisis. As a result, we opt for individual or collective decisions that express resilience by definition, allowing us to deliver the best possible service, despite the complex challenges, and that are able to change the status quo, without any losses."

Thinking, feeling, and acting based on this belief enabled an expansion of the vision, strengthening of the mission, and alignment of the purpose, in addition to new ways of reflecting on the management of people and activities, developing and improving skills and new behaviors that were decisive for the creation and expansion of services, improvement in stakeholder relations, and for turning processes and services more creative and innovative.

In times of uncertainty, facing the inevitable will be part of a new individual and corporate mindset. The crisis led to a great opportunity for advancement, such as opportunities for adjustments needed for individual and collective improvement, as well as learning about the importance of flexibility, resilience and, above all, solidarity and collaboration among different entities.

Multidisciplinary teams in the healthcare sector were put to the test, in an unprecedented scenario. The support from organizations, providing all the equipment, work methods, training, health, and emotional support, is essential for the success of the project. Project management, based on an agile model, leveraged by digital tools and non-exhaustive data support, proved in our experience to be the winning model to respond to the Covid-19 pandemic.

Change is the **essence of the management work**. Leaders must see change not as an occasional disruptor but as the very essence of their work. Setting goals and processes for achieving them, executing those processes, and carefully learning from them – should characterize the daily life of the organization across all levels.

1

Governments must keep the **preparation** for outbreaks preparedness on the public agenda.

- support the epidemiological response capacity
- maintain robust inventory of medical supplies and emergency supply chain mechanisms
- carry out regular outbreak drills

2

Most countries are far from realizing the **potential of integrating advanced data** and analytics to complement traditional, event-based surveillance in identifying infectious disease risks, so that **authorities can initiate efforts to interrupt individual transmission chains.**

- recognize that the threat of infectious diseases posed by one country is a threat to all nations
- build and maintain the capacity to investigate flexible, high-quality outbreaks across geographies
- develop strong pathogen surveillance, including through genome sequencing

3

While it is not possible to prevent all epidemics, it is possible to use all tools to prevent them, and **four approaches** to that can be highlighted.

- reduce the risk of zoonotic events by discovering unknown viral threats,
- reduce risk of zoonotic events by limiting interactions between humans and wildlife,
- limit antimicrobial resistance (AMR)
- administer vacciness on a wider scale

4

Prepare health systems to set detailed plans of how capacity can be focused on pandemic management and how incremental capacity can be added quickly. Peak capacity plans for pandemics must consider **the need to maintain essential health services.** It is becoming increasingly clear that the **secondary impacts of the COVID-19 pandemic** on population health are of a **similar magnitude to those directly ascribed to the disease.**

5

There is potential to trigger a renaissance in **infectious disease R&D**

- close gaps in vacciness and therapeutic arsenals against known threats, including influenza, for which effective R&D can provide significant advances
- scale vaccine manufacturing capacities for the production of 15 billion doses over a six-month period in order to provide sufficient coverage to immunize the global population
- invest in the development of new vaccines, antibodies, antivirals, and therapeutic platforms to combat emerging infectious diseases

Figure 2-17 Changes in the Health System – Part 2

> Overcoming any crisis will depend on the agility of the leadership in adapting its operating models to new circumstances

2.4 Excellence in Action: United Nations: Program Management for Humanitarian and Development Projects[8]

Introduction

Humanitarian and development-driven initiatives provide material, logistical and capacity building assistance to communities or a whole country in need of aid. Projects and programs are launched to address human-made disasters (conflicts, terrorism, technological hazards, etc.);

8 Material in this section was provided by Ricardo Viana Vargas, Farhad Abdollahyan, Carlos Alberto Pereira Soares, and André Bittencourt do Valle. Ricardo led the writing of this article when he was the director for Infrastructure and Project Management at the United Nations Office for Project Services (UNOPS). The opinions expressed in this case study do NOT represent those of the United Nations (UN), including its Member States and UNOPS. All rights reserved. Reproduced with permission of the authors.

At the time this section was written, all footnoted links were active.

disasters triggered by natural hazards[9] (earthquakes, floods, droughts, etc.) or to provide solutions to address development needs such as lack of health and sanitation, famine and poverty, illiteracy, lack of human resources qualification and lack of livelihood. In short, humanitarian projects are *"operations conducted to relieve human suffering, especially in circumstances where responsible authorities in the area are unable or unwilling to provide adequate service support to civilian populations"* (Reliefweb 2008), which means that the environment may be either unstable or even hostile to those who would provide the aid.

Humanitarian projects mainly focus on poverty-stricken regions of developing countries in Africa, Latin America, and Asia. These areas have been in poverty for a long time and the living conditions and quality of life are typically below international standards, as measured by the Human Development Index – HDI[10]. According to United Nations Development Program (UNDP) – Human Development Report, *"Overall human development levels continue to rise, but at a slower pace than before. The 2013 HDI value at the global level is 0.702, while the 2012 HDI was 0.700"* (UNDP-a 2014).

Humanitarian and development projects range from complex emergencies[11] (Reliefweb 2008) to small-scale, standalone, and output-oriented initiatives with limited scope and importance. While the former is relevant to program management, the latter is out of scope unless the projects are grouped together and managed as an emergent program as defined in the UK framework – Managing Successful Programs® (Cabinet Office 2011). This chapter covers those humanitarian and development projects that are undertaken in a post-disaster environment or in cases of chronic hunger and poverty. The international community contributes to these areas through the provision of funds, experts, food, medicines, farming and other technological tools, building materials, among other items. All these temporary engagements are projects managed with a varying scale of efficiency and effectiveness.

Humanitarian and development projects aim to produce outputs (deliverables) that contribute to results (i.e., provide capabilities leading to desired outcomes and impacts to the beneficiaries). Therefore, it makes sense to either group them together as emerging programs or launch them from the beginning as a vision-led program (Cabinet Office 2011). Although the scope and scale of these initiatives vary, there are principles such as transparency, mutual accountability, and governance; characteristics, such as complex multi-stakeholder environment; and critical success factors, such as the use of logical frameworks, benefit maps, and results-based planning that are common to all these projects.

Humanitarian and development programs are subject to external risks such as natural hazards, political and economic instability, and internal factors such as lack of transparent governance and local capacity that may interfere, change, or even halt their progress will be presented.

The characteristics, types, and the measurement of value of humanitarian and development programs will be presented. Some specific tools for this kind of programs, such as logical

---

9 A list of hazards is available in the UNISDR page at http://www.preventionweb.net/english/hazards/

10 The Human Development Index (HDI) is a summary measure of average achievement in key dimensions of human development: a long and healthy life, being knowledgeable, and having a decent standard of living. The HDI is the geometric mean of normalized indices for each of the three dimensions (for more detail access http://hdr.undp.org/en/content/human-development-index-hdi).

11 A multifaceted humanitarian crisis in a country, region, or society where there is a total or considerable breakdown of authority resulting from internal or external conflict and that requires a multisectorial, international response that goes beyond the mandate or capacity of any single agency and/or the ongoing UN country program. Such emergencies have, in particular, a devastating effect on children and women, and call for a complex range of responses (Relifweb, 2008).

frameworks, will also be exemplified. Finally, two cases of program management practices in this field will illustrate the adherence of the program management framework: Poverty reduction in Myanmar and urban rehabilitation in Haiti.

These cases vary in their nature, scale, and scope, and most importantly in governance and leadership. The Haitian government shared the management of its post-disaster reconstruction development program with foreign donors under the general coordination of United Nations agencies, while the government of Myanmar participated initially much less.

One of the programs of the portfolio of Haitian Reconstruction was the urban rehabilitation 16/6 Project (ONU-Haiti 2014). Its purpose was to design and execute the transfer of population from six camps, where they were temporarily allocated, back to 16 neighborhoods devastated by the 2010 earthquake.

Principles

Besides general program management principles, as described in Part 1 (Foundations of Program Management), some specific principles apply to humanitarian projects. These initiatives follow the global principles of humanity, neutrality, and impartiality (UN 1991). Adherence to these principles reflects a measure of accountability of the humanitarian community. Other principles also apply to humanitarian and development programs among which are transparency, national ownership and mutual accountability, and some crosscutting concepts such as sustainability, gender equality, and community engagement.

According to OECD[12],

> At the Second High Level Forum on Aid Effectiveness (2005) it was recognized that aid could – and should – be producing better impacts. The Paris Declaration [on Aid Effectiveness] was endorsed to base development efforts on first-hand experience of what works and does not work with aid. It is formulated around five central pillars: Ownership, Alignment, Harmonization, Managing for Results and Mutual Accountability.

The Accra Agenda for Action (2008) reaffirms the objective of strengthening partner countries' ownership of their development strategies and strengthening the ties, making governments accountable for their domestic constituents. It also broadens the concept to include engagement with the parliament, political parties, local authorities, the media, academia, social partners, and the broader civil society. The principles of Accra Agenda for Action are:

- **Ownership:** Countries have more to say over their development processes through wider participation in development policy formulation, stronger leadership on aid coordination and more use of country systems for aid delivery.
- **Inclusive partnerships:** All partners – including donors in the OECD Development Assistance Committee and developing countries, as well as other donors, foundations, and civil society – participate fully.
- **Delivering results:** Aid is focused on real and measurable impact on development.
- **Capacity development**: to build the ability of countries to manage their own future – also lies at the heart of the Accra Agenda for Action.

12 OECD, Paris Declaration and Accra Agenda for Action, http://www.oecd.org/dac/effectiveness/parisdeclarationandaccraagendaforaction.htm

Risk Context

- As the latest Human Development Reports (HDRs) have shown, the population in most countries has been doing progressively better in human development (UNDP-b 2014). Progress in technology, education and incomes may eventually lead to longer, healthier, and more secure lives. Nevertheless, there is also a widespread sense of instability in the world today – in livelihoods, in personal security, in environment and in global politics.
- More than 15 percent of the world's population are exposed to multidimensional poverty. High achievements on critical aspects of human development, such as health and nutrition, can be quickly undermined by a natural disaster or recession periods. Urban violence such as theft and armed robbery can leave people physically and psychologically impoverished. Corruption and impassive state institutions can leave those in need of assistance without alternatives.

There is a convergence among several frameworks that were considered as separate paradigms in the past. In fact, researchers, and practitioners in the field of development regard disaster risk (UNISDR[13] 2009), building national and local capacity, as well as having resilient infrastructure and communities, either as an integral part of humanitarian and development initiatives or as a prerequisite for their success (Figure 2-18). UNDP considers vulnerability and resilience through a human development lens in its latest Human Development Report (UNDP-b 2014).

$$\text{Disaster Risk} = \frac{\text{Hazard x Exposure x Vulnerability}}{\text{Resilience}}$$

Hazard	Exposure	Vulnerability	Resilience
A dangerous phenomenon, substance, human activity or condition that may cause loss of life, injury or other health impacts, property damage, loss of livelihoods and services, social and economic disruption, or environmental damage.	People, property, systems, or other elements present in hazard zones that are thereby subject to potential losses.	The characteristics and circumstances of a community, system or asset that make it susceptible to the damaging effects of a hazard.	The ability of a system, community or society exposed to hazards to resist, absorb, accommodate to and recover from the effects of a hazard in a timely and efficient manner, including through the preservation and restoration of its essential basic structures and functions. **Higher the Resilience = Lower Risk of disaster**

Figure 2-18 Disaster Risk and Its Relation to Resilience Adapted from UNISDR (2009)

13 United Nations International Strategy for Disaster Reduction, https://www.undrr.org.

To treat these uncertainties, UN agencies such as UNOPS[14], identify and respond to risks in the following risk categories when engaging their partners (governments, donors and implementing organizations) that may even require escalation to the Program Board level:

Government and regulatory – those events and conditions that may occur, changing the institutional framework governing the program and affecting its business case. A government may change its policies toward international cooperation and hence affect an ongoing program governance structure, for example.

Reputation – any incident that may affect the credibility of the implementing organization, as a transparent and politically neutral humanitarian and development agent should be managed.

Human resources – in humanitarian and development programs, HR risks are related to availability and skills necessary to run each project of the program, as well as manage the program itself. Most of the team members are contracted on a temporary basis and may not be easily replaced.

Implementation – risks may not only be related to design and execution of the solution but also to all aspects of procurement and logistics, working environment security or suppliers/contractors' quality and performance. These risks will affect capability delivery and may delay benefit realization.

Sustainability – humanitarian and development initiatives are oriented to sustainable outcomes, therefore all events and conditions that may impact environmental, economic, social, and national capacity dimensions of the program should be managed. An example of a sustainability risk would be related to assumptions of future maintenance requirements such as availability of operational budget for staff to run the capabilities and realize the benefits.

Complexity

In spite of all these coordinated efforts with the participation of rich donor countries and multilateral institutions and their poor country beneficiaries, specialists question the effectiveness of humanitarian and development aid (Figure 2-19). Fengler and Kharas (2010) observe that in the last two decades, although total aid has increased (from US$92 billion in 1992 to US$200 billion in 2008) and more players are in the game, aid has declined in *relative* importance in most countries. According to the same authors, a new and much more complex aid environment is emerging, in which direct foreign investment is higher than aid.

The donor landscape has also changed radically and new private players – International NGOs, foundations, and vertical funds[15] – are responsible for growing the share of aid volume (Figure 2-19). Private philanthropy alone now reaches US$60 billion annually, although most aid still comes from the members of the Development Assistance Committee (DAC), a club of 22 rich countries contributing with a total of US$120 billion per year. The "Third World" is no longer homogenously poor, and some countries, such as Brazil, not only reduce poverty at home but also become donors in Haiti and African countries. This "South-South" bilateral aid sums up to US$15 billion per year (Fengler and Harras 2010).

14 United Nations Office for Project Services, http://www.unops.org

15 Vertical funds are administered by multilateral agencies that focus on a single theme (such as Global Funds for AIDS, TB, and Malaria).

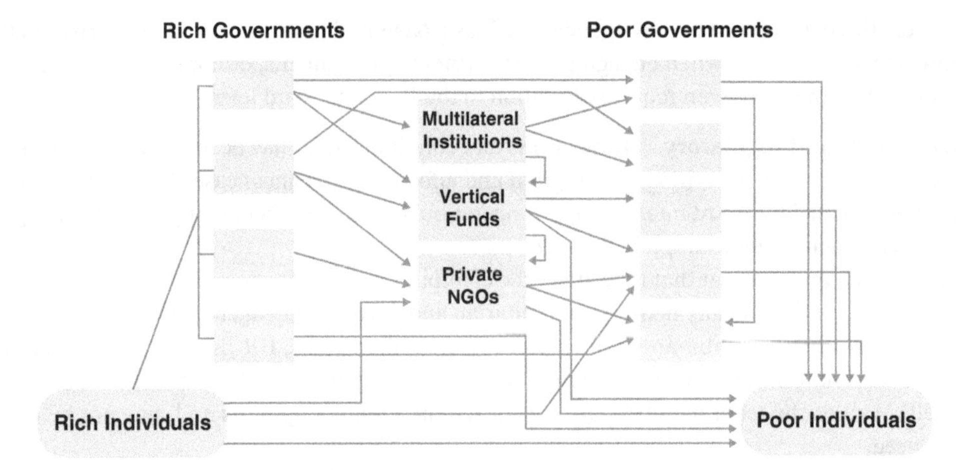

Figure 2-19 Complexity of Aid Flow Source: Fengler and Kharas 2010

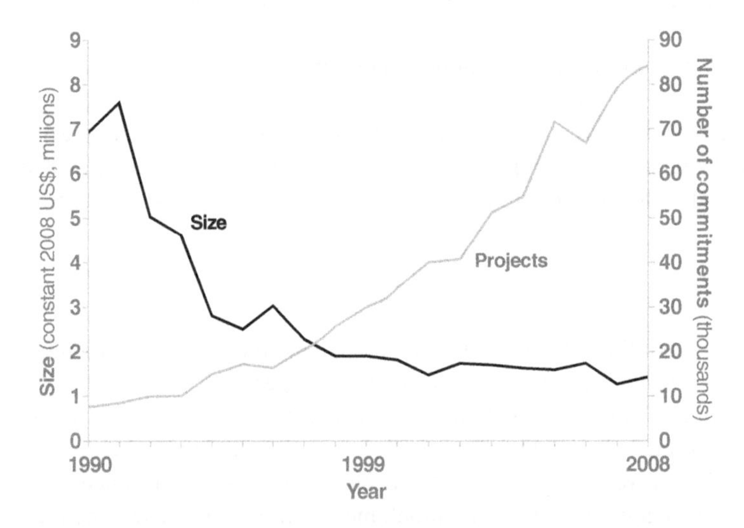

Figure 2-20 Project numbers and size, 1990–2008 Source: US Census Bureau; Banco de Mexico (in Fengler and Kharas 2010)

Although the volumes of aid are growing, the average project size – even of traditional donors – is shrinking (Figure 2-20). Smaller projects can deliver benefits to isolated communities, but they increase aid fragmentation that increases administrative costs and complicates donor coordination by recipient governments. Clustering small projects into programs can mitigate this tendency.

One of the recommendations of Fengler and Kharas supports the idea of emerging programs in the development aid context.

The focus should be on the dynamics of development. Individual project success does not always add up to systemic change. Projects that work well are not systematically scaled up. Donors have limited long-term engagement in or accountability for results in each area. A focus on dynamics can mean changing the institutional setup, with more aggressive monitoring, evaluation, and assessment of development results. Scalable and programmatic approaches commensurate with country needs have to be encouraged. It is important to identify needs, interventions, and gaps at a local and sectorial level, and to monitor progress in these areas systematically over time. (Fengler and Kharas 2011)

In other words, if a program management approach is applied there will be not only cost reduction and higher efficiency in use of aid funds, but also more effectiveness through the design of the right set of benefits to be achieved, planning and monitoring, and evaluation of outcomes and results.

Design, Monitoring and Evaluation Frameworks

Part Four of this book addresses the program life cycle, processes, methods, and tools that are applicable to most of the programs. Since the 1990s, most of the humanitarian and development programs have been managed based on the "Theory of Change" (ToC) model developed by Weiss (in Connell, Kubisch, Schorr, and Weiss 1995).

ToC is a rigorous yet participatory process whereby groups and stakeholders in a planning process articulate their long-term goals and identify the conditions necessary for those goals to be met. These outcomes are arranged graphically in a causal framework. ToC describes the types of interventions (projects) that bring about the outcomes depicted in the outcome's framework map. Each intervention is tied to an outcome in the causal framework, revealing the complex web of activities required to bring about change. It provides a working model against which hypotheses and assumptions can be tested on what actions will best produce the outcomes in the model.

Adherence to the ToC method keeps the processes of implementation transparent so that everyone involved knows what is happening and why. To be clear, every outcome in the theory is explicitly defined. All outcomes should be given one or more indicators of success. As implementation proceeds, organizations collect and analyse data on key indicators as a means of monitoring progress on the theory of change. Indicator data show whether changes are taking place as forecasted. Using the indicator data, program staff can adjust and revise their change model as they learn more about what works and what does not.

Rationales in a theory of change explain the connections between the outcomes and why one outcome is needed to achieve another. Assumptions explain the contextual foundations of the theory. Often, rationales and assumptions are supported by research, strengthening the likelihood of the theory and the likelihood, accompanied by a written narrative that explains the logic of the framework, that is, the chain of input, output, outcome, and impact.

ToC can be both a planning and problem-framing tool, and a monitoring and evaluation tool. In formulating long-term outcomes, preconditions, and interventions, ToC forms the basis of visioning papers, strategic and/or annual plans, and goal-setting processes. As an evaluation tool, ToC identifies the specific goals of the program and ties those goals to particular engagements. Data can then be collected to evaluate progress toward the stated goals, as well as the effectiveness of interventions in producing outcomes.

ToC maps out the initiative through stages:

- Identifying long-term goals and the assumptions behind them.
- Backwards mapping from the long-term goal by working out the preconditions or requirements necessary to achieve that goal and explaining why.
- Voicing assumptions made about what exists in the system without which the theory used would not work, and articulating rationales for why some outcomes are necessary preconditions to other outcomes.

- Weighing and choosing the most strategic interventions to bring about the desired change.
- Developing indicators to measure progress on the desired outcomes and assessing the performance of the initiative.
- Quality review should answer three basic questions: Is the theory used (1) plausible, (2) "doable" (or feasible), and (3) testable?
- Writing a narrative to explain the summary logic of the initiative.

Another popular model is the Logical Framework (known as LogFrame) originally elaborated by Rosenberg, Posner, and Hanley (1970) for United States Agency for International Development (USAID). The method has evolved into results-based management (RBM) and managing for development results (MfDR) used as a standard methodology in multilateral organizations like United Nations Development Program – UNDP and the World Bank.

The logframe is a project management tool that can be used to design, implement, monitor, and evaluate a project. It presents a wealth of information related to the project in a 4 × 4 matrix (Figure 2-21).

The logframe facilitates reflection on the basic elements of the project, such as its objectives, the activities that should be performed, the resources that are required, how to monitor the project's progress and results, and what risk could threaten the project.

The first column of the 4 × 4 matrix shows the project logic (also called intervention logic) – hence the name **logical framework.** The project's activities are captured on the bottom row. When the activities are completed, they are expected to lead to tangible outputs. All the different

Goals	Indicators	Verification Sources	Assumptions
Purpose	**Indicators**	**Verification Sources**	**Assumptions**
Outputs	**Indicators**	**Verification Sources**	**Assumptions**
Activities	**Resources**	**Means**	**Assumptions**

Figure 2-21 Logframe Matrix

results together will help to achieve the project's purpose (sometimes called specific objective). This is the main reason why the project was conceived in the first place. It is the problem that one wants to resolve. In a broader context, the project's purpose will help achieve one or more goals (or general objectives), which are captured in the top row. The term **project logic** means that one thing leads to another: the activities produce tangible outputs that lead to the project's purpose that contributes to one or more goals.

Both models use monitoring and evaluation to ensure that the outcomes and impacts are reached and that lessons learned are incorporated in the next phases of the program.

Lift Case Background

The Livelihoods and Food Security Trust Fund (LIFT) is a multi-donor consortium fund established in Myanmar in 2009 with the overall aim of reducing by half the number of people living in poverty and hunger. LIFT works in areas that account for about 90 percent of food poverty in Myanmar, including Ayeyarwady Delta, Dry Zone, Chin, Kachin, Shan, and Rakhine States. The purpose of LIFT is to increase food availability and the incomes of two million target beneficiaries through a network of implementing partners.

In the spirit of the Paris Declaration on Aid Effectiveness and the OECD/DAC guidelines on "Harmonizing Donor Practices for Effective Aid Delivery," the donors agreed on a multidonor trust fund approach. The program is driven by the conviction that pooling donor resources enables program coherence and leads to greater impact. UNOPS was engaged to administer the funds and provide monitoring and oversight.

At its commencement, donors and UNOPS faced the challenge of rapidly mobilizing funds to support livelihood recovery in the Delta, prone to cyclones. Operating within the framework of LIFT's overall goal, the $22 million Tat Lan (Way Forward) Sustainable Food Security and Livelihoods Program commenced implementation in 2013 to equitably and sustainably improve the livelihoods of 214 cyclone-affected communities.

The program management structure and processes have matured since its inception. The LIFT strategy places results at the center of its funding decisions. There is an overarching logical framework (logframe) with headline indicators that summarize LIFT's progress and performance against annual and program targets. The Implementing Partner logframes reflect in aggregate the overall LIFT logframe. The LIFT Fund staff monitors partner performance through bi-annual reports against an agreed logframe and six monthly field visits. LIFT has an agreed plan for M&E against the logframe, which, given the complexity of the program, comprises several layers. The logframe is reviewed periodically to reflect any changes in the program approach and targets.

According to DFID[16], who chairs the LIFT Donor Consortium, and as such owns the business case, results from 2010/11-year cyclone Nagis recovery activities funded by LIFT in the Delta have been very positive. As a result of the intensive training, demonstrations, and on-farm trials of new farming and storage techniques and inputs (tillage buffalo, improved seeds, appropriate fertilizer) supplied by the LIFT implementing partners, there was a 20–60 percent increase in the agricultural area cultivated and a 40–60 percent increase in crop yields. In addition to farming activities, LIFT has supported the creation of 9,579 small village-based businesses such as grocery shops, snack bars, soap house material production, mechanics, tailoring through training that assists

16 Department for International Development (DFID) leads the UK's work to end extreme poverty. https://www.gov.uk/government/organisations/department-for-international-development

villagers to explore potential opportunities, management skills and bookkeeping and revolving funds for business startup. Given its huge success, the LIFT program, which was originally was supposed to close in 2014, has been extended to 2018.

Identifying the Program

According to MSP®, the idea and the resulting vision that is driving the change generate the program mandate, which triggers the overall program management process.

In the case of LIFT, discussions began in 2008 amongst a group of donors on ways to help Myanmar make faster progress toward the achievement of Millennium Development Goals – MDG 1. After extensive consultations with key stakeholders from the Myanmar government, embassies, UN agencies, and NGOs, the LIFT Program Mandate is to assist Myanmar toward the achievement of MDG1[17], which aims to eradicate poverty and hunger. More precisely, LIFT's purpose was to increase food availability and the incomes of two million target beneficiaries.

In March 2009, LIFT was officially launched, stating that

> LIFT's vision is to be a collective and influential voice for innovation and learning, and to provide a platform for enhanced policy engagement on sustainable agriculture, food security and rural development. LIFT's goal is to contribute to sustainably reduce the number of people in Myanmar living in poverty and hunger. LIFT's purpose is to increase livelihoods' resilience and nutrition of poor people in Myanmar by focusing on interventions that increase income, food availability, utilization and stability of access to food.[18]

LIFT's goal and purpose is to be met through the achievement of four outcomes:

1) Increased incomes of rural households.
2) Increased resilience of poor rural households and communities to shocks, stresses, and adverse trends.
3) Improved nutrition of women, men, and children.
4) Improved policies and effective public expenditure for pro-poor rural development.

According to LIFT's Logical Framework, the delivery of eight key outputs will lead to the achievement of its four high-level outcomes. The outputs that will drive LIFT's projects are:

1) Increased sustainable agricultural production by smallholder farmers.
2) Improved market access and market terms for smallholder farmers.
3) Increased employment in non-farm activities for smallholders and landless.
4) Increased access to affordable financial services by smallholder farmers and landless.
5) Improved diets of women, men, and children.
6) Safeguarded access to and sustainable use of natural resources for smallholders and landless.
7) Strengthened local capacity to support and promote food and livelihoods security.
8) Generation of policy-relevant evidence regarding smallholder farmers and landless.

---

17 MDG 1: Reduce by half the proportion of people living on less than a dollar a day; achieve full and productive employment and decent work for all, including women and young people; reduce by half the proportion of people who suffer from hunger.

18 Livelihoods and Food Security Fund, https://www.lift-fund.org/en.

The narrative that connects these outcomes and outputs can be found in the way LIFT understands the structural dynamics of the rural economy and the implications for its target group – poor smallholder farmers and the landless. The strategy divides its target group into:

- Households with commercial potential to "step up" the value ladder and out of poverty. They have the opportunity because of access to land, labor, and markets to invest in achieving higher productivity agriculture and better market terms and LIFT can support them in doing this.
- Households that can productively "step out" of agriculture, and into more productive sectors of the economy over time. This could be a local "step out," finding better-paid employment in local non-farm activities. It can also be a "migration step out" to take advantage of opportunities further afield. LIFT can facilitate these processes and minimize adverse impacts on the household of making such a transition. Households without commercial potential or opportunities to leave agriculture or the rural economy and that have little option but to "hang in." For them, own account subsistence agriculture or working as agricultural labor is, in effect, a safety net. For those in this situation, LIFT can take direct steps to improve their food security and nutrition.

A program requires initial and ongoing top-level sponsorship to gain and maintain the necessary commitment to the investment, resources, time scales, and delivery of changes that are involved.

The sponsoring group in the LIFT program are representatives of contributing donors from Australia, Denmark, the European Union, France, Ireland, Italy, the Netherlands, New Zealand, Sweden, Switzerland, the United Kingdom, and the United States of America, forming the Donor Consortium that is the highest authority in terms of the Fund's governance and the ultimate decision-making body for the Fund. At the beginning, the Donor Consortium Board met at least twice per year. Since 2014, it has met annually.

The Donor Consortium and the Fund Board will be solely responsible for policy dialog, for the identification of the Fund's policy and strategy, and for approving activities to be implemented by the Fund Manager and by implementing partners (Figure 2-22).

Each donor to LIFT will have a bilateral contribution agreement with the Fund Manager. These contribution agreements are the main mechanism through which each donor exerts fiduciary and legal authority over the funds they provide for LIFT. The overarching objective of the **Donor Consortium** is to provide a structured forum for donors to LIFT's achievements and current challenges, as well as its finances and main priorities for the coming year.

The **Senior Consultation Group** – comprised of senior representation from the government of Myanmar, representatives from implementing partners, civil society, and the private sector – serves as an advisory body and sounding board for the Fund Board. The Senior Consultation Group has no formal decision-making authority within LIFT.

The **Fund Board** will act with delegated authority of the Donor Consortium, providing strategic leadership for and oversight of the implementation of LIFT. The Fund Board will focus on strategic decisions, policy decisions, donor coordination, building relations with the Government of Myanmar, and overall performance management of the Fund Manager. The Fund Board also selects the projects that receive LIFT funds with guidance from the Fund Manager. The United Kingdom (DFID) is the current Chair of the Fund Board (Senior Responsible Owner – SRO). Australia (AusAID), the European Union, Switzerland, the United Kingdom (DFID), and USAID are Fund Board members. There are also three independent members and one donor observer of the Fund Board. The Fund Board meets at least once every four months.

UNOPS acts as LIFT's Fund Manager (Program Manager role as defined in MSP®). The Fund Manager will be responsible for effective, transparent, and efficient management of the Fund on behalf of the Fund Board and will have delegated authority for the management of the Fund

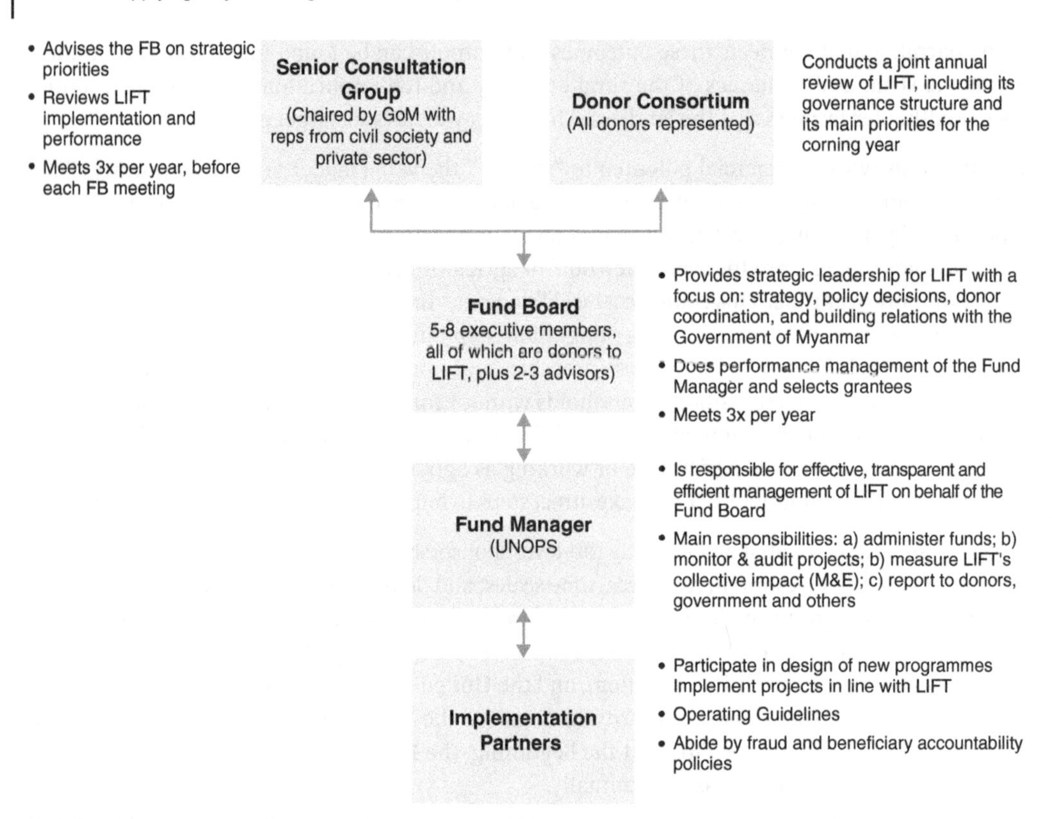

Figure 2-22 LIFT Program Governance Structure Source: Adapted from LIFT Fund.

under the rules and regulations of UNOPS, and in accordance with the strategy approved by the Fund Board (Figure 2-23). The Fund Manager is responsible for the overall management of the fund and the monitoring and oversight of the financial and technical performance of the implementing partners. The Fund Manager is also responsible for implementing monitoring and evaluation, generating knowledge and learning.

In order to use the funds effectively during implementation of projects, the Fund Board carefully selects the implementing partners using certain criteria. The current implementing partners include local, international and UN organizations.

The LIFT Program has a series of documents and agreements and a LIFT Concept Paper, which together cover items referenced by MSP® in the program brief. There was no formal document such as a Program Preparation Plan to guide the preliminary mobilization and planning of the program formally. However, the approach and schedule were reflected in the documents approved by the donor consortium.

It is highly advisable to conduct an independent formal review of the program brief to assess the scope, rationale, and objectives of the program. The independent review of LIFT, however, did not follow a structured gate/stage review such as the OGC Gateway Review,[19] which is applied for UK public sector programs, as well as in a number of other countries that adopted the same assurance

19 http://webarchive.nationalarchives.gov.uk/20100503135839/http:/www.ogc.gov.uk/ppm_documents_ogc_gateway.asp

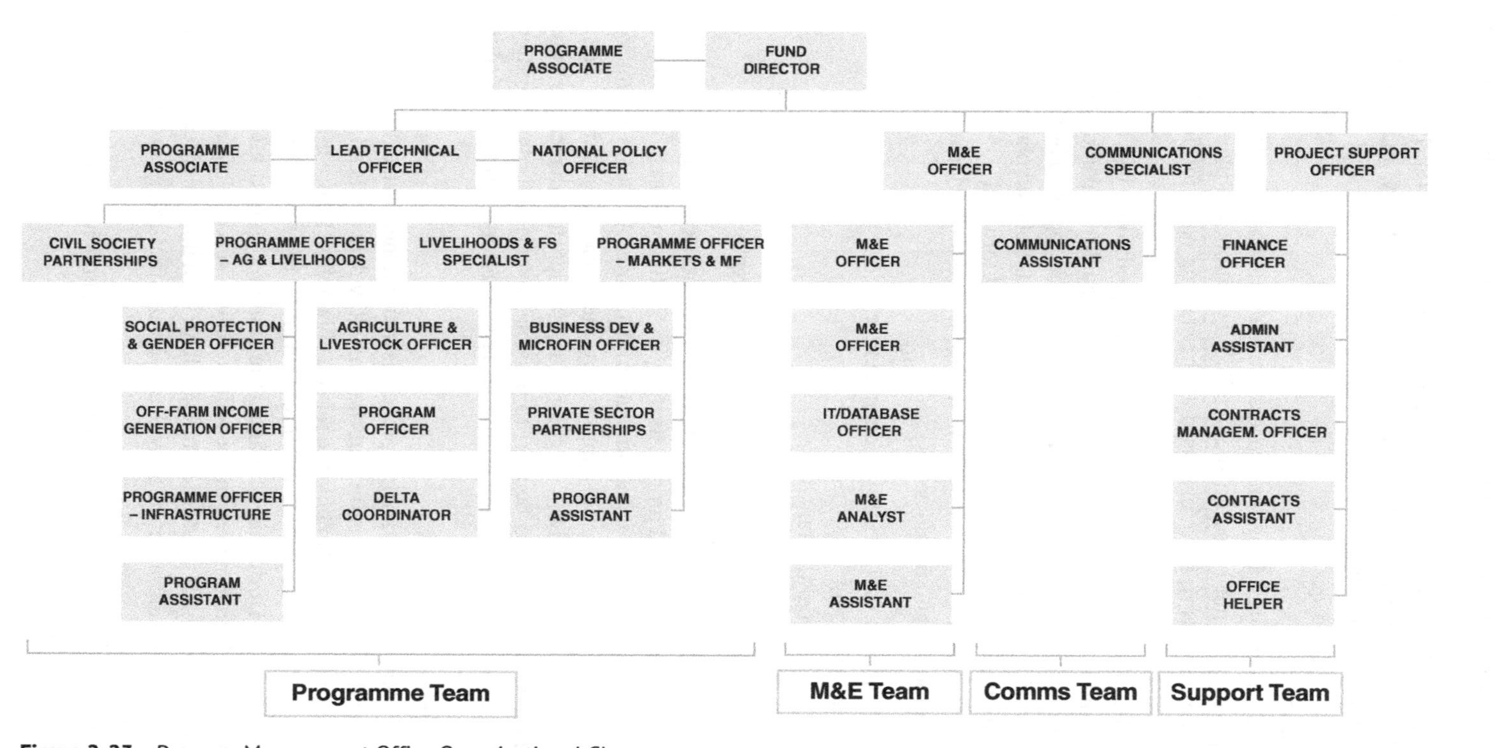

Figure 2-23 Program Management Office Organizational Chart

model before signing the agreements. However, LIFT was evaluated using another approach – the Value-for-Money assessment.[20]

UNOPS and the LITF Fund Consortium representatives signed the Program agreement on October 2, 2009. This event triggered the tranches, capability delivery, and benefit realization, in accordance with the MSP® methodology.

Defining the Program

According to MSP®, Defining a Program process provides the basis for deciding whether to proceed with the program or not based on the detailed definition and plans of the program. The Defining a Program process is, therefore, used as the starting point for developing the program definition in more detail.

The Program Definition document specifies how the organization will be changed by the successful program, and how it will be governed in terms of quality, stakeholders, issue resolution, risks, benefits, resources, and planning and control. It must be approved by the Sponsoring Group and the SRO before the program can be formally established. Detailed planning is undertaken at this stage to ensure the best – and most realistic – case is put forward.

The Program Plan document in this definition stage is vital to provide a structured framework for managing the program and includes:

- Timescales, cost, outputs, and the project dependencies
- Risks and issue management
- A detailed schedule of program tranches
- Transition plans
- Plans for monitoring and controlling performance and targets

DFID as chair of the Donor Consortium prepared a business case for LIFT. However, the program plan document was elaborated progressively by work streams based on arising and changing scenarios and events, such as cyclones and other natural hazards. A new LIFT Strategy (2014–2018) was finalized in September 2014 and approved by the Donor Consortium in October 2014. Throughout its program activities, LIFT monitors and attempts to enhance the impact of its projects upon household nutrition and be alert to possible adverse and often unintended impacts upon household and community resilience through a rigorous monitoring and evaluation process.

Managing the Program

The purpose of the Managing the Tranches process is to implement the defined program management strategies for the program, ensure that the capability delivery is aligned with the strategic direction of the organization, and enable the release of benefits (Figure 2-24).

A tranche is a portion of something – in this case, a group of projects and activities that deliver a step change in capability. During the Managing the Tranches process, after each tranche, End of Tranche reviews can be held to review the benefits achieved in accordance with the business case and the benefit realization plan. Once the strategies have been implemented, the main function of this process is monitoring all program activities to ensure that things are done correctly to the defined governance.

In October 2014, the program was halfway through and at the highest level. The end-of-tranche evaluation was carried out and a new tranche was initiated.

20 https://www.gov.uk/government/uploads/system/uploads/attachment_data/file/255126/value-for-money-external.pdf

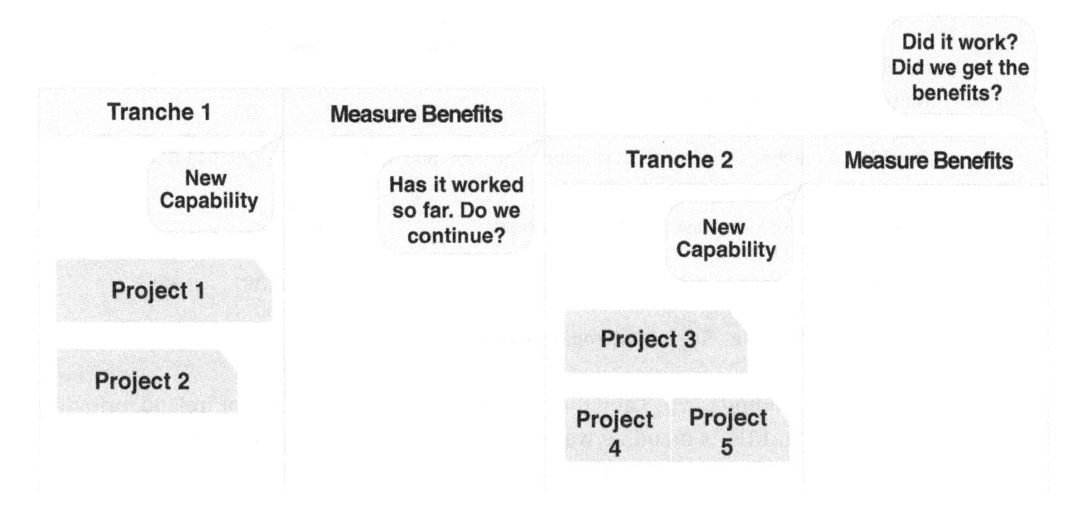

Figure 2-24 Managing the Tranches Source: MSP® (Cabinet Office 2011)

Of course, the situation in rural areas is fluid and sensitive to exogenous events. New roads and access to markets or new technologies can give households commercial opportunities that did not exist before. Or natural and man-made disasters of life-cycle crises can push some people into poverty.

LIFT devotes its resources to positively impacting the rural economy and enhancing the opportunities for all members of its target group through:

- Village-based interventions that are helping households to realize their economic potential as farmers (productivity raising activities), as landless (income-generating activities and jobs in the non-farm economy) or as potential migrants (making migration safe and productive), whilst protecting and enhancing the resilience of the poorest (through social protection measures).
- Supporting economic activities that play out at regional or national level and that can help to improve the overall environment and offer economic opportunities – for example, delivering a range of financial products, strengthening inclusive value chains, and making better use of information technology.
- Actively harvesting lessons and generating evidence to achieve systemic change through targeted advocacy that will inform the formulation of better, pro-poor policy and effective public expenditure.

Delivering the Capability and Realizing the Benefits

As shown in Figure 2-25, the results of the LIFT Program are impressive. By the end of 2011, 49 projects had received funds from LIFT, from which 22 were completed in 2011, and the 27 ongoing projects were being implemented in four distinct agro-climatic regions: Ayeyarwaddy Delta, Dry Zone (the low-lying central part of the country that includes large parts of Mandalay, Magway, and southern Sagaing Regions), Hilly Region (upland areas in Kachin, Chin and Shan States), and Coastal Region (coastal areas in Rakhine State). Including the support provided in 2010 and 2011, more than 200,000 households (more than one million people) received assistance from LIFT projects as direct beneficiaries.

In 2013, LIFT disbursed US$ 31.3 million to its implementing partners, which is 14 percent more than in 2012. The Fund supported 58 projects in 107 of the 330 townships across Myanmar.

	2010	2011	2012	2013
Number of projects	22	27	44	58
Number of townships where LIFT is active (out of 330 townships)	28	94	100	107
Cumulative number of households reached	153,808	223,229	372,528	511,505
Number of studies commissioned	0	2	8	15
Cumulative no. of loans disbursed to households (agricultural and non-farming purposes)	1,218	8.103	86.568	151,212
Cumulative number of CBOs strengthened	1,682	3,467	6,391	9,389

Figure 2-25 Activities, Reach, and Impact Source: LIFT Annual Report 2013

Donors increased their funding in 2013 and the Fund welcomed the Republic of Ireland, bringing donor membership to 11. LIFT's mandate was extended for two additional years, until the end of 2018.

Using implementing partner (IP) data, cross-referenced for accuracy with data from LIFT's extensive 2013 household survey, the Fund was able to track steady progress against its output and purpose indicators. By the end of the year, LIFT-funded projects had reached 511,505 beneficiary households, or about 2.5 million people. In addition:

- More than 290,000 beneficiary households reported that they had increased their food security by more than a month.
- Nearly 60,000 households reported higher incomes because of LIFT support.
- The number of households accessing affordable credit for agriculture doubled (since 2012) to 130,000 households.
- The Fund exceeded its progress targets in 17 out of the set of 22 indicators.

2.5 16/6 Project in Haiti

Case Background

The earthquake of January 2010 in Haiti killed 220,000 people, leaving 1.5 million people homeless and spread over a thousand camps. Many of them were already living in poverty and although their houses had not completely collapsed, they were living in precarious housing and neighborhoods of the capital, exposed to extreme climatic phenomena and risks of natural disasters, without access to basic services.

As low-income neighborhoods in Port-au-Prince were in the past often established on steep hillsides with little or no planning and using poor quality construction materials, many were severely affected by the earthquake.

It was imperative to initiate an improved reconstruction of Port-au-Prince, through the identification of sustainable housing solutions for displaced persons and original residents of neighborhoods, and the improvement of their living conditions.

To cope with the emergency situation, the president of the Republic of Haiti established, on 30 September 2011, the "16/6 Project," a two-year pilot program, with the United Nations' support, through four of its agencies: UNDP, International Labour Organization (ILO), International Organization for Migration (IOM), and UNOPS.

Figure 2-26 16/6 Project Rehabilitation

In addition to the rehabilitation of neighborhoods of the capital and provision of improved housing, this pilot program, funded by the Haiti Reconstruction Fund and the Canadian Government, sought to improve the living conditions of the inhabitants of affected areas, promoting the development of basic social services prioritized by the community and the creation of income-generating activities. This integrated solution included the closure of six priority camps and the Champs de Mars camp, and the relocation and improvement of the quality of life of the displaced, while meeting the urgent needs of physical infrastructure and social services selected by the communities.

Identifying the Program

More than half of the houses in the neighborhoods selected for rehabilitation suffered partial or heavy damages from the earthquake due to a lack of seismic standards. Many residents in these high-density neighborhoods had little choice but to resettle in the camps nearby, including Place St. Pierre, Place Boyer, Primature, Canape Vert, and the Stadia Silvio Cator and Mais Gate, which are now closed.

The pilot program aimed to facilitate the rehabilitation of 16 neighborhoods[21] through improved reconstruction of housing, access to basic services prioritized by the community, and the creation of income-generating opportunities (Figure 2-26). At the same time, the program aimed to facilitate the return of displaced people living in the six camps, which hosted about 5,000 households,

21 Morne Hercule, Morne Lazard, Nérette, Delmas 60 Argentina, Upper Pan, Pan Netherlands, and Morne Villa Rosa, Bas Canapé Vert Bois Patate, Jean Baptiste, Mapou / Mont Elbo, Spoiled maize 1 and 2, Barbancourt, Carrefour Clercine Backgrounds Delmas 31 and 33.

mainly from these 16 districts spread over the municipalities of Port-au-Prince, Delmas, and Petionville. This pilot program was named 16/6 Project (although "16/6 Program" would have been a better name, given its nature and characteristics).

The program's objectives were to ensure that displaced people in the six camps associated with the 16 districts found a solution to the housing problem. The return to the original neighborhoods was facilitated by the reconstruction of quality housing. Rehabilitation of the 16 districts of origin was undertaken on the basis of the priorities of its people. Rising incomes and access to employment of its people supports the sustainability of basic social services and access to credit ensuring adequate housing can be maintained. The capacity of Haitian stakeholders was strengthened to carry out reconstruction in Haiti by applying the model of the 16 districts.

The rehabilitation of neighborhoods and access to sustainable housing solutions included subsidies to rent or repair and the reconstruction of housing, according to the required standards of quality and safety. The program also provided support and training for building professionals in micro, small, and medium enterprises in the construction sector. Coordination and a smooth process in the districts were ensured by strengthening neighborhood committees, which created a platform for dialogue between the different levels of the government, the program, and the interests of the community. A communication strategy was designed and implemented to ensure good information and continuous interaction with the beneficiaries of the program.

The protection and equity dimensions played a key role in the program, particularly in its efforts to offer the best possible choice for neighborhood residents and displaced people regarding housing solutions and priorities for investment in the development infrastructure and access to services. The program also planned to invest an equivalent amount of resources in housing solutions, with interventions aimed at improving living conditions in the area, the development of income-generating opportunities and employment, and reducing risks.

The stakeholders of this program are at three levels – national, municipal, and local. The key stakeholders are the primary councils and CASECS / ASECS (the communal committees), other community leaders, and the beneficiary families in targeted neighborhoods and camps. The program provided frequent consultations with these stakeholders to ensure the achievement of the program objectives.

Given the multidimensionality of this program and the potential overlap with current or planned interventions, the program's strategy was to actively seek synergies with other projects and initiatives, including the support program for the reconstruction of housing and neighborhoods (Housing Support Project), which covers many areas of capacity building and data collection that will be useful for this pilot program, the project of shoveling fragments 1 and 2 of UNDP, the ILO training program, and the income generation for vulnerable women project of UNDP.

Program Planning and Controls

The 16/6 Project was structured around four projects or activities, each allocated to one of the UN agencies that supported the Haitian Government in the implementation of the program, according to their mandate and field of expertise. The projects were:

The Return Processes

In February 2011, the International Organization for Migration (IOM), with the support of the Civil Protection (DPC) began to register the displaced, providing the Haitian government and international agencies with reliable data on the needs and the situation of these people.

Return meant back to a safe habitat. In other words, returning to the homes and neighborhoods of origin without risk to the safety of people is linked to the quality of construction, given the natural hazards such as landslides, floods, heavy rains and winds associated with hurricanes and possible aftershocks and posing no consistency problem for urban operation.

It was necessary to focus on the original neighborhoods and camps associated with them. The engagement of individuals and communities was a key factor. This work stream of the program should also promote income-generating activities to meet a priority concern of affected populations. To avoid tensions between communities and to provide equitable support to affected populations, it was also necessary to promote equity in activities in support of return and a broad partnership with the economic and social agents.

Housing

The earthquake damaged or destroyed around 200,000 homes, at the same time increasing the deficit estimated at one million homes in Haiti, half of which corresponded to Port-au-Prince. Therefore, the program through its housing project established four modalities for repair and reconstruction of houses, with the intention to support the affected families in the neighborhoods of origin, while increasing the capacity of the rental stock.

The houses were repaired through two implementing rules: "agency driven" and "owner driven."

There were two modalities for the reconstruction of houses. A modality on condominiums in situ, with the owner on land allocation and another one based on available land, where houses were built on higher grounds. Most of reconstruction methods had been used by in Haiti since 2010.

Rehabilitation of Neighborhoods

The 16/6 Project began in November 2011 and helped reaffirm the local governance structures in each neighborhood, in the form of "community platforms." These platforms were composed of different leaders (15–20 members) present in each area, such as grassroots organizations leaders, religious leaders, representatives of women's groups, youth, or notables.

The Community Platforms became a single interlocutor valid speaking on behalf of the community. They would be inclusive groups representing the community of a given neighborhood and integrating all the different participating groups provided by law or not. A special effort was made to ensure that excluded or marginalized groups in general were properly included. "The space for dialogue" became an open discussion forum to facilitate dialogue between local and national governments for the expression of the will of the residents. The success of the neighborhood rehabilitation process depended on the active participation of communities. In this sense, their views were duly considered and incorporated.

Members of community platforms, who knew their neighborhoods, their needs, and potential needs, participated in the community planning exercise. Originally, these neighborhoods lacked urban planning. Leaders had been trained to read and understand maps, to establish the points, priorities and to define future interventions. Each of the districts and areas produced an accurate diagnosis, a management plan that would guide future actions and project briefs for each of the priorities identified by the community.

The management plans in question considered the zoning maps regulations, which were established through an exercise conducted by the 16/6 Project's technical experts together with the national authorities. These maps determine the level of risk faced by each of the districts in terms of landslides, floods, or earthquakes. This micro-zone analysis established which part of the districts can be considered a red or orange zone, which means that the return to these areas is discouraged because families are exposed to a very high risk and mitigation would not be possible; yellow areas were classified as considerable risk but that can be reduced through mitigation measures such as pipe, drainage, retaining walls and investments that make the areas safe to live; and finally the green parts where the living conditions are appropriate. The regulatory zoning maps are accompanied by a regulation that establishes the conditions of each area and recommends the work and investment to be made and the necessary technical guidance. This regulation could eventually be converted into a standard (law).

Monitoring, Evaluation, and Knowledge Management
This program represents a balance between humanitarian response and long-term actions required. For this, a monitoring and evaluation system was set up to monitor the implementation of the program and its impact on socioeconomic indicators, particularly in light of the wide range intentions of the program. The knowledge created in the process is captured by lessons learned documents that permit the relevant future interventions to use them and to move toward enhanced public proposals and institutionalization of the program as an urban planning and rehabilitation reference.

Lessons Learned

The landscape of humanitarian and development programs has changed considerably in the last decades. According to Fengler and Kharas (2010), more funds are being channeled to development aid, as well as emergency humanitarian initiatives; however, the value per projects is decreasing. Therefore, the value-for-money principles – economy, efficiency, and effectiveness – are even more required than before. Grouping projects and managing them as a program appears to be one of the approaches to achieve more value for the investment done by the international donor community. Disciplines such as fund management and monitoring and evaluation techniques ensure that the investment in this kind of temporary initiatives, successfully delivers outcomes and, eventually, the impacts expected by the beneficiaries.

The case studies demonstrated that when solid design, planning and governance are in place from the beginning, the chances of success increase considerably. In the case of LIFT, expressive results of poverty reduction are observed. The Haitian rehabilitation program shows that with the right mix of community engagement and ownership, resilient construction methods and proper governance, post-disaster reconstruction can be achieved within time, cost, and quality, while ensuring that the benefits are realized and social returns on investment are obtained.

2.6 Conclusions

Humanitarian and social projects are some of the most challenging and rewarding types of projects anyone can undertake. The combination of human aspects like empathy and motivation, if correctly associated with planning techniques, agility, and risk management, can deliver sustainable results for the beneficiaries and society in general.

References

Weiss, Carol (1995). Nothing as practical as good theory: exploring theory-based evaluation for comprehensive community initiatives for children and families in Connell, James, Kubisch, Anne, Schorr, Lisbeth and Weiss, Carol [editors] (1995). *New Approaches to Evaluating Community Initiatives: Concepts, Methods, and Contexts. Roundtable on Comprehensive Community Initiatives for Children and Families*. New York: Aspen Institute.

Fengler, Wolfgang, and Kharas, Homi (2010). Delivering Aid Differently – Lessons from the Field. *The World Bank Economic Premise Poverty Reduction and Economic Management (Prem) Network*. February 2011. Number 49. Available at: http://siteresources.worldbank.org/INTPREMNET/Resources/EP49.pdf.

LIFT (2014). LIFT Strategy. Available at: http://lift-fund.org/Publications/LIFT-strategy-(2014-18).pdf.

OECD (2008). Accra Agenda for Action in Brief. Available at http://effectivecooperation.org/files/resources/Accra%20Agenda%20for%20Action%20in%20Brief%20ENGLISH.pdf.

ONU HAITI (2014). Fast Facts 16 Neighborhoods 6 Camps. http://www.onu-haiti.org/wp-content/uploads/2011/12/16-6-factsheet-new.pdf.

Reliefweb (2008). Relief Web Glossary of Humanitarian Terms. http://reliefweb.int/report/world/reliefweb-glossary-humanitarian-terms.

UNDP-a (2014). *Human Development Report 2014,* UNDP: New York.

UNDP-b (2014). *Summary Human Development Report 2014,* New York: UNDP.

UNGA (1991). *United Nations General Assembly Resolution 46/182* (19/12/91). New York: UN.

UNISDR (2009). United Nations International Strategy for Disaster Reduction (UNISDR): Geneva available at http://www.unisdr.org/files/7817_UNISDRTerminologyEnglish.pdf.

Rosenberg, Leon J., Posner, Lawrence D. and Hanley, Edward J. (1970). *Project Evaluation and Project Appraisal Reporting System*. Fry Consulting Incorporated. Available at http://pdf.usaid.gov/pdf_docs/PNADW881.pdf.

3

Pillar #3: Project Management Is Creating Innovative Cultures

3.0 Background

"Culture eats strategy for breakfast" is a famous quote attributed to the legendary management consultant and writer Peter Drucker. This is increasingly fitting in the innovative cultures required for the success of tomorrow's strategic initiatives. Several attributes reflect this innovative culture: agility as a highly visible and valuable trend in how organizations conduct their business, the focus on enhanced safety to make mistakes, and the support for autonomy of work to drive creativity.

Executive management now appears to recognize and appreciate the contributions an innovative culture, that is anchored in project management principles and dynamic ways of working, could be bringing to their companies. The increasing urgency around project management is seen at the right levels of the future organizations' executive leadership, leading to the need for an integrated end-to-end implementation of its principles.

This chapter presents a model for the innovative culture. Each of its building blocks dimensions and foundational dimensions will be supported by various originations' excellence in action contributions to shaping this innovative culture. Key themes from these contributions clearly highlight some key learnings that tomorrow's organizations could consider building upon.

3.1 Introducing the Innovative Culture Model[1]

An innovative culture has a look and feel that is like no others. If we think of Amazon, Google, or a number of other organizations that embody the pace of continual development of creative solutions, that are customer-centric, it is no wonder that a few unique attributes to their culture of delivering with pace and excellence on their strategic initiatives will stand out.

The *Innovative Culture Model* shown in Figure 3-1 sets the tone for this chapter in addressing that look and feel. It shows four building blocks dimensions and three foundational dimensions. The

1 The Innovative Culture Model has been developed by Al Zeitoun and published as part of the Siemens Digital Industries Software white paper, Simcenter: the heartbeat of the digital twin, adopting a digital mindset to deliver and scale future innovative solutions.

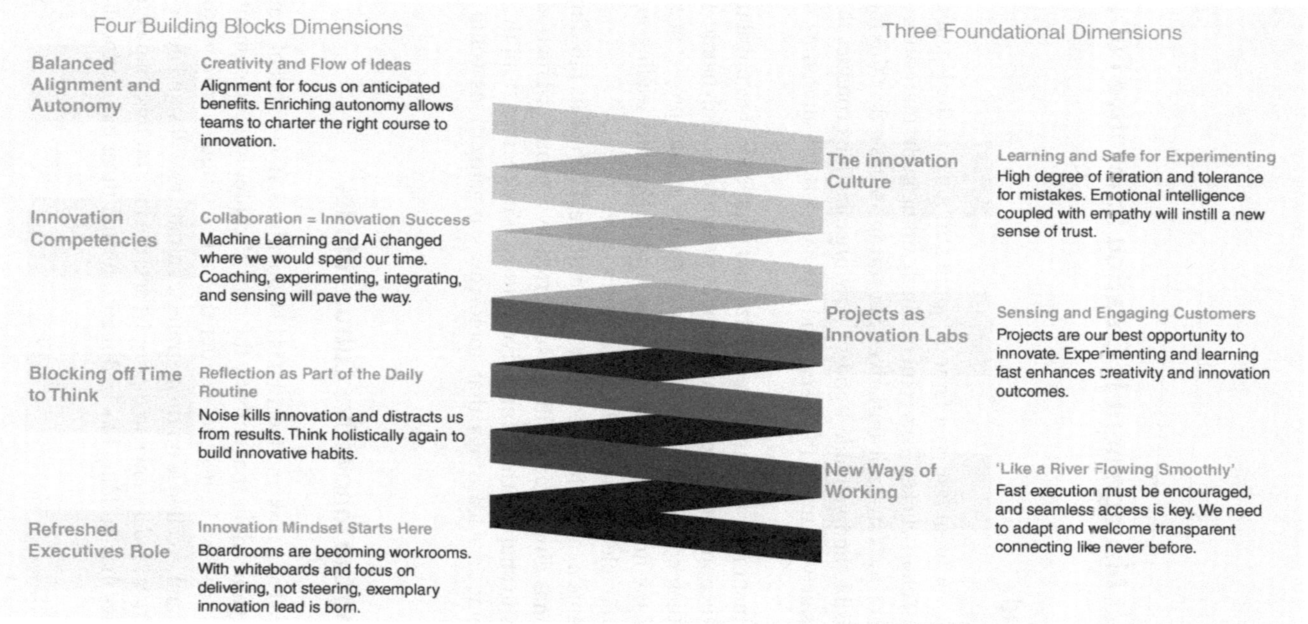

Four Building Blocks Dimensions

Balanced Alignment and Autonomy

Creativity and Flow of Ideas

Alignment for focus on anticipated benefits. Enriching autonomy allows teams to charter the right course to innovation.

Innovation Competencies

Collaboration = Innovation Success

Machine Learning and Ai changed where we would spend our time. Coaching, experimenting, integrating, and sensing will pave the way.

Blocking off Time to Think

Reflection as Part of the Daily Routine

Noise kills innovation and distracts us from results. Think holistically again to build innovative habits.

Refreshed Executives Role

Innovation Mindset Starts Here

Boardrooms are becoming workrooms. With whiteboards and focus on delivering, not steering, exemplary innovation lead is born.

Three Foundational Dimensions

The innovation Culture

Learning and Safe for Experimenting

High degree of iteration and tolerance for mistakes. Emotional intelligence coupled with empathy will instill a new sense of trust.

Projects as Innovation Labs

Sensing and Engaging Customers

Projects are our best opportunity to innovate. Experimenting and learning fast enhances creativity and innovation outcomes.

New Ways of Working

'Like a River Flowing Smoothly'

Fast execution must be encouraged, and seamless access is key. We need to adapt and welcome transparent connecting like never before.

Figure 3-1 The Innovative Culture Model

building blocks set the stage for how project management principles help create this innovative culture. The foundational dimensions are descriptors of how the look and feel value is demonstrated in these innovative cultures of future organizations.

3.2 Balanced Alignment and Autonomy

One of the most difficult decisions organizations make is the right level of governance that gives them the confidence in the achievement of strategic outcomes, while not overdoing it in exerting excessive controls. As we explore this first building block, we are influenced by the changing dynamics of the way or working and the continual new generations expecting a higher link to purpose and growing flexibility in how they do their work. This critical tension between alignment and autonomy had been repeatedly surfacing over the past few years. The future of work is hybrid, and this is exactly one of those critical questions to answer regarding that hybrid future approach.

The culture model shows the theme of this balance reflected in "creativity and flow of ideas." This flow of ideas allows for focus on value and achievement of benefits. The strength of the autonomy is tested when the teams continuously adjust their course back to creativity and achieving innovative solutions. The Sunrise UPC excellence in action example shown next illustrates a few of the shifts that organizations are doing now ahead of the future organizational excellence in living this delicate balance. This shift starts with the future project manager and how becoming a project leader sets the stage for a few other changes in team dynamics, communications, engagement, and building a change culture.

3.3 Excellence in Action: Sunrise UPC[2]

The Cultural Shift from Project Manager to Project Leader

The world today is changing at an incredible speed, much faster than in the past twenty years. The so-called *digital disruption* pushed the raise of new business models and many industries have been already put under pressure by this transformation. Just a few examples confirm that there is no space to go back to the *old times:* WhatsApp is connecting every day billions of people, without having a proprietary Telco infrastructure, Uber is the largest taxi company that owns no taxi, Airbnb is the largest accommodation provider that owns no rooms, Amazon Web Services is one of the most successful provider of computer infrastructure, without producing servers, and PayPal has a market value greater than all the listed banks in Germany. Even in the traditional auto sector, the revolution arrived and a company like Tesla is now sitting together with giants like Toyota or General Motors.

The recent pandemic just accelerated this pace. Lockdowns have encouraged growth in digitalization across many sectors, like online retail, digital finance, online entertainment, gaming and telehealth and it is expected that most of these changes will be long-lasting and new investments

---

2 Material in this section has been provided by Luca Gambini, Director of Business Program Management for Sunrise UPC, He has 20 years of experience in the Telco Industry and is an Expert Adviser at the London School of Economics.

will be done in the future to ensure they will stick. To stay competitive and win market shares in this challenging economic environment, companies must run new strategies and re-invent themselves. The role of Technology is becoming crucial in any sectors and being the real engine for innovation (not only an enabler for efficiency or cost reductions). A few years ago, ING CEO Ralph Hamers understood this trend and stated in an interview: "We want to portray ourselves as a tech company, with a banking license."

Therefore, in a society where the only certainty is the constant change and where the technology power is growing, the projects that are required to support this transformation are different from the past, and different will be the way of working to successfully deliver them. Especially in some industries, like telecommunications, a whole revolution has happened in recent years. The introduction of Agile methodologies and the scaling of them in companies like Sunrise UPC (the second-largest Telco player of Switzerland) improved many aspects of a project implementation:

- Reduced time to market to launch new products
- Increased customer satisfaction
- Better software quality, creation, and empowerment of autonomous "cross functional" teams
- Reduced barriers between IT and business units
- Removal of silos within the organization
- Creation of a new "explorer mindset"

These fit perfectly with what is requested in these times of disruption.

What about the profession of project manager (PM) in this picture? Is this changing as well? It looks clear that these transformational agendas have brought new challenges to the project management professionals and that, in order to adapt to this new landscape, the skills of a PM must also evolve.

Let's analyze in which direction is going this evolution. If we look at the Project Management Institute (PMI) triangle of competences, we see that in this complex and globalized environment, the technical skills are not enough anymore. The companies are seeking additional skills in the Strategic and Business Management and in the Leadership domain.

Leadership is at the core of PM profession and is changing as well with the times. It is no longer correlated with authority, title, or power of a single person. It is instead associated with empowering others in doing their job at the best, removing all the obstacles that they can encounter. This is particularly true in crises times that are characterized by uncertainty, with problems that are not familiar and where there is a lack of information and understanding. Therefore, it is crucial to build leaders who are setting clear priorities and naturally giving the power to their teams to discover and implement new solutions to accomplish those goals.

This is perfectly applicable to the project management profession, where less "command and control" style will be needed and where "servant leadership" will be more and more important to manage projects in such complex context. As already summarized by Kerry R. Wills some years ago in a paper presented at a PMI congress, "The Project Manager of yesterday, whose main focus was on building a project plan, telling people what to do, and communicating status is now a thing of the past and will not succeed in this current challenging environment. This is the new world in which the Project Leader will thrive."

The project leader role requires a bigger focus on the relationships with the team members and the stakeholders of the project. It also focuses much more on the personal qualities, which determine the extent to which people will follow you as a real leader, because there is no leader without followers.

The following leadership improvement actions explain what is fundamental to successfully build the transformation from project manager to project leader and how all this was applied in Sunrise UPC's "Business Program Management" department:

- **Team management** is one of the most important activities of a project leader. This means putting your team in the condition to perform at the maximum level, making sure all the people are focused on the project goal and works smoothly together to reach it. As a leader, you need to be able to identify the best performer for each different role, like a coach in a football team. From my experience in Sunrise UPC, I can suggest that every person in the team, internal or external to the company, must have the same level of empowerment, avoiding setting up just a simple customer–supplier relationship. If you also work with external vendors "near shore" or "offshore," it is very important that they perfectly know the final goal of the project and which strategic business objectives are supported. It looks obvious, but I have seen many projects fail just because the development team was ignoring the real scope of the software they were building.
- **Engagement and motivation** are also crucial in order to create a sense of trust among the different people that are part of the team. The project leader should have a clear vision and use it to engage people, motivating them and giving them the clues of their work. Be able to listen and take care of the needs of the employees should be at the top of a project leader priority list. In Sunrise UPC we always aim to create a friendly atmosphere at work, with a culture of no blaming or finger pointing and where people can work with passion and dedication. Celebrating the project success is also important to make sure that the hard work of both technology and business employees is equally recognized. Any sign of discouragement must be spotted immediately by the project leader, who should then implement actions in order to keep high the commitment of the people. This requires a deep understanding of the psychology behind the real motivation of your team.
- **Communication** is clearly a vital skill for a project manager because the essence of this work is all about being able to properly communicate at different level with different stakeholders. But the project leader has to do much more. A real leader cares for transparency in communication, without creating hidden agendas or messages behind the lines. This will increase the trust about not only what he/she says but also what he/she really means. Objectivity and fairness are also crucial, because conflicts will for sure arise in complex projects, and the leader must be able to mitigate them without picking favorites. Especially under pressure, even when the things are going wrong on a project, in Sunrise UPC we maintain a transparent and open communication toward management and sponsors. This guarantees that all the stakeholders are equally informed, can feel the right empathy with the project team, and are keen to help, when needed.
- **Solution mindset** is important as well to improve the performance of the team. It is too easy for a leader to drown in the negatives of unexpected issues or problems. It is more difficult to face them with positiveness and implement the best possible solution. Even harder is to analyze the root cause of a problem and remove it, but this is the approach to problem solving that we are steering at Sunrise UPC, in order to reduce the future workload of the teams.
- **Culture of change** is another key element that must be fully embraced by tomorrow's leaders since change is an inevitable part of our workplace. In my experience, quite a lot of time and effort are needed to prepare employees for a transition, and there is usually a natural pushback against change, especially if it is unexpected or outside of your comfort zone. But since we cannot avoid changes, we need to offer our employees the right training in the soft skills needed to make them mentally stronger. Resilience is not purely a factor of our traits or our surroundings – it can be developed, and nourished.

- **Servant leadership** is not a new concept. It was introduced to the business world in 1970 by Robert K. Greenleaf in his book *The Servant as Leader*, but represent today one of the most powerful frameworks that the project leader can apply to improve the performance of the team and make a project successful. The whole concept behind it is very simple: to be a leader, you need to be seen first as the best possible servant of the team. Only after having demonstrated that will the people follow you. This means that you commit in giving to your people all the resources and the support that will make them work at the best performance, removing any obstacles or impediments along the road, and protecting them from any unwanted politics or distractions. To use the words of Max De Pree in his book *Leadership Is an Art*: "In short, the true leader enables his or her followers to realize their full potential." This is the style of leadership that I am trying to apply in my department and that should guide any decisions of my managers.

In conclusion, the world is changing fast, and the role of the project manager cannot be anchored to someone who can tick off tasks as complete on a plan. PM role must shift toward a leadership position for successfully implementing any change needed to support the strategic goals of an organization. Will there be still the need of this profession in the next years? Difficult to predict, but even in a future where automation and artificial intelligence systems will gradually replace the work of many professionals, project managers will still provide an irreplaceably human combination of emotional intelligence, leadership, and ethical conduct.

3.4 Innovation Competencies

The mindset of future organizations is projectized. This requires a project revolution across all areas of an organization's ecosystem. There is an increasing trend in how coaching, experimenting, integrating, and sensing qualities are regarded by the next decade's excellent organizations. This decade has seen an increasing demand for collaboration, a unique competency required for operating with excellence. The next decade will see this competency continuing to grow in value as the ways of working center around teams and teams of teams working across complex and large global programs.

The theme for this second building block dimension of "collaboration = Innovation Success" is an illustration of the importance of this competency. Given the nature of increasing technology that dominate the future of work, such as the enabling effect of artificial intelligence (AI) that shifts where the project managers spend their time, a shifting from planning activities to engaging activities becomes a strong sign of successful change. The future excellence is highly engaging. The Bosch contribution example next highlights great ideas for promoting the value of project management at the right levels of the organization and create that mindset.

3.5 Excellence in Action: Bosch[3]

At the onset of the 2020s, we saw a next major step in thinking and acting in projects. Their contribution to success will be acknowledged by organizations and societies.

---

3 Material in this section has been provided by *Dr. Dieter Butz PMP®, Director, Corporate Project Management, Bosch (retired)*. ©2021 by Bosch. All rights reserved. Reproduced with permission.

"Project" has become a normal term for undertaking an endeavor limited in time. Other prerequisites are not always tied to the use of this term, although in most publicly available definitions more attributes are used to define a *project*. The most stated is the uniqueness of the endeavor, sometimes in addition to the degree of uncertainty in how to reach a successful end. Both are necessary aspects of a "project" in order to separate it from a "(repetitive) task" along a known and well-defined path of whichever size. Also, clarity in what to achieve with the endeavor is indispensable for starting a project. Even advanced development or basic research needs a rough idea of what to explore before starting.

Projects of whatever definition have become normal in most organizations, regardless of their purpose and size.

In recent history, especially in huge hardware projects, people started to describe their methods of predictive or plan-driven progress, where every detail until the end was fixed right at the start. This was sufficient as long as the focus lay on predicted success, using large teams from various functions in a hierarchical environment, and no real option to react to the unexpected. As soon as the latter did not hold true any longer, the degree of detail in planning was changed into the "rolling wave planning," where only the near-term future was planned in detail, the mid-term future was planned with much less detail, and the long-term future only in coarse milestones. When software became a regular component or even the deliverable itself, changes could be dealt with much quicker and easier. And we saw mostly small teams of highly capable experts in various software development domains, i.e., the same function. This was the birth of change-driven or Agile methods.

After initial trials to copy Agile to pure hardware or combined software-hardware projects, people found this not to be the silver bullet to speed up and reduce planning efforts in larger projects and in organizations that were used to a strict hierarchical thinking.

Nevertheless, both plan-driven and change-driven approaches have their sweet spots and should be combined to reach predictable deliverables for the customer and speedy adaptation to changing boundary conditions. This describes the hybrid approach, which draws from both sources to varying degrees. Early stages may be conducted using more Agile elements, whereas later stages may draw more from predictive methods. Thus, hybrid will be the choice of the future until further methods evolve and enlarge the methodological toolbox for conducting projects.

The more projects we see in an organization, the more they are understood as the most appropriate way to implement the organization's strategy. Such project portfolios need to be actively managed, with decisions made by the organization's senior executives. This puts projects and the people responsible for their setup and execution into an entrepreneurial position. Even the strategy itself may be developed using elements of project management.

Bosch started to set up a corporate-wide approach to project management around the year 2000 by coordinating the activities and organizational structures that had evolved in many units fairly independently. Since then, units have shared and learned procedures and knowledge, and have brought the topic further by a close and intense collaboration. Several important aspects will be covered in this section.

Methods and Tools in Project Management Spread across Sectors and Industries

As far as we have written sources, the origin of project management lies in the construction "industry" (e.g., completion of the Cheops pyramid around 2570 BCE, construction of the Great Wall of China around 210 BCE). Archeologists have found records of large teams erecting buildings in a planned and coordinated way.

In more recent history, project management methods were published for projects carried out in the field of research and development (e.g., predictive approaches during the Manhattan Project of the American military in the 1940s and the American Apollo space program in the 1960s, introduction of "rolling wave planning" -where only the next phase is planned in detail and those further out rougher and rougher- in the 1980s, Agile approaches for software development during the 1990s, culminating in the initial "Agile Manifesto" in 2001).

Today, project management is still widely considered as a concept relevant for new hardware or software products/services and for advancing infrastructure (e.g., creating buildings or transportation concepts, exploiting raw materials, shaping nature).

But the concept carries much further. Several organizations in the private and public sector as well as in private life (e.g., clubs and societies) are using it in strategy, management, administration, financial, or legal domains – and many more. They have concluded that applying project management on a professional level supports them in being successful. Smaller or very specialized organizations may select a single project management method and use it across their complete organization. In many – especially larger – organizations, multiple project management frameworks will and must exist in parallel or it will lead to intermeshed or hybrid approaches.

These approaches may be specific to the organization, as the needs of sub-units are diverse. This holds even more true if they serve multiple industries or sectors. The size of these subunits can range from well-established units in mature environments with tens of thousands of employees, to small start-ups in niches or new markets that only consist of a handful of people. Nevertheless, if an organization-wide approach is desired by top leadership, a certain degree of unification is needed to keep an overview of activities in and across all units. This will enable collaboration and inspiration between units and allow employees to move from one unit to another without having to relearn and readjust to a completely new environment and project mindset.

This unified basis includes three major aspects, which may be supported by professional standards:

- Toolboxes of diverse project management concepts and methods along with hints on their sweet spots and requirements for the organization's set-up. They often also offer tools and templates in quite some detail for easy adoption and tailoring by an organization.
- Compatible role definitions and their skillsets that fit into a bigger picture in the organization.
- A consistent and evolving mindset with an appreciation of project management across the whole organization.

Based on such pillars, existing concepts for the management of projects can be disseminated into areas of the organization that do not yet use them, and new concepts can be integrated into the overall approach, often after a piloting phase in a selected unit. Without this try-and-adapt mindset it is difficult to develop a unit or the complete organization continuously for better efficiency, effectiveness, and customer satisfaction.

A project is normally considered to be over as soon as the project deliverables have been accepted by the recipient. The involved people are disbanded, the project is forgotten, and the effect on the recipient is neither monitored nor followed up. This is one reason why many projects are considered a failure. They just do not realize their benefits. These might not only include a market success of a new product or service but also lead to a new way of working at the recipient (e.g., a new functional process, a new IT tool, a new organizational setup).

This will change in the future. Benefits realization management will play a much more prominent role and will require a highly active sponsor for a long period of time. Several Agile frameworks systematically support benefits throughout the project.

Many organizations will use projects to create desired results. To make this easy for their employees, they will install a dedicated team supporting the use of the selected and most beneficial methods/tools, roles, etc. This will help them recognize and realize the benefits of establishing a consistent project management governance and support structure.

Although a substantial number of projects will remain to create new products, services, etc., all functions will conduct or at least contribute to projects, often with a higher impact to the organization than any new offer. This will spread knowledge and professionalism in project management and lead to additional organizational change considerations in many projects.

Expansion of the Project Management Methods Toolbox

When software became more and more relevant as a project deliverable in the 1980s and 1990s, building and testing became much faster and cheaper, and you could react to changes to the project deliverable along the way – if the recipient of the deliverable and the realizing team collaborated closely. This new method needed no formal project manager – instead, it had a method master – and less formal planning for further-out details. It was called "Agile" and laid the foundation to the "Agile Manifesto," first published in 2001. It stated four items for better ways of developing software that were valued more than previous ones.

Consequently, what worked well in a (pure) software environment was challenging the "old school" approach in the hardware world, and even more so in the increasingly intertwined hard- and software world. A first attempt was to copy (ideally everything of) the Agile approach to the hardware world and to claim that this is the only viable and future-oriented way to conduct a project. But reality was a bit more complicated than mistakenly understanding the Agile items "we value . . . more than . . ." as "we do only . . . instead of . . .". Especially planning and documentation were considered overly burdensome, which got organizations into trouble later.

This origin of the Agile approach confirmed to many people that projects were smaller endeavors leading to a hardware or a software/IT product. And it limited the understanding of Agile mainly to R&D projects in a single location. With products becoming larger and more complex, and with more impact to the organization, collaboration among multiple functions became necessary. This simultaneous engineering (SE) involved engineering and manufacturing, engineering and purchasing, all three functions, or even additional functions or multiple locations, sometimes in other countries.

This showed that the requirements for a nonexistent project manager were now beyond the typical capabilities of a lead engineer and team organizer, method specialist, or subject matter expert. This role now needed at least basic knowledge in other functions, increasing business acumen, and leadership skills. With increasing team size, the domain specialists were members of the team and the project manager no longer had to be the best expert of the project subject. This was tackled by the newly evolving Agile scaling frameworks, which opened up a whole new world for individuals who were good leaders (especially when having no direct command of the team members) and entrepreneurs with a broad approach, who focused on maximizing the business benefit for their organization in the long run. This trend will persist and make project managers an essential element of implementing an organization's strategy for a successful future.

Many concepts and methods for managing projects accepted by professional communities are published by professional organizations like PMI, IPMA, and AXELOS. The common idea behind is described in ISO 21500. Originating from a plan-driven approach, the Agile world with multiple change-driven approaches emerged in the software domain. All these concepts include how to plan and conduct project work in phases or iterations, and how certain roles should collaborate on their way to realize the project deliverables. And all professional organizations offer descriptions of their concepts and many as well various certificates to prove familiarity with them. Since certification is a relevant revenue source for all providers, and since each considers their approach as the "best," certificates from different issuers are not truly comparable or cross-convertible. Even if someone had the best approach for specific situations, there is no one-size-fits-all approach for the complete range of projects.

Organizations that want to strengthen their project management capabilities often select a preferred concept, with certain additions or hybrids of others, for their internal setup of project management. By this they benefit from the further development of these concepts. Some even actively contribute to their development.

For project managers and team members, it will become normal and essential to be proficient in a multitude of methodological toolboxes such as predictive, Agile, and any hybrids between them, or in yet-to-come methods in project management. Sponsors and possible review or steering committee members need to have a basic understanding of all these methods. As digitalization spreads into areas, we might not even think of today, new software development approaches like "citizen development," where people without programming proficiency can code apps and smaller programs, will find more application areas.

Figure 3-2 illustrates the Stacey Matrix, a guide to selecting the initial approach. Knowing the degree of clarity about

- the path (know-how, process) to solve the task, no matter how big it may be, and
- the goal of the deliverable

identifies the starting point of a project. Reassessing this spot regularly along the project progress will very often show that from wherever we started, we will progress toward the lower-left corner.

The longer the trajectory of the project's path in this diagram is, or foreseeably will be, the more proficient project manager and team need to be in the approaches it suggests.

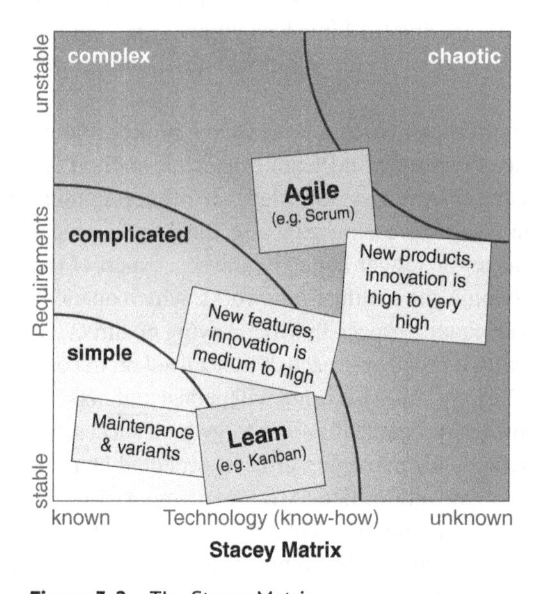

Figure 3-2 The Stacey Matrix

Special tasks along a known process will start and end in the lower-left corner. They will no longer be called a project, even though they have a time-constrained beginning and end.

The rise of artificial intelligence (AI) will strongly influence the landscape of methods and tools. In a first step, data analytics will support to establish and evaluate data on conducted projects. Accuracy, completeness, and consistency will be big challenges. In a next step, project reporting will be automated. Then AI will make suggestions for project plans, and finally – in the context of the complete project portfolio – scenarios for a single project or for the complete portfolio will be developed automatically. This will support project and portfolio managers and decision makers in most efficiently and effectively loading the organization.

Another aspect should not be overlooked: The larger a project is, the more it includes – or is even dominated by – organizational change or new ways of working. This intensifies the focus not only on the project deliverables, but even more so on project benefits. Therefore, the project sponsor must be aware of this, and the project manager must have a decent knowledge and understanding of these aspects and act accordingly, e.g., by a major focus on stakeholders and their management. All too often, this is neglected by both roles, predominately if they are used to smaller or mid-sized projects with mostly technical content.

In addition, the larger the project team and/or the touchier the project topic in the organization, the more leadership and political skills sponsors and project managers need to show. They act as influencers for the good cause of their project. Further detail will be covered in the section on roles.

Organizational development/change will become as important for achieving benefits as is expertise for the project deliverable. This will show in new (versions of) project management methods and tools. At least larger projects will need to take into consideration the influence of their deliverable on the way of working, the organizational structure, or new distribution of power in the affected units.

Line functions will stay the home base for functional excellence, whereas projects will be viewed as the leading temporary organizations to realize organizational targets utilizing the expertise of multiple functions.

After all this said, it is not a surprise that project performance metrics will include the contribution of a project to the organization's success, and the organization's performance metrics will reflect the combined impact of its projects. This will be split into project deliverables and project benefits.

For new products or services, benefits may be their expected performance on the market (turnover and result), the influence on the reputation of the organization, the legal position in its environment, customer satisfaction, or many more.

This will ultimately lead to a project-driven steering approach of organizations and, in turn, contribute to a new view on financial planning and administration. However, this will not be easy to implement, as existing systems of allocating budgets will no longer be adequate and position holders in a hierarchical management will resist giving up control of money and direct reports.

Supporting IT Tools

Today, a lot of projects are planned and report using standalone IT solutions like simple spreadsheets or slides. This is conceived as the least administrative effort. At the same time, this keeps all knowledge and experience in the project and in the heads of those involved. It prevents other projects from learning or the organization from having an overview or intervening if priorities change.

Most projects use IT tools provided by the organization. However, they are not always aligned and hosted on the same server platform and rarely data is stored in common data bases requiring a minimum data structure, completeness, and correctness of entries.

We will therefore see more IT tools that support the following:

- Planning, monitoring, controlling, and adjusting/terminating individual projects in a multi-project environment that uses a single or multiple methodological approaches, be it in parallel or in sequence. AI will reduce the administrative burden and even predict opportunities and threats for further proceeding in the project. This will be applied as an early warning system for the project manager and the sponsor.
- Steering project portfolios for maximizing portfolio benefit for the organization. AI will reduce administration burden here as well and will propose scenarios for the portfolio manager and the decision maker.

Project Roles

As important as proficiency in methods is having a good understanding and implementation of the corresponding roles needed to conduct projects successfully:

- The one initiating a project and responsible for realizing its benefits for the organization (e.g., sponsor)
- The one responsible for realizing the project deliverable (e.g., project manager)
- The team doing the actual project work
- Further roles, especially in scaled Agile frameworks
- Appropriate qualification of all role holders

Most Agile frameworks do not foresee a single individual responsible for realizing the project deliverable but spread this task over a complete self-managed team and someone responsible to know and adhere to the rules.

An important element of describing roles is to explicitly state their responsibility and their corresponding authority. Otherwise, role holders are "toothless tigers" who may be hindered to live up to their role.

Sponsor

The sponsor is the initiator and advocate of the project ensuring sufficient (financial and human) resources and is accountable for realizing project benefits for the organization. The project sponsor leads the project through the initiating processes until it is formally authorized. The sponsor serves as the escalation path for issues beyond the control of the project manager and may be involved in other important issues such as authorizing changes when risks are particularly high. Therefore, the sponsor needs to be at an adequate hierarchical level to ensure project success, reflecting the project impact. In addition, the sponsor needs to have sufficient time to live the sponsor role, especially if sponsoring multiple projects.

Sponsors are contributors to strategy implementation in the short term (project setup and realization of deliverables) and in the long term (lasting project benefits). This will strengthen their position. As facilitators for implementation of organizational strategy and achievement of related project benefits, their compensation will be strongly influenced by the success of "their" projects.

A challenge – at least for longer project durations (from identifying project targets to preparing project start to realizing project deliverables and thereafter to realizing project benefits) – is that the initial role holder might no longer be in this role at the end. It is not always easy to continue work started by a predecessor as long as the organization's governance and senior management does not help to keep interest in the project deliverables and benefits alive. Organizations need to find a way to secure consistency from initiation to benefits realization.

Sponsor's Responsibility (example)

- Visibly take ownership and accountability for the project targets, deliverables, risks, and benefits.
- Establish the project charter based on a business case.
- Monitor and control the business case.
- Focus on project deliverables.
- Support the project team and ensure full project backing by senior management and all contributing functions.
- Ensure long-term benefits realization of project deliverables.

Sponsor's Authority (example)

- Initiate the project; define the project objectives and business case.
- Determine the project impact category in line with the organization's regulations.
- Issue the project charter and approve changes to it.
- Assign and empower the project manager.
- Contribute to project manager's goal setting and performance evaluation (e.g., by participation).
- Install a steering committee and be a member of it.
- Sign the project management plan and changes to it.
- Traceably approve milestones/major releases, e.g., between project phases and at project closure.
- Terminate the project.

Project Manager

The project manager is the person assigned by the performing organization to lead the team that is responsible for achieving the project deliverables.

As project managers typically lead cross-functional or multidisciplinary teams, they must report to a management level responsible for the business success or benefits of their project deliverables.

In an Agile project, project managers shift from leading the team to serving the team. Thereby, they change their emphasis to coaching people who want or need help, fostering greater collaboration on the team and aligning stakeholder needs. As leaders, project managers encourage the distribution of responsibility within the team: to those people who have the knowledge and experience to get the work done.

Bigger projects typically become more challenging regarding their structure, complexity, and political sensitivity of the topic, and diversity in methods. This requires broader approaches to the project deliverables regarding their methods, specification/realization, and liability/compliance issues. It is no longer necessary that a project manager is the best subject matter expert in the project deliverable while still being able to understand the involved functional areas to see when and why things are not going well and to help team members not performing well. This may suffice in smaller and focused projects or in sub-projects. Many project managers do not recognize that this is an important aspect for rising in the ranks of their organization. In recovering a failed project – be it because of deficits in contents or "political" success – deep expertise in the project subject may even be hindering a restart.

Therefore, a project manager will need to act as a multi-talent and cover the full bandwidth of topics relevant for the sector or business of the project deliverable. In many cases this does not require being the single-best expert in the project topic. Such people might be essential on the project team, but not necessarily leading it. And to be honest, often times these experts are not seeking or preferring broad leadership positions.

At least in larger organizations the role of a project manager will not go away.

Project Manager's Responsibility (example)

- Lead the project team and the project; achieve project objectives.
- Establish the project management plan based on the project charter, integrating lessons learned from other projects.
- Implement the project management plan together with the project team and all other resources; assume target responsibility for the project team members to the extent of their involvement in the project.
- Appropriately apply all Project Management Knowledge Areas, respecting all relevant regulations.

Project Manager's Authority (example)

- Make decisions within the approved project management plan.
- Escalate decision needs beyond the project management plan to the sponsor.
- Appropriately contribute to the following for all project team members, respecting any prevalent legal restrictions.
  - Nomination
  - Goal setting and performance evaluation (e.g., by participating)
  - Determination of qualification measures

Every employee needs a skill or competence profile appropriate to the role and its impact on the organization. Four competence areas are relevant: Entrepreneurial, Leadership, Interpersonal, and Professional Competence (see Figure 3-3).

The organization can define how much of which dimension is required for a specific project manager position or position category. Such a procedure will help finding individuals who fit these requirements. On the other side, it will help individuals to assess themselves and prepare for such an assignment, e.g., by training.

Thereby, project managers will be considered and trained as leaders of the organization.

This will also lead to the situation, that the degree of achievement of project deliverables will decisively influence project manager compensation on more levels than today.

Project managers (at least of larger, long-lasting projects) will become direct superiors of their (core) team members. Team members will return into their line functions for updating their own or their organization's expertise.

Bosch established a career path for leading projects in 1989 (See Figure 3-4). Although originally described as a separate career, it is now understood as one of two variants of a single leadership career path: leading a project or leading a line function. For leading and inspiring their team(s), holders of either position require pretty much the same strengths and interpersonal skills. This is even more true the higher they rise in rank. What is different is the management

Competence Area	Competencies	Dimension
Entrepreneurial Competence	**Result Orientation**	Focus on target and benefits achievement
		Decision-making
		Scope management (incl. change management of requirements)
	Future Orientation	Strategic thinking & creativity; strategy implementation
		Continuous Improvement
		Business acumen
		Uncertainty management (threats & opportunities)
Leadership Competence	**Leading Myself**	Awareness of myself
		Curiosity and resilience
	Leading Others	Leadership principles
		Team management and facilitation
		Resource management
Interpersonal Competence	**Cooperation**	Intercultural competence (country, industry, enterprise, department. . .)
		Understanding customer needs / processes
		Stakeholder engagement (incl. customer relationship)
	Communication	Negotiation management / Conflict management
		Communications Management (incl. Networking)
		Language skills
Professional Competence	**Breadth of Experience**	Project management approaches (incl. PM methods like predictive/hybrid/Agile)
		Integration management (incl. coordination of project related processes and project change management)
		Procurement / Supplier management
	Depth of Knowledge	System and domain-specific know-how
		Schedule management
		Quality management
		Problem solving
		Cost management

Figure 3-3 Entrepreneurial, Leadership, Interpersonal, and Professional Competence

aspect: line functions develop functional expertise (e.g., for use in projects). At present, from the perspective of labor law, they are and stay the superior with hiring, firing, and compensation power. They often set targets and evaluate target achievement alone or jointly with the project manager. But we see more and more examples that this is conducted jointly with the project manager.

From a budget point of view, a project manager is responsible for the released project budget over project duration, which might be provided by multiple cost centers, whereas a line manager is normally responsible for a single cost center with the budget released for a single year. This follows the fact that in many companies the supervisory board or board of directors must release the company's annual budget plan. This is then broken down into cost centers for easier follow-up.

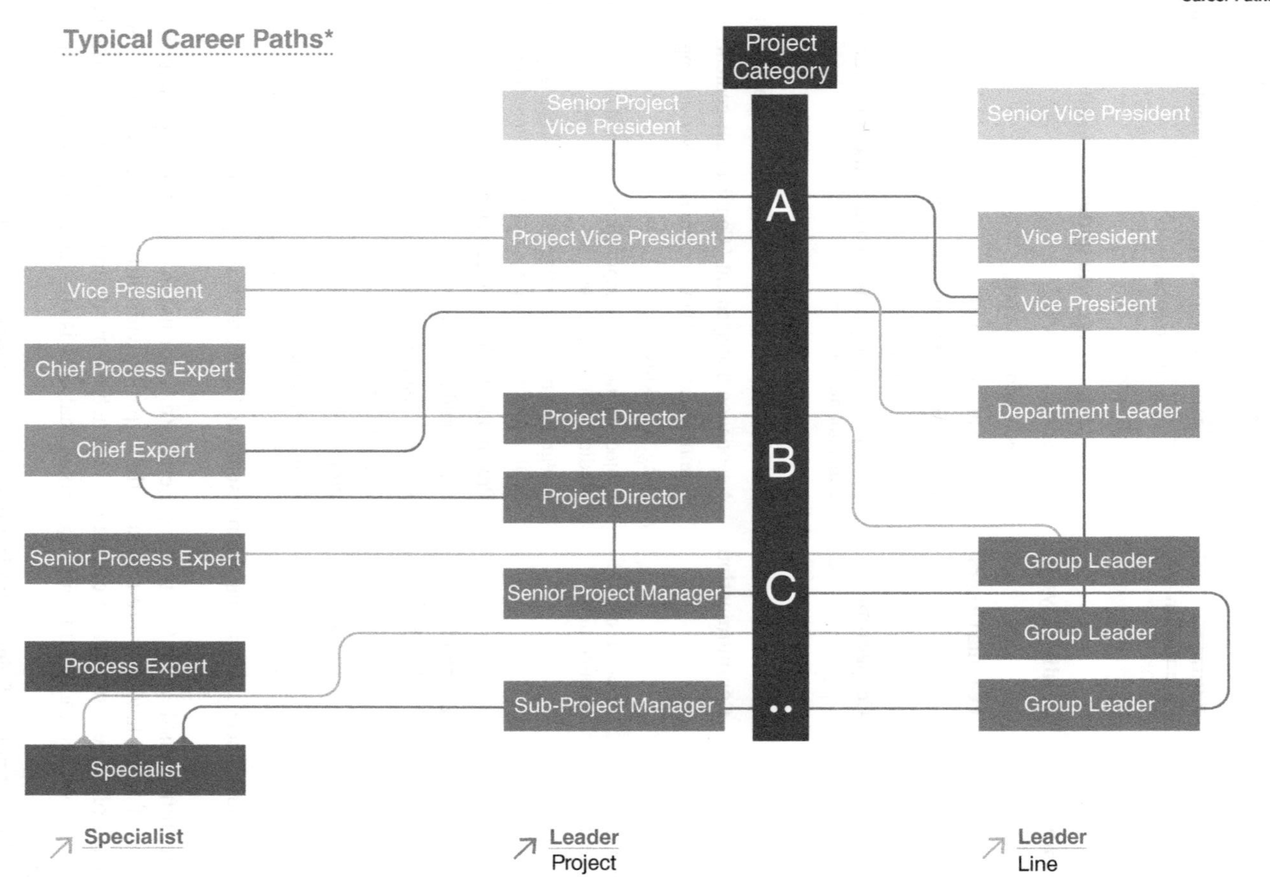

Figure 3-4 Career Path

Organizations set up along their projects might choose to base budget planning on multiyear project planning and select the next year (and maybe a few years ahead) for the release process. Projects then take priority for consecutive years unless they are consciously postponed or terminated. This is also feasible within a system of annual formal releases, and we will see more and more implementations of this project-based budget planning in the upcoming years.

Project Team

The project team (in an Agile environment, often called development team or implementation team) realizes the project deliverable according to the method selected for the project life cycle (phase). In large projects, this might be structured as a core team with multiple subteams, each of which may use a different PM approach. The project manager needs to install a suitable management structure to ensure success of the overall project.

In Agile teams, members can step in for one another and decide on which backlog item to realize and how. This limits team size and fields of expertise needed for the task. This will stay to be the sweet spot of applying Agile methods. In the future, we will see more hybrid approaches of predictive and Agile, e.g., Agile in early and predictive in later phases, requiring from team members to understand and apply such hybrids to the best benefit of a satisfactory and well-managed project deliverable.

Further Roles

Mastery in project management approaches will become a requirement for most management or leadership positions. Leadership will be a mandatory competence area for both, be it with or without direct reports. Line and project managers will form the joint leadership team of their organization.

Further roles (e.g., a product owner and a Scrum Master in Agile environments) will persist in mostly Agile environments. The same will happen to scaled Agile roles if frameworks like Scrum at Scale, SAFe, LeSS, Nexus, or others or even mixtures of some of them are used. Some of today's scaling approaches introduce roles at higher levels that remind us of traditional hierarchical structures in line organizations. All roles will be looked at in a less dogmatic way, and we will see adaptations and crossovers amongst methods.

The challenge will be to keep the oversight and to keep projects manageable, even if some use different approaches for a complete project or only for single phases of a project.

Qualification of Role Holders

All role holders will have some kind of qualification for their role, be it on the job or by a formal training, including external and/or internal certification. Training and certification will be structured in multiple layers like beginners, intermediate, and advanced. The larger and more impactful the project, the more we will see formally trained role holders familiar with the required hard and soft skills. These will be defined along the lines described for the project manager role. Training needs of individuals will be identified after comparing the required level per competence dimension of the designated role with their personal assessment.

At least on the more basic levels, formal training will rely on offers from the training market, both to have trainers proficient and certified in the approach of one or more professional associations and to have proficient trainers who make their living on training. Especially part-time

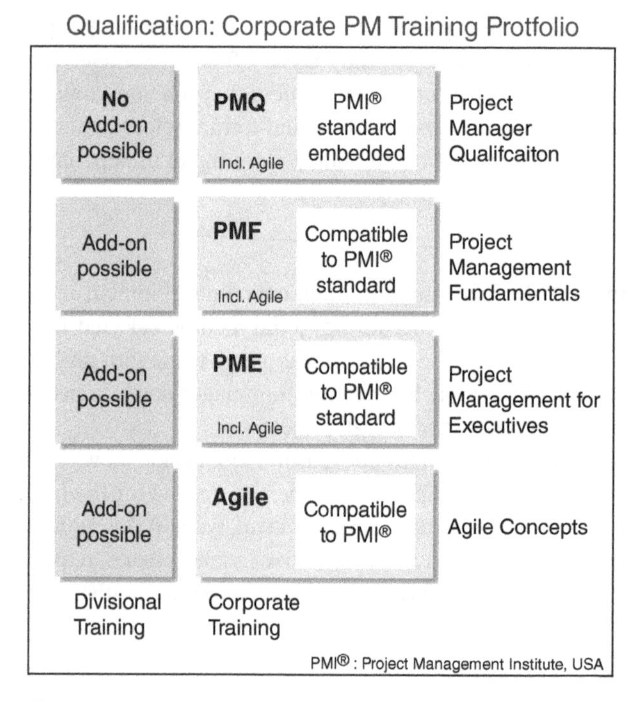

Figure 3-5 Corporate PM Training Portfolio

internal trainers might not be available on short notice due to urgent issues in their organization, or they might not further develop their training skills.

Internal training and certification of position holders will add further layers if needed by project size or specifics or any competitive advantages.

As an example, Figure 3-5 shows the principal training setup at Bosch.

Project Management Backbone in Organizations

Projects will be considered the standard way to implement the organization's strategy and its elements. Even strategy itself will be developed as a project. Organizations will establish overarching project management structures to support and expand their competitive situation.

Consistently successful projects require a supporting ecosystem. This is often underestimated. A vision and mission and a well thought-through support system are and will become even more important: the project management organization. It requires at least two roles: a senior executive responsible for the project management setup and someone (often called the project management office, PMO) implementing, monitoring, and improving this setup and consulting the executive suite and possibly substructures in subunits.

Project portfolios will be used as important elements of steering the organization. They will be managed by dedicated staff who prepare and evaluate options for decisions, while senior leadership makes the decisions. Project portfolios will focus on utilization of internal and external resources for external and internal deliverables, while product portfolios focus on the "products" serving a certain (mostly external) clientele. Both overlap but are not identical.

Project benefits will be decisive for project approval. They will be based on the value of the project deliverable and the benefits to their recipient.

Project Management Promoter

The promoter is a member of the highest leadership level and thereby the highest authority in project management in the organization. The promoter emphasizes the importance of project management for the organization and sees to its implementation, thereby supporting project success.

PM Promoter's Responsibility (example)

- Secure achieved level of expertise in project management in the complete organization.
- Shape project management in the organization.
- Ensure existence of a promoter in all subunits.
- Create and sustain a supportive working relationship with the PMO.

PM Promoter's Authority (example)

- Set all rules and regulations of project management in the organization along with any auditing mechanisms.
- Obtain and discuss targets for and results of maturity assessments from all subunits.
- Nominate the PMO head and see for appropriate resources.

Project Management Office (PMO)

The PMO is a management structure that standardizes the project-related governance processes and facilitates the sharing of resources, methodologies, tools, and techniques.

On the one hand side, for a complete organization this requires an appropriate set of governing rules (e.g., at least coarsely defining toolboxes of methods and from which toolbox to choose and when, how to report to whom and with which metrics). The latter should not only reflect the true and current status of the project, but as well the impact of the project deliverable to the unit. This helps the promoter to follow up on combined project benefits after the project manager and team have been discharged and the sponsor has taken over again. The larger these benefits are, the more organizational strategy is the guiding rail for the metrics to use. This procedure will attract even more interest by promoters in the future. The PMO will more often take the role of a navigator who guides the organization through "troubled waters." In split or large organizations, a concept of how to steer and review project progress on multiple levels needs to and will be defined.

PMO's Responsibility (example)

- Develop, implement, maintain, and improve all elements of project management.
- Facilitate learning across subunits.
- Communicate and represent project management internally and externally.
- Utilize external knowledge to further develop the internal project management.
- Establish and sustain a working relationship with the project management promoter.

PMO's Authority (example)

- Issue rules and regulations regarding project management.
- Audit subunit's adherence to corporate rules and regulations jointly with internal auditing.
- Design and further develop corporate project management qualification programs, and review implementation.
- Examine candidates and grant internal certificates in project management.

PMO offers are defined and communicated based on the PMO service portfolio given in Figure 3-6 and must be derived from organizational needs. Each subunit will have its own PMO.

PMO Service Portfolio

Target Group	Operational *Support*	Tactical *Optimization*	Strategic *Guidance*
Executives	• Act as the **interface** between project managers and executives.	• Provide a **neutral view** on the project protfolios and resources in the org unit.	• **Consult** and be a **sparring partner** for executives.
Projects	• Provide **operational support** for project reviews. • Provide **Project Office** support.	• **Coach and support** projects in case of escalation/crisis.	• **Manage** project/portfolio benefits.
Project Managers (PjM)	• Establish and maintain communities for PjM networking and knowledge transfer. • Maintain **overview of PjM positions and PjM.** • Provide **coaching** to project managers.	• Foster appropriate project leadership **Career Paths.** • Enable **Mentoring** of new PjM and agile role holders by experienced once. • **Consult** PjM and agile role holders in career opportunities.	• Act as **disciplinary superior** of project managers.
Organization	• **Coordinate** PM-related topics **within org units.** • Act as **interface** for PM-related topics **with other org units.**	• **Coordinate** all org unit-specific project life cycles. • **Promote** PM awareness throughout the organization. • Conduct **PM maturity self assessment** and project auidits.	• Define and provide org unit-wide **PM processes, tools, and methods** • Set and enforce **rules for staffing** project leadership positions. • Provide **PM qualification curricula** for all project roles.

Important note: This PMO service protfolio is an overview of what a PMO at Bosch **could** do.
Each PMO has to define its service portfolio according to the needs of the org unit.

Figure 3-6 PM Service Portfolio

Organizational Development

Looking at the importance of project management, it is obvious that it needs to be a decisive part of organizational development. Knowing where to start from some kind of maturity assessment can help. Applying a multilevel organizational maturity model (be it self-developed or from a professional association) helps to identify potential targets for the organization within its entrepreneurial and competitive environment. Monitoring and achieving such self-determined goals support identification with and proficiency in project management.

The roles of a promoter and a supporting PMO are instrumental in developing the organization toward a beneficial project orientation. They may exist on a corporate level and be cascaded down into the prevalent organizational setup, as can be seen for the example of Bosch in Figure 3-7. By that, units across complex organizational setups can still be differently mature while applying a comparable assessment procedure.

Such project management systems will become more widespread in organizations that support a coherent approach. It might even lead to making the executive responsible for organizational development the promoter and to tie the PMO and organizational development department together.

On the first layer below corporate, all divisions duplicate the corporate set-up and appoint one member of the executive leadership team as the divisional promoter, supported by the divisional process owner heading the divisional PMO. A further breakdown into business units, regional units, or plants is possible. A similar setup will be implemented in more organizations in many more sectors and industries.

Project Mindset and Appreciation in the Enterprise as an Important Element of Entrepreneurship

It is obvious that all that was described above will not happen if it is not supported by the senior leadership of an organization and implemented by all layers below.

Therefore, Bosch has decided that one member of the board of management is taking over the role of corporate promoter as the highest authority for project management in the Bosch Group. Together with the corporate PMO, they issue corporate regulations and descriptions on PM concepts and elements based on a vision and mission for PM at Bosch, which is well-suited to give guidance to the complete organization (see Figure 3-8). This includes further development of career paths, involvement of project managers in target setting and performance review of team members to whom they do not have a direct reporting line, and company-wide training offers jointly with corporate HR.

Adherence to the regulations is monitored by the corporate PMO and inspected by corporate auditing.

Conclusion

Implementing a future-oriented approach to project management in an organization is based on a project mindset on all levels, covers a lot of interconnected facets, and takes a long breath. Just announcing that you are following a certain method is not enough in the post-Agile era.

Organizations will put more emphasis on their project management system with aligned and clearly described roles and methods, and with an ecosystem that fosters project thinking as a part of entrepreneurial action. Project orientation will be the "new normal" in most organizations. This covers organizational setup, budgeting mechanisms, and leadership approaches.

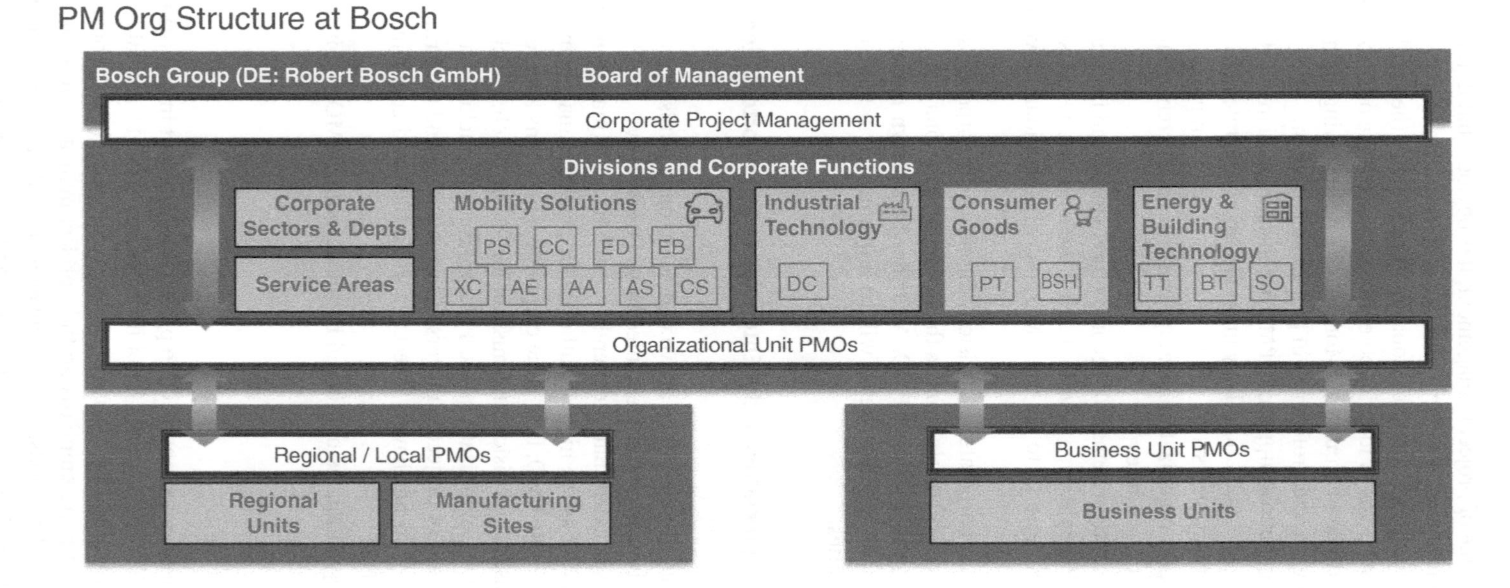

Figure 3-7 PM Organization Structure at Bosch

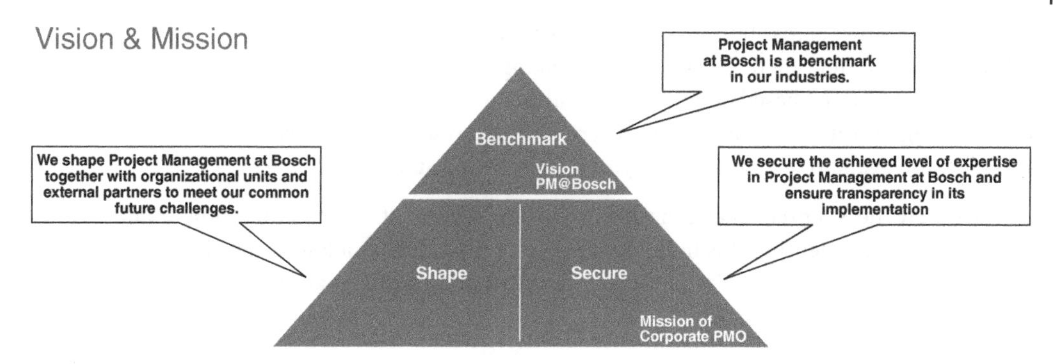

Figure 3-8 Vision and Mission

Organizational development will become a stronghold in making the organization project oriented. It supports or merges with the PMO to determine the best-suited methodological approaches and role concepts for conducting projects – whether one method, multiple methods in parallel, or a hybrid between two or more methods. Projects will be the future way of working with line functions providing the right amount of future-oriented expertise. Advanced project managers will be admitted to and even requested in the leadership ranks of organizations.

Substantially more automation will make day-to-day work (e.g., planning and reporting) of managing projects easier. Data automation and AI will help making predictions about suitable next steps or potential shortcomings in current projects. Success of projects will no longer be defined by achieving project deliverables. Long-term benefits will become decisive, so that organizational targets are supported by projects.

Organizations will therefore put more effort on training and certification in an ever-broader project management, be it when recruiting new talent or further developing existing talent by internal or external measures significantly beyond a pure method-oriented skillset.

3.6 Blocking Off Time to Think

If one would envision the difference maker for innovative cultures and focus on just one ingredient, this would be the one. With the huge distractions that have dominated the way of working over the last decade and the increasing set of competing demands, it became virtually impossible to do a good job in taking time to think. As basic as this sounds, the reality of success in carving this kind of time to think has become a luxury and the next decade must create a reversal of this phenomena. Future project managers will think again for a change!

The benefits of blocking off time off to think have been known by executive leaders for some time. These benefits will cascade across future organizations' lines of business and will contribute to likely excellence in a few areas:

- Performing more at a higher level of focus and motivation
- A creative approach to problem-solving
- An increase in customer-centricity and satisfaction
- Developing a clearer end-to-end view of the ecosystem
- An increase in learning new and effective practices
- Improvements in quality

- Better integration across business units and projects
- Higher quality and timeliness of decision-making
- Creating a solutions-based organizational culture
- Enhancing prioritization capacity of where time and resources are best utilized
- Having fun again in business and improving teams and staff moral

As we view the 3M excellence example in the next section, it demonstrates how small, simple, steps toward implementing this building block not only achieves a few of the above benefits but also changes the work dynamics in the environment and results in an improved level of energy across teams and the company.

3.7 Excellence in Action: 3M

Companies often struggle on how to motivate and reward the workforce into project management processes. In some companies, workers believe that only the R&D group or any other innovation groups are responsible for coming up with innovation ideas and exploiting them. 3M took a different approach and set the standard on how to involve the entire company.

Background

Companies that are highly successful at innovation and strategic project management use innovation as the driver for sustained corporate growth. 3M Corporation is prime example. Most researchers agree that 3M's success emanates from its corporate culture that fosters an innovation mindset. As stated by Irving Buchen [2000]:

> 3M announced that all employees were free to spend up to 15 minutes each working day on whatever ideas they wanted to work on. The only restriction was that it could not be at the expense of their regular assignments. They did not have to secure approval for their project. They did not have to tell anyone what they were working on. They could bunch their l-minute segments if they needed more solid blocks of time. They did not have to produce anything to justify or pay back the time taken.
>
> What was the result? There was electricity in the air. Employees came earlier and stayed later to extend their innovation time. Many walked around with a weird smile of mischief and even fun across their face; some even began to giggle. But they were also enormously productive. Scotch tape came out of this ferment; so, did Post-its.
>
> Perhaps equally as important, morale was given an enormous lift; general productivity was higher; teams seemed to be closer and working better together; the relationships between middle-level managers of different divisions seemed to improve. In short, it was a win-win situation. The innovative gains were matched by a new spirit that changed the entire culture.

Companies such as Google and Hewlett-Packard have programs similar to 3M's 15 percent time program. 3M's program was initiated in 1948 and has since generated many of the company's best-selling products, 22,800 patents, and annual sales of over $20 billion.

There are several distinguishing characteristics to the 3M culture, beginning with employee encouragement for innovation and idea generation. Employees are encouraged to follow their instincts and take advantage of opportunities. 3M provided forums for employees to see what others are doing and to get ideas for new products and to find solutions to existing problems. The culture thrives on open communications and the sharing of information. Employees are also encouraged to talk with customers about their needs and to visit 3M's Innovation Centers.

Strategic direction is another characteristic of the 3M culture. Employees are encouraged to think about the future but not at the expense of sacrificing current earnings. Using the Thirty Percent Rule, 30 percent of each division's revenues must come from products introduced in the last four years. This is tracked almost religiously and forms the basis for employee bonuses.

Funding sources are available to employees to further develop their ideas. Seed money for initial exploration of ideas can come from the business units. If the funding requests are denied, employees can request corporate funding.

Rewards and recognition are part of 3M's innovative culture. A common problem facing many companies is that scientists and technical experts believe that the 'grass is greener' in management than in a technical environment. 3M created a dual ladder system whereby technical personnel can have the same compensation and benefits as corporate management by remaining on a technical ladder. By staying on the technical ladder, people are guaranteed their former job even if their research project fails.

Similar to Disney and Apple, 3M created the Carlton Society, named after former company president Richard P. Carlton, which recognizes the achievements of 3M scientists who develop innovative new products and contribute to the innovation culture.

One of the most significant benefits of the 3M culture is in recruiting. Workers with specialized skills are sought after by most companies and the culture at 3M, which offers a significant amount of "freedom" for innovation, helps them attract talented employees.

Conclusion

3M's success set the standard that others have copied on how to involve the entire organization into innovational thinking. All employees must be made aware of the notion that they can contribute to innovation and their ideas will be heard.

3.8 Refreshed Executive Role

Excellence in future organizations will consistently start in the boardroom. This is the golden opportunity for project management to be seen with the right lens and to be reborn as a strategic competency as we highlighted in Chapter 1. These future boardrooms have executive leaders thinking and operating in a project system-like manner. They are no longer focused purely on steering, and they are more highly energetic and engaged. This is signaling the right accomplishment project managers are reaching by being the ones at the helm of many of tomorrow's companies.

This refreshed new role of the executive makes a big dent in the classical bureaucracy, hierarchy, and many of the past decades' siloed ways of working. By exemplifying the role of the product owners and providing proper sponsorship that is balanced and that is truly captain-like, these executives will demonstrate in action many of the Servant Leadership behaviors we highlight in Chapter 6.

The next section, the GM excellence example, showcases this fourth and final building block. The commitment to making a societal difference in the face of unprecedent challenges highlights the success story when cultures and their leaders are ready to exemplify resilience and adapt fast in the role they play while excelling in delivering benefits of major programs and projects.

3.9 Excellence in Action: General Motors[4]

General Motors and the Ventilators

When one thinks of project risk management, the thought of a pandemic usually isn't at the top of the list, if it is on the list at all. In 2020, things really turned upside-down when the COVID-19 virus hit the world. In March 2020, General Motors idled most of its manufacturing and assembly plants in order to slow the spread of the virus and protect its workers. So here was one of the largest companies in the world, with one of the most versatile workforces, stalled and waiting for the next bit of news.

What happened next was a brilliant pivot requiring the use of project management talent and application of manufacturing resources to help respond to the very pandemic that impacted production of its vehicles. GM's mobilization of the same project and program management resources that manufacture vehicles produced incredibly fast results for a corporate social responsibility initiative.

Companies like General Motors today maintain a self-regulated strategy called corporate social responsibility (CSR) that is integrated into the firm's business model and identifies the ethically oriented activities the firm will undertake for the benefit of consumers, society, ecology, and government regulations. Traditional CSR activities in many companies focus on the consumption of certain natural and renewable resources such as water, energy, and other materials. However, as General Motors (GM) has admirably demonstrated, there are other ways to demonstrate CSR.

Included in the description of a firm's CSR is usually the term *social,* which may be defined as improving or showing concern for human life without impacting the capacity of the supporting ecosystems. This leads us to the term *social innovation*, which is the creation of new or innovative products and services that support CSR. The outcomes of social innovation often focus more so on a concern for society instead of profitability.

GM's CSR effort regarding ventilators began on March 17 after they were approached by stopthespread.org, a coalition of CEOs, trying to organize companies to help stop the spread of the COVID-19 pandemic. The coalition contacted Mary Barra, GM's CEO, and suggested that GM team up with Ventec, a company located in Bothell, Washington, and a manufacturer of small portable ventilators. Within 24 hours, executives from GM and Ventec had their first conference call and the following day, a team from GM flew to Seattle to meet with Ventec.

> Senior stakeholders were quickly on board with the overall objective and remained in contact with all project workstreams on a daily basis. Empowered teams met daily and were loosely coupled to coordinate dependencies, understand the timeline, and resolve issues quickly. Issues were documented on shared folders, triaged, and resolved.
>
> Roadblocks did not exist for long. Decisions were made and the team moved ahead. Deliverables and approvals that would normally take weeks, only took minutes, due to the clear communication paths, team focus and executive priority. Capital was approved and contracts were put in place with minimal executive oversight/approval being required.
>
> All of this was possible due to the clarity of the GM vision established up front and unwavering commitment of GM people at all levels of the organization to deliver ventilators. There was plenty of Inspiration to get this done, as if someone in our family depended on it. People were excited to be involved. Each shipment was a celebration, down to the last one shipped just before midnight on the contract deadline.
>
> – Jeffrey J Hall, PMP, GM GPSC Sr Development Lead

4 Material in this section has been provided by Jeffrey J Hall, PMP, GM GPSC Sr Development Lead.

By March 20, GM had engaged its global supply base, and within 72 hours, suppliers had plans to get all the necessary parts. The UAW's national and local leadership signed off on the project and by March 25, and crews began preparing GM's 272-acre complex in Kokomo, Indiana, for production.[5] About 800 full-time and part-time GM employees were assigned to the project to produce 30,000 ventilators. According to a GM spokesperson:

> The Detroit automaker will produce and deliver the [30,000] ventilators to the government by the end of August, with the first 6,132 ventilators being delivered by June 1, [2020].[6]

Senior management at GM were committed to the CSR ventilator effort. "Our commitment to build Ventec's high-quality critical care ventilator, VOCSN, has never wavered," GM said. "The partnership between Ventec and GM combines global expertise in manufacturing quality and a joint commitment to safety to give medical professionals and patients access to life-saving technology as rapidly as possible. The entire GM team is proud to support this initiative."[7]

> *GM Information Technology (IT) group was a critical piece of the Ventilator production. While significantly scaling up the Ventec process to meet GM's volume expectations, each ventilator part had to be tracked through the manufacturing steps. GM reached out to Ventec's IT partners immediately and quickly worked through the contracting and architecture required to meet the business needs for quality, tracking, and throughput. This was done in parallel with setting up the initial manufacturing space to start building 'non-saleable' units within 3 weeks of GM's initial visit to Ventec.*
>
> – Jeffrey J Hall, PMP, GM GPSC Sr Development Lead

GM and its partner Ventec Life Systems delivered the last of the 30,000 ventilators owed the U.S. government as planned. GM said the full federal order was completed in 154 days, with one ventilator made about every seven minutes.[8] According to GM CEO Mary Barra:

> ..."the automaker's motivation to produce the critical care ventilators was fueled by thousands of people at GM, Ventec and our suppliers who all wanted to do their part to help save lives during the pandemic. It was inspiring to see so many people achieve so much so quickly."[9]
>
> *GM transitioned out of the Ventilator business after 5 months and enabled Ventec to take over the production line immediately upon completing the government contract. There was minimal down time in the transition due to careful planning. Ventec continued to build ventilators on that same production line until the ventilator demand diminished later in 2020.*
>
> – Jeffrey J Hall, PMP, GM GPSC Sr Development Lead

5 https://www.freep.com/story/money/cars/general-motors/2020/08/14/no-shortfall-supply-ventilators-gm-ventec-stockpile/3366678001/
6 https://www.cnbc.com/2020/04/08/gm-to-build-30000-ventilators-for-us-for-489point4-million.html?__source=sharebar|email&par=
7 Michael Wayland and Christina Wilkie, "Trump orders General Motors to make ventilators under Defense Production Act," CNBC, March 27, 2020, https://www.cnbc.com/2020/03/27/trump-orders-general-motors-to-make-ventilators-under-defense-production-act.html?__source=sharebar|email&par=sharebar
8 Jamie L. LaReau, "GM and Ford end critical care ventilator production," *Detroit Free Press,* September 1, 2020, https://www.freep.com/story/money/cars/general-motors/2020/09/01/gm-and-ford-deliver-last-ventilators-amid-coronavirus/3449490001/
9 Ibid.

GM's commitment to CSR did not end with the delivery of the last ventilator. GM was still making face masks. The company said it will donate 2 million face masks to Michigan public schools as part of the State of Michigan's MI Mask Aid partnership. GM's contribution includes 750,000 child-size masks for elementary students. Those will be ready for delivery by Sept. 14, GM said. Also, 1.25 million adult-size masks for high school students, faculty, and staff will be ready for delivery by September 28.[10]

Nobody knows what will happen in the future. There could another pandemic, acts of God, or wars that require companies to be able to pivot and redirect their resources from their primary purpose to one focused on the benefit of humanity as did GM with the ventilators and masks. This ability will bring more than just financial benefits to the company. People will turn to GM when innovation and ingenuity are needed to solve corporate social responsibility related challenges, even on the global scale.

3.10 The Innovation Culture

In order to attract Generation Z and following generations, future organizations will have no option but to create cultures of massive experimentation. Having the capacity to build this level of safety in the culture will make it more attractive for project work in the future. When we say that the future of work is shifting to a gig economy, we see this demonstrated by the ever-increasing trust currency that needs to be a clear area of investment in the future.

Innovation cultures are fun, highly networked, social, relaxed, and anchored in learning fast as the only answer for enhancing innovation. To get there, future excellence will require a much higher degree of flexibility around the use of the practices of project management and a literal application of the idea that these practices are only a guide. Context of these future gigs that project personnel will play will matter most and will allow for a level of confidence in how innovation dominates in the way of working, and results in a clear market differentiation.

When we think of this first foundational dimension of the culture model as previously shown in Figure 3-1, Apple comes to mind – possibly at the top of the list. The number of disruptions the company has created over the decades, including disrupting its own products, creates a crystal-clear example of the leaders' guts needed to create the relentless focus and zest for excellence in innovative cultures. The Apple excellence story, in the next section, highlights some of these continuous innovation culture attributes and offers us a reflection moment for what likely could dominate future organizations that look to substantial scaling and not just sustaining their progress. A strong message regarding ownership models is also highlighted in this example and directly reflects the potential for the next gen project managers.

3.11 Excellence in Action: Apple

When people think about companies that have a history of continuous innovation, they usually start with Apple. But what many people do not recognize is that the path leading to continuous innovation success may be strewn with some failures, roadblocks, challenges, and possibly lawsuits such as Apple encountered with Microsoft and Samsung over intellectual property rights.

10 ibid

Many of the nontraditional or strategic projects of the future will be accompanied by higher-than-usual levels of risk. There can be many different reasons for the failure and there could be lawsuits over intellectual property rights. Project teams must understand that this might be a way of life in the future and the corporate culture must have a tolerance for failure.

Background

Some of Apple's successes include the Macintosh computer, iPod, iPhone, iPad, Apple Watch, Apple TV, HomePod, Software, electric vehicles, and Apple Energy. But there is also a dark side of unsuccessful consumer products that were launched during the 1990s, such as digital cameras, portable CD audio players, speakers, video consoles, and TV appliances. The unsuccessful consumer products were not the result of an innovation failure but from unrealistic market forecasts.

Successful innovations increase market share and stock prices while unsuccessful products have the opposite effect. Apple was highly successful with the Macintosh computer from 1984 to 1991. From 1991 to 1997, Apple struggled financially due to limited innovation. Apple returned to profitability between 1997 and 2007. In 2007, Apple's innovations in mobile devices were a major step to its astounding success.

Innovation often creates legal issues over the ownership and control of intellectual property. The growth of the internet created a problem for piracy in the music industry. Apple's Steve Jobs commented on Apple's music business success, stating that "Over one million songs have now been legally purchased and downloaded around the globe, representing a major force against music piracy and the future of music distribution as we move from CDs to the internet."

Apple uses several different types of innovation on its strategic projects. Some products use incremental innovation, such as updated versions of cell phones, whereas other products appear as radical innovation. Apple also uses both open and closed innovation. For open innovation activities, which includes partnering with people outside of the company, Apple created a set of Apple Developer Tools to make it easier for the creation of products to be aligned to Apple's needs.

Apple also created innovations in its business model, which was designed to improve its relationship with its customers. Apple created a retail program that used the online store concept and physical store locations. Despite initial media speculation that Apple's store concept would fail, its stores were highly successful, bypassing the sales numbers of competing nearby stores and within three years reached US$1 billion in annual sales, becoming the fastest retailer in history to do so. Over the years, Apple has expanded the number of retail locations and its geographical coverage, with 499 stores across 22 countries worldwide as of December 2017. Strong product sales have placed Apple among the top-tier retail stores.

Apple created an innovation culture that gave people the opportunity to be creative. Unlike other cultures where executives assign people to innovation projects and then sit back waiting for results to appear, Steve Jobs became an active participant in many of the projects. Numerous Apple employees have stated that projects without Jobs's involvement often took longer than projects with it. This is expected to be a trend for the future, where executive and managers at all levels provide active support and assistance for projects rather than being bystanders.

At Apple, employees are specialists who are not exposed to functions outside their area of expertise. Jobs saw this as a means of having "best-in-class" employees in every role, including project management. In traditional project management, project managers often share the accountability for project success and failure with the functional managers that assign resources to the project.

Apple is known for strictly enforcing single-person accountability. Each project has a "directly responsible individual," or "DRI" in Apple jargon.

To recognize the best of its employees, Apple created the Apple Fellows program, which awards individuals who make extraordinary technical or leadership contributions to personal computing while at the company. This is becoming a common practice among companies that have a stream of innovations. Disney created a similar society called "Imagineering Legends" to recognize innovation excellence.

Conclusion

Continuous successful innovation is possible if the firm has a tolerance for some failures and recognition that the marketplace may not like some of the products. Also, the company's business model may change, as happened with the opening of Apple Stores. Innovation at Apple is an excellent example of how project managers must make both project and business decisions.

3.12 Projects as Innovation Labs

As we look to the future, and find ourselves in a place where agility and resilience have been dominating our workplace, we find the appetite for experimenting and learning at their highest. This is where this foundational dimension of the innovation culture is born. The idea of the innovation labs is a game changer in the future generations of project managers and teams. It ticks many boxes for how future generations will like to work. The excitement, the change, seeing outcomes fast, and witnessing the impacts of their work, and having the opportunity to play.

One natural element to this innovation lab concept is the outcome of the work of the project or program. Testing different ways to reach those outcomes in the form of the processes or ideas to get project teams to achieve value is a unique role in the creativity essential in innovation cultures. Yet, the elements of this innovation lab concept are not just related to the outcomes; they are heavily directed to the people. Experimenting with the right mix of skills, shifting roles in project teams, rotating leadership, figuring out the right autonomy and alignment mix, and at which intensity level and cadence of team interactions best results could be achieved, are all in play in tomorrow's labs. This is central to creating the learning organizations of the future where project management is seen at the core for how learning crosses organizational silos to create a tightly connected excellence in operation. As we view the Samsung excellence example in the next section, the idea of the lab is reflected in how it established Strategy and Innovation Centers as a key ingredient for innovation success and the effectiveness of knowledge management.

3.13 Excellence in Action: Samsung

There is a significant difference between introducing a few new innovative products occasionally and becoming a global innovation leader. The main difference, as Samsung has mastered, is by creating a culture that has a heavy focus on business value creation. Effective cultures are built around a set of core competencies that all members for project teams must support.

Background

Samsung and other companies have adopted a value innovation approach, which allowed them to create a corporate culture for becoming an innovation leader rather than an innovation copier and follower. Some characteristics of Samsung's culture include the following:

- Innovation-oriented sponsorship and governance from the top of the organizational hierarchy downward
- Line of sight about the executives' vision and the company's strategy and strategic objectives for all employees to see
- Use of open innovation practices as well as seeking out innovation ideas internally
- Establishing Strategy and Innovation Centers as well as Open Innovation Centers for better knowledge management
- Recognizing that knowledge management supports core competencies for storage and reuse of knowledge from R&D activities
- Recognizing that globalization in a turbulent environment requires nontraditional systems
- Maintaining customer-focused product innovations
- A willingness to accept innovation risk-taking
- Flattening of the organizational hierarchy
- Development of speed-focused innovation strategies and execution
- Decisions are being made quicker than before
- A reduction in cycle time from months and years to weeks
- Low-cost manufacturing

Conclusion

These characteristics have allowed Samsung to develop superior core competencies. The results of Samsung's value-driven culture have led to innovations in products, technology, marketing, cost reduction, and global management.

3.14 New Ways of Working

One of the advantages to the project economy, as is reflected by projects being central to how work gets done, is the fluidity of change in how we work. As the world has experienced needing to shift to being highly remote in times of unprecedented pandemics, a critical tension was created in the thought processes of organizations and their employees about how prioritization, offices use, teams' formulation, success measurement, outcomes achievement, and values that matter are viewed and handled.

One thing for sure became obvious: the ideas of back to normal, finding a new normal, and all types of debates around that, could drain a large amount of energy. In addition to the evident remaining resistance to change shown by some organizations, these debates confirmed that there is still a very vast amount of potential for project management in how organizations of the future work. Given that change is the core of what projects are about, projects drive resilience, adaptability, and the fluidity needed for many new ways of working, if we still need to continue to call it that. An ideal realization for this debate could be that organizations wanting to excel in the future need to embrace continual fluidity in their operating model and a fast tilting for where and how work gets done.

The risk appetite in the new way of working is much higher than we might have had to experience as an average across organizations. In addition, every excellent future organization would need to be a data-driven organization, where higher investment in the quality of date and the intelligence gathering form data is closely prioritized. The expression that might be most suitable to these future organizations is to think of themselves like a River Flowing Smoothly. This fluidity is a strong illustration of the linkages needed between, the culture, the practices, and the people. Leadership and the dominant skills in these future organizations will contribute to enhancing this fluidity. T-shaped skills where there are wide set of critical skills to enable that tilt across the enterprise will be more expected, and would likely consistently include data-related expertise. On the other vertical side of the "T" we will still depend on a specific expertise, although openness to reskilling and upskilling often will likely dominate the future way of working.

Fast incremental execution must be encouraged, and seamless feedback and access to value is key. We will need to adapt and welcome transparent connecting like never before which also means courageous leadership and ever-increasing respect for the vulnerability qualities in tomorrow's leaders.

3.15 Excellence in Action: Siemens[11]

Master Complexity and Deliver Reliability with an Agile Product Driven Project Management Capability

A new paradigm of project management delivers new innovation with speed, agility, and traceability (Figure 3-9).

There are many types of projects and programs that drive innovation with customers all over the world. The top five problems that I've seen many companies face:

1) **Lack of program discipline.** Companies do not invest in the proper planning and execute critical processes like bid/proposal with important suppliers in an ad-hoc way. No traceability.
2) **Lack of flexibility in methodologies.** Projects/programs are complex and require different methodologies to properly manage risk and accelerate delivery. One size doesn't fit all. Often projects are either under or over managed with checklists of steps.
3) **Assets are recreated instead of reused.** Many projects/programs are completed as if they are the first and only project the company has ever and will ever do. IP is not managed properly with the context and traceability that would enable asset reuse. Many projects waste massive effort on tasks that have already been done, if the company only knew where to find the knowledge and had the context to reuse it.
4) **Siloed processes and disconnected team members.** Disconnected teams with different plans that a PM brings together lack the collaboration to streamline work and execute

11 Material in this section has been provided by Suzanne Kopcha. She is the Vice President of Strategy at Siemens Digital Industries Software. Suzanne has 31+ years of experience in global strategy, business development and leading digital transformation and M&A initiatives. She has 2 US Patents and serves on 2 non-profit boards focused on the development of youth.

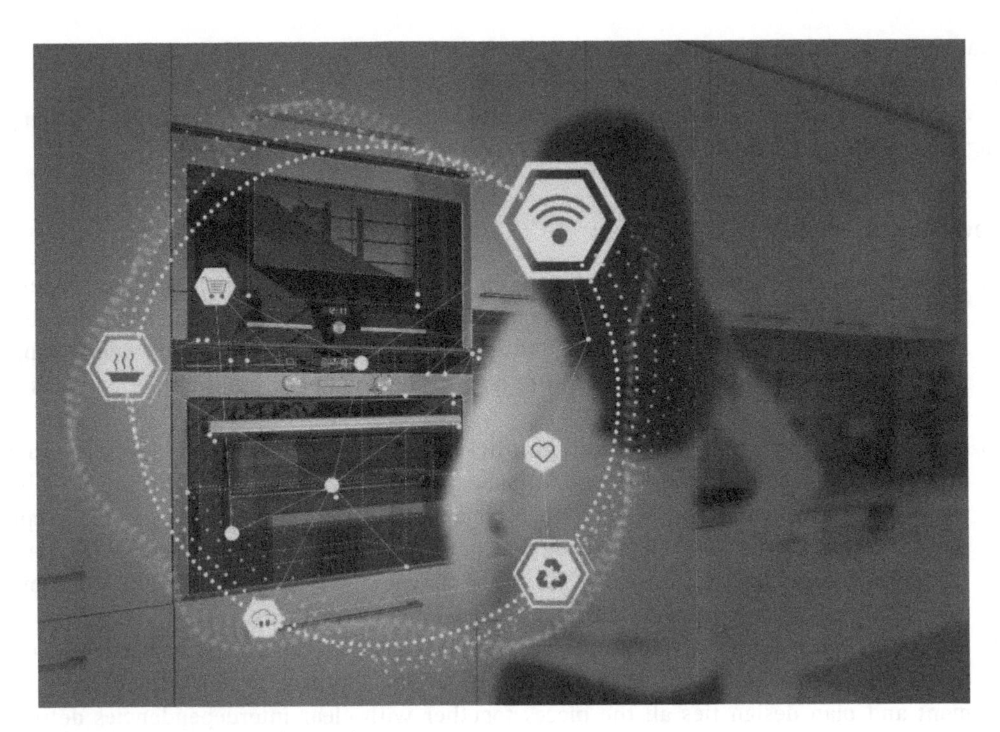

Figure 3-9 Accelerating Innovation

concurrently. Often times, these teams don't know when to stop and when to accelerate because the integrated plan comes together at status updates and is not available in real-time.

5) **Capability constraint of the right resources.** Resource allocation is not based on the requirements of deliverables and companies don't have the flexibility to understand the impact of changes to resource requirements.

Trends pushing the industry to a new way in an accelerated fashion:

- **Dynamic market:** Increasing complexity and ever-changing requirements.
- **Increased speed and complexity of innovation:** new, personalized, connected, smart.
- **Increased competition:** Lower barrier of entry from small, nimble competitors; online commerce.
- **Increasing importance of sustainability, the circular economy, and the lifecycle of a product;** not just its launch
- **Infinite data available** for after-market service revenue and closed-loop continual improvement.
- **Supply chain complexity:** Number of suppliers required to complete a project/program.

A Transparent, Systems Approach Enables a New Way

Building project/program management capabilities on the same platform, integrated with product lifecycle management capabilities, is enabling companies to accelerate innovation, drive a real-time system-driven approach that is Agile, flexible, and rigorous; delivering better innovations, faster, with lower risk and executed right the first time.

With this new approach, industries as diverse and aerospace and consumer products are changing the way they work to adopt these seven new best practices:

Seven Best Practices

1) **Product/Brand Master Data:** Bringing the commercial context to a company's product portfolio into the Innovation platform enables a company to trace the design and validation artifacts of their innovation portfolio to the product they are used in. This is critical when a change is required and facilitates the ability to propagate the change across multiple products in one step.
2) **Seamless integration of project and product milestones/artifacts:** Seamless integration of project and product milestones and artifacts on the same platform enables project milestones to be tied to actual release status of design deliverables. It is not about updates in a separate system. A company can no longer suffer unmet requirements or dependencies that were not fulfilled. All of the statuses are visible in real-time to every team member, and workflows can prevent mistakes often made with disconnected manual methods.
3) **Systems driven approach to requirements:** Products are becoming increasingly connected and increasingly more complex. A system-driven approach to requirements management and plan design ties all the pieces together with clear interdependencies defined and managed in the system – even across suppliers, including sourcing decisions and deliverables.
4) **Context and Traceability of Innovation / Reuse:** All of the steps of a product development project can be tracked with full context of what was required to deliver that step. With this level of record keeping and traceability, a company has the ability to systemically reuse innovation assets across multiple products and projects – accelerating new product projects. Imagine starting a project with all the design and validation for two-thirds of the project. This is a very real scenario in many companies – but most companies don't have the traceability required and end up redoing the same work for the next initiative, wasting precious time and resources.
5) **Smart Planning:** Includes dynamic templates based on the complexity of the product and project and leverage machine learning and AI to continually improve the system's ability to create the right plan that optimizes speed with managed risk. The system also enables a PM to define the right methodology for the right part of the project, leveraging Agile when appropriate and more traditional methods when that is appropriate to the level of complexity of the work. A smart planning process will also take into account the amount of reuse potential in a new product; eliminating work that is no longer needed.
6) **Closed loop change process:** Real-time visibility to changes and their impact enables better, more confident decision-making and fewer mistakes from change impacts not fully understood.
7) **Shift from sequential to concurrent execution:** Because of the real-time visibility and project and product statuses being tied together, companies are able to accelerate work and begin developing many more things in parallel with the knowledge of when to advance or stop based on other work progress (or delays).

Benefits

- Speed – plan tied directly to work enables fast, confident decisions on how to accelerate while managing risk.
- Quality – systems-driven requirements tied to deliverables eliminate the risk of disconnected processes and missed requirements.
- Eliminate non-value-added work – Smart Planning, AI-driven continual improvement, and artifact reuse can drive months (or years) of work out of a project.
- Flexibility and control to quickly adapt to changing market dynamics with full visibility to the impact of change.

3.16 Readying and Sustaining Tomorrow's Excellence Cultures

Creating momentum for excellence in the future will center on placing higher value on creative ideas, getting youth in the center of change, and using seamless digital-people models. Dismantling barriers to fluidity in cultures and mental barriers will be a strategic priority often discussed in boardrooms. Governance will be risk-centered. Continual assessments of the enterprise threats and opportunities associated with fluidity in operating and the correlation of these assessments to the missions of these organizations will make board rooms much more fun and intense places for creativity. We will see continual disappearance of the controlling role of boards and more evidence of the designing and the testing of ideas role.

The following three examples of excellence in organizations represent additional extended view of the ingredients highlighted thus far in this chapter and as demonstrated in the culture model. The first in section 3.17 provides such a fun view by Juan Carlos for the culture dynamics highlighted in this chapter about future organizations cultures. By looking at the role of project and program managers as holistic individuals who need to be energized to tackle the fluidity of the workplaces in the future, Juan Carlos draws a clear picture of a day in the life of that leader. He points to many of the practices that we have tested in this decade and how they will be a consistent norm for sustaining success in future organizations that have to operate as transparent and highly value-centered entities. This is a wonderful illustration of the connected cultures we need to see happen more frequently tomorrow, where every effort could be easily traced and mapped to what it contributes to, what strategic objective it influences, and ultimately what healthy results it creates.

Then the following section 3.18, Solvo360, shows the great opportunity ahead for project management in driving the future growth of emerging economies. Whether in Latin America as the example focuses or in the growing regions of Africa, or elsewhere across developing countries, the intensity of the likely impact of project management is still untapped. The example conducts a historic review of the changes witnessed and then takes us into today and the potential of the future value of project management in driving a portfolio of growth investments. It highlights with conviction what the application of many of the pillars in this book could have in terms of a fresh focused impact on many of the world countries and their youth.

The last section, 3.19, show cases what continuous innovation requires. The highlighted Texas Instruments example suggests the use of the success pyramid idea as a connector for the future organizations' way of working. The impact of this pyramid in creating entrepreneurial mindsets and the strong and clear ownership needed across project teams is very valuable.

3.17 A Future (Working) Day in the Life of the Program Manager[12]

This is a story about how the day of a program manager in an ICT company might look like in the future. I believe that we will have more programs than projects. The reason is that we will use a different approach to provide value and business benefits to the customers. The so-called project manager will be more of a program manager or business solution executive whose purpose is to maintain a sustainable flow of value to their customers.

In the future, it is likely that organizations will use a mix of Agile, traditional project and product management methodologies to implement the changes they want. In addition, there is a need to apply Lean and System Thinking approaches to the work they do to provide a sustainable flow of value.

There is growing need for frameworks that combine and integrate practices, knowledge, and competencies from several industries and professions within the organization. The organization as a system of systems consist of functions contributing to a purpose, and it needs an effective collaboration of those functions so that the flow of value will be delivered in timely and sustainable way.

The Agile methods provide the team a high level of autonomy to define how they will solve the challenges they are facing. The goal is to have self-organized and cross-functional teams that can deliver customer value independently, whenever it is possible. Team members will collaborate to make sense of the problem that they need to solve at the same time that they are starting to master the subject. The more the team investigates and learns, the more probabilities for creating better innovations increase. For this, the team needs time and space to innovate. It is the organization's leaders' responsibility to provide those resources. At the end, everybody wins. First, the team motivation grows as they learn more about their purpose and understand the alignment of their work to the big picture, by using a system thinking approach. Second, the leaders have time to define, develop, and communicate their vision. Finally, the leaders and the team can define the objectives and key results contributing to organization's vision.

Those methodologies and frameworks have an impact on the practices the project or program manager uses and how he acts. How will Lean, Agile, and System thinking impact the future project or work of the program manager? A day of the project or program manager could be like this.

As the day starts, he will meditate for few minutes. He will exercise for 30 minutes to wake up his body and get the stamina needed for the day's learning opportunities coming ahead. After that, he will check his journal or calendar to review the planned course of the day, and remind himself of his yearly, monthly, and weekly objectives. He will think about how the day could flow, the meetings to participate in, the people to meet, and the expected results.

After arrival at work, he will meet the team for a daily stand-up (of one of the teams within the program) to discuss the progress of the work in progress, upcoming activities, and what challenges and impediments are hindering their progress. The meeting is partially virtual, as part of the team members are around the world. The team decided to have those meetings twice per week, since this provides them more value than having those meeting daily. Those meetings twice a week enable him to get in contact with the teams, to see the objective realization status

12 Material in this section was written by Juan Carlos Guzmán Monet, PMP®, IPMA B-level®, SAFe SPC, MSc. Eng.

first-hand, know the team's feelings and understand and recognize what possible actions are needed to help the team members reach their objectives.

After the meeting, he will stay to talk with some of the team leaders or members to discuss solutions to the impediments expressed by the teams and how he can contribute to those solutions. After the meeting, a good cup of coffee or tea in the canteen is always welcome; here, he might run into other members of the organization and chat about life, what is going on in the organization, and whether there are weak signals of events that might impact the program objectives or otherwise need his attention. When enough information has been gathered and the coffee cup is empty, the manager will go to his place to review the current period status in the program dashboard.

In the new program setup, the teams plan their work for a three-month period and define the scope of work they can deliver within this time frame. The approach for managing scope has changed from the traditional project management triangle approach, where the scope is guiding the time, it takes to deliver results, and the amount of money needed, into an opposite direction, where the scope is determined by the time we plan for, and the amount of people available, and their capacity to deliver value and realize benefits within the available timeframe.

The program progress is measured by objectives and key results within certain time frame. The program objectives and key results are based on the organization's objectives and its purpose to the society. The objectives and key results are hierarchical and from each level key results the below level objectives are derived. This enable us to see the objective alignment to the organization strategy and how we are going to implement it.

From the dashboard report, he can drill down from the overall company strategy to specific parts of the organization's strategic goals, even to the business initiatives for the program and finally, to the team levels. He can see the results from several dimensions: delivery of value (expected business value vs. actuals), achievement of results (planned vs. achieved), team happiness, number of improvements planned vs. implemented, scope changes (number of planned features vs actual delivered vs number of new features requested), and the learnings. The key feature in this case, is the amount of functionality or problems which are solved during specific period, and their contribution to the value expected from the customers. Other information presented in the report, is the quality related information like percentage of test cases automated, number of defects found in the solution, and their resolution time. In addition, he can see the program and team's risks, impediments, and dependencies that needs his attention.

During the dashboard review, he makes notes in his journal to see what actions are needed to resolve the challenges he can see from the data, but making decisions based only on those facts is not enough. Is important to perform a qualitative analysis of the data to make more accurate decisions. For that reason, he will go and see the team members and other stakeholders and ask questions about how the program is progressing, what new innovations they are working with and where they need the program manager support.

In the corridor, he meets one team leader and discusses the challenges his team is facing. The team leader's highlights some impediments and dependencies the teams have, one internal and one from the external vendor. During the discussion, they realize that it is better to go into a room to discuss the subject without interruption. They look in the corridor's map to immediately book an empty room to discuss possible solutions. Since they need more people to find alternative solutions, they call a few subject matter experts and ask them to join the meeting for their input and help. As many of them are very busy, they are lucky to get two people onboard for their discussion. They start drawing the architecture on the wall and start defining the problem they need to solve, not jumping directly into solutions but really finding the reasons why they are at this stage.

The program manager would rather listen, facilitate the discussions, and guide the team to find the solution themselves. They are the experts and know possible solutions, but the program manager's role is to lead the people to find the solution themselves, not find the solutions for them. The team describes several alternatives solutions and options they believe they could implement, test, and review the results, to decide which recommendation to provide.

As lunch time arrives, he will have a lunch meeting with one of the steering group members, who is interested to hear about how the program is progressing and how he can support the program to implement the changes within the permanent organization. They discuss the program context and how the ongoing changes might impact it, the business value the program is delivering, how much the organization is investing on this program and its importance to the overall organization's future. Understanding this information is crucial as it defines a plan of action to achieve the program's goals, as well as program's objectives and key results.

After lunch, it is time for a good walk around the building to clear his head and reflect on what has happened so far during the day and not forget the program and personal's long- and short-term objectives, and what roadmap the program is implementing. The walk brings new ideas and clarification of the objectives and where the program is.

One of the teams is organizing a demo, and he has been invited to see what they have done. The team receives feedback about the presented solution and suggestions for improvement. Then they run a retrospective of what the team has done in their planning period. The team documents what they have learned, what was not good and what they plan to do in the future to improve it, what was good, and they should continue doing. The information goes into the database where the other teams could see each other team's results and planned actions in the future.

After the meeting, he will continue walking around the building discussing with other people about the program and their challenges: management by walking around. When coming back to his place, he recognizes that it is time to prepare for the next period planning meeting and will go to the periodic meetings with the business and product managers to prioritize the work to be performed during the next planning period (a time box of three months) and review the solutions' roadmap the program is supporting to realize the expected benefits and business value.

In the meeting, they recognize the changes to the scope, in the form of changed priorities, and delivery order, but these changes do not impact the funding provided to the business initiative the program belongs to. So he relaxes for a while before going home.

Before closing the day or the week, he does some journaling where he writes about:

- What did I learn today/this week?
- What should I change?
- What should I continue doing?
- How did I contribute to the achievement of my goals?
- How did I contribute to our organization's goals?

After answering those reflections questions in his journal, he writes and plans the upcoming two weeks. In the plan, he specifies the objectives and key results he wants to achieve in the next two weeks. Those results are measured quantitatively and qualitatively.

In the future, most people will have smart homes with high-speed internet connections up to 1 Gbps, so before leaving the office, he checks an app for his smart refrigerator to see what food he should buy from the grocery store. He places an online order and receives notification about when he can pick up the order on his way home. His car GPS recommends which route to take to fetch the groceries quickly and avoid the traffic congestions. He feels happy because realizes that his work has a purpose: improving people's lives.

What is your purpose? How would your typical (working) day as a project manager look like in the future?

3.18 Excellence in Action: Solvo360[13]

The Future of Project Management in Emerging Countries

It is difficult to understand the behaviors of the present, without considering what happened in the past. The study of history is a discipline that allows us to know and interpret the events of the past to provide elements that enable us to understand the present and thus guide the generation of information, for decision-making and actions for the future. This is extremely useful to avoid making the same mistakes and to be able to mitigate the impact of change management, which is a real barrier. By knowing where we come from, we can better visualize where we are going and understand what is happening, because the present is the result of our past and the source to seek the best solutions to the multiple difficulties that these changes produce.

The events and difficulties that occur daily have an impact and shape the development of human endeavor, and knowledge does not escape this. This dynamic generates new situations, with new premises and conditions and impacts to overcome. In an exercise of analysis and forecasting of these new situations, we seek the best options and the continuous improvement. One must consider identifying and describing the events that occur, the proper interpretation of that reality, and the foundation of the conceptual frameworks that support it to determine the best solutions.

At the PMI Conference in London,[14] the evolution of project management was associated with the events that occurred in each historical moment, and this is important for our reflection on the future of project management. Next, a series of facts will be presented that show that relationship to which I refer.

During the 1960s, a period marked by the discussion to conquer space travel, where the time factor was essential to achieve success, the discussion focused on analyzing and managing schedules and all the instruments that define time control. In addition, it reinforced the principle that planning is a powerful management tool.

In the 1970s, we observed the emergence of subcultures such as the hippie, the strengthening of political ideologies around the world, the development of philosophical and political thoughts that emphasize the human being. This brought with it the emergence of teamwork not as a process of organized people in the development of production lines, already implemented since the beginning of the 20th century, but as the ability to complement the skills and abilities of each person who works in the field project and integrating and harmonizing to consider the competencies of the team in the planning and execution of the projects.

In the 1980s, it began to be observed that the planning line could be modified by non-planned components that affect the results. The environment, nature, and unforeseen considerations, can

13 Material in this section has been provided by Ofelia Pérez Figueroa, Sociologist, Specialist in Business Economy and Project Management.
14 SHENHAR, Aaron J.: "Project Management Evolution: Past History and Future Research Directions". PMI Conference in London, United Kingdom, July 2004.

occur at high speed, generating effects on the results to be achieved, without giving time to seek solutions. Therefore, it was necessary to deepen in risk management practices and management of uncertainties. This would allow us, using models and tools, to create a focus of attention and program actions that mitigate the unwanted effects or catalyze the expected results with new actions, not initially proposed.

During the 1990s, as individual and corporate needs grew and multiplied, some markets expanded. The dilemma occurred in understanding how to maximize the ability to create formulas to handle *simultaneity*, that is, in the face of this whirlwind of situations. The challenge is how to develop coherent execution strategies, managed efficiently and in the short term, of a large volume of projects, for which the management of Programs and Portfolios acquire vital importance. It is a priority to develop methodologies for their management. Organizations add tools that guide, align, and facilitate the execution of projects.

On the first two decades of the twentieth century knowledge flows and multiplies. The appearance of new concepts and technologies originated new ways for people to connect. That made the companies adapt new strategies that could be effective in the globalization era. We saw how the big geographical distances get significantly reduced as time passed and the impact of the political decisions made themselves feel on every part of the world. Who would have said that political decisions in Venezuela generated such a migratory impact on all of the Americas?

In the year 2020, an iconic moment occurred that will last at least 10 years. The Covid-19 pandemic and quarantine appeared to help overcome its consequences. Quarantine became a catalyst of unthinkable proposals that decades ago would not be possible, the impacted it had on the human beings, its gregarious spirit, and its necessities to communicate and elicit emotion. This, along with the technological advancement, has placed "virtual" at the top of its usage in every field, and project management doesn´t escape this situation. In almost all companies, projects are a necessity and well-known project management techniques and tools continue to be used because they are recognized to provide benefits. But understanding how to be efficient in the face of this new expression makes us recognize that we are facing a change in the culture of humanity. Therefore, plurality must be present, and this will force maintaining the validity of the principles of adaptation and strategy, but now tolerance for diversity is incorporated.

The facts must always be evaluated since there is no single magical solution to a problem. It is the circumstances and the understanding of them which allows to overcome the existing barriers even in the implementation of more effective and efficient management models related to the degree of maturity of the organizational environments. Therefore, the future requires the existence of adequate pillars, becoming premises to guide actions in the uncertain, changing, and challenging future. This book describes very well these pillars that are factors of success to carry out the management of projects in the future.

The success of the solutions will largely depend on the incorporation of these factors. But countries and regions have different degrees of maturity in managing social and organizational behaviors. These countries can be classified as developed and emerging. In this case, we propose the evaluation of three socioeconomic indicators to classify them. These are:

1) *Gross domestic product (GDP)* at purchasing power parity (PPP) values per capita, which is the sum of all final goods and services produced by a country in a year, divided by the estimated population. It is measured in MM US $ and allows us to recognize the growth capacity of a country, reflected in greater consumption, greater purchasing power, greater employment, and greater capacity to formalize the economy of a country. The higher this indicator is, the more opportunity for growth is presented.

2) *Doing business,* whose results can range from 0 to 100, the higher the better the performance of the economy in a country. It offers a quantitative measurement of the regulations on opening a company, managing construction permits, property registration, obtaining credit, protecting investors, paying taxes, international trade, and other activities for the development of enterprises and companies. This will, in turn, generate institutional credibility, sustainable business development and growth in regions and countries.
3) *Human development* implies placing people at the center to achieve their well-being, to evolve their potential with greater opportunities and to live a life that they value. Human development is a concept that includes elements such as freedom, health, and education to promote the growth of the country. The result is between 0 and 1, the higher the result the better.

In the American continent, one can observe a great geographic dispersion and population volume. There are also geographical, ethnic, and cultural differences. The first indicator analyzed is the one that refers to the GDP at values of purchasing power parity (PPP) per capita.[15] The United States is the one with the highest value in the region and the 5th in the world, followed by Brazil and Argentina in the 8th and 28th in the world, respectively. The worst is Guyana, which is in position 158 in the world. Doing business[16] presents the United States with the ranking of the first country in the region and 8th in the world, followed by Chile with the 56th place in the world and the first in LATAM. The worst indicator is presented by Venezuela with position 188. No LATAM economy is among the top 50 stocks in the world, keeping them lagging in the world economic context. The best human development indicator[17] is presented by Canada, which has position 16 and the United States, position 17, while in LATAM it is presented by Chile with position 43 and Argentina with position 46. The worst position in the region is presented by Haiti, with 170th place. So, there is still a long way to go for LATAM. In emerging countries, the principle of sustainability is difficult to achieve.

By analyzing these three indicators, we can conclude that the US and Canada are developed countries, and these indicators have been sustainable over time, compared to the rest of the countries, which we will group as LATAM and which we classify as emerging countries. This last group is extensive and presents marked differences, but the countries coincide in the limited capacity to maintain trends in these three indicators, together with the instability of the institutions and the rules and regulations, which makes their development not sustainable over time. In the last five years these indicators have increased and decreased dramatically, thus allowing us to conclude that all LATAM countries are part of the emerging countries category.

On the other hand, while the United States and Canada have a population of more than 328.2 million[18] inhabitants, LATAM handles almost double the population of 630 million[19] inhabitants, so the well-being of its inhabitants is not sustainable, either. The development situation of projects in Latin America has suffered the same fate. Although several interesting initiatives have been developed, their vulnerability has been highlighted. An example of this was in Venezuela,

15 International Monetary Fund (IMF) (2018). "Report for Selected Countries and Subjects". Retrieved February 21, 2019.
16 World Bank: "The Doing Business 2020 Report", 17th edition, 2021.
17 United Nations Development Program (UNDP) "Human Development Index (HDI)". www.hdr.undp.org. Retrieved December 11, 2019.
18 U.S. Census Bureau. "2020 Census Apportionment Results." April 26, 2021. Accessed May 3, 2021.
19 CEPAL: "Updating of population estimates and projections for the countries of Latin America and the Caribbean." www.cepal.org. Retrieved April 22, 2020.

where we will talk about the case of Petroleum de Venezuela (PDVSA), which was on a straightforward path of achievements and, which after 20 years, today appears as an almost bankrupt company, with its corresponding impact on the country since it was their main source of income.

PDVSA was created on August 30, 1975, by government decree absorbing the operations and assets of the companies belonging to the Venezuelan Petroleum Corporation, and the functions of planning, coordination, and supervision of all oil operations in the country, under the tutelage of the Ministry of Energy and Mines, and under the legal framework established in the Organic Law of Hydrocarbons. In the beginning, it mainly carried out the activities of the headquarters of the nationalized operating companies. These subsidiaries were later joined by PEQUIVEN, who took over the petrochemical industry. CARBOZULIA exploited the coal beds in Guasare. BITOR (Bitumen del Orinoco), produced, transported, and commercialized the ORIMULSION. At the end of 1982, PDVSA began the Internationalization Program that lasted until 1998, the main objective being the acquisition of assets outside the country with its own resources or through partnerships with transnational companies. PDVSA was formed according to the following premises:

- The management team and operational performance of the corporation is based on the career management of the personnel and the growth in the company is done in terms of merits.
- Your staff are remunerated with competitive salaries, thus allowing the development of a career path.
- It is structured as "holding" and "political" appointments that are only made up to the board of directors of the parent company. The subsidiaries were corporations, with a board of directors elected based on merit, and which was accountable to the assembly of the parent company.
- PDVSA and its affiliated companies adopted a different legal form from the other existing state companies in the country. They were created as public companies under private law. Corporations (SA) with a sole shareholder, the Venezuelan state, represented by the government at shareholders' meetings.
- PDVSA's budget and financial management did not follow the approval process that of the rest of the public sector. This is carried out in business terms and oil policy under the approval of the government in the shareholders' meeting, following the rules of the Commercial Code.
- It was based on the existence of a set of laws that allowed a high degree of managerial and operational autonomy. As oil corporations, they would be evaluated on their own business and commercial merits and given leeway in these areas. This was part of the national and global success until 1999.

In 1975,[20] the Venezuelan oil industry produced 2,346 MBD (thousands of barrels per day), below its peak production of 3,708 MBD in 1970, and proven crude oil reserves were of the order of 18,390 MMB (million barrels).

In 1998,[21] Venezuelan production was 3,279 MBD of crude oil, 170 MBD of LPG (liquid petroleum gas) and 3,965 MMCFD (million cubic feet per day) of natural gas, which is equivalent to a total production of 4,133 MBPD of oil equivalent, with reserves of 76,108 MMB of crude oil and

---

20 ESPINASA, Ramón. "The rise and collapse of PDVSA thirty years after nationalization." *Venezuelan Journal of Economics and Social Sciences* 12 (1) Caracas Apr. 2006.

21 PACHECO, Luis A. "PDVSA 1998: Before the Storm." *PRODAVINCI* magazine in the Perspectives section. October 2018

146,573 BCF of natural gas. The country had the largest proven reserves in the world. In crude oil refining, 3,096 MBPD was reached, of which 1,620 MBD were in Venezuela (including the Isla Refinery in Curacao), 1,222 MBD in the United States, and 252 MBD in Europe.

All this evolution required technology and well-trained personnel. The technology was the responsibility of the subsidiary INTEVEP, the research laboratory that registered more than 300 patents, had more than 200 researchers with master's degrees and 160 with doctoral degrees. The education was carried out mainly by the CIED subsidiary, the corporate university created in 1995, which trained 2,000 people per day on average in its 12 headquarters nationwide, in face-to-face format and which made progress in development of training in e-learning. It included basic technical training up to advanced executive education.

Within PDVSA, investments in projects were quite important and it was necessary to be highly efficient in managing them. Therefore, it was necessary to follow strategies that would allow the alignment, management, and commitment of the personnel in its execution. The portfolio at the time had the potential for a great impact due to its magnitude, size, and nature. There were projects such as the expansion of the Paraguaná Refining Center to make it the largest refinery in the world, the creation of the José Antonio Anzoátegui Industrial, Petroleum, and Petrochemical Complex to promote the development of petrochemicals in the eastern part of the country, the development of the Cristóbal Colon Project destined to the exploitation of gas fields, and the proliferation of production plants in various areas of the country. For all of this, actions were needed, such as the following:

- Create the Capital Investment Guidelines (GGPIC), which became a reference for operational execution in the management of projects throughout their life cycle. Considered a best practice.
- Reorganize the project personnel, creating the Corporate Project Management, under the Engineering, Projects, and Maintenance Directorate, with four departments; the Project Definition and Development Management, where the project leaders were located who determined the structuring of each project and were integrated into the work teams from the early stages. The Planning and Control of Project Management that defined the plans and carried out the monitoring and control of them. The Contracts Management, which oversaw preparing the specifications and coordinating the contracts, since the execution of the projects was carried out by companies other than those firms in the local market. The Project Development Management organization oversaw investigating applicable practices to achieve continuous improvement and identify and develop the skills required for the staff.
- Establish a relationship framework with those involved by promoting the creation of the Venezuelan Chapter of the Project Management Institute (PMI-V), which in 1994, became the first Latin American Chapter.

For the year 1996, it was an obvious need to train, under the premises of resource optimization and management effectiveness, more than a thousand people who work in projects within PDVSA, subsequently personnel from the companies that provide service to PDVSA and finally to PDVSA itself and the general public. My objective as the Corporate Coordinator of Project Management training was to develop the skills of employees, especially leaders, executives, and work teams. The challenge was to Identify the areas of knowledge by role, to determine mastery levels and transmit the required competencies. Thus, the technical components and methodological skills were developed, but also the attitudes, behaviors, and knowledge of the business, within a framework of values and ethical guidelines.

It was my responsibility to lead the team that designed the strategy implemented in this area, which was based on:

- The development of training programs that was carried out through courses and workshops for the training of personnel who presented specific weaknesses in the management of concepts and tools according to the training requirements of the audiences. In 2002[22] alone, 56 workshops and project management courses were held, training 1,120 people.
- The professionalization of project leaders through the certification programs such as Project Management Professional of PMI, more than 12 preparation courses for certification were conducted, training approximately 240 people. The necessary steps were taken to carry out the test in CIED facilities in Caracas. In 2002[23] more than 100 people were certified as PMPs.
- The "Postgraduate Project Management" was designed at the level of specialization and master's degree in an agreement with the Andrés Bello Catholic University (UCAB) and with the permission of the Project Management Institute (PMI), made up of professors and trainers from both institutions. By 2002[24] there were 21 cohorts with an average participation of 20 people per group. For this reason, more than 880 people carried out fourth level studies in Project Management in Venezuela at the national level.

Training became part of the corporate culture. Thus, employees were perceived as the main asset, as individuals and as a work team, promoting their power of competitiveness and ability to share information, experiences, and individual and collective knowledge. Ultimately, the strengthening of the staff's competencies was sought and their alignment with the established strategic objectives was ensured. Adding value through the systematic offer of learning maps for the development of professional action and management in the planning and development of teaching-learning processes. All this was fostered by the corporate guideline that each employee should attend 80 hours of training per year.

In 1999, a new president of the Republic wins the elections, bringing with him prejudiced opinions about the oil industry, as well as advisers made up of historical opponents of PDVSA. This created an atmosphere of mistrust and in 2002, tensions exploded in the context of a generalized social and political crisis. From this point forward, PDVSA would no longer recover. Among the most outstanding facts were:

- The Venezuelan government seeks a new strategy to free itself from dependence on the US market and the dollar in its economy.
- Dismissal of more than 2000 skilled workers in 2003
- Disappearance of CIED in 2007
- In 2007, it was decreed the conversion of all the concessions in the Belt to a mixed company, where the State, through PDVSA, would become the majority shareholding.
- Dramatic decrease in the country's oil production, even though between 2004 and 2008 oil prices increased, reaching $140 a barrel.
- The management of PDVSA was now in the hands of unprepared military personnel and unqualified and unprepared operational personnel.

---

22 PÉREZ, Ofelia: Data issued by CIED systems for the preparation of the 2002 CIED Management Report.
23 Idem
24 Idem

- Lack of investment in exploration and maintenance of refineries
- The economic sanctions of the North American government against the PDVSA Company aimed at ending the company's corruption.

The results of this management, according to the information issued by OPEC[25] for the year 2019, reported that the national oil production fell to 680,000 B / day and the average production in February 2021 was 521 thousand barrels per day. In the refining area, gasoline and diesel are imported for internal consumption. As a result, it can be concluded that state-owned companies are prone to failure, either due to inefficiency or as victims of political interference, and PDVSA was no exception.

The successful future of project management in emerging countries is only possible if the strengthening of staff skills is achieved, and this is closely linked to Knowledge Management and Change Management. The competencies model and its strengthening emphasize developing training mechanisms associated with knowledge and know-how. This is within the context of knowledge management and its direct effect on the management of appropriate SKILLS for operational and managerial positions both in the technical and business management. Change management will strengthen the competencies of knowing how to be and knowing how to manage the required ATTITUDES and facilitate decision-making within appropriate contexts.

In conclusion, today predicting the future without looking like a utopia or an expression of science fiction is sometimes difficult to understand for many executives in these emerging countries. This means breaking paradigms, creating an inclusive and tolerant culture, more openness, and without forgetting an ethical framework in which to act. The general principle is growth, seeking the greatest well-being, and without damaging the environment.

The future will seem uncertain and turbulent, but that does not mean that it is challenging and attractive. We all have no choice but to go through it. The complexity of the realities, the breadth of knowledge and putting a group of people often with contradictory principles to act as a team, puts us in the dilemma of understanding that there is no single and absolute solution, but the most suitable solution. As we see opportunities to apply concepts like those outlined in the pillars discussed in this book, organizations will be able to move towards excellence. For this to happen, companies must hire managers and leaders capable of managing the ability to guide work teams on the best path, with firm and effective solutions and establish lines of innovation, creativity, and development for the management of enterprises. Being essential in emerging countries, we must look toward the future with more precise steps. And if you cannot do it, give yourself the right to be wrong and learn your lessons to do better next time.

3.19 Excellence in Action: Texas Instruments[26]

When companies recognize the need for continuous innovation for new or strategic projects, they are often at a loss as to where to begin. Some companies tend to focus on technology, tools, processes, and organizational restructuring. Texas Instruments quickly identified that by starting with people, effective project management leadership, and an organizational culture that supports strategic projects can accelerate the process.

---

25 OPEC: "Report on the oil production of the oil-exporting countries", March 2021

26 Part of this section has been adapted from Harold Kerzner, *Advanced Project Management: Best Practices on Implementation*, 2nd edition (Hoboken, NJ: John Wiley, 2004),46-48.

Background

Most people seem to believe that the future of project management begins with the development of a new or revised project management methodology. While this may be true in some circumstances, companies that appear to be excellent in project management realize that people execute methodologies and that the best practices in innovation and traditional project management might be achieved quicker if the focus initially is on the people rather than the tools. Therefore, the focus should be on the culture.

One way to become good at project management is to develop a success pyramid. Every company has their own approach as to what should be included in a success pyramid. Texas Instruments recognized the importance of focusing on people as a way to accelerate innovation project success. Texas Instruments developed a success pyramid for managing global innovation projects. The success pyramid is shown in Figure 3-10.

A spokesperson at Texas Instruments describes the development and use of the success pyramid for managing global projects at Texas Instruments:

> By the late 1990s, the business organization for sensors and controls had migrated from localized teams to global teams. I was responsible for managing 5–6 project managers who were in turn managing global teams for NPD (new product development). These teams typically consisted of 6–12 members from North America, Europe, and Asia. Although we were operating in a global business environment, there were many new and unique difficulties that the teams faced. We developed the success pyramid to help these project managers in this task.

Figure 3-10 The Success Pyramid

Although the message in the pyramid is quite simple, the use of this tool can be very powerful. It is based on the principle of building a pyramid from the bottom to the top. The bottom layer of building blocks is the *foundation* and is called "understanding and trust." The message here is that for a global team to function well, there must be a common bond. The team members must have trust in one another, and it is up to the project manager to make sure that this bond is established. Within the building blocks at this level, we provided additional details and examples to help the project managers. It is common that some team members may not have ever met prior to the beginning of a project, so this task of building trust is definitely a challenge.

The second level is called "sanctioned direction." This level includes the team charter and mission as well as the formal goals and objectives. Since these are virtual teams that often have little direct face time, the message at this level is for the project manager to secure the approval and support from all the regional managers involved in the project. This step is crucial in avoiding conflicts of priorities from team members at distant locations.

The third level of the pyramid is called "accountability." This level emphasizes the importance of including the values and beliefs from all team members. On global teams, there can be quite a lot of variation in this area. By allowing a voice from all team members, not only can project planning be more complete but also everyone can directly buy into the plan. Project managers using a method of distributed leadership in this phase usually do very well. The secret is to get people to transition from attitude of obligation to a willingness of accepting responsibility.

The next level, called "logistics," is where the team lives for the duration of the project and conducts the day-to-day work. This level includes all of the daily, weekly, and monthly communications and is based on an agreement of the type of development process that will be followed. At Texas Instruments, we have a formal process for NPD projects, and this is usually used for this type of project. The power of the pyramid is that this level of detailed work can go very smoothly, provided there is a solid foundation below it.

Following the execution of the lower levels in the pyramid, we can expect to get good "results," as shown in the fifth level. This is driven in the two areas of internal and external customers. Internal customers may include management or may include business center sites that have financial ownership of the overall project.

Finally, the top level of the pyramid shows the overall goal and is labeled "team success." Our experience has shown that a global team that is successful on a one- to two-year project is often elevated to a higher level of confidence and capability. This success breeds added enthusiasm and positions the team members for bigger and more challenging assignments. The ability of managers to tap into this higher level of capability provides competitive advantage and leverages our ability to achieve success.

Conclusion

At Texas Instruments, the emphasis on culture not only benefited their innovation and strategic initiatives but also resulted in best practices that supported other initiatives. Texas Instruments focused on people rather than tools and processes. This may very well be the starting point for other companies that recognize the benefits achievable from cultural considerations, social leadership training, and other people-oriented skills. It is unfortunate that more companies do not realize the importance of this approach.

4

Pillar #4: Digitalization Is Central to Delivering Projects' Promises

4.0 Background

Digitalization is here to stay. Whether we have been concerned about artificial intelligence (AI) and whether this current wave will dominate the next decade and beyond, or we are energized by its promise, the disrupting effects on the organizations' business models and ways of working have been witnessed in the last few years and the next decade will encounter a much higher acceleration rate given the maturity potential of digitalization. Some believe that the AI effects are similar to the invention of electricity. How much autonomous working and life interactions would this create is amongst many of the open questions on the table.

The pandemic and related disruptions have closed the gap on many of the delays in digitally transforming organizations. Digitalization was finally the obvious solution. Many of today's project managers are dealing with decisions around the enhanced use of digitalization that could lead the way to a much higher level of accuracy in the achievement of projects' future promises. For the project manager, classic topics of prioritization will center around the flow of information, data analytics, cybersecurity, impact on product development speed, and the multitude of impacts in managing across the lines of a typically matrixed set of organizational boundaries.

The achievement of future projects' promises will directly contribute to organizational excellence. These promises will entail a closer link to strategic objectives of organizations and being consistently outcomes based and not stay tangled in being outputs based. The final shift from the high focus on controlled ways of project delivery coupled with extreme focus on value will be enabled by expanded digitalization. This promise will scale in potential when projects play a stronger role in connecting the business to technology. This is a great opportunity for a multitude of positive behavioral project shifts and for freeing time for the project managers to think again for a change and operate as true entrepreneurs.

According to PMI Pulse of the Profession study,[1] Beyond Agility, conducted with 3950 project professionals, there have been a multiple number of patterns and changes related to digitalization and its impacts on tomorrow's organizations. As an example of key points addressed in that study, is the impact of the pandemic on the pace of digitalization. The study shows Digital Transformation as the biggest change organizations witnessed during the year following the pandemic (68 percent of the surveyed professionals). This was followed by Business Strategy and Organizational adaptability at 64 percent and 61 percent, respectively. There is no doubt that this has cascaded and has

1 Pulse of the Profession® 2021 Study by the Project Management Institute (PMI). All rights reserved.

directly correlated to affecting the way of working of the various project teams within these organizations.

Another interesting area addressed in the study is the growing adaptability in how organizations work which confirms some of the patterns addressed regarding flexible ways of delivery. The term *gymnastic organization* was used in the study to reflect the openness associated with making project delivery methods choices and not be limited to sticking to one format of delivery or even a few. This added flexibility will likely dominate the future organizational excellence. The study also shows a wider adoption of digitalization such as in cloud solutions, the Internet of Things, and AI.

What is critical to the next generation project managers is going to depend greatly on the degree of integration of these varied technologies. Digital is still, to a large extent, untapped and that is changing rapidly in the next decade when coupled with the many specific revolutions that are taking place across a few of the growing industry sectors.

The next section highlights an excellence example by ASGC, sheds the light on many of the digitalization principles, and show cases their applications in the construction industry. The points made by the ASGC example challenge some of the prevailing views about digital, address culture elements, and begin the dialogue around the digital talent revolution ahead.

4.1 Excellence in Action: ASGC[2]

Derailers of Digital Transformation in the Construction Industry

The world is witnessing a historical extraordinary industrial disruption, a revolution that entails a transformation of humankind. Unlike any predecessor with a linear pace of expansion, Industry 4.0 is spreading with exponential velocity. It is renovating all traditional strategies and aiming to transform business models across countries, governments, companies, industries, and societies as a whole, distinct from other sectors. The construction industry has struggled to take advantage of technology in a significant way.

The construction industry is facing unprecedented transformation since the last decade; The industry is always known as slow, fragmented, labor intensive, late in technology adoption and low-profit margin; Nevertheless, with the development of different sectors across the globe.

The improvement became a mandate, especially with the complexity, constricted budgets and squeezed delivery time frame; rethinking the construction projects delivery became vital to plan, operate, deliver, and measure success.

Digital transformations provide industries with unparallel opportunities for value creation; understanding the industry 4.0 concept in construction terms would mean adopting intelligent technologies to transform construction processes and projects delivery to be more efficient, effective, and agile.

It will embrace the automation of construction delivery processes, ultimately fostering technological advancements to increase productivity, efficiency, and value (King 2017); creating a new profit pool by improving customer experience and expectations will drive business success in the

2 Material in this section has been provided by Maged Elhawary, Ph.D., PMP®, EVP,MRICS, Group Chief Information Officer at ASGC. He has held managerial positions in key international companies and has managed major projects in Residential & Commercial, Oil & Gas, and Industrial sectors. Maged held multiple leading roles with AACEI in the UAE.

industry. However, digital transformation is not well understood yet by the different stakeholders in the construction and built environment.

Xu et al. 2018; summarized the different waves of the industrial revolution in four major stages. The first one is the age of steam, where water power was used to operate machines; mass production started with the discovery of electricity, which lead to the second wave of the industrial revolution.

The existence of computers and automation was the base for the third Industrial Revolution; lately, with the development of internet speed and IoT concepts, the fourth wave of industrial revolution elaborated with the idea of cyber-physical systems (CPS), which is the near future for most of the industries including construction.

The three terminologies of **digitization, Digitalization** (capital D), and **transformation** are used when companies are willing to change the delivery method and depend more on technologies; clearing the doubt and defining the difference between those terminologies are essential (Figure 4-1).

With a lowercase D, digitization concerns transforming information from any format to another easily readable and understandable by computers to allow machines to deal with the data management, process, and analysis; a clear example is converting handwritten subject to a digital format.

On the other hand, Digitalization uses digital technologies to improve the delivery and change the business model while adding new values and opportunities to the process and operations through analogy to a digital system that implements digitization through automation to increase transparency and accuracy.

Nevertheless, digitization and Digitalization are correlated terminologies with a close meaning. Digital transformation is an environment that implements the digitalization concept to change the delivery method. Companies might digitize the information, digitalize the process, and then use IoT applications to transform information management and communication.

Figure 4-1 describes the primary differences of the four used terminologies of the complete digital transformation journey. As an example for capturing and recording labor productivity from the preliminary stage of rerecording by handwriting (Analog) then the using spreadsheets to record (Digitization) until the migration of using electronic document management system (EDMS), which is an advanced step toward (Digitalization) and finally using IoT & Radio Frequency Identification (RFID) technologies with its link to Enterprise Resource Planning (ERP) and human capital management (HCM) different systems to enable complete digital transformation journey.

Digital Transformation is always promising to improve the performance of different industries while many companies could fail for various reasons; the below discussion considers some derailers of transformation initiatives.

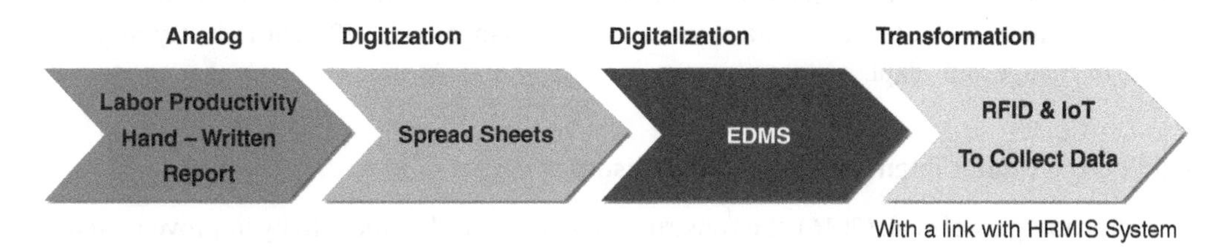

Figure 4-1 Digitalization Flow (Author's development)

Fuzzy Definitions

Lacking a clear definition of digital transformation, companies fail to connect digital strategy to their business growth, leaving them drifting in the fast-churning waters of digital adoption and change.

Lessons learned from telecommunications development and smartphones over the past decade should only push construction companies as no industry will be isolated from the game and competition change. Defining the meaning of digital transformation, some superficial views as the upgraded terminology for what their IT professionals used to produce. Others focus on digital marketing, software, or sales, but very few have a holistic vision of digital transformation.

Confusion of the Economics of Digital Transformation

Many professionals of different industries know a set of basic financial principles which learned and adopted many years ago and realized the results of its implementation; this built perception might contradict the fact of the new era and economics of the digital economy, such as Digitalization ends economic rent

Economy principles such as heavy assets are better than rent, and spending in CAPEX has better financial results than OPEX are one of the early lessons industry professionals learned in economics study. Digital is puzzling the best-laid tactics to capture surplus to add more value to the customer and end-user and changing the method of calculating the return on investment.

Over Indexing on the Usual Threats

Most corporates are concerned about the burdens of moving into the digital world because of the nature of the digital environment, unusual suspicious actions, data storage and security. Adopting cutting-edge technology that isn't yet mature and casting the concentration only on the unusual suspects is a critical concern that might affect the digital transformation of companies.

Different industries have sensitive data, but digital transformation never stops as cybersecurity and data management are highly developed to maintain trust and confidentiality.

Missing the Duality of Digital

One of the shared responses to digital threats when discussing with construction industry professionals is the following: "If disrupted, then I need to create something completely new." However, this is not a correct argument and approach; thinking of a fundamental two-by-two matrix to understand the challenge depends on two main factors, degree of change and pace of change, affecting different stakeholders in all industries.

The pace and degree of evolution differ from one sector to another as significant and rapid change puts slow payers at risk in business continuity plans such as telecommunications, marketing, and financial institutes. In contrast, low and slow changes give different industry players a chance to change and adapt.

Building the Digital Talent and Required Skillsets

According to Agarwal et al. (2016), the construction industry lacks productivity improvement due to the lack of talent development and shortage of training as the industry struggles to acquire and maintain the right talent and skillset.

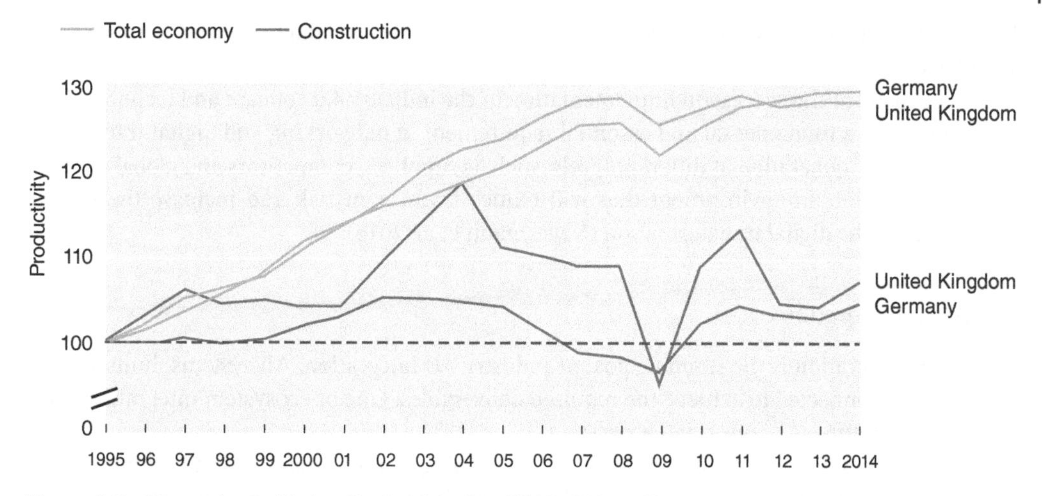

Figure 4-2 Comparison of Labor Productivity Bet (1995–2014). *Source:* Agarwal et al. 2016.

This issue is critical when discussing the industry 4.0 concept and technologies as the new skill set is mandated to adopt and implement digital culture. Technology awareness and high training requirements were ranked as the significant challenges of digital transformation strategies related to individuals and upgraded skillset.

Agarwal et al. (2016) compared performance and productivity of the construction industry with total economic performance in the period between years (1995, 2014) in the UK and Germany as mentioned in Figure 4-2.

It shows the seasonality of construction productivity performance over the years. It is lower than the average economic performance all the time, which they related the vast majority of inadequate performance to the absence of training and talent gap.

Cultural Resistance

Culture is one of the biggest challenges to change. People stick to what they used to do without having the desire to leave their comfort zone. Fear of change, the inability to deal with the new technology, and the fear of losing control are the significant drawbacks of the transformation in different industries.

Cybersecurity Concerns

According to Kusiak (2018), with unprecedented technological advancement, information and data security became the primary concern to everyone as the industry 4.0 concept is primarily about technology, connectivity, duality and data exchange in cyber-physical systems.

Securing data exchange, cyber risk, and safe storage are critical business risks; the vulnerability of data caused by the implementation of IoT technology is one of the new risks facing digital transformation. Adopting industry 4.0 technologies includes cybersecurity investment due to data security, an essential and critical rule in business safety and continuation plans.

Collaboration Environment

Due to the broad scope of digital transformation in the construction industry and its nature, full cooperation and integration between different stakeholders, supply chain and involved parties in different project life cycles, including governmental entities, is essential to succeed in transformation plans.

The backbone of the successful implementation of the industry 4.0 concept and technologies is collaboration as a fundamental and essential requirement in networking and digital transformation initiatives. Cooperation at different levels, such as suppliers, competitors and global corporations, will develop an environment that will reduce investment risk and increase the success possibilities of the digital transformation (Kagermann et al. 2016).

Ecosystem Integration

Ecosystem integration is the ultimate goal of industry 4.0 integration. All systems, humans, and machines are connected to achieve the required deliverables. One of ecosystem integration is the factory of the future as a connected ecosystem.

The multiple dimensional connectivity through mobility, supply chain, smart buildings under the umbrella of intelligent government will drive productivity and optimize the resources in a lean environment.

Similarly, in the construction ecosystem, integration is essential, especially with the increasing complexity of construction projects with time and cost constraints which requires a high level of control and integration.

Inadequate Identification of Corporates Strengths and Gaps

Many corporates fail to identify their skill set and analyze their strengths and gaps in people, processes, and technology which are considered one of the biggest pitfalls of change and improvement.

Analyzing core capabilities and potential improvement areas in different business deliverables and operations dimensions is essential to building a solid foundation for adopting technology and exploring future opportunities.

Investment Competitiveness and Low-Profit Margins

Agarwal et al. (2016) mentioned that the construction industry lacks other sectors in the allocated fund for R&D as it hardly reaches 1 percent compared to 4.5 percent in aerospace and 3.5 percent in automobiles, compared to telecommunications and pharmaceutics figures are in-comparable.

Lower investment in the construction industry R&D is a massive challenge for change and improvement, especially with low profitability margins. The positive side of such a dilemma is that industry could benefit from the borderless nature of different sectors and adopt ready-made technologies from others in addition to its initiatives of R&D approaches.

Focusing on Outputs Rather Than Business Outcomes

Casting the concentration of technology output only and missing the link with the business outcome is one of the major critical factors in adaptation failure. Technology link with business outcomes is crucial in deciding and selecting new initiatives and updating business processes.

Tracking and articulating business and customer targets are essential to kick-off with developments plans as accountability of technology outcomes should be shared between technology leaders and different business process owners (BPO).

Configuration/Change Management

Change management is always a critical task in all organizations in different industries, especially in construction, due to multiple reasons that vary between cultural changes, people resistance, and silos nature.

Ignoring the change management process is a considerable challenge that puts business continuation plans at risk. Innovation culture is a mandate before adopting digital transformation strategies, which requires management, processes, and people support to achieve the necessary objectives.

The three pillars of any transformation change are technology, processes, and people. People resistance will remain a critical factor in the transformation as they may feel insecure to adopt new methodologies or find it challenging to change the way they used to do.

Scalability and Strategic Alignment

Digital transformation projects and technology advancement projects will gain management support when stockholders can evaluate its value and impact on different business outcomes.

One of the core outcomes is the scalability of new proposals, which can be used on a wide-scale and gain the opportunities offers by economies of scale concept, not just a kind of ad-hoc implementation or single project with limited business outcomes.

Alignment between IT plans and corporate strategic objectives are essential for success; missing the connection between IT development plans and corporate strategic goals is critical for technology implementation objectives.

Such missed links could present overpromising plans for massive changes while ignoring the incremental improvement and organic expansion of ongoing projects and current achievements of recent developments.

Depending solely on new projects and ignoring capitalizing on current strength might result in losing management and stakeholders support due to the financial impact of new plans and business process owners' interest in participating.

Table 4-1 summarizes key challenges toward digital transformation and recommends the required actions toward successful implementation.

Table 4-1 Digital transformation challenges and solutions

	Key Challenge	Solutions
1	Ecosystem integration	-Digital strategy alignment -Transformation architecture -Applications and software usability -Availability of data lakes -Integration of services and technical support
2	Building Digital Talent (levels)	-C-Suite buy-in -Proper selection of change agents -Experts' involvement -Digital awareness -Proper talent acquisition

Table 4.1 (Continued)

	Key Challenge	**Solutions**
3	Standardization	-Standards, procedures, and processes -Innovation panel and auditors
4	Alignment between IT and Business strategy	-Corporate strategic objectives -IT strategy
5	Digital Environment	-Digital business plans -Change management plans -Upskilling individuals' talents -Top management support
6	Business added value	-Benefits realizations -Added values -Project management principles -RODI
7	Digital Transformation Continuation	-Innovation as a culture -Considerations of the technology hype cycle

In conclusion, digital transformation is an exciting journey that requires different corporates to be equipped with the adequate skillset and sufficient resources with clear criteria for accepting new initiatives and incorporating IT objectives with corporate strategic goals.

4.2 Digitalization and Projects Framework

A possible framework to address the correlation of digitalization to the achievement of future project's promises is shown in Figure 4-3. This framework highlights the six critical elements that will likely affect the ways of working of the next gen project managers the most. Naturally, there are multiple overlaps amongst these six elements of experimenting capacity, context drive planning, co-creation, virtual teams' engagement, outcomes-focused work, and the ever-changing ways of working.

As highlighted in the Insights Series of the PM World Journal,[3] opportunities ahead for digitalization are massive and the many examples highlighted in this chapter confirm that its application to the world of projects and programs are immense. Freeing the program and project teams to focus on creativity, true collaboration, and integrating amongst learning themes will be a much better use of people's time. Integrated enablement of digitalization takes away many of the stresses associated with poor estimates, wrong assumptions, communication deficiencies, amongst other classical obstacles toward an improved future way of working.

3 Insights Series, Author: Al Zeitoun (2021, August), *PM World Journal* (ISSN: 2330-4480) Vol. XI, Issue VIII, www.pmworldjournal.com

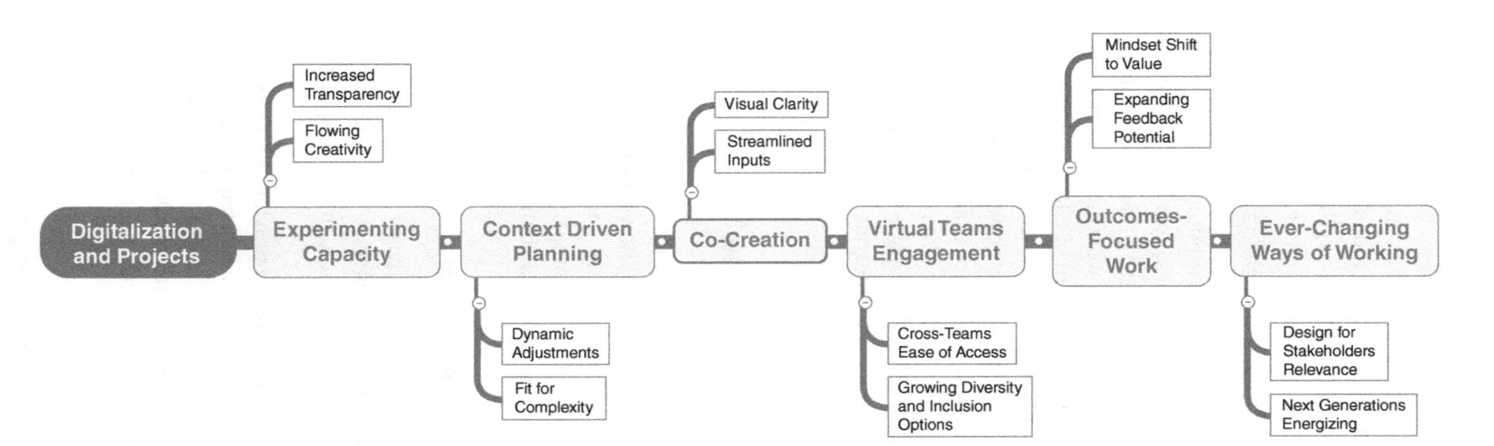

Figure 4-3 Digitalization and Projects Framework

4.3 Experimenting Capacity

One of the growing advantages of digitalization in projects is the ability to experiment. Projects are major investments for organizations and thus any of the savings we could create with experimenting, are greatly appreciated by the originations' leaders. Let us take the digital twin as an example and how the ability to create a physical replica of a product or a solution to gather closer insights and debates early on in the investment cycle and ensure a much higher closeness to the anticipated strategic value of these investment initiatives.

As shown in Figure 4-3, the two sub-elements picked under this first of the six critical elements are *increased transparency and flowing creativity*. Future projects' promises and associated organizational excellence are closely aligned with these two. These two sub-elements create a different organizational model that provides safety for the project managers to come up with the most fitting solutions to tomorrow's challenges, while working comfortably and closely with their teams. When transparency increases, given the amount of effective data we could gain from digital technologies, and the organization culture gives the open room for creative dialogues and critical discussions, as possibly facilitated by a PMO, many growth metrics are typically achieved.

In the next section, the ServiceNow excellence example highlights the impacts of automation and quality of data in driving effective decisions. The example also clearly describes the changing role of the PMOs as digitalization creates the space for stronger cross-functional collaboration, contributes to a culture of transparency, and integrates data in the best possible way to drive better understanding of business trends and their impact on organizational outcomes.

4.4 Excellence in Action: ServiceNow[4]

Digitization is the process of automating business processes – the circulatory system of any enterprise – and turning them into digital workflows. Digitizing the manual, repetitive parts of these processes improve customer experiences and empowers employees to perform more meaningful work. Automation can also help companies gain insights from the data collected as part of digitization efforts, drive organizational change, and create new business models.

Digital Workflows Are Essential to PMO Success

Digitization is core to the PMO's role as an advisor to business leaders on resource allocation across the enterprise portfolio. Digitization enables PMOs to adopt a strategic role in guiding enterprise agility and outcome realization, through the insight they gain across projects and programs.

The differentiating capability that digitization provides PMOs is insight into value and outcome realization at all levels of the company. PMO leaders require sufficient visibility and insight across their portfolio to make the right recommendations to business leaders to respond to change. Digital workflows are essential to identify outcome dependencies across projects, help manage across methodologies, and create excellent customer and employee experiences that position

4 Authors: Simon Grice (Senior Director, Innovation), Doug Page (Senior Manager, Product Management), Rani Pangam (Senior Director, IT Project Management), Tony Pantaleo (Director, Product Success)

Acknowledgment: Used with permission from ServiceNow, Inc.

PMOs for success. Digital workflows can also liberate PMOs from repetitive, manual, lower value tasks and enable them to shift energy toward driving enterprise-wide digital transformation.

Unifying Systems to Drive Capabilities

Fundamentally, the drive to digitally transform gives PMOs the ability to elevate from the tactical level of an individual project to overseeing the strategic capabilities to achieve change. With the right digital capabilities, project leaders can manage project portfolios at scale to ensure they are optimizing their tactics for the best result. The emphasis shifts from what work gets done, not on how.

Leaders can choose the right project methodology to meet business demands, and then connect it to both demand and outcomes through a digital workflow. With this "closed loop" in place, the PMO can capture and measure end-to-end strategic delivery across business lines, communicate where the business is developing new capabilities, and identify where efforts have stalled.

To do this, the digital capabilities that PMOs employ should support five needs, as shown in Figure 4-4:

1 Flexibility

Digitization should provide the ability to accommodate and adapt to multiple ways of working – both for projects that require traditional Waterfall-based project management approaches and those that use Agile methods. The mix of Waterfall and Agile will vary, but every business will need digital tools that enable PMOs to manage and communicate across multiple approaches and methodologies. The ability to use a single, closed-loop work management platform to plan, prioritize, and execute projects of all types – as well as measure outcomes and map them back to organizational strategies – is essential for most large enterprises.

2 Data-Driven Automation

To manage at a more strategic level, PMOs should automate manual, lower-level processes. Project management processes can be converted into digital workflows that standardize, automate, and optimize manual tasks. For example, chatbots are already being used to offload project administration processes, like creating demands and sending status updates, task reminders, or issues to be closed. As AI matures, it brings new insights into scope, time, and risk management. It also optimizes the science of project management as more project data is captured and ingested.

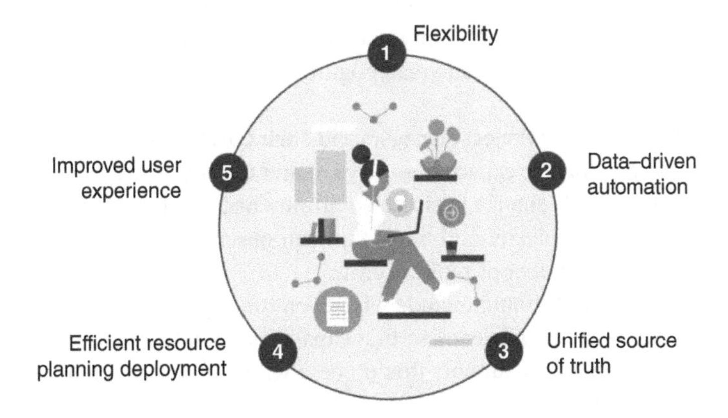

Figure 4-4 Digital Capabilities Drive PMO Excellence

Leading PMOs will bring capabilities like these into a hyper-automation strategy – the combination of previously siloed automation technologies to generate step changes in the efficiency and effectiveness of project management. Technologies such as machine learning, business process mining, and robotic process automation can be brought together to help automate the end-to-end workflow for a standard project management activity, like the development of work breakdown structures.

These advances promise to free up project managers leaders from the routine, administrative aspects of project management to focus on more strategic activities, such as organizational change management and portfolio investment alignment.

3 Unified Source of Truth

Having a single repository of data is critical to link projects and programs back to measurable business outcomes. Data from systems of record needs to be accessible in a platform that maps project dependencies into a company-wide view for business leaders, a perspective that no individual business unit can provide. This approach supports more effective communications, knowledge sharing, and insights across systems and approaches. To achieve this, digitization must deliver real-time data and analytics showing progress against business outcomes, as well advisability and connection to related projects that support common business outcomes.

With a unified view of resources, bottlenecks, change requests, and other factors across the entire portfolio of programs, resource allocation becomes much easier. Project managers can avoid over-committing and underdelivering resources and ensure greater efficiency. Project leaders should be able to see the end-to-end impact of a project or program – from the original demand, its design and build, then through its implementation and its outcomes, or its return on investment.

4 More Efficient Resource Planning Deployment

As the PMO aligns its portfolio of work to corporate strategies, the business will need a single planning system to support resource allocation.

This will become critical to success as businesses move closer to a 'gig economy,' where flexible and temporary jobs are performed on demand. Business leaders will need seek out specific skill-sets from different areas of the business to solve short-term demands.

Enterprise-level project planning will need the capability to source the skills to "swarm" or use ad-hoc, cross-functional teams to quickly complete a large project or program and drive faster time to market. Strengthening this capability will become even more important as business face new operating models that demand even greater agility.

5 Improved User Experience for Project Managers and Their Customers

Excellent customer and employee experiences drive higher topline growth, productivity, and user satisfaction. Especially in high-change business environments, a well-designed user experience is essential to reducing the complexity and noise that can obscure effective project management, especially as the scope and number of projects grow.

The digitization of the PMO should include a focus on the experience of project managers and stakeholders, and actively solicit feedback so that capabilities can be rapidly scaled, modified, or improved. Digital workflows can support this process with timeframes that embrace a deploy, learn, and adjust approach.

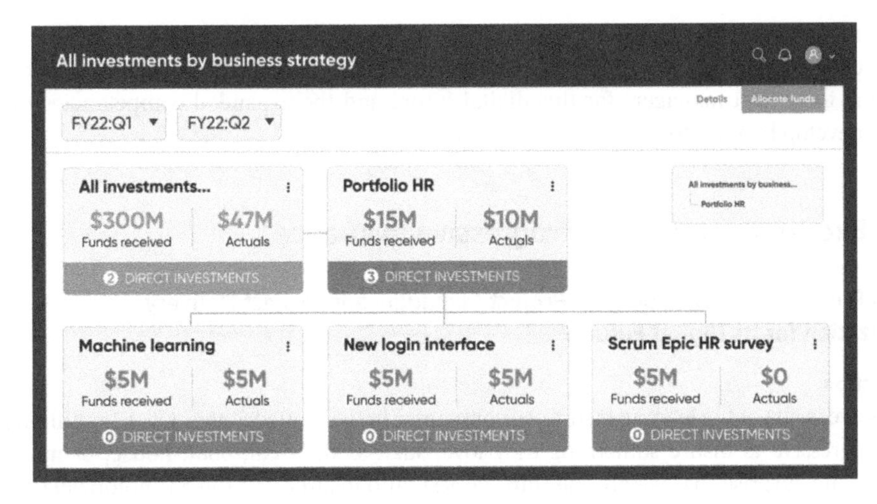

Figure 4-5 Align Technology Investments to Value

Digital Transformation beyond the PMO

Digital transformation of the PMO and project management is a key prerequisite for the PMO to enable the broader digital transformation of the enterprise. With its vantage point across business lines and functions, the PMO can compare digital progress and maturity using common yardsticks, such as velocity, intelligence, or experience, and enable cross-functional and executive dialogue around digital transformation priorities, using tools like heatmaps to assess digital maturity across business functions and capabilities.

Dashboards, as shown in Figure 4-5, can provide transparency to executives on the performance of their investments. By shifting its own focus toward outcomes and strategic agility, the PMO can position itself to advise the rest of the enterprise on the fastest and best paths to digital transformation.

4.5 Context-Driven Planning

This second critical element of digitalization and projects has driven much of the dialogue and decision-making process on the project delivery frameworks for the last decade. A good number of cases and results have been gathered across organizations over the past few years and confirm a movement toward a hybrid waying of working with few growing pains surrounding the right organizational agility requirements.

The first of the two sub-elements, *dynamic adjustments,* addresses the growing need for an iterative approach to how project planning is conducted, is now the norm. Building on the progressive elaboration concepts that have dominated the practice of project management for the past decade, organizations will be expected to develop stronger adaptive muscles and to continually adjust plans without sacrificing the integrity of clear and socialized project expectations. The second sub-element closely aligns with the first and centers around fit. This fit is driven by closer understanding of the complexity of initiatives and by ensuring expanded dialogues and acceptance for the tailoring that should happen to reach the right and reasonable comfort and control level in the degree of planning.

The Progressive Insurance example in the next section is a good illustration of the value of this context driven planning. This example uses the organization's maturity journey as a backdrop for preparing its project managers for this digital future and the expanded competencies they will need to develop in order to succeed in that future.

4.6 Excellence in Action: Progressive Insurance[5]

How Is Progressive Developing Its Project Managers and Project Delivery Organization for Its Digital Future?

Introduction

Progressive is one of the leading insurance companies in the United States. Over 18 million people trust Progressive to insure something they love, such as their car, boat, house, or motorcycle, making it no surprise that we are the third largest auto insurer and the #1 motorcycle/specialty RV/truck insurer. Customers can get insurance for almost anything, and most products can be purchased online in a few minutes. Our Fortune 100 company was started in 1937 with a business philosophy to approach auto insurance in an innovative way – like no other company had. In 1997, Progressive was the first company in the industry to offer consumers the ability to buy (not just quote) an auto insurance policy in real-time online and started earning industry awards for our site design and customer experience. Being at the forefront of digital transformation enabled us to have one of the largest Information Technology departments in the insurance industry, including the Enterprise Project Management Office (EPMO). As of 2021, we have over 220 project managers in the EPMO delivering products to our customers.

A few decades ago, the digital transformation was seen only in the Information Technology sectors with tech giants like IBM, Microsoft, and Google. Today, insurance companies are significantly impacted by technological advances. The first phase of digital adoption across the industry focused on automating processes and improving work efficiencies. A later stage of adoption concentrated on enhancing customer experience and collecting more data from customers. At present, insurance companies are focused on meeting customer demand and beating the competition within and external to the industry. Insurance companies use the latest technologies, including cloud computing, big data, artificial intelligence, and live chat, to make customer service more personalized and efficient.

These phases of change in the competitive insurance industry have changed the way business partners view project management. We can no longer sustain our traditional methodologies and project management skills focused on delivering projects to a defined scope on time and within budget. Delivering projects with agility and business value is of utmost importance to our business and customers in the current fast-paced, ever-changing technology environment.

Progressive's Evolution of Project Management

Mandatory 60-hour work weeks for months at a time to meet project deadlines was very common at Progressive in the 1980s and early 1990s. We all asked, how can we do better? Project Management in the IT world was still relatively new, and managers did the best they could with

5 Material in this section has been provided by Ganesh Kumaraswamy, PgMP, PMP; Kim Newton, PMP; Joe Anastasi, PMP, PMI-ACP, DASM.

limited tools and unrealistic expectations. Project Management was not recognized as a job title; instead, it was a role different people would play in their spare time, including service managers, programmers, business analysts, etc. Availability of training was very ad hoc, and some embraced the concept of estimating, assignments, and dependencies to create and manage to a critical path. Others continued to manage with tasks lists in spreadsheets.

In the late 1990s, small PMOs began popping up in IT departments with each doing something different. We recognized a desktop scheduling tool did not support the cross project work many people were engaged in which continuously led to overallocated people and missed dates. The adoption of an enterprise tool did not bring the results all had expected. Pockets of PMOs were having some success, but that success was not realized across the board. Again, we all asked, how can we do better?

That recurring question led to the implementation of a single IT Enterprise Project Management Office (EPMO), which was one part of a much larger organizational change to align most IT people into a functional matrix in the early part of 2010. Processes and standards, governance and portfolio oversight were now enforced across projects, with most following a standard waterfall methodology. Some projects used an in house developed software development methodology which incorporated Agile practices into the Waterfall methodology. Were we becoming more predictable? Yes, all would agree we were more predictable and had more transparency, but many believed we were still not living up to expectations. We were not working closely enough with our business partners and our structured processes were holding us back from delivering what was most important more quickly. Again, we all asked, what can we do better?

During the middle of the 2010 decade, many small pockets of developers began finding better ways of working and Scrum rapidly became what all deemed was the silver bullet as a grass roots effort. The experience and learnings were invaluable and that, along with the adoption of the Lean and Agile mindset, showed there was much more we could do to increase flexibility in the delivery of new and improved applications that would enable us to retain and grow our customer base. The evolution continues as we welcome and embrace better ways of working. As one can see, continuous improvement is embedded in our culture. We are now asking ourselves; how can we be more Lean and Agile?

Figure 4-6 is a timeline of activity.

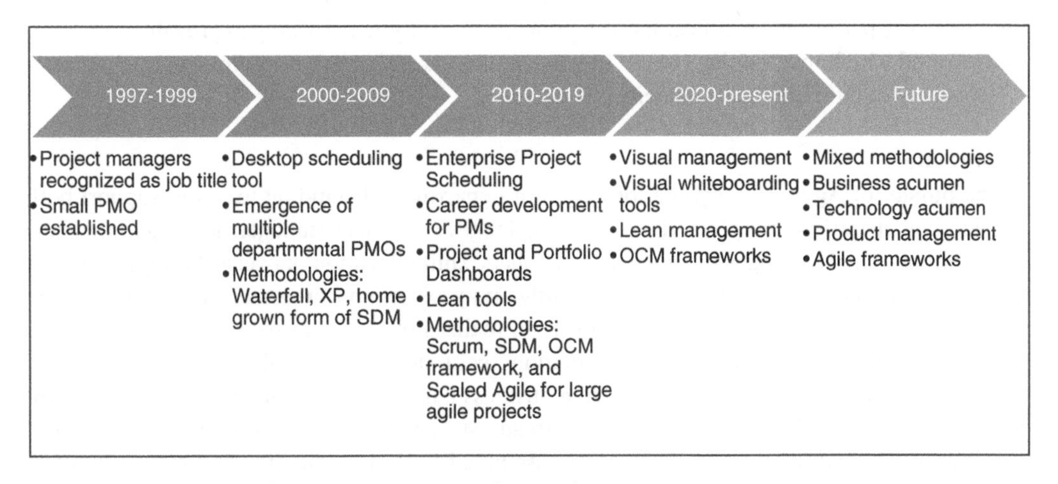

Figure 4-6 Evolution of Project Management at Progressive

Current State

As a result of response to this change in demand, close to 70 percent of projects at Progressive are now delivered using Agile methodologies like Lean, Scrum, and Kanban. The project manager's scope of responsibilities is blended with other roles as we transition to a new way of managing projects. The renewed expectations for our methodologies have created an environment where project managers find their roles evolving to change methodologies and industry practices. Organizing a multiyear project scope and delivering the project on time and budget is no longer deemed the only success criterion for completing a project. The EPMO organization and leadership have flexed their expectations on achieving the traditional project metrics on labor hours and duration. The focus has shifted to breaking the projects into small customer value deliverables and releasing the products incrementally. Project managers are expected to define the minimum viable products (MVP) with a business value measured and agreed on by the business or product owner. With this change, project managers and sponsors have embraced new leadership mindsets like Agile and servant leadership.

With a business focus on agility and speed to market, multiple departments within Progressive have transformed their organization to become more agile by organizing their group into standing Scrum teams. The departments have framed their operation and project work to be managed by these Scrum teams. The project managers working in these groups play non-traditional roles (e.g., Scrum Master, impediment remover, integration management with other teams, etc.) to align with the functions defined by the methodologies. In some cases, the standing teams are not leveraging people outside of EPMO. Project managers eliminate distractions that could impede the work of Scrum teams. They coordinate and/or integrate workflows across teams, departments, and portfolios. They also gather, consolidate, and interpret disparate data sources (work requests) to help prioritize.

With future programs and projects focused on speed to market business value delivery, the Enterprise Project Management Organization (EPMO) has evolved to support more experimentation regarding how teams are organized and with new methodologies that align with the type of projects delivered: implementing brand new technology, upgrading the existing technology, new business line product, enhancing the current development in the market, compliance projects, and operational efficiency projects. The different execution approaches create a new challenge for the project managers leading the projects or programs working in a matrixed environment with multiple methodologies. In addition, the different methodologies have disrupted our traditional enterprise key performance indicators. The project organization is working to provide more guidance around which methodology to use in what situations and how we continue to track metrics effectively with different methodologies.

Future State

In the pursuit of Agile delivery, the formation of departmental standing teams to support both projects and operations has given rise to new challenges for project managers leading in this new environment. The challenges include team members' split allocation between the support of the products and executing on projects concurrently and projects and programs needing to execute across mixed methodologies. As the organization realizes the value of being agile and lean, we anticipate more departments within Information Technology teams will embrace a method of delivery ensuring alignment and buy-in from business partners. Project managers initiating projects in a matrix organization will be faced with adjusting and flexing their project management leadership style and processes to lead teams assembled from different standing teams. In a few situations, we anticipate project manager's to adapt their strategies to get the work done by placing project work in standing team backlogs versus getting specific people assigned to their projects.

The mixed methodology or hybrid framework environment leads us to adjust the project delivery process and metrics to track progress and manage dependencies and the flow of information between teams. The traditional project manager role and responsibility must be flexed and must reflect the new delivery context. In future projects, the project managers will take on roles like release manager (responsible for coordinating efforts from multiple standing teams for a single product release), Scrum Master (who ensures the team adheres to the theory, practices, and rules of Scrum and maximizes the value created by the Scrum Team), integrator (responsible for coordinating internal and external dependencies, vendor engagement, and consolidated reporting for Scrum teams).

Another future experimentation we anticipate for project organizations is to align project managers to a specific product line to gain more profound knowledge and experience to deliver improved products and solutions. In addition, project managers must develop strong business and technical acumen to lead innovation projects. Project managers who are not equipping themselves to stay current with the technology or with deep business acumen to facilitate business discussions will encounter difficulty supporting organizational direction and demand.

How Is Progressive Preparing Its Project Managers for the Future?

With the rapid evolution of Agile methodologies and product management, the Enterprise Project Management Organization assessed the viability of the project manager in the new solution delivery context of digital business. A few strategic investments were made to support the development of project manager competencies for the future.

First, the organization partnered with internal Agile coaches and external vendors to mature the Agile methodologies practices and developed the team's capability to scale for larger teams. The capabilities were developed through the Agile methodology and Agile leadership training, Scrum training, and coaches dedicated to the teams.

Second, the organization invested in starting up the College of Enterprise Project Management. A senior dean leads the college. A group of associate deans with diverse project management delivery and service experience who are passionate about learning and development round out the leadership team.

The college conducted multiple project management skill assessments with project managers from all job levels. The assessments included rating the individual project managers on project management technical competencies and a focus group session with key stakeholders and sponsors to assess their leadership competencies. After consolidating the data, the group prioritized investments in project managers' skill development as follows:

- **Business acumen**: The deans, in partnership with business teams, conduct business acumen training, roundtables, and business presentations to increase the business knowledge of project managers.
- **Technology acumen:** The college works with enterprise architecture organizations to impart the latest technology trends and educate project managers on foundational technical knowledge. This is done through panel discussions and workshop sessions.
- **Leadership competencies:** The assessment revealed the need to strengthen the capabilities of project managers to manage ambiguity, communication, leadership styles, and change

management in their future assignments. The team is leveraging its partnership with online educational portals to provide short-burst experiential learning. Project managers can take training as a group for a few hours, apply their understandings, and come back to share their experiences, enriching their knowledge in a short time. The deans also have set up peer-to-peer support systems for project managers new to the role to increase the pace of learning and improve their performance.

As with any profession, project management is evolving quickly to keep pace with the rapid change in the business environment. As the environment becomes more global, digitalized, uncertain, and volatile, companies need to adapt and bring services and products to market with more speed and agility than ever before. Project managers are the people who can make that happen, and Progressive's project management organization is prepared to take on the challenges the future demands.

4.7 Co-Creation

One could not talk about achieving projects' promises without the art and science of co-creation. This refers to the interactive involvement of customers, users, suppliers, partners, various experts, and many other stakeholders as needed to shape and create the solution or expected product or service of a given initiative. The future is bound to see an increased appetite and expectation around this principle. The two sub-elements in the framework that were chosen as most relevant here are: *visual clarity and streamlined inputs.*

Visual clarity is addressed multiple number of times in examples and ideas throughout this book and highlights an ever-increasing need for clear stories and ways of connecting teams and stakeholders around simple and concise visuals. The second sub-element, streamlined inputs, is also addressed throughout the book's chapters. Of special mention here, the ideas of design thinking and how the continual inputs gained could shape up the right solution especially if powered by digitalization's wider capacity for rich data to confirm how the gained inputs align with previous successes and future potential.

The discussions and points raised across the next sections 4.8 through 4.18, which include four sections contributed by Dundas Data Visualization, strongly support the importance of this element of the framework and support the value of becoming a knowledge centric organization in this digital era.

4.8 Growth in Information Warehouses

The expansion and use of project management into more business applications and nontraditional projects have forced project managers to learn new skills to navigate in these unchartered waters. There were now new challenges related to problem-solving and decision-making. All of this required more information than was currently available in the earned value management system.

As project management applications grew, emphasis was initially placed on creating a project management information system (PMIS) based almost entirely on earned value management efforts using just time, cost, and scope metrics. Additional information, if necessary, came from personal contacts. Most decisions were based on information related to the triple constraints and the impact it might have on the well-defined requirements listed in the statement of work. Today, project managers are expected to have access to the necessary information, tools, and processes to support complex problem analysis and decision-making. Advances in technologies, the need for competing constraints, the expansion of project management into strategic projects, and the growth of information warehouses are driving companies toward business intelligence (BI) systems.

For more than 20 years, the project manager's prime source of current information related to project management excellence and continuous improvement efforts came from the PMO or the Best Practices Library managed by the PMO. Business-related information was provided in the business case or by the project sponsor. We believe today that we are managing our business by projects and that project management is the delivery system for sustainable business value. Therefore, project managers are expected to facilitate business decisions as well as make project decisions and need direct access to a great deal of project and business information that does not appear in the earned value measurement system. The challenge has been in selecting the best way to access this information.

Throughout the life of a project, especially strategic projects as discussed in previous pillars, there is a significant amount of data that must be collected including information related to the project business case, project benefits realization plan, project charter, project master plan, customer interfacing, and market analyses. This information is necessary for innovation decision-making. The knowledge contained in information warehouses, as well as the amount of information and speed of access, provides companies with a source of competitive advantage.

If the intent of a strategic project such as one requiring innovation is to create a commercially successful product, then the team members must understand the knowledge needed in the commercialization life cycle even though many of the team members may not be active participants during commercialization. Making the wrong decisions in the early project stages can create havoc during commercialization efforts. Team members should review commercialization data records of previously introduced products so that they understand the downstream impact of business decisions. This information should be included in information warehouses.

4.9 Knowledge Repositories

Companies are now creating knowledge management systems and knowledge repositories, as shown in Figure 4-7. However, most companies have significantly more categories than what is represented in Figure 4-7. Included in these systems are metrics/KPIs libraries that show what metrics are available and how they should be displayed.

It can be extremely difficult to get people to use a knowledge management system correctly unless they can recognize the value in its use. With the amount of information contained in knowledge management systems, care must be taken in how much information to extract. A mistake is often made whereby the only information extracted relates to your company and the targeted, end-of-the-line customer base. While it is true that not all customers and stakeholders are equally important, to increase the value created by the project, we must also consider new customers that might eventually buy into our products and services. Focusing on diverse needs of a larger customer base may prove rewarding especially if some customers no longer are important and new customers enter the market with needs that are different than current customers.

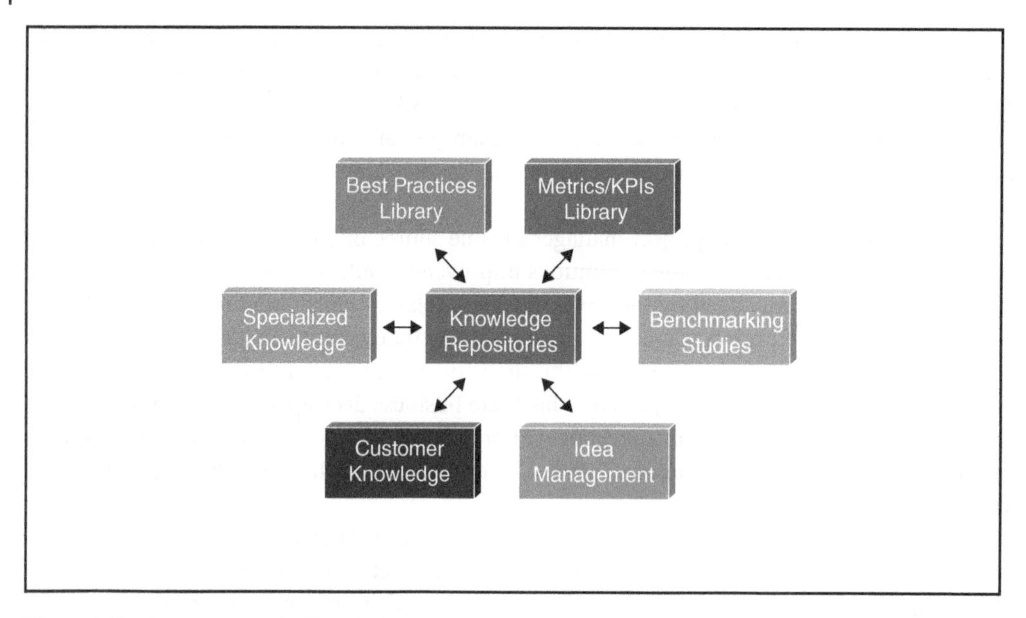

Figure 4-7 Components of a Knowledge Repository

Companies invest millions of dollars in developing information warehouses and knowledge management systems. There is a tremendous amount of rich but often complicated data about customers, their likes, and dislikes, and buying habits. This knowledge is treated as both tangible and intangible assets. But the hard part is trying to convert the information into useful knowledge.

The Benefits of a Knowledge Repository

The use of a knowledge management system is expected to become a necessity for all future project teams. Project teams should first map out the mission-critical knowledge assets that are needed to support the project's strategy. The next step is to determine which knowledge assets to use and exploit. By mapping the knowledge assets, you put boundaries around what the project is designed to do. Unfortunately, the only true value of a knowledge management system is the impact on the business. Simply stated, we must show that the investment in a knowledge management system contributes to a future competitive advantage.

Knowledge management can increase competitiveness, allow for faster decisions and responses to disruptive changes, and rapid adaptation to changes in the environment. Knowledge management access is critical during design thinking. The growth in information has created a need for cloud computing.

Intangible Intellectual Capital Assets

The information contained in a knowledge repository is often referred to as intellectual capital. As shown in Figure 4-8, intellectual capital is frequently considered as intangible assets categorized as human, product and structural capital. These are knowledge-related assets normally not identified on the balance sheets of companies, but they can be transformed into value that leads to a

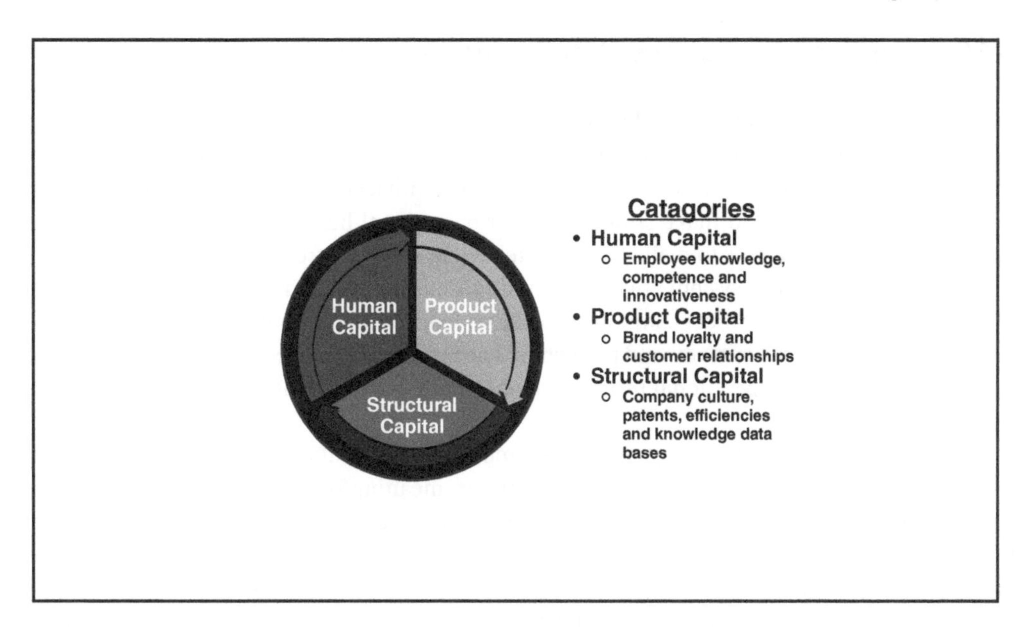

Figure 4-8 Three Critical Intangible Components of Intellectual Capital

Table 4-2 Sources of knowledge

Source of Knowledge	Description
Explicit	Encoded knowledge that can be found in books, magazines, and other documents
Implicit (or Tacit)	Knowledge in the heads of people. Also, knowledge retained by suppliers and vendors. Knowledge may be difficult to explain.
Situational	Knowledge related to a specific situation, such as a specific use of a product.
Dispersed	Knowledge that is not controlled by a single person.
Experience	Knowledge obtained from experiences or observations of clients using the product; must understand user behavior.
Procedural	Detailed knowledge on how to do something

sustainable competitive advantage. Knowledge data bases and information warehouses are needed to support intellectual capital components. These intangible assets that are used to define intellectual capital could be strategically more important to the growth and survival of the firm than its tangible assets. Project teams are becoming more knowledgeable into the importance of intangible assets.

Categories of Knowledge

In Figure 4-7, we showed the components of a knowledge repository. The knowledge in each component can come from multiple knowledge sources. There are several sources of knowledge, and they are not mutually exclusive. Table 4-2 lists some ways to classify knowledge sources:

Although we have stressed in this pillar the importance of using a knowledge management system, project teams must understand that part of their job will include updating each of the knowledge categories.

4.10 The Need for Business Intelligence Systems

For many companies, simply having knowledge repositories and information warehouses will not be sufficient to support future business objectives and strategic projects. BI systems are often considered as the next step to combine business information with technologies in a manner that allow executives and managers to make strategic and/or operational business decisions. Similarly, BI can enable project managers to make decisions related to strategic projects.

The components of a BI system are data gathering, data storage, and knowledge management. Metrics information is a critical component of BI systems. The information contained in the BI systems can be historical, current, and predictive. The information can come from several sources including strategic and operational project management benchmarking studies often conducted by PMOs.

BI technologies are designed to handle large amounts of "big data", whether structured, semi-structured, or unstructured, and present the data on meaningful dashboards to enable better business decisions and take advantage of business opportunities, especially when implementing strategic projects. The technologies used in BI systems allow companies to look at external data (i.e., information from the markets in which the company operates) and internal data (i.e., financial and operational data) together and create business intelligence information to support strategic, tactical, and operational projects. BI systems facilitate corporate decision support systems by transforming raw data into meaningful and competitive business intelligence. However, there are still companies that believe that BI systems are merely the growth of business reporting systems.

Project managers will need to learn new decision-making tools, including digitalized economics, artificial intelligence, and the Internet of things (IOT). With large amounts of data, teams may have to rely on analytical statistics, which include:

- Descriptive data analytics: analysis of historical data including past successes and failures
- Predictive data analytics: analysis of the data to make predictions of what might happen
- Prescriptive data analytics: look at the reasons why things may happen, options for risk mitigation of future work, and options to take advantage of opportunities

4.11 Big Data

The growth of big data will most likely impact most companies worldwide. For effective analysis of the data, project teams will need workers that possess data science capability. The skills will include statistical methods, computational intelligence, and optimization techniques.

There are numerous mathematical models that currently exist to support project decision-making efforts using big data. A list includes:

- Financial models (ROI, IRR, NPV, payback period, benefit-to-cost ratio, breakeven analysis)
- Time (scheduling models)
- Money (cash flow models)
- Resources (competency models)
- Materials (procurement models)

- Work hours (estimating models)
- Environmental changes models
- Consumer tastes and demand models
- Inflation effects models
- Unemployment effects models
- Changes in technology models
- Simulation and games models
- Mental models

The expected benefits from using big data effectively include:

- Detection of patterns and trends related to time, cost, and scope
- Comparison to other projects as well
- Identification of the root causes of problems
- Better use of "what-if" scenarios
- Better trade-offs on competing constraints
- Better tracking of assumptions and constraints
- Better tracking of VUCA and the enterprise environmental factors
- Better response to out-of-tolerance situations
- Better capacity planning decisions involving resource utilization
- Ability to make strategic rather than just operational decisions
- Ability to make change management decisions
- Decision-making can be pushed down the organizational hierarchy, but there will be "rules for decision-making" established
- Emphasis on long-term perspectives rather than just short-term
- A reduction in the risk of making the wrong decision because of a lack of information

Project teams seem to focus on the knowledge management portion of the BI system. This includes:

- How performance metrics are created and reported
- How benchmarking information can be extracted
- Statistical and predictive analytics
- Data visualization techniques and dashboard design
- Business and project reporting for executives and stakeholders

PMOs and project teams are now participating in designing and managing BI systems projects. For some companies, this will bring additional challenges related to understanding BI systems and data visualization practices, including new ways of designing BI dashboards and interfacing with stakeholders. Therefore, it is incumbent on project team members to adopt the new skills identified in earlier sections of this pillar.

For companies just starting to implement BI projects, selecting existing knowledge bases as a first BI project may be a good way of evaluating the effectiveness of, and organizational readiness for, BI technology. It also can provide the dual benefit of controlling scope (i.e., starting small) and allowing project managers a suitable learning curve.

In the next four sections are white papers graciously provided by Dundas Data Visualization that address several of these challenges.

4.12 Top Seven Things to Consider When Choosing a BI Tool[6]

Choosing your Business Intelligence (BI) and analytics tool can be a lengthy process. There are countless vendors to choose from that all seem to offer similar features and make similar promises. How do you choose the right one for your organization, from the many? Here we've laid out the most important factors to take into consideration when vetting and selecting a business intelligence tool, so you can make the right choice for your organization.

Visual Functionality and Presentation

When creating your data visualizations, is the tool easy to use and does it support a variety of user types? Does the data visualization software allow for the level of customization and flexibility that you need to support your objectives, and the requirements of the various functional groups in your organization? Your business intelligence reporting tool is the face of the whole operation. It's what your users will refer to when making business decisions. So, while visuals may not seem all that important when choosing a BI tool, if your dashboards are difficult to create or to understand, or feature poor-quality visuals, then it will be infinitely more difficult to get people to actually use the tool in the first place.

Integration

Before selecting your BI tool, you need to figure out if you're looking for a standalone solution, or an integrated solution. A standalone solution can only be accessed and viewed in its own application, whereas an integrated solution can be accessed and viewed through any pre-existing applications, websites, and services within your company. Integration allows you to put the benefits of BI directly into a service that your users already use and are familiar with.

Now, if you've decided to go with a standalone solution, you have your pick of the lot. Nearly every BI solution can operate as a standalone program. But the biggest problem you run into with standalone solutions, is that they have the lowest adoption rates. It all goes back to that old adage: out of sight, out of mind. Standalone solutions exist in a separate application that needs to be opened every time it's used. And let's be frank here, most people will forget. Integrated solutions bring your analytics tool forward to a location that can't be ignored.

Integrated tools come in two types: partially integrated and fully integrated. And the difference between the two is important. While aesthetically, there isn't much of a difference – both partially and fully integrated solutions will have the same basic look. Where the big difference lies is in data security; with partially integrated solutions being far less secure than fully integrated (to the point where they actually present a security risk to your company's data).

If you choose to go with an integrated solution, be sure to find out if that solution is partially or fully integrated. Most companies will advertise integration, but not specify.

Customization

Every organization is unique in its operations and needs, so you need to ensure that you select a BI vendor that can accommodate your specific requirements – for today and in the future. Can the BI vendor build a solution that fits with your business and data requirements? Is the solution

6 This section was provided by Dundas Data Visualization, Inc.

flexible in that it can be extended or altered to meet future needs? Many vendors only provide minimal customization abilities, and some don't even offer the ability to expand. While this might work for some companies, most will need a custom setup – one that can integrate flawlessly into their operations and grow as they grow. Business users can be very particular about the way they consume their data making customization and extensibility critical factors to take into account when selecting your BI vendor and tool.

Mobility

A mobile BI tool is an easy, and powerful way to keep everyone connected. Mobility features are crucial to factor in for any company that has off-site workers, board members, or partners that need access to company data in order to be effective and informed. Determine what you need from your mobile BI application. Do your users need to simply view dashboards and reports, or also create or edit them on-the-go? Mobility has come far in the BI space, and more and more vendors are offering mobile solutions. Just be sure that they offer the features that are necessary for your operation's success.

Training Requirements

Any good quality BI tool will require some product training. This is all a part of the necessary investment you need to make in order to reap the full benefits of any BI solution. BI has the capacity to radically change your organization for the better, but only if you are willing to make the upfront investment in time and energy. Decide early on to take the time to train your users on the BI product, or risk missing out on some of the biggest benefits BI has to offer. Another thing you need to keep in mind is the quality and flexibility of the training offered. Does your vendor train both advanced business and technical users? Some vendors only offer training for the technical group, which means you now have to expend your own resources training your advanced business users. Look for a solution that trains all users, so your organization can gain the biggest rewards.

Pricing

There are three major things to consider in regard to cost: subscriptions, growth, and hidden fees. Get the full details on all costs for your solution now, and as your needs change and grow. Does the vendor require quarterly or yearly subscriptions on top of any upfront costs? This is something that may not be advertised however may come into play later.

As your business grows and expands, so will your BI needs. Will the vendor allow you to scale your solution with a flexible pricing model? Are there any additional fees that aren't listed in the pricing model, but necessary for full implementation or scale up of your BI reporting tool (ex. hardware, server use, development/testing environment etc.)? Make sure to find out what you will be spending on your BI solution up front to avoid any surprises in the future.

IT Support

An easy way to make both your business and technical users happy is to go with a self-service BI tool. Self-service BI allows business users an easy BI experience; freeing up your IT department to focus on other tasks. Make sure that while your selection provides self-service capabilities, it still allows for IT to easily control the deployment and prevent *end-users chaos.* Not all users would want to use the self-service options (some will just be simple consumers) and not all users should have the same level of data access and control. Make sure your vendor also has a fully capable support team to accommodate any issues or questions your technical users may have.

You want to make sure you're relying on an expert, flexible vendor who will help innovate solutions custom to your organization's needs or problems. Vendor support teams that go above and beyond really make a difference to your entire BI experience.

A BI tool will radically improve your company's operations, but only if it's the right tool for your organization. This includes not only the product itself, but the training, support, and expertise your vendor provides for you. You can easily reap all of the benefits BI has to offer, by choosing the right BI vendor and reporting tool for your organization's needs, today and in the future.

4.13 Stop Treating Business Intelligence Projects as IT Projects[7]

In our modern society, companies and employees want to be data-driven and analytical, because it implies that they'll perform better and be more successful. Hundreds – probably even thousands – of new business intelligence initiatives are started every single day.

But in reality, most of these newly started initiatives will fail. According to a Gartner forecast, only 20 percent of BI initiatives will deliver actual results thought 2020, meaning the rest will fail. I believe that the main cause of such high failure rates are a series of common mistakes and misunderstandings that I will try to address below in this blog.

Business Intelligence Is Not an IT Project but a Strategic Endeavor

Before undertaking a business intelligence (BI) initiative, it helps to understand what exactly BI is. According to Gartner, business intelligence is defined as:

> *"An umbrella term that includes the applications, infrastructure and tools, and best practices that enable access to and analysis of information to improve and optimize decisions and performance."*

One of the most common mistakes companies make, is treating their BI initiative as an IT project, when in actuality, it should be treated as a strategic business project. Yes, data and analytics projects will always rely on technical expertise and IT support, however, they should primarily be driven by a business need. When evaluating the success of a BI project, the primary benefits should always be seen from the businesses side. The BI project results you should be expecting and striving for would be elevated business performance, better decision making, increased competitive advantage, and optimized return on investments. If your BI project didn't impact your business, it either wasn't a BI project or it was a BI project that failed.

A Business Intelligence Project Is a Change Initiative

Let's imagine we have two fictitious companies: **Company A** and **Company B.** Each of these companies are rolling out very similar BI initiatives with the objective of enhancing their overall

7 This section was written by a Senior Consultant at Dundas Data Visualization, Inc.

business performance. Each company employs workers with very similar skillsets, maintains similar budgets for their projects, and are experiencing similar business problems they'd ideally like to remedy.

With that information, it's safe to assume that both companies would end up with quite similar results. If **Company A** were to enjoy a huge return on investment and experience a massive jump in how effective their business was, you'd naturally assume **Company B** enjoyed the same success, right? But what if **Company B's** return on investment was negligible? What if their BI initiative was a complete waste of time? Where did they go wrong? What were the differences? Well, in most cases of failed business intelligence initiatives, I've found the difference-maker – the crucial factor in tipping the scales between success and failure – to be an understanding that BI projects are strategic business projects that requires proper change management.

Change Management Is Key to Business Intelligence Success

Analytics initiatives often begin with someone recognizing the importance of data-driven decisions and a desire to know more and do better – hence a pressing need for change. As in any other business project, if support comes only from upper management but employees don't understand the value of the change, they will continue to use old processes, old methods, and old tools where possible. That's why it is so important to lead with a change of attitude and behavior rather than with a change of tools and processes, meaning we need to follow best practices in change management first, and best practices in business intelligence second. To argue this point, let's take a look at the steps needed to make a business intelligence project successful:

1) Cross-level and cross-departmental awareness of the importance of the project and the benefits it'll deliver.
2) Your staff must display the required knowledge and skills to support the initiative.
3) The project has an internal support group along with champions/advocates across all levels to address any issues and questions.
4) The project starts with small, easily achievable goals to prove its benefits and enforce a culture of small wins.
5) You must remain positive and transparent throughout the project and ensure employees are involved in the ongoing initiative.
6) The project must be sustained and embedded within the company's culture.

Yes, the six steps above are basically describing any change initiative within an organization, meaning that a business intelligence project should be treated like a strategic change project.

Successful Dashboards Are Not Data Dumps

Another key to a successful analytics initiative, is ensuring a change in your stakeholder's mindset. Often times, while implementing BI, project leaders and stakeholders tend to get stuck in the *old Excel world* and still want to see a bunch of tables or wish to present as much data as possible on a single screen, because in their mind every single row and column is *critically* important. That is not true, and this approach to business intelligence must be changed. A proper BI project begins with an understanding of the businesses needs and problems that require solving, but its success hinges on understanding that the way we consume data is more important than seeing it all at once.

What Should We Take from This?

In conclusion, I would like to highlight that we should stop treating business intelligence projects purely as IT projects. BI projects are business projects that involve the evolution of data consumption, the enhancement of old processes and tools, the advancement of business performance, and the improvement of analytical decision-making. The days of going through multiple excel spreadsheets and tons of unnecessary data in the hope of discovering insights are long gone.

As Leonardo Da Vinci once stated, "Simplicity is the ultimate sophistication." Now we strive for an easier and quicker way to make more effective decisions. This is why we start with change, and that is why BI projects should be treated like strategic change initiatives as a whole.

4.14 Dashboards vs. Reports: Which One Should You Go With?[8]

What Exactly Is Reporting?

Reporting. . .businesses live and die by it; employees live and die by it. So why is there such confusion over reports and dashboards? The term reporting has often been used to describe both reports and dashboards, with users commonly referring to simple tables placed within dashboards, as reports. Traditionally, most IT led business intelligence (BI) tools have been focused on reports that provide users with very little interactivity and visuals. However, modern data discovery tools revolutionized this and introduced tools that allow for exceptional interactive visualization, often delivered to greater audiences by means of dashboards. Today, the majority of BI tools administer either reports or dashboards (or have tendencies to focus primarily on one and egress limited capabilities to the other).

The issue that arises when deciding between using reports or dashboards is that both terms are often construed as being synonymous with each other. The differences between both deliverables are often difficult to interpret. Oftentimes, there's disconnect in regard to reporting and dashboards.

What Are the Differences Between Reports and Dashboards?

The best way to distinguish between the two concepts is to highlight each terms' disparities. It's important to understand that each medium is generalizable, as there are always exceptions to each respective case.

Enterprise Reporting

Enterprise reporting, or production reporting, is a method by which reports are distributed within an organization to provide users with support for their work. Generally, in traditional BI tools, reports are primarily text and table based, and periodically include visual components. More often than not, these reports are scheduled for automatic distribution, as they tend to be used on a more regular basis. These reports can be arranged via pagination, which is the exact placement of elements on a page. This technique is capable of complex display logic for creating printed reports or online operational reports. Even though losing popularity, paginated reports will continue to be a standard reporting method when it comes to displaying raw level data or pixel perfect data views

8 This section was prepared by Marketing at Dundas Data Visualization, Inc. ©2021 by Dundas Data Visualization. All rights reserved.

designed for printing. It's important to note that reports can be lengthy and span multiple pages. As a result, the creation of reports can be a bit more complex requiring a more technical user. Reports are, in summation, commonly comprised of background data and other decision informing information.

Dashboards

Conversely, dashboards present the most important information required to achieve specific business objectives, at a glance. They are designed for at-a-glance monitoring and decision making. Generally, a dashboards' objective is to present all necessary content on one screen, and drive decision-making without overwhelming the eye. A dashboard will focus on visual, interactive features and will allow users to physically 'drill' into information. Dashboards predominantly focus on decision driving, rather than informing, and are targeted at Key Performance Indicators, or other decision driving information.

Now, these are all generalizations. Every dashboard does not need to be interactive, and every report does not need to be static. A good example of a non-tabular report is a multi-page, pixel perfect report, which incorporates maps, data images, data labels, and other report elements that can be manipulated to a very precise degree:

Ad-Hoc Reporting

Ad-hoc reporting is definitely a part of the reporting world but is different than enterprise reports and dashboards as it is usually designed for very particular purposes, often without much planning in advance. Ad-hoc reporting is self-service, which allows for reports to be easily built with the exact information users require without consuming developer resources. Users are able to create those reports in order to analyze explicit business questions that are not answered within the organization existing set of enterprise reports and dashboards. Typically, with ad-hoc reports, the BI solution is connected to the desired modeled data sources, parameters are established, and frameworks for what the end-user will see are specified. Ad-hoc reporting is designed to capacitate end users and allow them to create their own solutions usually via drag and drop. Therefore, this method is more desirable when there is an excess number of end users, who yearn to act on data and information independently, while still having access to similar data sets.

In traditional BI tools, Ad-hoc reporting is often referred to as Ad-hoc querying as it allows users to generate a dataset (usually in the form of a table with different columns of data) simply by selecting the data fields required (the table columns) without having to write SQL or other queries code.

In modern BI tools, that includes data discovery capabilities, as well as allowing for Ad-hoc queries to be created via drag and drop, but the outcome is usually much more visual based. Using different types of visualizations rather than tables allows the user to reveal trends, patterns, and data relationships. Often, the visuals created using those tools serve as the starting point for a dashboard. As mentioned above, this is often a source of confusion, since those can also be simple tables included within dashboards.

So, Which Should I Use?

What's important to take away, is that there is room for all types of reporting. Putting Ad-hoc reporting aside, both enterprise reporting and dashboards are used to communicate data and information to executives and information workers who need to be able to monitor data and quickly make decisions that allow them to get back to do their job. It's integral that objectives can be delivered on the pair and the user is able to integrate the two easily. The ideal situation involves starting with a dashboard and drilling down into a more detailed report that is already filtered on previously identified issues located on the dashboard. On the other hand, ad-hoc reporting will serve business analysts and executives that are trying to further analyze a new business situation and find new insights.

As different users have different skills and needs, it is important to have a BI environment that enables all users to interact with their data as needed – ideally in a single, front-end solution that allows for all those types of reporting in one place, rather than having multiple tools that typically increase your data inconsistency and silos concerns.

4.15 Mapping Dashboards to Objectives[9]

Overview

Dashboard solutions allow an organization to visualize and monitor their data in a meaningful manner, while enabling them to make better operational and strategic decisions.

In order to build a functional dashboard, we need to answer the following key questions:

- What are the business objectives that the organization needs to achieve?
- What is the gap between where we are now and what we are trying to achieve?
- What are the key success factors that will help us achieve our objectives?
- What are the current issues that are preventing us from achieving our objectives?
- What are the key performance indicators that will enable us to monitor if our key success factors have been met?
- What are the key performance indicators that will enable us to be proactive to mitigate any risks in achieving our objectives?
- Who are the stakeholders that will help us define our key performance indicators?

The To-Be, the As-Is, and the Gap

A dashboard solution is the means to achieving the organization's objectives. Therefore, before building the dashboard we need list the objectives that the organization needs to achieve.

Objectives need to be specific, understood by the whole organization bounded by time, and measurable. Once the objectives have been set, we need to define how far we are from achieving them. To do so, we need to ask ourselves the following questions:

- Have we met our objectives?
- If not, then why? List the problems that are currently preventing us from getting there.
- For each identified problem, what can we do to solve it?
- For each identified solution, what do we need to do to implement it?
- For each implementation, how do we know that it is working properly?

When we answer these questions, we can know where we are and what we need to do get to our "To-Be" situation.

9 This section was provided by Dundas Data Visualization, Inc. ©2021 by Dundas Data Visualization. All rights reserved.

How to Get There

Now that we know what we want to achieve, how far we are from achieving it, and the solutions that needs to be implemented to get us there, we need to list the measures that have to be monitored in order to know that our solutions are working correctly.

The key performance indicators (KPIs) are the set of measures that focus on the critical areas that we have to monitor in order to make sure that we are on the right track.

An effective key performance indicator should incorporate the following:

- The numeric value(s) that we are monitoring
- The dimensions that we are monitoring these values against (e.g., are we monitoring these values over a 12-month period, by department, by region, etc.)
- The filters that we're using to slice the data (e.g., are we filtering these values by date range, by employee, etc.)
- The targets/benchmarks that we are using to compare these values against (e.g., this year versus last year)
- The states that we need to know if the values are good or bad (e.g., the values between 0 and 100 are bad, 100 and 200 are OK and above 200 are good)

Validation

Once we have the KPI list defined, we need to make sure that these KPIs will actually contribute to achieving our objectives. So, for each KPI defined we need to ask the following questions:

- Is the KPI understood by all the stakeholders?
- Is the corrective action known when the KPI reflects negative values?
- Can the KPI be traced back to the success factors and identified solutions for the organization?
- Is the KPI linked to certain departments so that we know who is responsible when any issues are identified?

4.16 Virtual Teams Engagement

This element of the framework is directly corelated to the enhanced productivity global organizations witness when they mobilize the right talents, support their growth culturally, and sponsor them with the technologies and infrastructures that break down barriers for successfully managing projects across the organization. The world events that led to higher experimentation with digitalization to connect and engage teams have created unprecedented and unexpected growth in the potential of the first sub-element which is the *cross teams ease of access*.

Not only did the team manage to connect with folks they seldom engaged with in a given world region, but also, they managed to expand their reach and almost seamlessly collaborate with great energy across global boundaries. This supported beautifully the second sub-element in engagement that resulted in the *expansion of diversity and inclusion options*.

There has been a decade leading to the need of evidence in progress for this sub-element, yet the timing and excelling in using digital, coupled with the cultural shifts that organizations needed to make, set us up nicely for the next decade with sponge-like ability to respect and absorb views that are different and unique. When we finally listen with empathy, rather than with our own opinions getting the way, this is bound to lead to a clear maturity trend for the future project leaders and their teams.

The IBM multifaceted excellence example in the next section covers a number of the elements in the framework, including this element of engagement. The highlighting of the skills storm ahead is quite intriguing. The future digital project team members, PMOs, and the likely several impacts on talent, certifications, and the future of work, are a great collection of predictions that support the complexity experienced when many of the pillars in the book come together to shape these ever-changing ways of achieving organizational excellence.

4.17 Excellence in Action: IBM[10]

Abstract

2020 was an important lesson for every business in the importance of agility and resilience to business success. Operations can be devastated by unexpected events whether pandemic, foodborne illness, severe weather disruptions, geopolitical change, or even international trade policy changes. Executives know this all too well and continue their efforts to create more resilient operations across enterprise functions.

Today's operations must by dynamic, responsive, and interconnected to an organization's ecosystem and workflows. This requires end-to-end enterprise visibility, real-time insights, and decisive actions – particularly in escalating situations. Businesses that transform their core business processes into intelligent workflows with the help of exponential technologies like AI and automation are poised to address today's workforce shifts, supply chain challenges, and customer service disruptions – and to thrive in tomorrow's recovering market.

Rise of Intelligent Automation

Over the decades, automation has touched most industries – from the factory floor to banking transactions and oil refineries. But intelligent automation enables change at a whole new level. Artificial intelligence (AI) and automation—together they become intelligent automation – alter the way humans and machines interact, in terms of how data is analyzed, decisions are made, and tasks and activities within a workflow or system are performed. In addition to potential costs savings, intelligent automation can dramatically enhance an enterprise's ability to respond, adapt, and thrive in a challenging market. Organizations that build a robust automation program combine a broad set of technologies, including robotics, bots, and devices, with AI capabilities such as machine learning, natural language processing, augmented intelligence, and computer vision and hearing. A blend of the appropriate technologies for the task at hand supports intelligent workflow efficiencies but can also help drive revenue and profit.

The IBM Institute for Business Value, in collaboration with Oxford Economics, conducted a comprehensive study to discover more about the impacts of intelligent automation initiatives today (2021) and in the near future. They posed a broad range of questions related to automation investments, priorities, benefits, and impact to 1,500 executives around the world, representing 21 industries in 26 countries.

Executives surveyed reported that in the next three years, the two most important elements in defining competitive advantage will be the customer experience and workforce skills

10 Material in this section was provided by Jim Boland, Project Management Global Centre of Excellence Leader, IBM Global Services Ltd.

Figure 4-9 Customer Experience and Workforce Skills

(see Figure 4-9). They also tell us that digital initiatives can greatly influence those elements: When asked where their digital initiatives deliver the highest value to their organization, 75 percent of respondents point to customer experiences, while 64 percent say workforce management.

The Skill Storm

Throughout economic history, talented humans have been a source of innovation and advancement – their skills the impetus for economic growth. Today, however, multiple factors, including continued – and rapid – technological developments and business and operating model innovation, have contributed to market shifts that are redefining industries. Combined with various economic and market disruptions, as well as significant demographic shifts in many countries, these factors have created a perfect storm that is affecting the value of, need for, and availability of workforce skills. The result is a looming global talent shortage with the power to severely impact individuals and economies worldwide.

The digital era has provided the opportunity and the need for speed – and that, in turn, has led to new ways of working. Remote working, always-on access, transparency, less hierarchy, pop-up teams operating across functional and organizational boundaries, and organizations operating within an ecosystem of partners all require a cultural of agility and, in turn, new skills for the workforce. In a recent IBM research study executives' views regarding the priority of critical skills have taken a turn from digital and technical to behavioral. In 2018, soft skills dominated the top four core competencies global executives seek (see Figure 4-10).

Executives' responses indicate workers require a blend of both digital skills and soft skills – also called behavioral skills – to be successful in the workforce. Why the growing importance of behavioral skills? Several factors are likely at play. The last few years have been marked by significant investment in technical skills. Indeed, entirely new areas of expertise, such as data science and machine learning, have saturated nearly every industry in a new business environment laden with powerful technology.

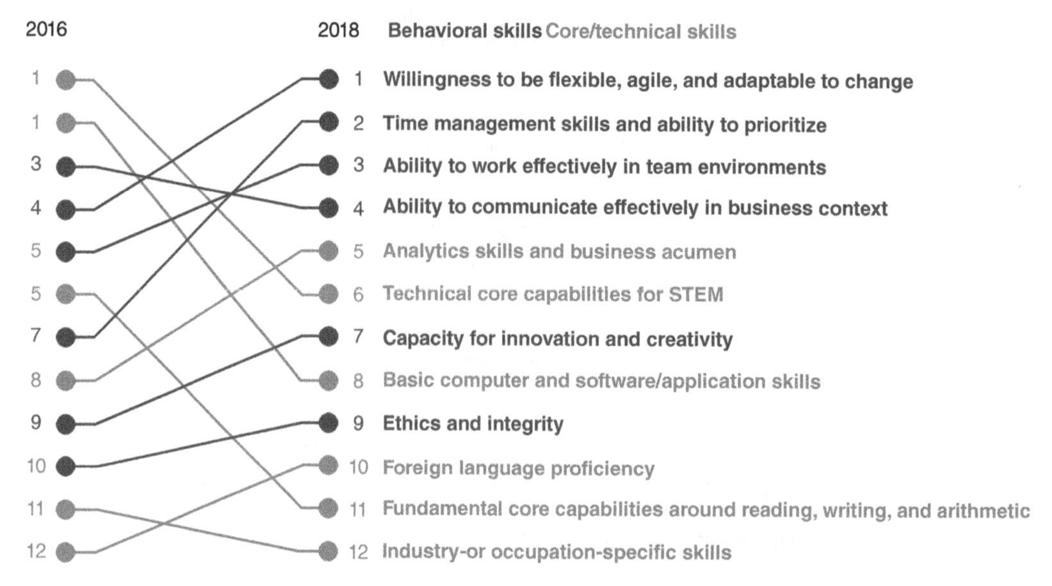

Figure 4-10 Behavioral Skills as the Most Critical

While organizations still struggle to address gaps in technical skills, there have been significant efforts and investments to address these gaps at multiple levels to lessen the impact on organizations. Executives are now tasked with continuously innovating and succeeding in this constantly evolving landscape. And they recognize that navigating it requires individuals who can communicate effectively, apply problem-solving and critical-thinking skills to drive innovation using new technologies, and draw and act on insights from vast amounts of data. It also calls for creativity and empathy, an ability to change course quickly, and a propensity to seek out personal growth. Expectedly, teamwork and organizational flexibility top executives' list of most important attributes for successful innovation.

Future of Project Management

Data from the Project Management Institute's 2020 *Pulse of the Profession®* reveals that 11.4 percent of investment is wasted due to poor project performance. And organizations that undervalue project management as a strategic competency for driving change report an average of 67 percent more of their projects failing outright. The data also showed that most organizations place nearly an equal emphasis on developing leadership skills as they do technical skills (65 percent and 68 percent, respectively).

IBM is a leading global hybrid cloud and AI, and business services provider. We help clients in more than 175 countries capitalize on insights from their data, streamline business processes, reduce costs, and gain the competitive edge in their industries. Nearly 3,000 government and corporate entities in critical infrastructure areas such as financial services, telecommunications and healthcare rely on IBM's hybrid cloud platform and Red Hat OpenShift to affect their digital transformations quickly, efficiently, and securely. IBM's breakthrough innovations in AI, quantum computing, industry-specific cloud solutions and business services deliver open and flexible options to our clients. All of this is backed by IBM's legendary commitment to trust, transparency, responsibility, inclusivity, and service.

IBM has been acutely aware of the strategic importance of project management since the inception of the IBM Project Management Center of Excellence (PMCOE) in 1997. Project management, leadership, and inter-personal skills – often referred to as Power or Behavior Skills – together with knowledge of emerging technologies are seen as essential for our project management professionals.

According to Orla Stefanazzi, IBM's Global PMCOE Communications program lead "*To maximize the benefits of disruptive technologies, project professionals must embrace how artificial intelligence and game-changing tools can empower them for the future. The IBM Project Management Center of Excellence is embracing disruptive technologies to empower and support our global project management community in the delivery of their projects.*"

The focus on personal development and acquiring future skills is paramount across IBM.

According to Jim Boland, IBM's Global PMCOE Leader

> *Fundamental to the success of IBM's transformation is the skills and abilities of its workforce. Every IBMer is being asked to regularly evaluate their skills, and be ready for the marketplace of tomorrow. There is a culture of continuous learning across the entire enterprise, enabled by IBM's Your Learning platform. We all need a broad range of skills to be successful, for example, project managers need to understand AI and Hybrid Cloud, Consultants, and Architects need to understand project management and so on.*

1. Embracing Intelligent Automation

Project managers need to embrace what were once emerging technologies, such as Cloud, AI, and Automation and think how they can best utilize these enablers across the full lifecycle of their projects.

As a Cloud and AI organization at its heart, IBM is at the forefront in embedding these technologies both internally and externally.

Our project managers need to consider:

- How they can use AI to assist their decision-making process, identify and solve risks and issues etc.
- How they can use automation in their project teams to empower people and free up their time for higher-value work e.g., capturing of meeting minutes, assigning, and following up on actions, escalations, etc.
- How can digital bots and other automation techniques be embedded in their solution set, e.g., reengineering a business process with a combination of human and machine resources

The following are some notable examples of intelligent automation in action across IBM

I. IBM's Project Health Dashboard

This dashboard helps our project managers identify where a project presents early signs of a risky "delivery." It uses a Predictive Risk Score, developed by our data scientists, to extract the learning from historical data by analyzing projects that both went well and those that did not and identifying attributes that might have contributed to this result. This is achieved through Correlations Data Analysis (Descriptive Statistics, regression, multi-Variate statistics, time series), identification of attribute of interest and model definition using Random Forest and Gradient Boosting regressor. The output of the prediction provides a list of the most contributing factors to the risk and why.

For example, do we have a good track record to deliver projects on target with this customer? Have we been successful over the last years in delivering this type of solution in this geography or market area? At this stage of the project, are our financials aligned with past successful delivery of similar project size?

II. Digital PMO Worker (Bot)

A project manager's work is never done. Managing clients, team members, vendors – and then there are operational tasks. From various project reporting, on and off boarding, plan tracking, task assignment – the hours spent on manual data extraction, preparation, review, and follow up are staggering.

Digital PMO will autonomously complete most of these operational tasks with the click of a button, providing valuable tracking and planning support to the project manager. Digital PMO has led to significant saving of effort and time for the project managers enabling them to focus on client and team management tasks and critical issues. Digital PMO is picking up new skills to become more proficient e.g., automatic task tracking and assignment, staffing forecast and risk prediction.

The Digital PMO implements the workflow of each PM operation skill with set of automation components orchestrated in runtime to perform a standard set of tasks. One or many skills are configured specific to an account requirement. The project manager can configure the Digital PMO to schedule automatic execution of specific skills when desired or execute on demand.

Several different styles of work can be supported by automation. These styles can be defined by the combination of "uniqueness of work" and "volume of work" as in Figure 4-11.

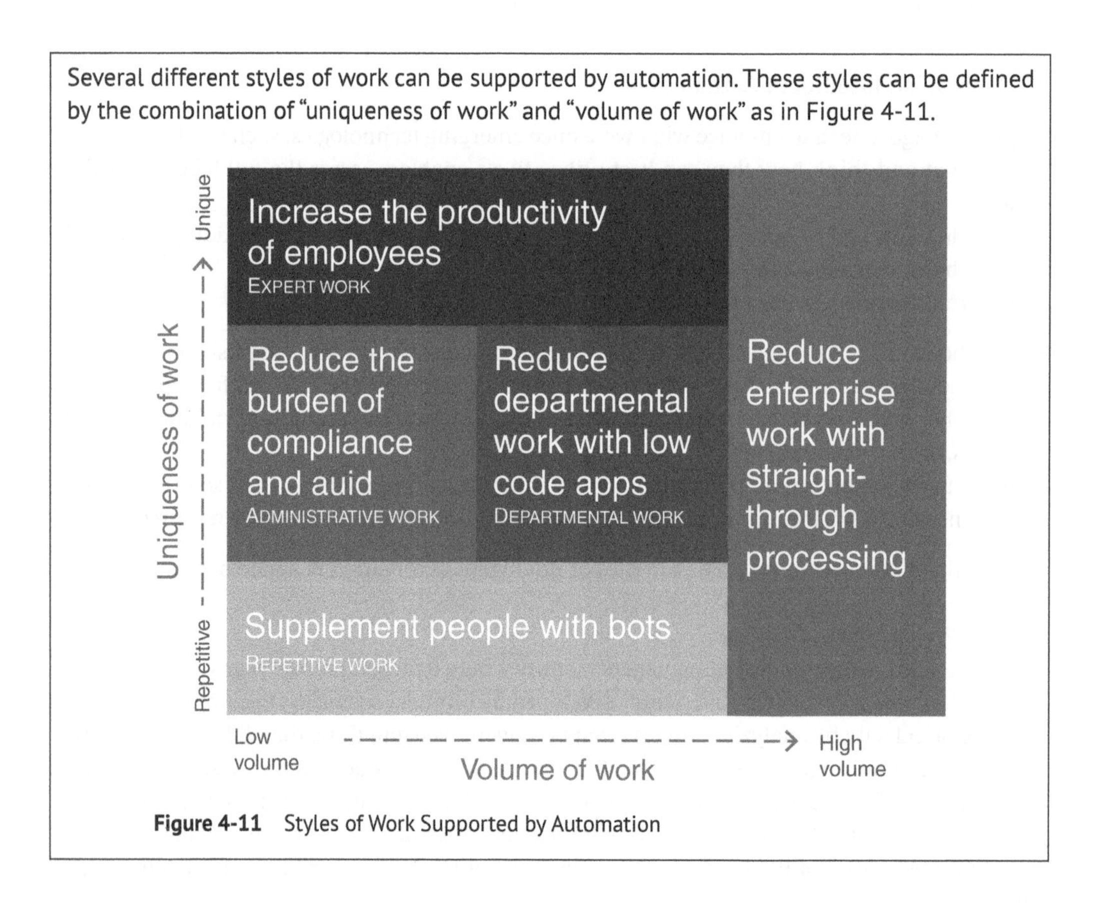

Figure 4-11 Styles of Work Supported by Automation

III. Dispute Manager Digital Worker

This is a Digital Worker or bot, which augments the skills of a human in processing invoice disputes. It collects and digitizes disputes with relevant supporting documents, identifies missing data, validates dispute cases, performs dispute approval/rejection, or need based escalation, and processes the dispute before finally completing/closing a dispute case.The Digital Worker collaborates with a human for critical decision-making, enabling the human counterpart to interact with it via a human in the loop (HITL) interface. It provides a unified dashboard for real-time monitoring of the dispute management process, and utilizes data collected throughout the process to generate reports and KPIs.

IV. IBM Project Management Center of Excellence AI-Powered Virtual Assistant (Hugo)

Hugo is an AI-powered virtual assistant built using IBM Watson that assists our global project managers in all their questions and queries relating to project management. It is a self-learning engine that resides on top of various knowledge repositories across IBM. Our project managers interact with the portal and search and locate the most relevant information to suit their requirement – text, video etc. The portal will interpret the questions asked and answers provided to continuously enhance its capabilities.

2. Transforming Process, Methods, and Tools

Successful implementation of projects and programs requires a management system that addresses all aspects of planning, controlling, and integrating with business and technical processes.

IBM's structured project management system addresses delivery challenges to reliably deliver business commitments to its clients:

- Risk is managed more effectively because the project is properly defined, within the client's business environment, with risks clearly identified and managed.
- Productivity is increased by a clear definition of roles, responsibilities and deliverables resulting in faster startup through the use of knowledge management, less rework, and more productive time in the project.
- Communication is easier and clearer because client and IBM project teams form more quickly and use common terminology.
- Client visibility to the project plans, schedule, and actual performance against the project objectives is enhanced, helping to increase client satisfaction.

IBM's comprehensive project management solution has three dimensions: coverage, depth, and scope applicable to projects and programs.

The first dimension is **Scope**. IBM has developed the enablers and professionals needed to manage the delivery of projects and programs of all sizes and complexity. These enablers include a full scope PM method, a PM Tools Suite, PM management systems and a staff of PM professionals that are trained and experienced in these enablers. The enablers are integrated so that they complement and support each other.

The second dimension, **Coverage**, ensures the enablers (method, tool suites, and processes) are comprehensive and scalable to appropriately serve the requirements of the enterprise's management team, from projects to programs and portfolios. IBM's PM professionals also have a range of skills and experience from project manager to executives.

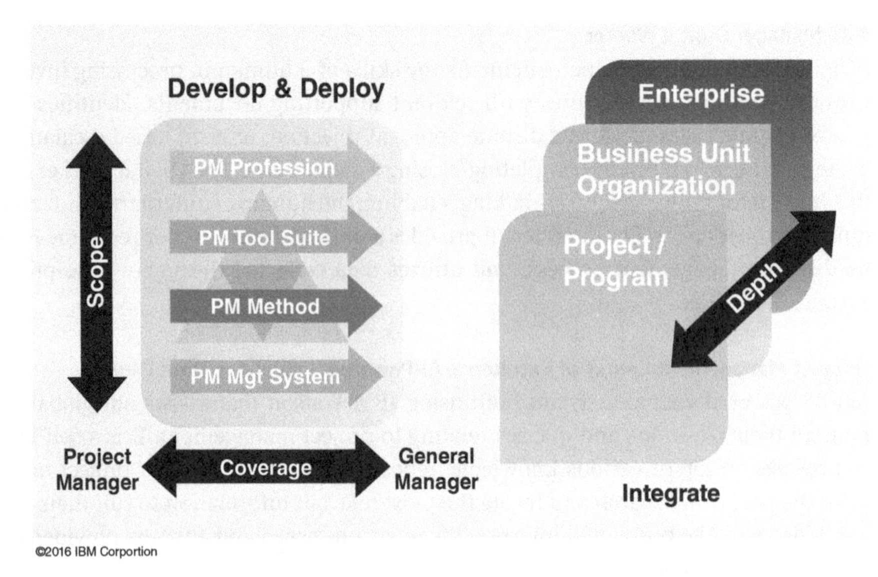

Figure 4-12 The Three Dimensions of IBM's Project Management System

The third dimension is **Depth**. Depth addresses the integration of project/program management disciplines and data with the management systems of the enterprise at all levels.

In summary, the project management approach involves building PM deliverables that have the full scope of items needed to implement and control the delivery of a project or program, have the coverage to be applicable from the top to the bottom of the organization and have the depth to be integrated into the very essence of the enterprise.

IBM's structured approach to managing projects and programs includes understanding and adapting to meet our clients' needs and environment. A Project Management System is the core of this structured approach.

I. IBM's Project Management Method (WWPMM)

IBM Methods provide guidance on how to build and deliver client solutions in a proven and consistent manner. They provide market differentiation by leveraging our Intellectual Capital, unify capabilities for integrated propositions and value creation. Methods embed our unique culture and approach directly into day-to-day work and make solutioning and delivery more robust. Applying Agile, DevOps & Design Thinking practices, as well as Cognitive tooling helps accelerate the value delivered to our clients.

To provide its project teams with consistent methods for implementing project management globally, IBM developed the Worldwide Project Management Method (WWPMM), which establishes and provides guidance on the best project management practices for defining, planning, delivering, and controlling a wide variety of projects and programs. The goal of IBM's project management method is to provide proven, repeatable means of delivering solutions that ultimately result in successful projects/programs and satisfied clients.

WWPMM covers the core domains of People, Process, and Business Environment in depth and specifies PM work products contents. WWPMM is comprised of three distinct PM Methods – WWPMM, WWPMM (Program) for the management of complex programs, and WWPMM (Agile)

for the management of Agile or hybrid engagements. It includes work products that align with Agile approaches.

WWPMM describes the way projects and programs are managed in IBM. It is documented as a collection of plans and procedures that direct all project management activity, and records that provide evidence of their implementation. To be generic and applicable across IBM, the project management method does not describe life-cycle phases but, rather, PM activity groups that can be used repeatedly across any lifecycle. This allows the flexibility for the method to be used with any number of delivery approaches (Predictive, Agile, and Hybrid) and life cycles.

An essential element of WWPMM is that it is dynamic, responsive, easy to consume and adaptable to meet the requirements of today's projects and to be of benefit to our global project managers.

There are key benefits in helping and enabling teams to adopt methods in a simplified manner including:

- More consistency in how we solution and deliver client projects.
- More efficient project delivery leveraging methods and common work practices.
- Better integration across business units for multi-offering and complex projects.

The Project Management Method is just one of many technical and delivery methods that exist across IBM. Another great example of a delivery method is the IBM Garage, which is discussed later in this section.

WWPMM is hosted on IBM's Method Workspace platform presenting approved method content in a highly structured format, enabling IBMer's to efficiently find, tailor, and adopt the content to fit their project profile (scaling up and down as needed). Key to this is the ability for users to easily consume best practice work practices and products.

> *The Method Workspace provides IBM's project managers with a central source for project best practices. The project manager can choose amongst business areas or offerings to find the right delivery method, which has IBM's Project Management Method embedded within it. The workspace provides project managers the flexibility to tailor both delivery and project management methods to meet geography, business line, or client specific requirements while still preserving consistency in method adoption.*
>
> *– Orla Stefanazzi, IBM's Global PMCOE Communications program lead"*

II. IBM's Project Management Tools

In today's complex hybrid environment, the requirement for a suite of tools that are easy to adopt, use and that can scale up and down according to your project profile has never been greater.

IBM's Strategic Project Management Tool is the IBM Program Work Center (IPWC). IPWC incorporates best practices in the form of tried and tested templates for various project and program types from across the business, whilst enabling customization to specific client needs. In addition to comprehensive reporting capabilities, IPWC has a built-in project scheduler along with specialized work centers to manage risks, assumptions, issues, dependencies, changes, business benefits and documents. IPWC has interfaces with other tools such as IBM's Project Financial Tools, Jira, Box, and many others. IBMs WWPMM is fully integrated with IPWC.

The Project Journey Guide (PJG) developed and used by IBM Global Business Services is another great example of an application that has been designed with the project manager in mind.

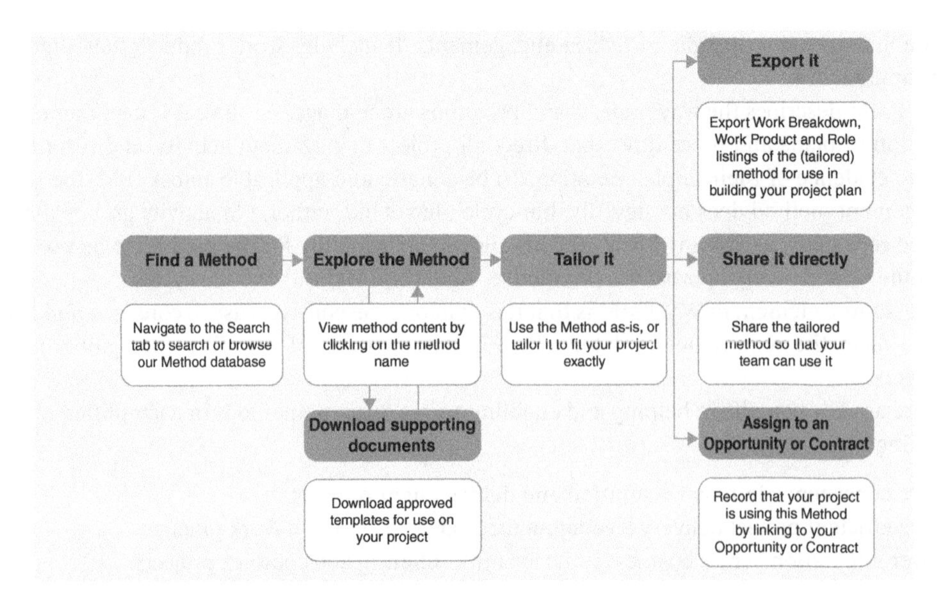

Figure 4-13 Using IBM's Method Workspace

The application helps the user easily navigate through the complex requirements involved in launching and closing their project.

It provides a step-by-step guide on

- What needs to be done
- When it needs to be done
- Which steps to follow to complete the tasks, and how the tasks relate to each other
- What the status/progress is of the project's launch/closure

The PJG provides the project team with a series of activities that need to be completed combined with target completion dates based on the project's start / end date. The PJG's radar view allows teams to focus their attention on the activities that need to be done first, and then continuing to those that can be addressed later. It is easy to identify if the project's launch or closure is on track or falling behind.

The PJG provides several levels of details on how to complete each activity. New and less experienced PMs can learn on the job using the guide to help them navigate through the critical phases of their project.

III. IBM's Garage Methodology

The IBM Garage Method is an approach that adopts flexible methodologies to meet the changing nature of projects.

The IBM Garage embodies the DNA of a startup to create a new way of working. In Garages around the world, companies both large and small are empowered to create engaging solutions. Garage teams help you start your transformation in workshops, where you design and build real working MVPs that meet your business needs and scale those solutions across the company.

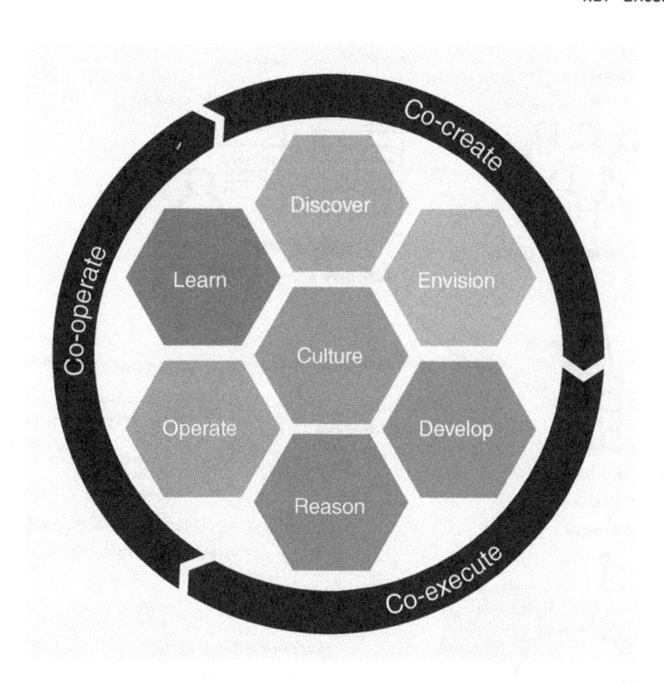

These integrated practices are grounded in experience – from implementation at global scale through culture change – into a single, seamless end-to-end approach: the IBM Garage Methodology (Figure 4-14). The Methodology drives Enterprise Design Thinking at scale, is built on Agile principles for co-located and distributed teams, leverages DevOps tools and techniques for continued delivery and operations, fosters digital talent and culture change, and enables Site Reliability Engineering.

At its roots, the Garage Methodology is a cultural movement; it's all about people.

The key principles of the methodology are:

01	Design and deliver solutions that delight consumers and achieve measurable business outcomes.	06	Business value is achieved when the solution is in reliable, secure production.
02	Learn from your users by testing hypotheses and adjust based on your results.	07	Autonomous teams are more productive.
03	Deliver value iteratively and rapidly.	08	Diverse, multidisciplinary teams innovate and solve problems faster.
04	Increase quality and reduce toil through automation.	09	Effective communication is transparent and enjoyable.
05	Don't over plan and don't anticipate. Do just enough to deliver a meaningful outcome.	10	Principles guide practices. Adapt your practices to your culture and situation.

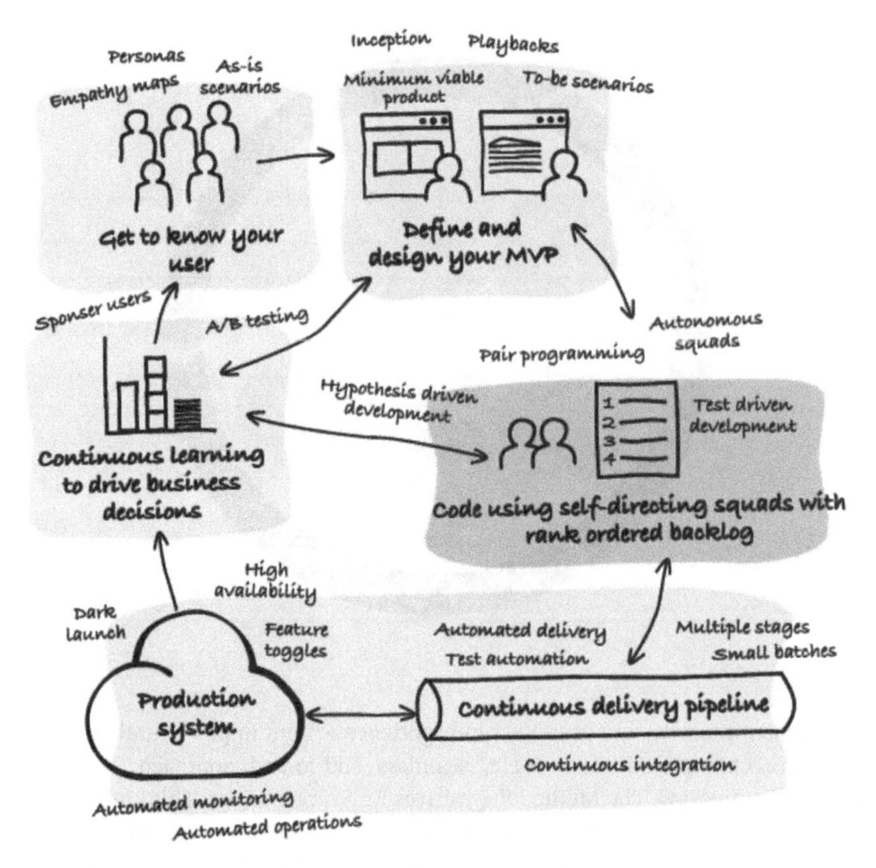

Figure 4-14 Using IBM's Garage Methodology for Cloud Applications

3. Project Management Certification

The IBM Project Management profession has established an end-to-end process to "quality assure" practitioners' progress through their project management certification journey.

Professionals can enter the profession at different levels, depending on their experience and skills. Validation of a practitioner's skills and expertise is accomplished through the validation process. The validation process, outlined in Figure 4-15, is composed of accreditation (at the entry and mid-range levels), certification (at the higher, experienced levels), and at recertification (to ensure profession currency).

Each badge requires a combination of education and experience, demonstrating that the applicant has the required skill levels and capabilities across all project knowledge areas. Certification levels also include a series of validation interviews with subject matter experts (SMEs) as well as evidence of giveback to the PM profession. The PMCOE supports IBMers in their certification journey, including bootcamps, webinars as well as mentoring and coaching.

Education is not limited to project management content – there is growing focus on skills and education in areas such as design thinking, Agile mindset, leading teams, collaboration, and other behavioral skills that have become fundamental to their success.

PROJECT MANAGEMENT – ACCREDITATION AND CERTIFICATION

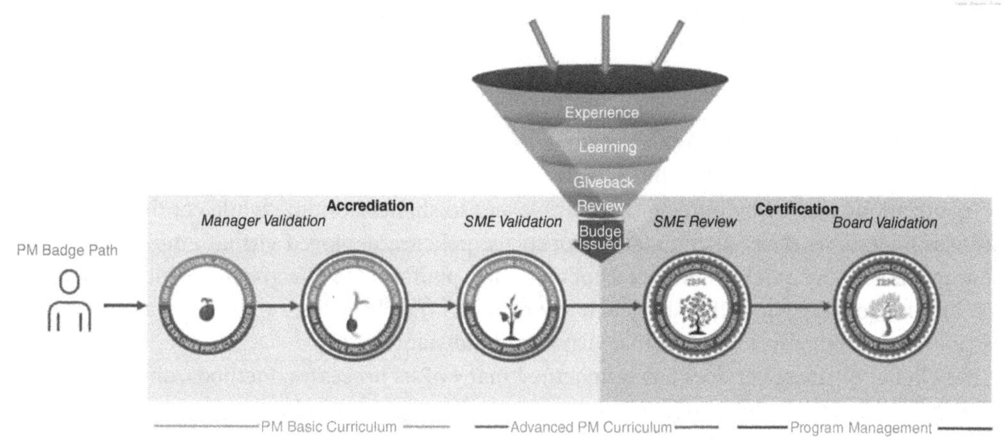

Figure 4-15 Validation Process

Coaching and/or Mentoring is also key to the ongoing success of project management capabilities across IBM. IBMers applying for project management certification are required to have a PM mentor/coach assigned to support them in managing their way through the process.

IBMers who are successful in achieving certification are then encouraged to mentor/coach other practitioners on their project management journey – in fact, mentoring fellow IBMers is a mandated prerequisite to the recertification process, as well as other ways of giveback.

4. Skills Development

We draw an important distinction between project management learning and learning for project managers. Employees in many job roles will benefit from an understanding of project management and for the project manager, strong skills in project management alone are not sufficient for success.

Consequently, our core curriculum exists to teach the discipline of project management to any employee who will benefit, irrespective of their job role and over 50 percent of the learning is taken by employees who are not job coded as project managers. Beyond this core of structured learning, we offer a wide curriculum of content covering areas such as additional project management techniques, personal development, leadership, industry insights, tools, and technology.

We provide the structured learning path around our core curriculum to ensure that our project managers have a solid grounding in the breadth of the PM discipline. Project management roles are however very varied in their specific requirements and so we also provide a wide breadth of learning that can be accessed to best meet the needs of the individual and the role.

The future will not be a polarized world of predictive project management vs Agile. Successful project managers will be able to select the right approaches for their specific context and bring them together into a customized project management system. In our curriculum, we increasingly seek to blend different approaches together and equip our project managers for this hybrid world they will inhabit.

Perhaps most importantly of all it is essential to encouraging a growth mindset and a culture of continuous learning in the PM Community. This, combined with a breadth of available learning, is what will ensure that PMs can embrace change in the profession and flourish.

5. The Future of Work

As already stated, COVID-19 has had an unprecedented impact not only on health and humanity, but also created a ripple effect on businesses across industries.

Organizations reacted by moving their employees to shelter-at-home while continuing critical business operations. They also updated workplace policies, deployed virtual employee engagement processes, and quickly embraced collaboration platforms to effectively execute work virtually. In this evolving world, the physical barriers of an office have been removed with the shift to distributed, self-empowered teams in a virtually equidistant world.

IBM Global Business Services has reimagined many of its processes, methods, and tools under this service delivery model in this new era to not only prepare our clients to help absorb shocks from unforeseen events but also be able to help accelerate digital transformation despite disruption.

According to a 2020 IBM Institute for Business Value study, 62 percent of C-Suite executives now plan to accelerate digital transformation efforts over the next two years.

However, as organizations adjust to the new normal, the need for accelerating Digital Reinvention® and cloud adoption has become a crucial lever for growth.

IBM's perspective on Digital Reinvention envisions a Cognitive Enterprise™ powered by intelligent workflows, underpinned by hybrid multi cloud. However, navigating the journey in the uncharted waters of remote, contactless work requires a resilient delivery model designed for both virtual and physical work patterns.

IBM Global Business Services has evolved and strengthened its services delivery model to help their clients address the new requirements and continue to drive sustained value. They call this holistic model *Dynamic Delivery*. It integrates technical foundations and humans in the network with virtualized methods, enhanced with AI and automation, to support rapid adaptation to changing business conditions.

Going beyond just virtual collaboration or automating select processes, it represents the next generation of service delivery necessary to help businesses realize value in the new normal.

Dynamic Delivery comprises three key components, each with defined set of capabilities:

1) **Delivery foundation:** Fulfills the technical requirements necessary for a flexible service delivery model.
2) **Humans in the network:** Provides the roadmap for building and managing virtual teams, developing and deploying new engagement methods, enhancing leadership models, and enabling cloud-based knowledge management.
3) **Contactless delivery:** Uses the necessary methods and tools framework enabled by extreme automation and seamless virtual collaboration.

These components and capabilities are based on insights IBM Global Business Services gained from virtualizing its own operations across the globe and successfully delivering multiple client engagements since the onset of the pandemic.

> *"Bringing all of this into a holistic model, with a robust and scalable set of tools, processes and methods enables our project managers get on with the job of delivering."*
>
> *Jim Boland, IBM's Global PM Centre of Excellence Leader.*

Summation

To deepen project leaders' understanding of the major developments reshaping our world, the Project Management Institute has identified five megatrends, which all share a common thread: they are exacerbating the endemic exclusion, disruption, and discontent that have crept into our society. These megatrends are Covid-19, Climate Crisis; Civil, Civic, and Equality Movements; Shifting Globalization Dynamics and Mainstream Artificial Intelligence. Each of these trends impacts project management.

IBM's continued commitment and focus on empowering project managers to understand and embrace disruptive technology and on developing the necessary skills for the future will enable our project managers to be successful in tomorrow's projects.

They must become acutely aware of the environment in which they find themselves.

They must

- Manage remote teams that have been virtual for over a year and know how to adapt post Covid-19, to a new post-pandemic reality.
- Lead ambitious, innovative projects to tackle the climate crisis and understand climate impacts on project decisions.
- Foster diversity and inclusion in their project and client teams. Speak out where individuals are marginalized or excluded.
- Embrace and support youth and talent across project initiatives.
- Leverage intelligent automation to enhance project delivery and realize significant benefits for both IBM and our clients.

In a world of rapid technological and disruptive change the project manager of tomorrow will certainly be faced with challenges. . . But who better to lead this change than those that understand the scale and what's at stake? Those who have proven their capability to be agile, flexible, and committed and who deliver strategic, financial, commercial, and operational benefits to their organizations time and time again. The future of the project manager – and project management – really does look bright.

The Authors

Jim Boland – Leader, IBM Project Management Center of Excellence

Orla Stefanazzi – Communications Program Lead, IBM Project Management Center of Excellence

Noel Daly – Project Management Profession Lead, IBM Project Management Center of Excellence

Michael Coleman – Project Management Learning Program Manager, IBM Learning

About IBM

IBM is a leading global hybrid cloud and AI, and business services provider. We help clients in more than 175 countries capitalize on insights from their data, streamline business processes, reduce costs, and gain the competitive edge in their industries. Nearly 3,000 government and corporate entities in critical infrastructure areas such as financial services, telecommunications and healthcare rely on IBM's hybrid cloud platform and Red Hat OpenShift to affect their digital transformations quickly, efficiently, and securely. IBM's breakthrough innovations in AI, quantum computing, industry-specific cloud solutions and business services deliver open and flexible options to our clients. All of this is backed by IBM's legendary commitment to trust, transparency, responsibility, inclusivity, and service.

For more information, visit www.ibm.com

References

Karen Butner, Tom Ivory, Marco Albertoni, and Katie Sotheran (2020). Automation and the Future of Work. IBM Institute for Business Value.

Annette LaPrade, Janet Mertens, Tanya Moore, and Amy Wright (2018). Enterprise guide to closing the skills gap RESEARCH INSIGHTS. – IBM Institute for Business Value.

Project Management Institute (2020). *PMI Pulse of the Profession.*

Project Management Institute (2021). *Megatrends 2021.*

Appendix: IBM Project Management Center of Excellence History

In the mid-1990s, industry dynamics such as worldwide competition, resource pressures, rapid change in customer segments, and technology drove IBM to a different organizational structure than its traditional hierarchical management approach. Additionally, IBM identified the lack of strong project management as a major contributor to project failure and client satisfaction issues across the corporation. These factors led IBM to decide to become a project-based business at that point in time, applying and integrating project management disciplines across its core businesses, processes, and systems.

In November 1996, IBM consolidated its efforts to become a project-based enterprise and established the IBM Project Management Center of Excellence (PMCOE). The PMCOE charter was to drive IBM's transition to and support of professional project management worldwide, a competency deemed necessary to effectiveness and success within a matrix enterprise. Since its inception, this charter is periodically reviewed to ensure its ongoing value and relevance.

4.18 Outcomes-Focused Work

The execution of projects in the future organization will have a much closer connection to the business value. The learning and maturing of project management processes and principles over the past decade have shown a much higher readiness to challenge the classic busy work project teams have focused on and have questioned every aspect of how success is viewed. The biggest shift has been around value streams and the hyper focus on outcomes rather than outputs. This shift has been well supported by digitalization that enabled much more effective simulations and drove the success of the incremental delivery mindset that started to dominate project work.

The future will continue to see this mindset shift to value dominate project work. This is also likely to be the case across the decision-making levels of the organization and be exemplified by organizations' boards as a way of excelling in governance in the future. Digitalization has already set us up nicely also for the second sub-element here too, with expanding the potential of feedback as technology has simplified the access to that feedback and the richness of data analytics has prepared us to look deeply behind the obvious problems causes, surface numbers, and trends, and go deeper to root causes and the complex multi-factors analyses in order to expedite delivery of meaningful results across the portfolios of initiatives.

The excellence example in the next section by Dubai Customs is an illustration of the disruptions that are taking place and help shape the future mindset shifts and the enhancements needed for organizational understanding and execution of collaboration.

4.19 Excellence in Action: Dubai Customs[11]

Dubai Customs is disrupting Project Management roles by developing "AI Virtual Assistants."

Digital transformation, and the Fourth Industrial Revolution, have been gathering momentum across all industry sectors, driven by a new wave of disruptive technologies in the future. At Dubai customs, we believe in continuous improvement and constant innovation to capitalize on the future opportunities that technological disruptions can create. In many ways, be the pioneers in the global arena by effectively utilizing new technologies as they emerge.

As digital becomes an ever-increasing part of Dubai Customs DNA, we consider improving our portfolio and project management practices maturity by shifting gears from reactive to proactive by embracing disruption with adaptability and lifelong learning in future project management roles. In line with our next steps toward improvements and innovation, Dubai Customs is currently exploring using artificial intelligence in creating a new era where humans and machines will collaborate to deliver projects, execute strategies, and provide insights to our key stakeholders.

Using our standardized project delivery capabilities today, we can improve collaboration between our essential processes, increasing efficiency and effectiveness. In addition, the project management tools utilized today assist our program and project managers in optimizing their efficacy and tracking critical metrics against KPIs and project milestones. However, these current tools provide information on where the project-related information will need to be extracted. They cannot currently predict a "what if" and future scenarios or proactively alert our project managers before a significant issue arises.

In line with the desired future scenario, the Dubai Customs project delivery department has set out to conceptualize the need for creating an AI Virtual Assistant with a customized knowledge base that will work together with our project managers providing the required tools, which push out project-related information to proactively think and do on behalf of our project managers and provide on-demand real-time project information with efficiency and effectiveness (Figure 4-16).

The idea exploration used machine learning to learn from our historical project's data, enabling the project management tools to provide further support and accuracy to the decision-making process. This set of algorithms would learn and adjust for the future, connecting all the required project activities to plan, budget, execute, and track projects effectively to manage a project from start to finish while having AI Virtual Assistant at hand helping our project managers to navigate their project forecasts better.

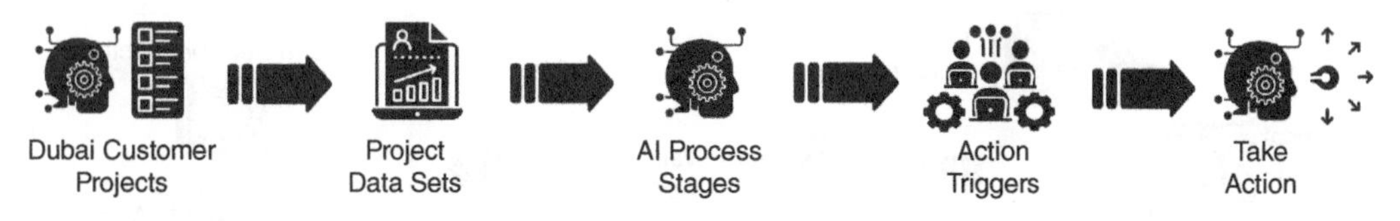

Figure 4-16 AI Disrupting Project Management

11 Material in this section has been provided by Ajith Nair, Senior Research Head / Excellence Specialist, Project Delivery Department, Dubai Customs.

Utilizing AI Virtual Assistant in Project Management

In the proof of concept, the Project Delivery department, collaborating with the innovation department, using the accelerated exploratory lab approach, collected massive amounts of project-related data from the PMO related to status and progress reporting, developed for different stakeholders and top-level management. The AI Virtual Assistant would assist the project managers by performing the administrative and repetitive tasks on their behalf, enhancing their skills (Figure 4-17).

Necessary real-time actions the AI Virtual Assistant would perform are as follows:

- Data Management: Entry, Collection, and Analysis.
- Project Schedules: Prepare and update schedules from multiple input sources
- Status Reports: Generate status reports by also flagging exceptions to alert the project managers.
- Interactive Medium: Active interaction with project managers promptly informs them of overdue tasks, schedule constraints, and requests from project stakeholders while also updating their current workload and task management.
- Auto-Assignment: Assign tasks to people with set target dates, sending out actions and follow-ups, organize meetings, analyze variations, and send reminders to the respective teams.

The collaborative teams analyzed the project reports data sets for the AI Virtual Assistant to provide better insights on trends, forecasts, and project managers' performance. The project management tools (PPM) database of risks and issues, change requests, schedules with baselining history and project managers, and resources utilization information also provided significant input into the AI system.

The developed AI Virtual Assistant provided a natural language interface to allow users to query the data in a 'human-like way, responding to project-related information, improving usability, enabling access to the information in real-time with suggested recommendations of actions to be taken by the project manager and the Project Delivery director. The AI Virtual Assistant reduced the project manager's operational workload by providing critical real-time information access to any executive/depts at an exceptional level of speed and accuracy.

At Dubai Customs, we believe in continuous improvement and constant innovation embedded in our DNA. In 2020 Dubai customs also bagged the World's largest professional award for Project Management Offices, their organizations, and leaders, winning both "World PMO of the Year" and the "Asia Pacific PMO of the Year" award titles (Figure 4-18).

The AI virtual assistant prototype is just a steppingstone on the potential forward to utilizing disruptive technologies in project management and demonstrating how project managers and AI can work together in the digital era to cope with future challenges.

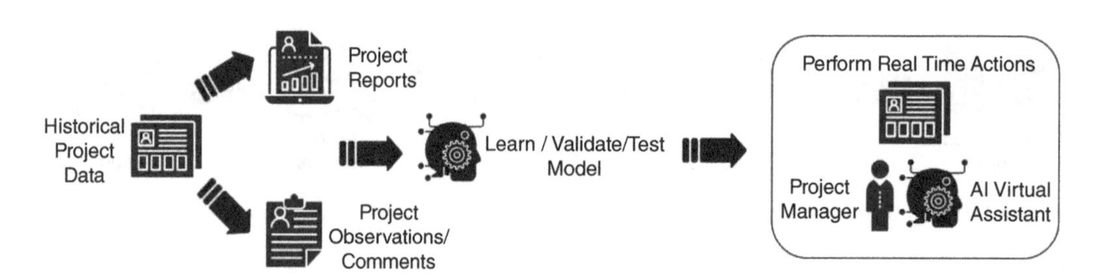

Figure 4-17 Building the AI Model

Figure 4-18 The PMO Global Award. *Source:* PMO Global Alliance.

4.20 Ever-Changing Ways of Working

Any discussions of normal are not going to be fitting in the next decade. It is an absolute must in the digital future to adapt at an increasing rate and to train ourselves to continuously reinvent ourselves and our interactions. Two sub-elements are addressed in this element of the framework. The first one is the *design for stakeholders' relevance*. Digitalization is empowering us to have a very different and a much higher appreciation for the closeness to stakeholders. No more in the next decade will stakeholders' analysis be an afterthought. AI will tremendously enable the finding of the patterns around effective engagement in order to achieve repeatability of the successful ones and build a model of organizational excellence never before possible.

Another key impact on project work with digitalization, is the *energizing of the next generation* of talent and workers. The digitally native mass of future workers that will be seen leading in the next decade will make this increasing shift to digital solutions a must have. This will contribute to a higher level of motivation and energization that links to the potential of continuously trying different things and to immediately seeing the results of the playing and testing enabled by technology.

The example by Wuttke & Team in the next section is an excellent illustration to the stretching in the project management universe into the field of software development under the umbrella of the term *citizen development*. It is a great disruption in the world of work and complements many of the shifts that organizations are bound to encounter on the road ahead.

4.21 Excellence in Action: Wuttke & Team[12]

Citizen Development Project Management

There is a new initiative in the wide world of project management. This initiative is called Citizen Development and has an admittedly strange name at first glance. The interesting thing about this initiative is first of all that it is not a new project management method or any variation of an existing approach. Citizen Development is a software development approach. So, in the first moment it is about software and not about project management at all. Project management comes into play later.

12 Material in this section has been provided by Thomas Wuttke, PMP, CSM. He is President and CEO of Wuttke & Team, a Gita brand. Thomas is Author, Speaker and Senior Trainer for Project and Risk Management ©2021 Gita GmbH, Germany. All rights reserved. Reproduced with permission.

What Exactly Is Citizen Development?

Citizen Development is the development of IT products by non-experts. Whereby the laymen are not called laymen, but citizens. In other words, Citizen Development sounds better than software development by people who have no idea on IT. It is a kind of democratization of IT development, so that in the end everyone involved can program with the help of "LowCode/NoCode" application systems.

At first glance, this sounds surprising and not promising to the ears of the IT departments of this world. Not to say as a threat. Just yesterday, the IT departments scooped up all these private developments from the business departments and centralized them again. These private individual developments also had the nice epithet "ShadowIT" and referred exactly to the uncontrolled growth beyond the control of the IT managers. Who doesn't know the tons of private spreadsheets that performed many miracles with creative macro programming and crashed many a database? Seen in this light, Citizen Development sounds like a legalization of the bad habits of recent years.

However, it is not like that. The big difference between the proliferation of some half-baked programming and other shadowy growths and the new Citizen Development idea is that this time IT is fully involved. Ideally, the introduction of Citizen Development will also happen under the leadership of the IT department.

ShadowIT

But let's take another quick look at ShadowIT. It wasn't all that bad, otherwise it wouldn't have spread so rapidly. A number of workarounds in the specialized departments were successfully implemented with the help of increasingly better software components, and they also unfolded their effective blessings.

That was the good side. What was clumsy was the uncontrolled proliferation: Hardly anything was documented. Development hung on individuals and the more powerful ShadowIT became, the more powerful monopolies of heads became. Due to fluctuation, however, it was then very quickly "back to zero" again. And if the ShadowIT not only read out data but also changed or processed it, it was only a matter of time before inconsistencies caused more or less damage.

These problems were countered, in part radically: There was an attempt to eradicate ShadowIT with stump and stick. It was banned, closed down. Only centrally approved modules, functions and/or apps were allowed. The weeds in the garden were fought with Agent Orange, so to speak.

Unfortunately, the advantages were also lost with it. ShadowIT was exactly where it was needed. Used by people who knew exactly what they needed. No scope statement had to be written, nothing had to be handed over. The creation process was "lean" in the best sense of the word. And it was necessary because the challenges did not decrease.

Citizen Development = Advantage IT

Those who nevertheless tried to implement their local departmental requirements by the professionals in the sense of integrated IT development got deadlines (in the best case) that were completely unacceptable.

What would be to think of reducing the disadvantages and further using the advantages? Voilá: That's exactly what Citizen Development is.

IT professionals can focus on the really big challenges and concentrate their resources on tasks that would not be solvable by non-professionals. Because Citizen Development does not mean that from now on everything has to be done by laymen. Of course not. But we still have the problem that IT specialists are in short supply, and an ever-growing backlog of what they see as trivial issues is piling up that are simply not being tackled for capacity reasons.

And in most cases, the implementation of these topics – regardless of whether they are trivial or not – requires intensive communication between the requester and the person performing the work.

What would you think if the requesters didn't have to describe their tasks and communicate them to IT in a complicated manner, but instead handled them themselves? And this even though these requesters do not know how to program? And then with the blessing of IT?

Programming for the Ordinary Folks

This is made possible by program generators that get by with little or no code understanding. The technical term for this is "LowCode/NoCode" and for the sake of simplicity, we will abbreviate these code generators as LoCoNoCo in the following.

The goal of Citizen Development is to solve simple requirements from the business department. These can be quite simple, such as the design of a form with integrated plausibility. However, many tools can already do this on their own. But the mapping of a workflow with integration of data from the central databases is also Citizen Development. Here, the gradual complexity already becomes visible. The transition from a simple plausibility check form to a workflow with Citizen Development to a professional application is fluid. There is no hard interface and the spheres and boundaries of influence probably have to be negotiated on a company-specific basis.

Digitization Needs Citizen Development

In the next few years, there will be so many changes on the agenda that it will be impossible for the IT departments' staff shortages to handle the entire volume. In order to keep pace with the ever faster spinning world, there must be approaches that at least partially place the power of development in the hands of the experts or even users. According to conservative estimates by leading consulting firms, the number of apps will increase tenfold (!) in the next few years. And worse: according to Gartner, about 70 percent of apps in 2025 are not even developed today. This gigantic mountain of digitization projects is waiting to be implemented and processed.

Currently, however, we do not even consider the interface between citizens and professionals. It is sufficient if we stay with the citizens. So, as stated above, the professional department now has the theoretical ability to develop and implement their own applications in a straightforward manner. The providers of LoCoNoCo solutions also suggest exactly this aspect. It is enough to think hard about the end result once and we have a ready-made app with a supplier quality control on the cell phone. Well, not quite.

Citizen Development and Citizen Development Project Management

It may be that through the LoCoNoCo platforms the development is relatively easy and can be done quite well. However, this does not mean that the underlying project – even if it is comparatively small – is as easy to manage.

If we take a look at typical project management processes, it quickly becomes apparent that scope has always been and remains a challenging issue. If requirements are not properly elicited in a LoCoNoCo environment, citizens will soon have no desire to do the IT work themselves.

And here lies the real challenge, one could also say a key Citizen Development risk. Because the technical implementation is or seems to be so simple, this simplicity is also projected onto the project ("What do we need a project for now?"). However, it would be more than a shame if this unnecessary confusion between product development and the implementation process meant that the opportunity of Citizen Development was not properly exploited.

Citizen Project Management Is Agile

Only Agile approaches are suitable for the implementation of projects in the Citizen Development environment. The stakeholders involved would not understand the idea of trying to solve problems with a waterfall approach. Rather, quick successes are needed, which are continuously expanded and extended. On the other hand, it makes little sense to use an approach like Scrum for the implementation of Citizen Development projects. The necessary framework conditions are not even available for this. What is needed is an Agile approach, not Scrum, but closely based on the good ideas and experiences that have been gathered so far.

Characteristics of Citizen Development Projects

If we assume that the business departments now create their applications themselves to some extent with the help of a construction kit and with the support of central IT, then this environment typically has the following framework conditions for the implementation of Citizen Development projects:

1) Runs alongside day-to-day business and thus always has priority
2) Very rarely will the entire department work on development.
3) No exclusive development teams.
4) Changing roles within the team.
5) No dedicated "Agile" or "Scrum" Master.
6) Product owner and team can also overlap. There is not the crystal-clear separation.
7) Depending on the size of the application, these can also be simply one-person projects.
8) Great danger of multitasking coupled with the temptation to develop a handful of apps simultaneously.

The question now is what an ideal project setting for Citizen Development applications might look like.

Although rather unpopular now, we would divide a Citizen Development project into at least three to four time periods. In the past, these time periods were called *phases*. And since we are not attached to fashion trends, we also say phases here.

The first phase can become shorter and shorter as experience progresses, and in the end, it can be omitted altogether, since this phase involves getting acquainted, familiarized, trained, and experienced with the Citizen Development environment.

Phase 1 Getting Familiar

Let's assume that the infrastructural requirements are already in place. A suitable LoCoNoCo platform has been selected, and collaboration with IT has been established and agreed on. For the first projects, an increased effort due to training and adaptation has to be added and considered. Perhaps it also makes sense in the department to agree on the ground rules for using the tool and also to ensure that not everyone wants to try out the new toy at the same time.

Phase 2 Scoping the Idea

Here the central question is in the background: What do we want to achieve? Experience from thousands of projects has taught that this stage cannot be set aside or skipped too quickly.

What scope should the application actually take on? The questions that arise here are typical for project management and are actually completely independent of the project method. Or also: Just because we are developing the whole thing agilely, we are still allowed to use our head shortly beforehand.

Very typical questions: What are the requirements of the user groups? What will the world look like after the successful implementation? Which framework conditions have to be considered? Which assumptions have been made? When must the system be available? Who is affected by the implementation? What changes for people? Does the use need to be trained? Who can all make a meaningful contribution?

If you think at first that this is all far too bureaucratic, you run the risk of repeating the same mistakes that many generations of accidental project managers have experienced again and again over the past 50 years. Resist the temptation to start coding immediately.

Ask yourself the critical question: what exactly do you want the system to improve? Engage colleagues and use the most important tool in the creative process there is: Sticky Notes. And it doesn't matter if this session takes place online or onsite. Sticky notes are available real and virtual. Let the ideas flow freely in the brainstorming session. Maybe there are approaches that you don't even see and that are even better than your first thought.

Build the backlog of your future application from these ideas. What is a backlog? Let's make it simple and say it is the consolidated collection of your ideas. Classically, it is required that the individual backlog items should also always promise a benefit and should also follow a clear syntax.

We would not be quite so dogmatic about it in the Citizen Development environment. It is sufficient to consolidate and summarize the list of ideas. The summary then leads to a backlog element, which in Scrum is called a user story. You are welcome to call it that, we call it Backlog Element. Describe this element from a user's perspective and describe the functionality. What should happen when certain actions are taken?

Example: When all criteria for acceptance are met, a log should be printed.

Now if your backlog has 10, 20, or even 50 such elements, that's not a problem. There is no upper limit, there is no closing time. You can always and at any time develop, continue, rewrite, or collapse this list. What would be a good idea, though, is for you to come up with some sort of minimum scope. Which of your let's say 36 backlog items would need to be done in order for an initial scope to work in a meaningful way?

That this sum of backlog items also unfolds a benefit, that the world becomes a bit better when the first version is deployed and running. To stay with our example: Let's say you have identified 36 backlog elements, but would be satisfied if the most important 11 from this list were already working.

Phase 3 Step into the Cycle

Phase 2 is then completed, but we can resume work on the backlog at any time and continue. Or also: Phase 2 is completed, but we can return to it at any time.

In Phase 3, we now consider the following:

1) Which of the 11 elements mentioned in our example do we want to start with?
2) Where are perhaps "low-hanging fruits"?
3) How much time do we want to take for the first swing?
4) Who will help and can work on the development on the side without neglecting the daily business?
5) When do we consider the work to be completed? This is a very important question to ask at the very beginning. In the Agile professional field, this is called the definition of done and by defining it early, you save yourself quite a bit of discussion in the end.

To stay in our example, let's say we choose 4 backlog items for the first swing and allow a month for this. We can already create a meeting and invite interested stakeholders. This also makes the whole thing a bit more binding.

We assume that during this month, during these four weeks, we will work on these four elements again and again, depending on how the daily business allows us to have this fun.

We stay in the third phase and repeat this iteration until the 11 elements are reached. And if it has become 13 in the meantime because something else has been added, that's no big deal, either.

The important thing is: We always determine the duration of an iteration at the beginning of the iteration, and we don't stick slavishly to the four or three weeks. It can also be 1 week or six. Even one after the other. Fixed iteration durations are for fixed teams and professional environments. We keep it flexible.

Phase 4 Breathe Life into It

In our example, once we have the 11 elements done at some point, it is now time to transition our application into live operation. This transfer to live operation can be done implicitly and without major inquiries because it is self-explanatory. Or we may have to organize a training session for a small group of participants, or a workshop, or a short online meeting or webinar. In any case, this is not to underestimate the communication for the operation. Just because we were busy with this application the whole time, does not mean that our colleagues in the immediate vicinity understood the last few dodges.

4.22 Digitalization and Projects Path Forward

The future of work will continue to see digital transformation initiatives becoming the norm to the point that the term might not even be mentioned much anymore. As mentioned in the *Insights Series of the PM Work Journal*,[13] when the pressure of 10X scaling remains to be a strategic priority for world organizations, it is vital to have an end-to-end well connected ecosystem approach for the success of digital transformations. The Success model shown in Figure 4-19 is a proposed model for achieving such success. It is going to be critical that we have the right balance of culture, system, and processes closely checked and aligned around the aspirations of excellence in the future organizations.

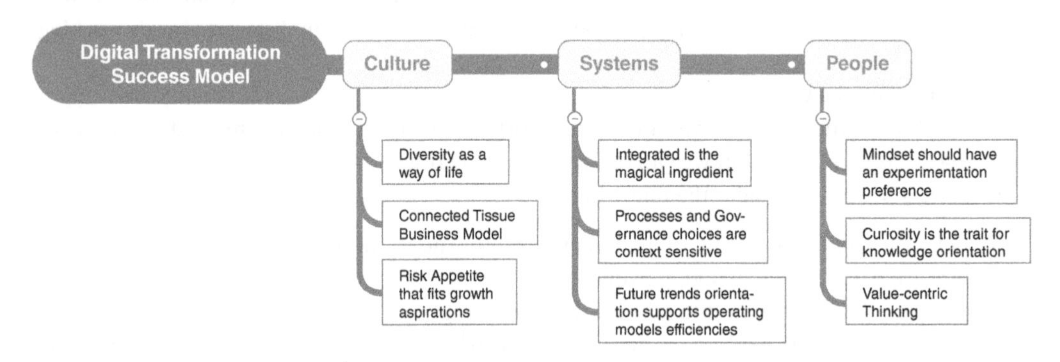

Figure 4-19 The Digital Transformation Success Model

13 Insights Series, Author: Al Zeitoun (2021, August), *PM World Journal* (ISSN: 2330-4480) Vol. XI, Issue VIII, www.pmworldjournal.com.

5

Pillar 5: Evolving Project Delivery Skills

5.0 The Changing Landscape

The environment in which the project managers perform is changing significantly, especially those involving strategic projects in the global marketplace. Our projects are becoming more complex. The importance of time and cost are no longer the only success criteria in the minds of clients and stakeholders. Clients want to see the business value identified in the projects they are funding.

All new processes, such as flexible methodologies needed for strategic projects, are accompanied by advantages and disadvantages. Sometimes, the advantages and disadvantages are not readily apparent at first, and this creates problems when designing educational programs for new skills for potential project team members. An example would be the improvements in requirements management practices resulting from using flexible methodologies.

In traditional projects, using a one-size-fits-all approach, there existed a heavy reliance upon the statement of work accompanied by the belief that, once defined, it would remain the same throughout the duration of the project. Flexible methodologies made it clear that validation throughout the project would be necessary because of evolving client needs. Collaboration with the client and stakeholders would be ongoing rather than just at the initiation of the project and at closure.

Flexible methodologies made it apparent that project managers are managing part of a business rather than just a project and are therefore expected to make business as well as project-based decisions. Project teams now have access to information warehouses and knowledge management systems, accompanied by a line-of-sight to senior management such that they can now perform most of the duties conducted previously by project business analysts. Business-related project metrics are now part of project management information systems. This becomes important when analyzing the evolving scope. Requirements management practices are now seen as a core competency for project management teams. This leads to new skills needed for continuous problem solving and decision-making.

Project managers must learn new skills to prepare for challenges of future projects. Project decisions are no longer a single person endeavor. Historically, critical decisions related to business value were made by project sponsors and senior management. Today, project managers are expected to form and lead problem-solving and decision-making teams. Part of this is being driven by pillars discussed previously, namely the growth in strategic projects, the need for team empowerment, and new project leadership styles.

Most project managers have never been trained in problem solving, brainstorming, creative thinking techniques, and decision-making. They relied on experience as the primary teacher accompanied by well-defined business cases and statements of work. While that sounds like a reasonable approach, it can be devastating for ineffectively prepared project managers that end up learning just from their own mistakes rather than the mistakes of others.

5.1 Problem Solving and Decision-Making

Project problem solving and decision-making skills could become the greatest challenge facing the next generation of project managers. On traditional projects, problems were solved, and decisions were made, based on the information in a well-defined business case and statement of work. Additional information was found in the standard forms, guidelines, templates, and checklist dictated by the one-size-fits-all methodology. Project managers relied on experience and, if necessary, assistance from the project sponsor.

Today, the environment in which the project managers perform is changing significantly. Our projects are become more complex, and many have strategic implications. The importance of time and cost has reached new heights in the minds of clients and stakeholders. Decision-making in the future may become value-based decision-making. All of this is creating challenges for project managers in how they identify and make decisions to resolve problems. To make matters more complex, project managers are now seen as managing part of a business where decision-making may be in areas unfamiliar to the project manager. Project teams must be educated in problem identification and resolution techniques. The failure to do this can have a serious impact on strategic project outcomes.

Challenging Project Management Scenarios

Below are six situations that illustrate how even some simple projects can have complex problems requiring difficult decisions to be made, especially the types of decisions that project managers are unaccustomed to making.

Situation #1: The Phoenix Project

Scott had heard rumors about how much trouble the Phoenix Project was in. People were leaving the Phoenix Project like rats deserting a sinking ship. Scott was grateful that he had nothing to do with the Phoenix Project. Unfortunately, all that was about to change.

Scott was a well-disciplined project manager. He was a PMP® and was excellent in following the enterprise project management methodology his company had developed. The forms, guidelines, templates, and checklists provided by the methodology made managing projects easy. Sometimes, Scott even believed that it was like a "no brainer" to function as a project manager. Everything was laid out for him, and all he had to do was to follow protocol.

On Friday afternoon, Scott was informed that the project manager on the Phoenix Project was reassigned and the Scott would be taking over the project. Furthermore, the executive made the following assertions:

- "We are thinking about cancelling the project, but first we want to see if you can turn the project around."

- "We know that we need a new business case because we will never be able to achieve all of the business value expected from the original business case."
- "You'll have to come up with a different solution, one that hopefully will work."
- "I'm not sure how much support the executives can give you. You are out there all by yourself, except of course for your team."
- "I expect you to have a recovery plan to show me in less than a week."

The projects that Scott was used to managing started out with well-defined expectations, a clear business case, and a good statement of work. Time was no longer a luxury or even a simple constraint; it was now a critical constraint. The forms, guidelines, templates, and checklists in the enterprise project management methodology did not account for this type of situation. Furthermore, Scott had never been in any meetings that required brainstorming, problem solving, and critical decision-making. It was now quite apparent that Scott was not qualified to function as a recovery project manager.

Situation #2: Brenda's Dilemma

Brenda had never been placed in such a position before while managing projects. In the past, Brenda believed that she was expected to make project-based decisions predicated upon technology. But on this project, Brenda was expected to make both business-related and project-related decisions. Previously, Brenda relied heavily on her project sponsor for business related decisions. But now, it appeared that most of the decisions rested on her shoulders.

The sponsors on her projects no longer wanted to hear about problems without additional data. In the past, project managers had the tendency to send their problems to their sponsors, and eventually the sponsors would resolve the problems. Soon, sponsors were spending more time solving problems on projects rather than performing the activities dictated by their job descriptions. To resolve this problem, the sponsors were now telling the project managers, "Do not come to us with problems unless you also bring alternatives and recommendations!"

Brenda was experienced in making project-based decisions where alternatives were developed based on the constraints of time, cost, scope, quality, and sometimes risk. But making business decisions would require consideration of additional constraints, namely safety, image, reputation, goodwill, stakeholder relations management, culture, future business, and customer satisfaction.

Brenda's project team was composed mainly of engineers many of which had never taken any courses in business. They understood technology and how to develop technical alternatives. They knew little about marketing and sales activities. Many of the engineers on her team were also prima donnas who believed that their opinion is the only opinion that counts.

Putting all these people in a room and asking them to develop and evaluate alternatives to the problems would certainly be difficult. Brenda did not know where to begin.

Situation #3: The Brainstorming Meeting

Paul was delighted that all the subcontractors were willing to send technical representatives to the brainstorming meeting. Paul's company won a contract from one of their most important clients to develop a new product. The contract involved state-of-the-art technology that was unavailable to Paul's company except through subcontracts.

The client worked with Paul in the selection process of the subcontractors. The client knew right from the start of the project that there were risks on the project and that the product may not be able to be developed without significant tradeoffs. The client's original statement of work was more of a "wish list" of deliverables with little chance of being accomplished.

As the problems began to mount, Paul and the client jointly agreed that the direction of the project must change to salvage as much value as possible. The subcontractors had to participate in the brainstorming meeting because they possessed the expertise regarding what could and could not be done. Paul had great expectations that everyone in attendance could agree on a new direction for the project.

Paul had been in other brainstorming sessions and knew that all ideas should be listed but not evaluated or criticized until sometime later, perhaps even after the session was completed. But since several of the suppliers were not geographically local, Paul opted to conduct the meeting using the Nominal Group Technique whereby each person would present their ideas and be subject to immediate evaluation and criticism.

Each of the subcontractors presented their arguments for why their approach would be best. Even though the ideas of several subcontractors could possibly be combined into a workable solution, the subcontractors refused to budge on their position. The subcontractors' adamant position made it appear that they were more interested in how much follow-on business they could get rather than what was in the best interest of the client.

Paul now had a problem. How could he get the subcontractors to work together to come up with an agreed upon direction acceptable to all? Paul had never been in this type of situation before. Obviously, this was not what Paul had expected as an outcome of the problem-solving meeting.

Situation #4: The Lack of Information

John was an experienced project manager; at least he thought that he was. His company had an enterprise project management methodology that contained forms, guidelines, templates, and checklists for just about any situation. It was a one-size-fits-all approach. Unfortunately, there were no instructions for John's current dilemma.

The statement of work was reasonably clear as to what direction the project should take. Everyone knew that the technical approach was optimistic and may not be achievable to satisfy the business objectives. When the project plan was prepared, the primary planning objective was "least time" so that the product could be introduced into the marketplace quickly. But since the technical approach could not work, it was necessary to redirect the project and create a new project plan.

John did not know whether the planning objective would still be "least time," or whether it would change to "least risk" or "least cost." John also realized that he needed additional information from Engineering, Manufacturing, Marketing, and Sales to solve his problem and plan.

John's first attempt to collect the problem-solving information was met with resistance. John recognized quickly that information was a source of power and these functional areas were unwilling to give John the information he needed. The lack of critical information was placing John in an impossible. Engineering would not confirm what new technical approaches were possible. Manufacturing could not provide any information on manufacturing costs without knowing the engineering design. The Sales personnel could not provide any information without knowing the manufacturing costs. Marketing wanted to make sure there was a market need for the new design before providing information.

Repeatedly, John went to the functional areas asking for help. The answer was always the same; "Let me think about it and I'll get back to you." Not willing to throw in the towel yet and concede failure, John went his sponsor, the vice president of Marketing, and explained the situation. After explaining his dilemma, the project sponsor replied, "I can get you some information from our marketing personnel, but you are on your own with the other functional areas. Our functional silos have brick walls around them, and I have no authority over any of the resources in the other functional areas. You'll have to do the best you can." John went back to his office and began contemplating his future with the company.

Situation #5: The Information Overflow Dilemma

Anne was placed in charge of a project to create a new product. Although Anne was experienced in project management, many of her newly assigned team members had little experience working on projects. Anne knew she could help them once the project was under way, but the greater problems would most likely occur during problem solving and decision-making when developing the project plan.

During the project's kickoff meeting, Anne went through the statement of work in detail. She was convinced that everyone understood what had to be done. Anne instructed everyone to reconvene in a week with information regarding their specific efforts on the project. The information Anne needed from everyone was:

- Hours needed for each work package
- Grade level of the workers needed for each work package
- The cost of their efforts for each work package
- The time duration of each work package
- The anticipated risk of each work package

When the team reconvened, most of the team members came with alternatives. Many of the alternatives came with optimistic, most likely, and pessimistic estimates. Some of the team members came with historical data from five or six previous projects that were completed successfully.

This is not what Anne expected. She was now overwhelmed with information. To make matters worse, Anne knew her limitations and believed that she could not make a decision based on the massive documentation that the team members were providing. Anne had to find a way to limit the information overflow.

Situation #6: Impact of the Assumptions

Karl was an exceptionally talented engineer whose experience was restricted to engineering project management activities. Karl understood the constraints on the project he managed but was never provided with any assumptions, especially assumptions related to business decisions on projects.

For the projects that Karl had managed in the past, Karl made primarily project-based decisions related to technology. All decisions related to the business side of the project were made by the project sponsor. But for the project that Karl was now managing, Karl was expected to make both project and business decisions. The business decisions required and understanding of the assumptions.

Karl realized that the technical approach selected, and the expected technical breakthrough, might not be achievable. Selecting a new technical approach would certainly elongate the project and increase the costs. But changing the direction of the project would certainly have an impact on marketing and sales activities, especially on a long-term project. Some of assumptions that were likely to change over the duration of a project, especially on a long-term project, might include:

- The cost of borrowing money and financing the project will remain fixed.
- The procurement costs will not increase.

- The breakthrough in technology will take place as scheduled.
- The resources with the necessary skills will be available when needed.
- The marketplace will readily accept the product.
- Our competitors will not catch up to us.
- The risks are low and can be easily mitigated.
- The political environment will not change.

Karl looked at these assumptions and wondered how these would impact problem solving and decision-making.

The Need for New Skills

As stated previously, decisions are no longer a single person endeavor. However, the project manager and project team must possess the necessary skills to put the pieces together to arrive at the best decision. There are numerous books available on problem solving and decision-making. Unfortunately, they look at the issues from a psychological perspective with applications not always relevant to project and program managers. In this section, we will extract the core concepts of problem solving and decision-making that would be pertinent to project managers and assist them with their jobs.

Some books use the term problem analysis rather than problem solving. Problem analysis can be interpreted as simply looking at the problem and gathering the facts, but not necessarily developing alternative solutions for later decision-making. In this book, problem solving is use throughout reflecting the identification of alternatives as well.

Problem solving and decision-making go together. Decisions are made when we have issues or choices to make. In general, we must have a problem prior to making decisions. But there is a strong argument that decision-making is needed and must be used as part of identifying the problem and developing alternatives.

The Need for Information

There are several facts or generalities that we consider when discussing problem solving and decision-making, and they all are related directly or indirectly to the information we need. They include:

- Projects today have become much more complex than before, and so are the problems and decisions that must be made.
- Problem-solving techniques are used not only to solve problems but to take advantage of opportunities.
- Today, we seem to be flooded with information to the point where we have information overload and cannot discern what information is needed or useful for solving problems.
- Lower to middle levels of management are often valuable resources to have when discussing the technical side of problems. Senior management is valuable in the knowledge of how the problem (and its solution) relates to the overall business and the impact of the enterprise environmental factors.
- Problem solving today has become a project management core competency, yet most companies provide little training for their employees on problem solving and decision-making.
- The project team may be composed of numerous subject matter experts, but the same people may not be creative or capable of thinking creatively for solving problems.

- The people that created the problem may not be capable of solving the problem they created.
- Few people seem to know the relationship between creativity and problem solving.

Today, there seems to be an abundance of information available to project teams thanks to advances in information system technologies and information warehouses. Our main problem is being able to discern what information is critical and what information should be discarded or stored in archives.

For simplicity's sake, information can be broken down into primary and secondary information. Primary information is information that is readily available to us. This is information that we can directly access from our desktop or laptop and may come from information warehouses or business intelligence systems. Information that is company sensitive or considered as proprietary information may be password protected but is still accessible.

Secondary information is information that must be collected from someone else. This is generally the norm on strategic projects. Even with information overload from information warehouses, project managers generally do not have all the information they need to solve a problem and make a timely decision. This is largely due to the complexity of our projects as well as the complexity of the problems that need to be resolved. We generally rely on a problem-solving team to provide us with the secondary information. The secondary information is often more critical for decision-making than the primary information. Many times, the secondary information is controlled by the SMEs, and they must tell us what information is directly pertinent to this problem.

Collecting the information, whether primary or secondary information, can be time-consuming. Information overload often forces us to spend a great deal of time searching through information when this time should be spent on problem solving. Project plans in the future must allow for the time needed for problem solving and decision-making.

The project manager's challenge is not just getting the information but getting the right information in a timely manner. Sometimes, the information that the project manager needs, especially secondary information, is retained by people that are not part of the problem-solving team. An example might be information related to politics, stakeholder relations management, economic conditions, cost of capital and other enterprise environmental factors. This information may be retained by senior management and stakeholders.

Because timing is essential, project managers should have the right to talk directly with anyone they need to converse with to obtain the necessary information to solve a problem. Having to always go through the chain of command to access the information creates problems and wastes valuable time. Information is often seen as a source of power, which is one of the reasons why some companies control the chain of command.

Behind every door in a company is information of some sort. Project managers must be able to open those doors as needed. If project managers do not have access to those doors, then there are two options; follow the chain of command and hope that the information is not filtered by the time it gets to you or invite the person with this information to attend the problem-solving meeting. Where the people with the information reside in the organizational hierarchy often determines their availability to attend the meeting. The higher up they reside, the less likely they will be able to attend your meeting in the near term. The project manager's accessibility to information is critical.

Even though we have information overload and access to secondary information sources, there is no guarantee we will have readily available all the information we need. People that need to make decisions must accept the fact that they generally will not have all the information they need on hand. This can happen at all levels of management, not just on projects. We must be willing to make the best possible decisions based on the information we have at that time, even if it is partial information.

Too often, we rely on the chain of command for getting the information to help resolve a problem. If people believe that "possessing information is power," access to the needed information can be a problem, especially if they withhold some of the information. Because of the criticality of the constraints, time is not a luxury. Project managers must have the right or authority to access those people that possess the information. This assumes, of course, that the project manager knows where the information resides. This is sometimes the greater challenge, especially if the needed information is nowhere to be found with the company. We must go outside the company to get the critical information.

Problem Solving and Decision-Making in a Project Environment

Project managers today believe that each project that they are managing is part of the business and therefore they are managing part of a business rather than just a project. As such, project managers are expected to make business decisions as well as project decisions. However, there is a difference between project decision-making and business decision-making:

- Project decision-making focuses on meeting baselines, verification, and validation. Business decision-making focuses on market share, profitability, customer satisfaction and repeat business.
- Project decision-making involves the project team whereas business decision-making may include marketing, sales, and senior management.
- There are multiple tools that project managers use for project decision-making but most of the tools used for business decision-making are mainly financial tools such as ROI, NPV, IRR and cash flow.
- The focus of project decision-making is project performance whereas the focus of business decision-making is financial performance.
- The results of project decisions appear quickly whereas the results of business decisions may not appear for years.
- Most of the problems addressed as part of project decision-making are to maintain the baselines whereas business problems focus on alignment or changes to the business strategy.

Understanding decision-making in a project management environment requires that we understand how the project management environment differs from the traditional environment we are all used to. The project management environment is an interaction between people, tools, processes, and routine business work that must be accomplished for the survival of the firm as well as for project work. Project management activities may be considered secondary to the ongoing business activities. Because of the high levels of risk on many projects and the fact that some of the best resources are assigned to ongoing business activities, the decision-making process can lead to suboptimal or ineffective decisions.

There are other factors that make decision-making quite complex in a project environment:

- The project manager may have limited or no authority to make the decisions even though they may have a serious impact on the project's outcome.
- The people on the project team may not be able to make meaningful contributions to the decision-making process.

Because of the project manager's limited authority, some project managers simply identify alternatives and provide recommendations. These recommendations are then brought to the executive levels of management, or the project sponsor, for the final decision to be made. However, some people argue that the project manager should have the authority to make those decisions that do not alter the deliverables of the project or require a change to the constraints and baselines.

By now, you should have a reasonably clear picture of what decision-making and problem solving will be like in the future project environment. Some other important factors to consider include:

- There are numerous constraints imposed on the project.
- The constraints can change in relative importance over the life of the project, and new constraints can appear.
- The assumptions made at the beginning of the project may no longer be valid as the project continues.
- The project manager may not know all the constraints, even though some of them are listed in the project charter or business case.
- The project manager most likely does not have a command of technology in the area where the problem exists.
- The project team members may likewise not have a command of technology in the area where the problem exists.
- The project manager and the team are expected to make the necessary decision in a timely manner even though they may not have complete information.
- The client and all the associated stakeholders may not agree with the final decision.
- Expecting to always make the right decision is wishful thinking.

Real Versus Personality Problems

We generally believe that most problems are real and need to be resolved. But that is not always the case. Some problems are created based on the personalities of the individuals. Some people create problems unnecessarily so that they can somehow benefit, perhaps by being the only person capable of solving the problem. Some examples include:

- Resolution of the problem will get you more power.
- Resolution of the problem will get you more authority.
- Resolution of the problem will diminish the power and authority of others.
- You are the only one with the capability to resolve the problem, and it will improve your image and reputation.
- You will be regarded as a creating thinker.
- It will look good on your resume.
- It will look good during performance reviews.
- It will guarantee you employment.

It is important to discover at first whether the problem is real and if a simple solution exists. Many years ago, a department manager was afraid that downsizing would take place and that he would lose his position as a department manager. To protect himself, he gave the workers assigned to the project team conflicting instructions, knowing that problems would occur and that several projects might suffer. This resulted in rework and created problems for several projects. The department manager then called a problem-solving meeting with those project managers that were falling behind on their projects. The department manager stated in the meeting that almost all his employees were poor workers that needed constant supervision, and that the problems will be resolved by the department manager. He would provide these poor workers with much closer supervision. The department manager provided alternatives to the fictitious problems and stated that he would have the problems resolved within a few months.

The department manager believed that his position was now secure. But the project managers were not fooled. The project managers found out the truth, and eventually the department manager was fired for what he did. The project managers discovered that these were not real problems that needed to be resolved using problem-solving and decision-making techniques.

Deciding Who Should Attend the Problem-Solving Meeting

Problems are not resolved in a vacuum. Meetings are needed and the hard part is to determine who should attend. If people are not involved in the problem or the problem is unrelated to the work they do, then having them attend these meetings may be a waste of their time. This holds true for some of the team members as well. As an example, if the problem is with procurement, then it may not be necessary for the drafting personnel to be in attendance.

For simplicity's sake, we shall consider just two types of meetings: problem-solving and decision-making. The purpose of the problem-solving meeting is to obtain a clear understanding of the problem, collect the necessary data and develop a list of workable alternatives accompanied by recommendations. More than one meeting will probably be required.

Sending out an agenda is important. The agenda should include a problem statement that clearly explains why the meeting is being called. If people know about the problem in advance, they will have a chance to think about the problem and bring the necessary information, thus reducing some of the time needed for data gathering. It is also possible that the information gathered will identify that the real problem is quite different from what was at first considered to be the problem.

It is essential that subject matter experts (SMEs) familiar with the problem be in attendance. These SMEs may not be part of the original project team but may be brought in just to resolve this problem. The SMEs may also be contractors hired in to assist with the problem.

The people brought in for the identification of the problem and data gathering usually remain for the development of the alternatives. But there are situations where additional people may participate just for the consideration of alternatives.

Deciding Who Should Attend the Decision-Making Meeting

The decision-making meeting is different from the problem-solving meeting. In general, all the participants that were involved in the problem-solving meeting will most likely be in attendance in the decision-making meeting but there may be a significant number of other participants. Project team members should have the ability to resolve problems but not all the team members have the authority to make decisions for their functional units. It is normally a good idea at the initiation of the project for the project manager to determine which team members possess this decision-making authority and which do not. Team members that do not possess decision-making authority will still be allowed to attend the decision-making sessions but may need to be accompanied by their respective functional managers when decisions are required, and voting takes place.

Stakeholder attendance is virtually mandatory at the decision-making meetings. The people making the decisions must have the authority to commit resources to the solution of the problem.

The commitment could involve additional funding or the assignment of SMEs and higher-pay-grade employees. Project managers are responsible for the implementation of the solution. Therefore, the project manager must have the authority to obtain the resources needed for a timely solution to the problem.

Creating a Meeting Framework

For the problem-solving meeting, it is important to create a mental framework of the problem beforehand, including what should be accomplished in the meeting, and the limitations. Not all the people that will attend the problem-solving meeting will be familiar with the details of the problem. Some may have just a cursory understanding of the problem and others may not have known that the problem even existed prior to the meeting.

The mental framework should include all the information known thus far about the problem. Additional information will most likely be forthcoming in the meeting. If possible, the framework should be included in the invitation for the meeting and/or the agenda. Informing people about the problem prior to the meeting will get them to think about it and possibly even perform some preliminary research prior to the meeting. When people understand the framework prior to the meeting, they usually come to the meeting better prepared and may even recommend to the meeting organizer other people who should be invited.

Setting Limits

Problem solving and decision-making can go on for a long period of time. Limitations must be established early on. Some of the limits include:

- How much time is available to address the issue?
- How much money is the project willing to commit to resolve the problem?
- How many resources can be assigned to resolve the problem?
- Will the assigned resources have the required skills?
- Which facilities can be used for testing or other activities?
- How important is the project to the company?
- How important is the project to the client?
- How important is the project to the stakeholders?
- What is the criticality of the project and the problem?

It is not uncommon for the limitations to be identified in the problem statement or even in the agenda. When people understand the limitation early on, decisions are usually made in a timely manner.

Establishing Boundary Conditions

The limitations discussed previously were limitations on the meeting. Limitations or boundary conditions must also be established for the solution to the problem, and the limitations can impact the alternatives that will be selected.

We know for a fact that, when solving a problem and identifying alternatives, we do not have an infinite sum of money to implement a solution or as much time as we may want. We consider these as constraints, but they are also boundary conditions. Boundary conditions can be established by the client, stakeholders, and the eventual users of the deliverables. A partial list of boundary conditions might be:

- Staying within the project's constraints on time, cost, quality, and scope
- Without increasing the risks on the remaining work on the project
- Without altering the scope of the remaining work on the project
- Without altering the company's normal flow of work
- Without gold-plating the solution
- Without including unnecessary functionality
- Knowing that only a limited number of additional resources are available for solving the problem and implementing the solution
- Without violating Regulatory Agency requirements such as those established by OSHA and EPA
- Without driving up the selling price of the product beyond what the customers will pay for it

Adaptive Decision-Making

There are several decision-making techniques that project teams can use. The boundary conditions and type of problem often determine which technique may be best.

Adaptive decision-making may require some degree of intuition. The problem is usually well understood, and the project team may be able to make the decision without outside support or sophisticated tools and techniques. Adaptive decision-making is the most common form of decision-making used on projects. Examples might include determining the following:

- The number of tests that should appear in the test matrix
- When an activity should begin or end
- How late an activity can start without delaying downstream work
- How late we can order raw materials
- Whether the work should take place on regular shift or overtime
- Whether a risk management plan is necessary, and if it is necessary, how much detail should appear in the place
- How often testing should take place to validate compliance to quality requirements
- The resource skillset needed, assuming there are choices
- The best way to present both good news and bad news to the stakeholders
- Ways to correct unfavorable cost and schedule variances
- The leadership style to be used to motivate certain team members
- How to best reward superior performance by team members

Innovative Decision-Making

Many strategic projects will include some degree of innovation, and project teams must then be prepared to make strategic innovative decisions. Innovation is generally regarded as a new way of doing something. The new way of doing something should be substantially different from the way it was done before, rather than a small incremental change such as with continuous improvement activities. The goal of innovation is to create hopefully long-lasting additional value for the company, the users, and the deliverable itself. Innovation can be viewed as the conversion of an idea into cash or a cash equivalent.

Innovative decision-making is most often used on projects involving R&D, new product development, and significant product enhancements. These decisions involve SMEs that may not be part of the project team and may require the use of more advanced decision-making tools and techniques. These decisions may require a radical departure from the project's original objectives. Not all project managers possess the skillset to manage projects involving innovation.

While the goal of successful innovation is to add value, the effect can be negative or even destructive if it results in poor team morale, an unfavorable cultural change, or a radical departure from existing ways of doing work. The failure of an innovation project can lead to demoralizing the organization and causing talented people to be risk-avoiders in the future rather than risk-takers.

Pressured Decision-Making

Time is a critical constraint on most projects, and this can have a serious impact on the time necessary to understand the problems and find solutions. As an example, let us assume that a critical test fails, and the client says that they will be meeting with you the day after the failure to discuss how you will correct the problem. They are expecting alternatives and a recommendation.

Typically, you might need a week or longer to meet with your team and diagnose the situation. However, given the circumstances, you may have to make a decision, right or wrong, based on the time available. This is high-pressured decision-making because resources may be standing by idling waiting for directions on what to do next. Given sufficient time, we can all analyze or even overanalyze a problem and come up with a list of viable alternatives.

High-pressured decision-making can also be part of adaptive and innovative decision-making as well. Being pressured to make a decision can have favorable results if it forces the decision makers to look at only those attributes that are critical to the problem. But, often, high-pressured decision-making leads to suboptimal results.

Given that these situations will happen, you must expect that you will not always have complete or perfect information to make a decision. Most decision-making teams must deal with partial information.

Decision-Making Meetings

While some decisions that are routine or adaptive in nature can be handled during regular team meetings, in general problem resolution team meetings should be set up as separate meetings. The attendance at a problem resolution meeting can be quite different from the attendance at regular team meetings. Stakeholders and clients may be required to attend problem resolution team meetings since they are the people most likely affected by the decision. Functional managers and SMEs may also be invited to attend. Outside consultants with critical expertise may also participate.

There is a wide variety of decision-making tools and techniques that project teams can use for decision-making during these meetings. The selection of the best tool or technique to use can be based on the complexity of the problem, the risks associated with the decision, the cost to make the decision and the impact if the decision is wrong, to whom the decision is important, the time available to make a decision, the impact on the project's objectives, the number of people on the project team, the relative importance to the customer or stakeholders, and the availability of supporting data.

More than one meeting may be required. The purpose of the first meeting may be to just understand the problem and gather the facts. The problem-solving team may then require additional time to think through the problem and identify alternatives. It is highly unlikely that a decision will be made at the first team meeting.

There are several models available for how project teams make decisions. A typical four-phase model might include:

1. **Familiarization stage.** This is where the team meets to understand the problem and the decision(s) that must be made.
2. **Options identification phase.** This is where the team performs brainstorming and lists possible alternatives for a solution.
3. **Option selection phase.** This is where the team decides upon the best option. The team selecting the preferred option may have a different makeup than the team developing the list of options.
4. **Justification phase.** This is where the team rationalizes that they made the right decision and possibly evaluates the results.

Advantages and Disadvantages of Group Decision-Making

As stated before, there are situations such as routine decisions where the ultimate decision is made by just the project manager. Groups are not needed in this situation. But many problems that appear on projects require group thinking.

There are several advantages to group decision-making:

- Groups provide better decisions than individuals.
- Group discussions lead to a better understanding of the problem.
- Group discussions lead to a better understanding of the solution.
- Groups make better judgments calls on the selection of alternatives.
- Groups tend to accept more risks in problem solving than do individuals.
- Clients appear less likely to question the decision of the group compared to the decision of an individual.
- People are more willing to accept the final decision if they participated in the decision-making process.

There are also several disadvantages to group decision-making:

- The discussions can be dominated by the personality of one person, regardless of whether that person is regarded as a subject matter expert.
- Groups may accept too much risk knowing that a failure would be blamed equally among all the members of the group.
- There is pressure to accept the decision of the group, even though you know that other decisions might be better.
- Too much time might be spent arriving at a consensus.
- Groups tend to overthink problems and solutions.
- It might be impossible to get the proper people released from other duties so they can attend the meeting.
- Finding a common meeting time that satisfies all parties may be difficult.
- If external people are involved, the costs associated with traveling could become quite large, especially if more than one decision-making meeting is needed.

Decision-Making Mental Roadblocks

Not everyone wants to make decisions or can make them. Some people would prefer to have others make all decisions, especially critical decisions. Reasons for this behavior might include:

- A previous history of making the wrong decisions
- Emotionally afraid of making the wrong decision
- Afraid of the associated risks
- Lack of conviction in one's own beliefs
- Having high levels of anxiety
- Unable to cope with the politics of decision-making
- Unfamiliar with the facts surrounding the problem and not willing to learn
- Unfamiliar with members of the team
- Possessing poor coping skills
- Lack of motivation
- Lack of perspective
- Being brought into the discussion well after the discussion began
- Unable to work under high levels of stress and pressure
- Afraid of working with unions that are involved in the problem
- Afraid of working with certain stakeholders involved in the problem
- Afraid of contributing for fear of being ridiculed
- Afraid of exposing one's inadequacies
- Afraid of damaging your career and/or reputation

These roadblocks are often categorized into five areas:

- Emotional blockages
- Cultural blockages
- Perceptual blockages
- Intellectual blockages
- Expressive blockages

Decision-Making Personal Biases

Biases can creep into our decision-making processes. A partial list might include:

- Believing beforehand that your solution is the only possible solution
- Ignoring evidence that supports a conclusion other than yours
- Neglecting to understand the root cause of the problem
- Refusing to search for supporting data for a decision
- Neglecting to understand how the wrong decision can impact the project
- Being afraid to state your opinion and siding with the person whom you believe will provide the best approach
- Being afraid to decide for fear that you may make the wrong decision
- Being fearful of having your ideas criticized
- Unwilling to think differently or out-of-the-box
- Adopting a wishful thinking approach to making a decision
- Adopting a selective perception approach and looking at only the information and alternatives that are in your comfort zone

- Making the decision that others expect you to make even when you strongly believe it may be the wrong decision
- Making a decision that is in your personal interest rather than the best interest of the project
- Spending too much time on small or unimportant things that are in your comfort zone rather than focusing on what is critical

The Danger of Making Hasty Decisions

The project's constraints often place the project manager is a position of wanting to make hasty decisions. Making hasty decisions is sometimes a necessity, but the results can be detrimental. Hasty decisions can lead to:

- Additional problems surfacing later in the project
- Rework that leads to cost overruns and schedule slippages
- Excessive overtime
- Customers and stakeholders that will lose faith in your ability to manage the project correctly
- Lack of faith in the problem-solving and decision-making process
- Manpower curves with peaks and valleys rather than a smoother out manpower curve
- Greater hands-on involvement by the governance committee
- More meetings
- An increase in reporting requirements
- Deliverables that are rejected by the client

Simply stated, speed in decision-making is a risky business.

Decision-Making Styles

Not all decisions are easy to make. Sometimes you must make a decision whether you are ready or not, and when you have partial rather than complete information available to you. Also, the decision to do nothing differently may be the best decision under certain circumstances. If the team believes that they can live with the problem at hand, then the team may wait and see if the problem gets worse before making a decision.

Project managers all have their own approach to decision-making, and this may vary from project to project. The style selected is based on the definition of the problem and the type of decision that must be made. Although some approaches work well, other approaches do more harm than good.

Textbooks on decision-making provide several different styles. The five styles most common for project managers are:

- The autocratic decision maker
- The fearful decision maker
- The circular decision maker
- The democratic decision maker
- The self-serving decision maker

The autocratic decision maker usually trusts nobody on the team and dictates the decision even though the risks are great and little time was consumed discussing the problem. Team members are often fearful of presenting alternatives and recommendations because the project manager might ridicule suggestions, believing that his/her decision is the only one. Team members may not contribute ideas even when asked.

The autocratic style can work if the project manager is regarded as an expert in the area in which the decision must be made. But in general, project managers today seem to possess more of an understanding of technology than a command of technology. As such, using the autocratic style when you have limited knowledge about the technology of the problem and the solution can lead to a rapid decision but often a decision that is not the optimal choice.

Most of the time, autocratic decision makers feel better making a decision by themselves without any input from others. They make a decision on the spot based on a feeling in their gut. This is often a hit or miss approach.

While the autocratic decision maker thrives on making the decision, right or wrong, and in a timely manner, the fearful decision maker is afraid of making the wrong decision. This is often referred to as the "ostrich" approach to making a decision. In this case, the project manager will bury his/her head in the sand and hope that the problem will disappear or that people will forget about the problem. The project manager also hopes that by waiting, a miracle solution will appear by itself, such that a decision may not have to be made at all.

Sometimes the fearful decision maker adopts a procrastination attitude, which is waiting for enough (or at least a minimum amount of) information so that a decision can be made. This does not necessarily mean avoiding a decision. The fearful decision maker knows that a decision must be made, eventually.

The fearful decision maker is afraid that making the wrong decision could have a serious impact on his/her reputation and career. The team may not be invited to provide alternatives and recommendations because that would indicate that a problem exists and that a decision must be made. Information on the problem may even be withheld from senior management, at least temporarily.

The project manager may try to get others to make the decision. The project manager may prefer to have someone act as the moderator of the decision-making group and, if a decision must be made, the project manager will always argue that it is a group decision rather than a personal decision. The project manager will avoid, if possible, taking personal accountability and responsibility for the decision.

As stated previously, time is a constraint on projects, not a luxury. Taking a wait-and-see approach to making a decision can lose precious time where the problem could have been easily resolved. Also, the longer we wait to make a decision, the fewer the options are.

The circular decision maker is like the fearful decision maker. The project manager not only wants to make the decision but wants to make the perfect decision. Numerous team meetings are held to discuss the same problem. Each team meeting seems to discuss the problem and possible solutions from a different perspective. The team members are given actions items that keep them scurrying about looking for additional information to support the perfect decision.

The circular decision maker is willing to make a decision but sacrifices a great deal of time looking for the perfect decision that everyone will agree to. The decision maker is willing to violate the time constraints on a project to accomplish this. The decision maker may also believe that the problem may disappear if they think about it long enough.

The project manager can adopt the circular decision-making style even if he/she is an expert in the area in which the problem exists. The project manager needs reinforcement from the team, and possibly superiors, that the best decision was made. In the eyes of the project manager, the decision may be deemed more important than the outcome of the project.

The democratic decision maker allows the team members to participate in the final decision. Voting by the group membership is critical and may even be mandatory. The company may even have a structured approach for this using guidelines or templates. This can happen even if the project manager is the expert in the area where the problem exists and even if the project manager has the authority to make the decision alone.

Democratic decision-making can create long-term problems. Team members may feel that they should be involved in all future decisions as well, even those where they might have limited knowledge about the problems. Asking team members to take an early vote on the solution to a problem can lead to apprehension if the team members are uncomfortable with making a decision based on incomplete information. Waiting too long to make the decision can limit the options available and frustrate the project team because of the time that was wasted overthinking the problem and the solution.

Democratic decision-making is a strong motivational tool if used properly. As an example, if the project manager believes that he/she already knows that decision that should be made, asking the team for their opinion, and giving credit to a team member for coming up with the same idea is a good approach. This encourages people to participate in decision-making and makes them believe that they will be given credit for their contributions.

Everyone sooner or later is placed in a position where they must decide when making a decision what is more important: their individual values or the organizational values. This situation often forces people to make decisions either in favor of themselves or the organization. A compromise might be impossible.

These types of self-serving conflicts can permeate all levels of management. Executives may make decisions in the best interest of their pension rather than the best interest of their firm. One executive wanted to be remembered in history books as the pioneer of high-speed rapid transit. He came close to bankrupting his company in the process of achieving his personal ambitions at the expense of the projects he was sponsoring.

Self-serving decision makers seem to focus on what is in their own best interest in the short term and often disregard what might be in the best interest of the project. In a project environment, this can become quite a complex process if the team members, stakeholders, the client, and the project sponsor all want the decision made in their own best interest. Suboptimal solutions are reached with several parties being quite unhappy with result. Unfortunately, self-serving decisions are almost always made for what is in the best interest of the largest financial contributor to the project for fear that, if the financial contributor removes support from the project, the project may be cancelled.

Decision-Making Tools

There are numerous complex tools that can be used for more complex decision-making tools. SWOT analysis looks at the strengths, weaknesses, opportunities, and threats in each situation. SWOT analysis was originally created as a strategic planning tool but has now been adapted to complex problem solving on a project or in a business venture. It involves specifying the objective of the problem or project and identifying the internal and external factors that are favorable and unfavorable to achieve that objective.

A SWOT analysis must first start with defining a desired end state or objective for the problem at hand:

- **Strengths:** characteristics of the team that give it the ability to solve the problem. This could include technical knowhow and expertise.

- **Weaknesses:** characteristics that may prevent the team from solving the problem. This could be the team's lack of technical ability.
- **Opportunities:** *external* opportunities if the problem is resolved.
- **Threats:** *external* risks or elements in the environment or with stakeholders that could cause trouble for the project or business.

Strengths and weaknesses are internal strengths and weaknesses and look at the capability of the internal resources to solve the problem. Opportunities and threats are external results that may occur if the problem is or is not resolved. Strengths and weaknesses indicated what you "can do," and this must take place before you look at the opportunities and threats that indicate what you "should do." Having a great alternative that will appease the stakeholders is nice if you have the qualified resources to accomplish it.

Pareto analysis is a statistical technique in decision-making that is used for selection of a limited number of tasks that produce significant overall effect such as the solution to a problem. It uses the Pareto principle – the idea that by doing 20 percent of work, 80 percent of the advantage of doing the entire job can be generated. In terms of quality improvement, a large majority of problems (80 percent) are produced by a few key causes (20 percent). In problem solving, 80 percent of the desired solution can be obtained by performing 20 percent of the work.

Pareto analysis is a formal technique useful where many possible courses of action are competing for the solution to a problem. In essence, the problem-solver estimates the benefit delivered by each action, then selects a few the most effective actions that deliver a total benefit reasonably close to the maximal possible one.

Pareto analysis is a creative way of looking at causes of problems because it helps stimulate thinking and organize thoughts. However, it can be limited by its exclusion of possibly important problems which may be small initially, but which grow with time. It should be combined with other analytical tools such as **failure mode and effects analysis** and **fault tree analysis** for example.

This technique helps to identify the top 20 percent of causes of a problem that needs to be addressed to resolve the 80 percent of the problems. Once the top 20 percent of the causes are identified, then tools like the **Ishikawa diagram** or fishbone analysis can be used to identify the root causes of the problems.

Multiple Criteria Decision Analysis (MCDA) or **Multi-Criteria Decision Making (MCDM)** is a discipline aimed at supporting decision makers faced with making numerous and sometimes conflicting evaluations. MCDA aims at highlighting these conflicts during problem-solving and deriving a way to come to a compromise. It is a combination of intuition and a systematic approach

Unlike methods that assume the availability of measurements, measurements in MCDA are derived or interpreted subjectively as indicators of the strength of various preferences. Preferences differ from decision maker to decision maker, so the outcome depends on who is making the decision and what their goals and preferences are. Since MCDA involves a certain element of subjectivity, and the morals and ethics of the persons implementing MCDA play a significant part in the accuracy and fairness of MCDA's conclusions. The ethical point is especially important when one is making a decision that seriously affects other people, as opposed to a personal decision.

There are many MCDA / MCDM methods in use today. However, often different methods may yield different results for the same problem. In other words, when the same problem data are used with different MCDA / MCDM methods, such methods may recommend different solutions even for quite simple problems (i.e., ones with very few alternatives and criteria). The choice of which model is most appropriate depends on the problem at hand and may be, to some extent, dependent on which model the decision maker is most comfortable with. A question with all the above methods, and methods not included in this list or even future methods, is how to assess their effectiveness.

In **paired comparison analysis,** also known as **paired choice analysis,** decision alternatives are compared two at a time to see the relative importance. The alternatives are compared, and the results are then tallied to find an overall winner. The process begins by first identifying a range of plausible options. Each option is compared against each of the other options, determining the preferred option in each case. The results are tallied and the option with the highest score is the preferred option. This technique may be conducted individually or in groups. It may include criteria to guide the comparisons or be based on intuition following an open discussion of the group. A paired choice matrix or paired comparison matrix can be constructed to help with this type of analysis.

A decision tree is a decision support tool that uses a tree-like graph or model of decisions and their possible consequences, including chance event outcomes, resource costs, and utility. It is one way to display an algorithm for decision-making. Decision trees are excellent tools for helping you to choose between several courses of action. They provide a highly effective structure within which you can lay out options and investigate the possible outcomes of choosing those options. They also help you to form a balanced picture of the risks and rewards associated with each possible course of action.

Decision trees are commonly used in operations research, specifically in decision analysis, to help identify the optimum approach most likely to reach a goal. Amongst decision support tools, decision trees (and influence diagrams) have several advantages:

- Are simple to understand and interpret. People can understand decision tree models after a brief explanation.
- Lay out the problem such that all options can be analyzed.
- Allow us to see the results of making a decision.
- Have value even with little hard data. Important insights can be generated based on experts describing a situation (its alternatives, probabilities, and costs) and their preferences for outcomes.
- Use a white box model. If a given result is provided by a model, the explanation for the result is easily replicated by simple math.
- Can be combined with other decision techniques such as probability trees.

An **influence diagram (ID),** also called a relevance diagram, decision diagram or a decision network, is a compact graphical and mathematical representation of a decision situation. It is a simple visual representation of a decision problem. Influence diagrams offer an intuitive way to identify and display the essential elements, including decisions, uncertainties, and objectives, and how they influence each other. It is a generalization of a Bayesian network, in which not only probabilistic inference problems but also decision-making problems (following maximum expected utility criterion) can be modeled and solved. Influence diagrams are extremely useful in showing the structure of the domain, i.e., the structure of the decision problem. Influence diagrams contain four types of nodes (decision, chance, deterministic, and value) and two types of arcs (influences and informational arcs).

ID was first developed in mid-1970s within the decision analysis community with an intuitive semantic that is easy to understand. It is now adopted widely and becoming an alternative to decision tree which typically suffers from exponential growth in number of branches with each variable modeled. ID is directly applicable in team decision analysis, since it allows incomplete sharing of information among team members to be modeled and solved explicitly. Extension of ID also finds its use in game theory as an alternative representation of game tree.

As a graphical aid to decision-making under uncertainty, it depicts what is known or unknown at the time of making a choice, and the degree of dependence or independence (influence) of each variable on other variables and choices. It represents the cause-and-effect (causal) relationships of a phenomenon or situation in a nonambiguous manner, and helps in a shared understanding of the key issues.

An affinity diagram is a technique for organizing verbal information into a visual pattern. An affinity diagram starts with specific ideas and helps you work toward broad categories. This is the opposite of a cause-and-effect diagram, which starts with the broad causes and works toward specifics. You can use either technique to explore all aspects of an issue. Affinity diagrams can help you:

Organize and give structure to a list of factors that contribute to a problem.

- Identify key areas where improvement is most needed.

This technique is useful when there are large amounts of data. The affinity diagram is a business tool used to organize ideas and data. The tool is commonly used within project management and allows large numbers of ideas stemming from brainstorming to be sorted into groups for review and analysis.

The benefits include:

- Adding structure to a large or complicated issue
- Breaking down a complicated problem into broad categories
- Gaining agreement on the solution to a problem

Game theory models or games, as applied to project management problem-solving and decision-making, allow us to address a problem in which an individual's success in making choices depends on the choices of others. Simply stated, this technique considers responses of outside participants. It can be used to address how the client and stakeholders might react to certain alternatives selected.

It is used not only in project management, but also in the social sciences (most notably in economics, management, operations research, political science, and social psychology) as well as in other formal sciences (logic, computer science, and statistics) and biology (particularly evolutionary biology and ecology). While initially developed to analyze competitions in which one individual does better at another's expense (zero sum games), it has been expanded to treat a wide class of interactions, which are classified according to several criteria. This makes it applicable to project management, especially on complex projects where multiple stakeholders exist, each with competing needs.

Cost-benefit analyses are most useful for problems involving financial decisions. The alternatives to a problem are usually those where the value of receiving the benefits outweighs the costs of obtaining them. Factors considered in cost-benefit analyses include:

- Return on investment
- Net present value

- Internal rate of return
- Cash flow
- Payback period
- Market share

Other parameters to consider that are more difficult to quantify include:

- Stockholder and stakeholder satisfaction
- Customer satisfaction
- Employee retention
- Brand loyalty
- Time-to-market
- Business relationships
- Safety
- Reliability
- Reputation
- Goodwill
- Image

Work groups or nominal work groups, as applied to project management, can be an interdisciplinary collaboration of researchers or SMEs that have convened to identify and/or solve a problem. The group may be external consultants or contractors. The lifespan of the working group can be one day or several weeks. Such groups have the tendency to develop a *quasi-permanent existence* once the assigned task is accomplished; hence the need to disband (or phase out) the work group once it has provided solutions to the issues for which it was *initially* convened.

The work group may assemble experts (and future experts) on a topic together for intensive work. It is not an avenue for briefing novices about the subject matter. Occasionally, a group might admit a person with little experience and a lot of enthusiasm. However, such participants should be present as observers and in the minority.

It is imperative for the participants to appreciate and understand that the working group is intended to be a forum for cooperation and participation. Participants represent the interests and views of stakeholders from disparate sectors of the community that happen to have a vested interest in the solution to a problem. Therefore, maintaining and strengthening communication lines with all parties involved is essential (this responsibility cuts both ways – stakeholders are expected to share what information, knowledge, and expertise they have on the issue.)

Each member of the work group may be asked to present their solutions to the rest of the group for analysis and be willing to accept constructive criticism. Work groups often have the advantage of arriving at a reasonably rapid decision but also suffer from the drawback that possibly not all the alternatives were considered.

The Delphi technique is a structured communication approach, originally developed as a systematic, interactive **forecasting** and problem-solving method that relies on a panel of experts. The experts may not know who else is a member of the panel, and all responses are provided anonymously.

In the standard version, the experts answer questionnaires in two or more rounds. After each round, a facilitator provides an anonymous summary of the experts' forecasts (or solutions to a problem) from the previous round as well as the reasons they provided for their judgments. Thus, experts are encouraged to revise their earlier answers considering the replies of other members of their panel. It is believed that during this process, the range of the answers will decrease, and the group will converge toward the "correct" answer. If convergence does not take place, the panel may be asked to select the five best alternatives for the next round. Then in the next round, select the three best alternatives. Then in the following round, select the two best alternatives. Finally, the process is stopped after a pre-defined stop criterion (e.g. number of rounds, achievement of consensus, and stability of results) and the **mean** or **median** scores of the final rounds determine the results.

Delphi is based on the principle that forecasts (or decisions) from a structured group of individuals are more accurate than those from unstructured groups. This has been indicated with the term "collective intelligence". The advantage of this technique is that the participants will provide their answers without being biased by others or openly criticized. People are free to state their opinion. The downside factors are that the process takes time (which may not be a luxury that most projects have) and that the best approach may be the combining of two or more alternatives rather than forcing people to select just one alternative.

Other decision-making tools, some of which may take longer to perform, include:

- **Linear programming applications:** These include the application of management science and operations research models for decision-making.
- **Trial and error solutions:** These are useful for small problems when the cause-and-effect relationships are reasonably well known.
- **Heuristic solutions:** Like trial-and-error solutions, heuristic experimentation is done to reduce the list of alternatives.
- **Scientific methods:** Problem-solving methods involving scientific issues; additional experimentation may be done to confirm the problem and/or hypothesis.

Problem-solving sessions normally involve only one decision-making tool. Most of the more complex tools are time-consuming or costly to use, and combining several of these together may be prohibitive.

Evaluating the Impact of a Decision

Anybody can make a decision, but the hard part is making the right decision. Decision-makers often lack the skills in how to evaluate the results or impact of a decision. What the project manager believes was that the correct decision may be viewed differently by the client and the stakeholders.

Part of decision-making requires the project manager to predict how those impacted by the decision will react. Soliciting feedback prior to the implementation of the solution seems nice to do. But the real impact of the decision may not be known until after full implementation of the solution. As an example, as part of developing a new product, marketing informs the project manager that the competition has just come out with a similar product and marketing believes that we must add in some additional features into the product you are developing. The project team adds in a significant number of "bells and whistles" to the point where the product's selling price is higher than that of the competition and the payback period is now elongated. When the product was eventually launched, the consumer did not believe that the added features were worth the additional cost.

It is not always possible to evaluate or predict the impact of a decision when making a choice among alternatives. But soliciting feedback prior to full implementation is helpful.

A useful tool for assisting in the selection of alternatives is a consequence table, such as shown in Figure 5-1. For each alternative, the consequences are measured against a variety of factors

Competing Constraints

Alternative	Time	Cost	Quality	Safety	Overall Impact
#1	A	C	B	B	B
#2	A	C	A	C	B
#3	A	C	C	C	C
#4	B	A	C	A	B
#5	A	B	A	A	A

A = High Impact
B = Moderate Impact
C = Low Impact

Figure 5-1 A Consequence Table

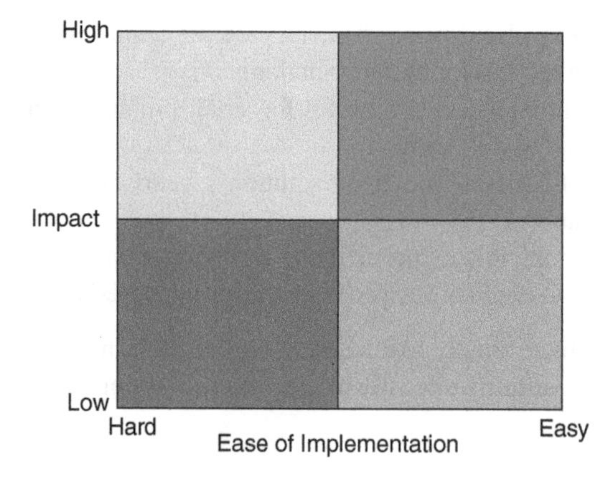

Figure 5-2 Impact Analysis

such as each of the competing constraints. For example, an alternative could have a favorable consequence on quality but an unfavorable consequence on time and cost. Most consequence tables have the impacts identified quantitatively rather than qualitatively. Risk is also a factor that is considered, but the impact on risk is usually defined qualitatively rather than quantitatively.

If there are three alternatives and five constraints, then there may be 15 rows in the consequence table. Once all 15 consequences are identified, they are ranked. They may be ranked according to either favorable or unfavorable consequences. If none of the consequences are acceptable, then it may be necessary to perform tradeoffs on the alternatives. This could become an iterative process until an agreed-upon alternative is found.

The people preparing the table are the people who make up the project team rather than possible outsiders that were brought in as SMEs for a particular problem. Project team members know the estimating techniques as well as the tools that are part of the organization process assets that can be used for determining impacts.

It is nice to have several possible alternatives for the solution to a problem. Unfortunately, the alternative that is finally selected must be implemented, and that can also create problems.

One of the ways to analyze the impact is to create an impact-implementation matrix, as shown in Figure 5-2. Each alternative considered could have a high or low impact on the project. Likewise, the implementation of each alternative could be easy or hard.

Each alternative is identified in its appropriate quadrant. The most obvious choice would be the alternatives that have a low impact and are easy to implement. But we often do not find very many alternatives in this quadrant.

The Time to Implement a Solution

Too many times in my life I have sat in problem-solving sessions and listened to team members come up with good (often brilliant) solutions to a problem. Everyone becomes enamored with the brilliance of the solution, but nobody seems concerned as to how long it will take to implement the solution. Significantly more time is consumed in implementing a solution than in decision-making.

Questions that should be considered include:

- Must we change our plans/baselines and, if so, how long will it take?
- How long will it take to get the necessary additional funding approved?
- Will the resources with the required skill levels be available when needed?
- Is overtime an option?
- How long will it take to procure those materials we need?
- Are additional reviews and meetings necessary before implementation can begin?
- Are additional reviews and meetings necessary as we implement the solution?

Simply stated, decision-making is easy; implementation is often difficult and time-consuming.

5.2 Brainstorming

Introduction

How often have we heard executives and project managers say, "We need to come up with some new ideas," or "We need to find some elegant solutions to this project problem"? The project manager then sends out emails to all the team members with the hope of filling every available seat in a conference room with the belief that the more people that attend, the greater the number of ideas generated. Brainstorming sessions are usually components of problem solving and decision-making. The brainstorming session may be virtual or onsite. Yet most project teams frequently fail at brainstorming.

Although brainstorming has become a popular group technique, when applied in a traditional group setting, researchers have not found evidence of its effectiveness for enhancing either the quantity or quality of ideas generated. However, it may still provide benefits, such as boosting morale, enhancing work enjoyment, and improving teamwork. Thus, numerous attempts have been made to improve brainstorming or use more effective variations of the basic technique.

Brainstorming sessions will become an integral part of project management in the future as we begin applying project management practices to more strategic type projects. Brainstorming is not just coming up with ideas to solve a problem. Brainstorming also includes a decision-making phase where the ideas are evaluated, and the best idea is selected.

Throughout the life cycle of any project, the team will be tested on their ability to find the best possible solution to a problem within the imposed limitations and boundaries. Brainstorming sessions can exist anytime whether a large or small decision must be made. This could occur in the planning phase of the project where we must come up with the best possible approach for a plan or it could happen in any later phases where problems arise, and the best solution must be found.

When the sessions work as planned, participants may feel elated in the creativity that they participated in. However, for some, they may experience feelings of anxiety, a loss of empowerment and even a hatred of the upcoming meeting based on past experiences in such sessions.

Although we normally discuss brainstorming as a means for identifying alternative solutions to a problem, brainstorming can also be used for root cause identification of the problem. These are situations where brainstorming techniques may not be appropriate. Most people seem to have heard about brainstorming, but very few have been part of brainstorming teams.

Onsite Brainstorming

Before we had advancements in technology, most brainstorming sessions were onsite. The onsite brainstorming session begins with the project manager spending less than five minutes discussing the problem or goal, and then asking the participants for ideas and solutions. For most participants, this may very well be the first time they heard about the problem or goal. They look at each other in amazement, and begin thinking about quick, rather than the best, solutions. For others, this may appear as a session for them to vent their emotions and believe that they are being asked to support a decision already made by senior management or project governance. Some people might believe that the session is their chance to exploit their hidden agenda.

Some people refuse to speak, even when asked, for fear of being criticized. Others might love the attention and want to spend an enormous amount of time in control of the discussion and defending their position. There may be people that have no ideas to contribute and throw their support behind the first idea that is brought up. It is not uncommon for some people to get hung up on details that may be irrelevant. Some people just look at the clock waiting for the meeting to end or are multitasking using their cell phones or laptops. Those people who may have good ideas may feel intimidated to speak if there are people in the room that are recognized as SMEs.

Some people may wish to express their ideas but need to think about the problem a bit longer.

The session, which was probably scheduled for one hour, finishes at the end of two hours or longer with just a few ideas, many of which may need further evaluation. The project manager then prepares a list of action items and tells everyone that another email will be sent out for the follow-up brainstorming session(s).

Reasons for Failure

Most brainstorming session simply do not provide the expected results, despite starting out with admirable intentions. This is a fact regardless of if you are using the Waterfall approach, Agile or Scrum. It is not because the brainstorming process does not work but because the sessions are conducted poorly. Understanding the reasons for brainstorming failure often serve as a motivational force for corrective action. Some of the most critical reasons for failure include:

- **Lack of training for the facilitators and attendees:** Most project management training programs discuss brainstorming but never fully train people on the right way for the sessions to be conducted. Project managers can do more harm than good by facilitating brainstorming sessions without being adequately trained. There are people professionally trained in brainstorming practices. In such cases, the professional should conduct the session and the project manager may simply be a participant, taking notes if necessary, and answering questions. Ideally, everyone should be trained in brainstorming techniques, so they have a good understanding of expectations when attending such sessions.

- **People spend too much time on solutions:** People tend to focus quickly on solutions without fully understanding the problem, goal, or question presented. While having people come prepared with ideas and solutions seems a good idea, the focus must be on the right question or problem before generating ideas. Sufficient time must be allowed for people to understand why the meeting was called. Even if this is explained in the invitation email, it should be reinforced at the onset of the meeting for alignment to the issues at hand. Solving the wrong problem is a waste of precious time and money.
- **Poorly trained facilitators begin the meeting by immediately asking for ideas:** The meeting should begin with an understanding of the ground rules (such as no distractions or interruptions), creating the right mindset, explaining the expectations on behavior of the participants (following directions), how the meeting will be conducted, and a clarification of the purpose of the meeting. Even though people may have attended brainstorming training sessions previously, taking a few minutes to explain the ground rules for the session is helpful.
- **Failing to consider the fears and apprehensions of the participants:** Some people have an inherent fear of brainstorming sessions, and this includes experienced personnel. The apprehensions might include fear of being criticized, fear of being drawn into a conflict, and fear of change if the implementation of some of the ideas might remove one from their comfort zone.
- **No control over criticism:** In brainstorming, criticism of ideas creates conflict and wastes valuable time needed to generate the maximum number of ideas. When people see ideas being criticized, they tend to withhold their own ideas to avoid being criticized.
- **Too long of a meeting:** Brainstorming sessions in some companies may be as short as 15 minutes. Meetings that go beyond an hour make people edgy and looking at the clock, hoping for adjournment. A study conducted at the University of Amsterdam showed that, when people work alone, they tend to come up with more ideas than when working in a group. It may be best to ask people to work first alone, or in small groups, to come up with their best ideas and then share the information in larger groups for evaluation.
- **Large groups can stifle creativity:** There is a valid argument that small groups of 5–10 people yield better results than large groups. Jeff Bezos, formerly Amazon's CEO, calls it the two-pizza rule: If the group can eat more than two pizzas, it is too big. The sessions may be composed of SMEs and may not include employees that might later be assigned to the implementation of the desired solution. Having a large group does not mean that more ideas will be forthcoming. Some people may feel intimidated by the size of the group and contribute only a limited number of ideas if at all. With large groups, people worry about how others will view their ideas and may be afraid to contribute for fear of criticism. People tend to contribute more meaningful results when in small groups. If the group is large, it may be best to break the participants into smaller groups at the start of the session for idea generation and then bring them together in a large group near the session's end for idea evaluation. Strong leadership is necessary to prevent the loudest voices and largest groups from drowning out the smaller voices and individualism.
- **Not building on the ideas of others:** Sometimes combining ideas may generate the best possible solution. For this to work correctly, people must be given ample time to express their thoughts and digest what they heard. New ideas should be encouraged from the combination of ideas already presented. This is the reason why smaller groups at first are often best. The best possible solution may be a combination of ideas.
- **Having the wrong balance of experience and knowledge in the session:** People should be invited to attend based on the contribution that they can make rather than simply because of their rank, title, or availability. Inviting people that have a valid interest in the topic, even if they are not part of the project team, may bring forth good ideas.

- **Not having a diverse group:** Having a diverse brainstorming team may be advantageous. More information may come forth that leads to a different and better solution to a problem. Diverse teams usually do a better job challenging the assumptions, looking at problems and solutions differently, thinking "outside the box," and dividing up the work requirements. Each member of the diverse group may come up with good ideas for parts of the solution and, when all the pieces are assembled, a good solution may result.
- **Allowing one person to dominate the discussion:** Some people like to hear themselves talk and try to dominate the discussion. This can be demoralizing to others and the frustration can prevent others from wishing to speak.
- **Information overload:** Brainstorming sessions run the risk of information overload. This is particularly true if multiple brainstorming sessions are needed. Information overload can be demoralizing but can be controlled using idea management software.
- **Premature evaluation:** The intent of a brainstorming session is to maximize the number of ideas, both good and bad. The assumption made is that the greater the number of ideas, the greater the chance of finding the optimal solution to a problem. Unfortunately, some groups tend to quickly jump on the first acceptable idea and run with it without proper evaluation. Forcing participants to vote without due consideration of the facts can result in the implementation of a suboptimal solution that everyone will question.

The Need for Brainstorming Structure

Simply putting a group of people in a room and saying, "Let's come up with great ideas," does not work well. Professional facilitation and session structure are necessary to maximize performance expectations. Conducting a brainstorming session is not the same as holding a weekly or monthly team meeting.

Some of the things that a company can do include:

- First and foremost, it is best to have a professionally trained facilitator conduct the brainstorming session to get people to contribute ideas, bring order from chaos, and limit distractions.
- Send out an agenda early that clearly states the purpose of the meeting, the ground rules, and the topic(s) to be discussed. The purpose of the meeting must be clear, not too big, and captured in a specific question or issue. If the problem is too big, the facilitator should break it into smaller components, each with its own question.
- If handouts will be used in the session, it may be best to provide the handouts with the agenda so people can review them prior to the meeting and then come prepared to ask the right questions and possibly make decisions.
- Clearly articulate the reason for the meeting and the goal and make sure that the goal is reachable in a reasonable time frame.
- While asking people to "think outside of the box" seems like a good idea, the best solution may be when the participants think "inside the box" instead.
- Invite participants that may have an interest in the topic even though they are not part of the project team.
- If meeting in person, ask people not to bring distractions such as cell phones, notepads, or laptops.
- Do not criticize any ideas no matter how bad they sound.
- If market research is required, ask the participants to obtain information from the end users rather than others.

- Encourage everyone to come prepared to speak and to share their ideas, whether good or bad.
- Document all ideas because some ideas may be valuable later for issues on other projects. There are several excellent software packages that deal with idea management and brainstorming activities.
- Some people are combative and continuously fight for their belief or their dislike for someone else's position. These people must be controlled to prevent the meeting from losing the intended purpose.

Virtual Brainstorming Sessions

The previous discussion assumed that the people attending the brainstorming session were in the same building or location. Generally, workers spend a minimum of 25 percent of their time in virtual communications. This percentage is increasing as companies are now embarking on virtual brainstorming sessions. More people are working from home because of several factors, including COVID-19, the cost of office rental space, and the fact that workers needed for the sessions might be dispersed geographically across multiple continents.

Virtual brainstorming is somewhat more difficult than onsite brainstorming because the virtual environment may require a different set of tools and software for communication, viewing, recording, and displaying of ideas, and interaction among participants. If the group must be broken down in smaller groups, multiple concurrent virtual sessions may be necessary. The proper use of virtual brainstorming tools can overcome the productivity loss encountered in onsite brainstorming, bring out more creative ideas per person, and generate a higher degree of satisfaction among the team members.

Virtual brainstorming has advantages and disadvantages, many of which are like onsite brainstorming. The benefits include:

- The workers are under less peer pressure and may not be intimidated by others on the call.
- It may be easier to put together a diverse team of participants.
- People are working alone or in small groups and may come up with more fruitful ideas than in larger groups.
- Large groups can participate virtually, and it is less likely that someone will want to dominate the discussion with their ideas.
- Large groups can be subdivided into smaller groups without worrying about title, rank, and expertise.
- There is less wasted time in virtual sessions than onsite sessions.

There are several disadvantages, including:

- Facilitators must ensure that the proper virtual tools are in place.
- It may take more time at the onset of the meeting to make sure that everyone is on the same page.
- Sharing documents may be difficult virtually; facilitators must ensure that all participants have the appropriate handouts.
- The way that communication takes place may make it difficult for workers to build on the ideas of others or to combine ideas.
- It may be difficult to break large groups into smaller groups virtually.
- Virtual participants may be less likely to ask questions than if they were in the conference room with the other team members.

- Professional facilitators are trained in emotional intelligence and how to read body language. They can observe the expressions on people's faces and watch what they do with their hands or the way they are sitting as an indication of whether they are upset or in agreement with the discussion. Their fears and apprehensions can be visible by how they act. This is difficult to observe virtually.
- Having an open dialogue where everyone gets to speak may be difficult to enforce.
- Having too large a group may prevent or discourage members from providing input.
- People may be multitasking or distracted, and the facilitator has limited control over the meeting.

Decision-Making Characteristics

Brainstorming is not just about generating ideas for others to evaluate and select. Usually, the group itself will, in its final stage, evaluate the ideas and select one as the solution to the problem proposed to the group. Based on the criticality of the choices, decision makers that were not involved in the brainstorming sessions may be invited to participate in the decision-making phase.

Decision-making has some degree of structure to it, as did the brainstorming phase. Decision-making involves the following:

- Objectives must first be established or reconfirmed.
- Objectives must be classified and placed in order of importance.
- Alternative actions must be developed for each reasonable solution.
- The alternatives must be evaluated against all the objectives.
- The alternative that will achieve all or most of the objectives is the tentative decision.
- The tentative decision is evaluated for more possible consequences.
- The decisive actions are taken, and additional actions are taken to prevent any adverse consequences from becoming problems and starting both systems (problem analysis and decision-making) all over again.

The decision-making activities are often more time consuming and costlier to perform than the problem-solving activities of brainstorming. This is largely due to the number of alternatives that can be identified, as well as the methods used to evaluate and prioritize them. Having a significant number of reasonable alternatives may seem nice, but being unable to arrive at a decision on which one to adopt can be troublesome.

There are several types of decision-making styles that people can use. There are also numerous tools that can assist in the decision-making processes.

People must understand prior to attending both the problem-solving and the decision-making meetings how the decisions will be made. There are several options available, and the approach taken to agree on the problem can be different from the decision on which alternatives were selected:

- The solution should not require resources or skills the members of the group do not have or cannot acquire.
- If acquiring additional resources or skills are necessary, that needs to be the first part of the solution.
- There must be a way to measure progress and success.
- The steps to carry out the solution must be clear to all, and amenable to being assigned to the team members so that each will have an important role.

- There must be a common decision-making process to enable a coordinated effort to proceed, and to reassign tasks as the project unfolds.
- There should be evaluations at milestones to decide whether the group is on track toward a final solution.
- There should be incentives for participation so that participants maintain their efforts.

Recommendations

Effective brainstorming practices increase the chances of success and foster teamwork. Any company can get lucky and occasionally come up with one or two brilliant ideas. But if a company wants a continuous stream of great ideas, then educating project teams in correct brainstorming practices is mandatory. Brainstorming education should be a requirement for all employees if a company wants a sustained competitive advantage. The training must consider the challenges of virtual brainstorming sessions, as well as onsite sessions.

5.3 Design Thinking[1]

Many future projects will be approved based on an idea, goal, or a strategic need but not fully described. One of the most important skills for managing these projects may be design thinking. Design thinking is a structured process for exploring ill-defined problems that were not clearly articulated, helping to solve ill-structured situations, and improving desired outcomes. Design thinking can help resolve performance challenges. "Design thinking helps structure team interactions to cultivate greater inclusiveness, foster creativity, deepen empathy, and align participants around specific goals and results" (Mootee 2013; 63).

Design thinking is a collaborating approach to creative problem solving in rapidly changing markets where breakthrough ideas may be necessary. Design thinking also mandates a close and trusting relationship with the team members and the stakeholders throughout the life of the project. The focus is on customer needs and "thinking outside of the box."

As part of design thinking, you must know who your customers are or will be. This information may come from an understanding of competitive research such as a SWOT analysis and may include identification of:

- The most profitable and demanding customers
- The less-demanding customers
- The new customers that are willing to accept a "good enough" product or service

Interviewing customers may be highly advantageous. The information can then be mapped as shown in Figure 5-3.

The team and the PM must also have a line-of-sight as to what are the strategic goals and growth objectives. There are no clear-cut tools or paths for identifying these goals. Some people argue that improper goal setting can change the intended direction for a project, whereas others prefer the need for some ambiguity with the argument that it creates space for creative and innovative ideas, more fallback options are available, and the team may have an easier time converting ideas to

---

1 Adapted from Harold Kerzner, *Innovation Project Management*, Hoboken, John Wiley Publishers, 2019, 135–138.

Stakeholders

Stakeholders Issues	Customers	Shareholders	Government	Management	Employees
Product Quality	A	C	B	B	B
Product Safety	A	C	A	C	C
Product Features	A	C	C	C	B
Product Cost	B	A	C	A	C
Delivery Date	A	B	C	A	A

A = High Stakeholder Importance
B = Somewhat Important to Stakeholder
C = Low Stakeholder Importance

Figure 5-3 Prioritizing Stakeholder Needs

reality using design thinking. When deciding to "kill" an idea because the value is not there, you must first put yourself in the users' shoes and think about the first user, not yourself. Future PMs must know the firm's tangible and intangible assets (capabilities and resources). The design-thinking approach to redefine value for the customers begins with people, not products.

Design thinkers seek to understand the cultures not only of others but also of themselves, recognizing that their own emotions, practices, and belief systems inform what, how, and why they do what they do.

Embedding design thinking into a business means embedding it into the company's strategy, corporate culture, processes and practices, systems, and structures. According to Mootee (2013; 60), "Applied design thinking in business problem solving incorporates mental models, tools, processes, and techniques such as design, engineering, economics, the humanities, and the social sciences to identify, define, and address business challenges in strategic planning, product development, innovation, corporate social responsibility and beyond." Unfortunately, most of these topics are not covered in traditional project management training programs but are a necessity for educating future project managers.

There are several benefits to design thinking:

- Greater focus on the customers' needs
- Discovering new customer insights (i.e., with engagement project management)
- More creativity through idea generation and prototyping
- Solutions created as quickly as possible
- Creates a culture of learning
- Better handling of ambiguity
- Better understanding of complex connections
- Incrementally little bets as we go along rather than big bets

There are also causes for design thinking failure:

- Not understanding the problem and/or lack of definition
- Lack of information
- Poor communication channels, especially with stakeholders
- Rushing into prototype development

- Expecting a final solution from using one prototype
- Looking for a quick solution
- Team members are closed-minded to ideas of others
- Relying too much on history and past customer behavior
- Overthinking the information at hand
- Failing to consider parallel paths
- Corporate gravity (i.e., removal from comfort zones)
- Cynicism

Project managers will face several challenges, beginning with a fuzzy front end (FFE). A well-managed FFE usually leads to better outcomes. Unfortunately, because of up-front uncertainty and a need to make quick choices, especially based on incomplete information, managing the FFE is a challenge.

Prototyping is another challenge. In linear prototyping, which occurs in traditional project management, the prototype is developed near the end of the project. In nonlinear prototyping, which is most common in strategic projects, several prototypes are developed, beginning upfront during design thinking. "Fail fast, fail cheap, and fail early" must be replaced by "Learn fast, learn cheap, and learn early." Full validation of an innovation may require several prototypes and possibly world testing.

Organizational friction will occur during design thinking and can be advantageous and create internal competition surrounding intangible assets such as the interpretation of intellectual property and the rationale for some decisions. People may then bring additional information to the table to support their position.

If the outcome of a project is to create customer value, then there is a need to bring design principles, methods and these new tools into organizational management and business strategy development (Brown 2008). Design thinking activities and project management are both evolving rapidly as transformation factors and processes in firms and the economic landscape change. Both fields are anchored in a practice characterized by methods and tools, but they are moving beyond that operational perspective toward a strategic one" (Ben Mahmoud-Jouini et al. 2016). A primary reason for this is because an output of design thinking may not be a product but instead a new business model that includes significant changes in customer management practices, innovation activities for the future, regulatory and social considerations, organizational considerations, and financial considerations.

There are more than 100 tools that can be used as part of design thinking (Kumar 2013). Some common design thinking tools include:

- Storytelling (providing narrative info rather than dry facts)
- Storyboards (depicting the innovation needs through a story with artwork)
- Mind maps (connecting all the information)
- Context maps (uncovering insights on user experience)
- Customer journey maps (stages customers go through to purchase the product and use it)
- Stakeholder maps (visualizing stakeholder involvement)
- Personas (determining who are the users and nonusers)
- Metaphors (comparison with something else)
- Prototyping (testing different ideas)
- Generative sessions (looking at stakeholder experience)

There are other tools that can be used, such as information warehouses and business intelligence systems. These tools help identify customers and their needs.

5.4 Excellence in Action: Disney[2]

Introduction

The future of project management in many companies may require deviations from standard practices and traditional project management skills because of the firm's unique products and services. New types of projects will require deviations from standard practices. Project teams may need to learn new skills. Disney University was created as a means for preparing project teams for the future challenges at Disney as well as present assignments.

Not all project managers that manage traditional projects are happy with their jobs and often believe that changing industries might help. The project managers that are probably the happiest with challenging and complex projects are the Imagineering project managers that work for the Walt Disney Company. Three of their Imagineering project managers (John Hench, Claude Coats, and Martin Sklar) retired with a combined 172 years of Imagineering project management work experience with the Walt Disney Company. But how many project managers in other industries truly understand what skills are needed to be successful as an Imagineering project manager? Is it possible that many of the Imagineering project management skills are or will be applicable to other industries and we do not recognize it?

The ***PMBOK® Guide*** is, as the name implies, just a guide. Each company may have unique or specialized skills needed for the projects they undertake above and beyond what is included in the *PMBOK® Guide*. Even though the principles of the *PMBOK® Guide* still apply to Disney's theme park projects, there are other skills needed that are significantly different from a lot of the material taught in traditional project management courses. Perhaps the most common skills among all Imagineering project managers are brainstorming, problem-solving, decision-making, and thinking in three rather than two dimensions. While many of these skills are not taught in depth in traditional project management programs, they may very well be necessities for all future project managers. Yet most of us may not recognize this fact.

Walt Disney Imagineering[3]

Walt Disney Imagineering (also known as WDI, or simply Imagineering) is the design and development arm of The Walt Disney Company, responsible for the creation and construction of Disney theme parks worldwide. Founded by Walt Disney to oversee the production of Disneyland Park, it was originally known as WED Enterprises, from the initials meaning "Walter Elias Disney," the company founder's full name.[4]

The term Imagineering was introduced in the 1940s by Alcoa to describe its blending of imagination and engineering, and used by Union Carbide in an in-house magazine in 1957, with an article by Richard F. Sailer called BRAINSTORMING IS IMAGINation engINEERing. Disney filed for a copyright for the term in 1967, claiming first use of the term in 1962. Imagineering is responsible for designing and building Disney theme parks, resorts, cruise ships, and other entertainment venues at all levels of project development. Imagineers possess a broad range of skills

2 ©2021 by Harold Kerzner. Adapted from Disney (A): Imagineering Project Management, *Project Management Case Studies*, 6e, Hoboken, John Wiley & Sons, 2021. Neither the Walt Disney Company nor any personnel from the Disney theme parks participated in the preparation of this case study.
3 Part of this case study has been adapted from Wikipedia, the Free Encyclopedia: Walt Disney Imagineering.
4 Wright, Alex; Imagineers. *The Imagineering Field Guide to Magic Kingdom at Walt Disney World*. New York: Disney Editions; 2005.

and talents, and thus over 140 different job titles fall under the banner of Imagineering, including illustrators, architects, engineers, lighting designers, show writers, graphic designers, and many more.[5] It could be argued that all Imagineers are project managers and all project managers at WDI are Imagineers. Most Imagineers work from the company's headquarters in Glendale, California, but are often deployed to satellite branches within the theme parks for long periods of time.

Project Deliverables

Unlike traditional projects where the outcome is a hardware or software deliverable, Imagineering project outcomes for theme park attractions are visual stories. The entire deliverable is designed to operate in a controlled environment where every component of the deliverable has a specific meaning and contributes to part of telling a story. It is visual storytelling. Unlike traditional movies or books that are two dimensional, the theme parks and the accompanying characters come to life in three dimensions. Most project managers do not see themselves as storytellers and tend to focus in two dimensions.

The intent of the theme park attraction is to remove people from reality once they enter the attraction and make them believe that they are living out a story and possibly interacting with their favorite characters. Theme park visitors are made to feel that they are participants in the story, rather than just observers, and this includes visitors of all ages.

While some theme parks are composed of rides that appeal to just one of your senses, Disney's attractions appeal to several of the senses thus leaving a greater impact when people exit the attraction. "People must learn how to see, hear, smell, touch and taste in new ways."[6] Everything is designed to give people an experience. In the ideal situation, people are made to believe that they are part of the story. When new attractions are launched, Imagineers pay attention to the guests' faces as they come off of a ride. This is important for continuous improvement efforts.

> *All I want you to think about is when people walk through or have access to anything you design, I want them, when they leave, to have smiles on their faces. Just remember that. It's all I ask of you as a designer.*[7]
>
> *Walt Disney*

The Importance of Constraints

Most project management courses emphasize that there are three constraints on projects, namely time, cost, and scope. While these constraints exist for Imagineering projects as well, there are three other competing theme park constraints that are often considered as more important than time, cost, and scope. The additional constraints are safety, quality, and aesthetic value.

Safety, quality, and aesthetic value are all interrelated constraints. Disney will never sacrifice safety. It is first and foremost the primary constraint. All attractions operate every few minutes 365 days each year and must therefore satisfy the strictest of building codes. Some rides require special effects such as fire, smoke, steam, and water. All of this is accomplished with safety in mind. Special effects include fire that actually does not burn, simulated fog that one can breathe

---

5 Ibid.

6 John Hench, *Designing Disney*, Disney Editions (2008), p. 2.

7 *Walt Disney Imagineering*, Disney Editions (1996), p.18.

safely, and explosions that do not destroy anything. Another special effect is the appearance of bubbling molten lava that is actually cool to the touch.

Reliability and maintainability are important quality attributes for all project managers but of critical importance for the Imagineers. In addition to fire, smoke, steam, and water, there are a significant number of moving parts in each attraction. Reliability considers how long something will perform without requiring maintenance. Maintainability discusses how quickly repairs can be made. Attractions are designed with consideration given to component malfunctions and ways to minimize the down time. Some people may have planned their entire vacation around specific attractions, and if these attractions are down for repairs for a lengthy time, park guests will be unhappy.

Brainstorming

With traditional projects, brainstorming may be measured in hours or days. Members of the brainstorming group are small in number and may include marketing for the purpose of identifying the need for a new product or enhancement to an existing product, and technical personnel to state how long it takes and the approximate cost. Quite often, traditional project managers may not be assigned and brought on board until after the project has been approved, added into the queue, and the statement of work is well-defined. At Disney's Imagineering organization, brainstorming may be measured in years and a multitude of Imagineering personnel will participate, including the project managers.

Attractions at most traditional amusement parks are designed by engineers and architects. Imagineering brainstorming at Disney is done by storytellers that must visualize their ideas in both two and three two dimensions. Brainstorming could very well be the most critical skill for an Imagineer. Brainstorming requires that Imagineers put themselves in the guests' shoes and think like kids as well as adults in order to see what the visitors will see. You must know your primary audience when designing an attraction.

Brainstorming can be structured or unstructured. Structured brainstorming could entail thinking up an attraction based on a newly released animated or non-animated Disney movie. Unstructured brainstorming is usually referred to as "blue sky" brainstorming. Several sessions may be required to come up with the best idea because people need time to brainstorm. Effective brainstorming mandates that we be open-minded to all ideas. And even if everyone agrees on the idea, Imagineers always ask, "Can we make it even better?" Unlike traditional brainstorming, it may take years before an idea comes to fruition at the Imagineering Division.

Imagineering brainstorming must focus on a controlled themed environment where every component is part of telling the story. There are critical questions that must be addressed and answered as part of Imagineering brainstorming:

- How much space will I have for the attraction?
- How much time will the guests need to feel the experience?
- Will the attraction be seen on foot or using people movers?
- What colors should we use?
- What music should we use?
- What special effects and/or illusions must be in place?
- Does technology exist for the attraction, or must new technology be created?
- What landscaping and architecture will be required?
- What other attractions precede this attraction or follow it?

Before brainstorming is completed, the team must consider the cost. Regardless of the technology, can we afford to build it? This question must be addressed whether it is part of a structured or blue sky brainstorming session.

Guiding Principles

Imagineers are governed by a few key principles when developing new concepts and improving existing attractions. Often new concepts and improvements are created to fulfill specific needs and to make the impossible appear possible. Many ingenious solutions to problems are Imagineered in this way, such as the ride vehicle of the attraction Soarin' Over California. The Imagineers knew they wanted guests to experience the sensation of flight but weren't sure how to accomplish the task of loading the people on to a ride vehicle in an efficient manner where everyone had an optimal viewing position. One day, an Imagineer found an Erector set in his attic, and was able to envision and design a ride vehicle that would effectively simulate hang gliding.[8]

Imagineers are also known for returning to ideas for attractions and shows that, for whatever reason, never came to fruition. It could be years later when they revisit the ideas. These ideas are often reworked and appear in a different form – like the Museum of the Weird, a proposed walk-through wax museum that eventually became the Haunted Mansion.[9]

Finally, there is the principle of "blue sky speculation," a process where Imagineers generate ideas with no limitations. The custom at Imagineering has been to start the creative process with what is referred to as "eyewash" – the boldest, wildest, best idea one can come up with, presented in absolutely convincing detail. Many Imagineers consider this to be the true beginning of the design process and operate under the notion that if it can be dreamt, it can be built.[10] Disney believes that everyone can brainstorm and that everyone wants to contribute to the brainstorming process. No ideas are bad ideas. Effective brainstorming sessions neither evaluate nor criticize the ideas. They are recorded and may be revisited years later.

Imagineers are always seeking to improve on their work – what Walt called "plussing." He firmly believed that "Disneyland will never be completed as long as there's imagination left in the world," meaning there is always room for innovation and improvement.[11] Imagineering also has created many ideas that have never been realized, although some, such as Country Bear Jamboree, do take form in one way or another later. Ideas and eventually future attractions can also come from the animated films produced by the Walt Disney Company or other film studios.

"The brainstorming subsides when the basic idea is defined, understood, and agreed upon by all group members. It belongs to all of us, keeping strong a rich heritage left to use by Walt Disney. Teamwork is truly the heart of Imagineering. . . In that spirit, though Imagineering is a diverse collection of architects, engineers, artists, support staff members, writers, researchers, custodians, schedulers, estimators, machinists, financiers, model-makers, landscape designers, special effects and lighting designers, sound technicians, producers, carpenters, accountants, and filmmakers – we all have the honor of sharing the same unique title. Here, you will find only *Imagineers*."[12]

---

8 George Scribner and Jerry Rees (Directors) (2007). *Disneyland: Secrets, Stories, and Magic* (DVD). Walt Disney Video.
9 Ibid.
10 Marling, Karal (1997). *Designing Disney's Theme Parks*. Paris, New York: Flammarion.
11 Scribner and Rees, 2007.
12 *Walt Disney Imagineering*, Disney Editions (1996), p. 21.

If I could pick any job here, I'd move my office to the Imagineering building and immerse myself in all that lunacy and free-thinking.[13]

Michael D. Eisner, CEO, Walt Disney

Imagineering Innovations

Over the years, Walt Disney Imagineering has been granted over 115 patents in areas such as ride systems, special effects, interactive technology, live entertainment, fiber optics, and advanced audio systems.[14] WDI is responsible for technological advances such as the Circle-Vision 360° film technique and the FastPass virtual queuing system.

Imagineering must find a way to blend technology with the story. Imagineering is perhaps best known for its development of Audio-Animatronics, a form of robotics created for use in shows and attractions in the theme parks that allowed Disney to animate things in three dimensions instead of just two dimensions. The idea sprang from Disney's fascination with a mechanical bird he purchased in New Orleans, which eventually led to the development of the attraction The Enchanted Tiki Room. The Tiki Room, which featured singing Audio-Animatronic birds, was the first to use such technology. The 1964 World's Fair featured an Audio Animatronic figure of Abraham Lincoln that actually stood up and delivered part of the Gettysburg Address (which was incidentally just past its centennial at the time) for the "Great Moments With Mr. Lincoln" figure exhibit, the first human Audio-Animatronic.[15]

Today, Audio-Animatronics are featured prominently in many popular Disney attractions, including Pirates of the Caribbean, The Haunted Mansion, The Hall of Presidents, Country Bear Jamboree, Star Tours: The Adventures Continue and Muppet*Vision 3D. Guests also have the opportunity to interact with some Audio-Animatronic characters, such as Lucky the Dinosaur, WALL-E, and Remy from Ratatouille. The next wave of Audio-Animatronic development focuses on completely independent figures, or "Autonomatronics." Otto, the first Autonomatronic figure, is capable of seeing, hearing, sensing a person's presence, having a conversation, and even sensing and reacting to guests' emotions.

Storyboarding

Most traditional project managers may be unfamiliar with the use of the storyboarding approach as applied to projects. At Disney Imagineering, it is an essential part of the project. Ideas at Imagineering begin as a two-dimensional vision drafted on a piece of white paper. Storyboards assist the Imagineers in seeing the entire attraction. Storyboards are graphic organizers in the form of illustrations or images displayed in sequence for the purpose of previsualizing the relationship between time and space in the attraction. Storyboards have also been used in motion pictures, animation, motion graphics, and interactive media. The storyboard provides a visual layout of events as they are to be seen by the guests. The storyboarding process, in the form it is known as today, was developed at Walt Disney Productions during the early 1930s, after several years of similar processes being in use at Walt Disney and other animation studios.

A storyboard is essentially a large comic of the attraction produced beforehand to help the Imagineers visualize the scenes and find potential problems before they occur. Storyboards also

13 *Wall Street Journal*, Jan 6, 1987.
14 Walt Disney Imagineering website.
15 Ibid.

help estimate the cost of the overall attraction and save development time. Storyboards can be used to identify where changes to the music are needed to fit the mood of the scene. Often storyboards include arrows or instructions that indicate movement. When animation and special effects are part of the attraction, the storyboarding stage may be followed by simplified mock-ups called "animatics" to give a better idea of how the scene will look and feel with motion and timing. At its simplest, an animatic is a series of still images edited together and displayed in sequence with a rough dialogue and/or rough sound track added to the sequence of still images (usually taken from a storyboard) to test whether the sound and images are working together effectively.

The storyboarding process can be very time-consuming and intricate. Today, storyboarding software is available to speed up the process.

Mockups

Once brainstorming has been completed, mockups of the idea are created. Mockups are common to some other industries such as construction. Simple mock-ups can be made from paper, cardboard, Styrofoam, plywood, or metal. "The model-maker is the first Imagineer to make a concept real. The art of bringing a two-dimensional design into three dimensions is one of the most important and valued steps in the Imagineering process. Models enable the Imagineer to visualize, in miniature, the physical layout and dimensions of a concept, and the relationships of show sets or buildings as they will appear.

As the project evolves, so too do the models that represent it. Once the project team is satisfied with the arrangements portrayed on massing models, small-scale detailed-oriented study models are begun. This reflects the architectural styles and colors for the project.

Creating a larger overall model, based on detailed architectural and engineering drawings, is the last step in the model-building process. "This show model is the exact replica of the project as it will be built, featuring the tiniest of details, including building exteriors, landscape, color schemes, the complete ride layout, vehicles, show sets, props, figures and suggested lighting and graphics."[16]

Computer models of the complete attraction, including the actual ride, are next. They are computer generated so that the Imagineers can see what the final product looks like from various positions without actually having to build a full-scale model. Computer models, similar to CAD/CAM modeling, can show in three dimensions the layout of all of the necessary electrical, plumbing, HVAC, special effects, and other needs.

Aesthetics

Imagineers view the aesthetic value of an attraction in a controlled theme environment as a constraint. This aesthetic constraint is more of a passion for perfection than the normal constraints that most project managers are familiar with.[17]

Aesthetics are the design elements that identify the character and the overall theme and control the environment and atmosphere of each setting. This includes color, landscaping, trees, colorful flowers, architecture, music, and special effects. Music must support the mood of the ride. The shape of the rocks used in the landscape is also important. Pointed or sharp rocks may indicate

16 *Walt Disney Imagineering*, Disney Editions (1996), p. 72.

17 Some people argue that the aesthetics focus more on creating a controlled environment than on reality, thus controlling your imagination.

danger whereas rounded or smooth rocks may represent safety. Everything in the attraction is there for the purpose of reinforcing a story. Imagineers go to highly detailed levels of perfection for everything needed to support the story without overwhelming the viewers with too many details. Details that are contradictory can leave the visitors confused about the meaning of the story.

A major contributor to the aesthetics of the attraction is the special effects. Special effects are created by "Illusioneering," which is a subset of Imagineering. Special effects can come in many different forms. Typical projected special effects can include:[18]

- Steam, smoke clouds, drifting fog, swirling effects
- Erupting volcano, flowing lava
- Lightning flashes and strikes, sparks
- Water ripple, reflection, waterfall, flows
- Rotating and tumbling images
- Flying, falling, rising, moving images
- Moving images with animated sections
- Kaleidoscopic projections
- Liquid projections, bubbles, waves
- Aurora borealis, lumia, abstract light effects
- Twinkling stars (when fiber optics cannot be used, such as on rear-projection screen)
- Spinning galaxies in perspective, comets, rotating space stations, pulsars, meteor showers, shooting stars, and any astronomical phenomena
- Fire, torches, forest fire
- Expanding rings
- Ghosts, distorted images
- Explosions, flashes

Perhaps the most important contributor to the aesthetic value of an attraction is color. Traditional project managers rely on sales or marketing personnel to select the colors for a deliverable. At Imagineering, it is done by the Imagineers. Color is a form of communications. Even the colors of the flowers and the landscaping are critical. People feel emotions from certain colors either consciously or subconsciously. Imagineers treat color as a language. Some colors catch the eye quickly and we focus our attention on it. "We must ask not only how colors work together, but how they make the viewer feel in a given situation. . . . It is the Imagineer's job to understand how colors work together visually and why they can make guests feel better."[19]

"White represents cleanliness and purity, and in many European and North American cultures . . . is the color most associated with weddings, and with religious ceremonies such as christenings. Silver-white suggest joy, pleasure, and delight. In architecture and interior design, white can be monotonous if used over large areas."[20] ". . . We have created an entire color vocabulary at Imagineering, which includes colors and patterns we have found that stir basic human instincts – including that of survival."[21]

18 See "Bill Novey and the Business of Theme Park Special Effects"; BloopLoop.com. The paper provides an excellent summary of various special effects used by Illusioneers. In addition to the projected special effects, the paper also describes laser effects, holographic images, floating images, mirror gags, gas discharge effects, and fiber optics.

19 Hench, p.104.

20 Ibid, p.135.

21 *Walt Disney Imagineering*, Disney Editions (1996), p. 94.

Aesthetics also impacts the outfits and full body costumes of the cast members that are part of the attraction. The outfits that the cast members wear must support the attraction. Unlike animation where there are no physical limitations to a character's identity or mobility, people may have restricted motion once in the costume. Care must be taken that the colors used in the full body costumes maintain the character's identity without conflicting with the background colors used in the attraction. Even the colors in the rest rooms must fit the themed environment.

Imagineers also try to address queue design by trying to make it a pleasant experience. As people wait in line to see an attraction, aesthetics can introduce them to the theme of the attraction. The aesthetics must also consider the time it takes people to go from attraction to attraction, as well as what precedes this attraction and what follows it. "For transition to be smooth, there must be a blending of themed foliage, color, sound, music, and architecture. Even the soles of your feet feel a change in the paving explicitly and tell you something new is on the horizon."[22]

The Art of the Show

Over the years, Imagineering has conceived a whole range of retail stores, galleries, and hotels that are designed to be experienced and to create and sustain a very specific mood – for example, the mood of Disney's Contemporary Resort could be called "the hello futuristic optimism," and it's readily apparent given the resort's A-frame structure, futuristic building techniques, modern décor, and the monorail gliding quietly through the lobby every few minutes. Together, these details combine to tell the story of the hotel.[23]

Imagineering is, first and foremost, a form of storytelling, and visiting a Disney theme park should feel like entering a show. Extensive theming, atmosphere, and attention to detail are the hallmarks of the Disney experience. The mood is distinct and identifiable, the story made clear by details and props. Pirates of the Caribbean evokes a "rollicking buccaneer adventure," according to Imagineering Legend John Hench, whereas the Disney Cruise Lineapos;s ships create an elegant seafaring atmosphere. Even the shops and restaurants within the theme parks tell stories. Every detail is carefully considered, from the menus to the names of the dishes to the Cast Members' costumes.[24] Disney parks are meant to be experienced through all senses – for example, as guests walk down Main Street, U.S.A., they are likely to smell freshly baked cookies, a small detail that enhances the story of turn-of-the-century, small-town America.

The story of Disney theme parks is often told visually, and the Imagineers design the guest experience in what they call "The Art of the Show." John Hench (one of Disney's Imagineering Legends) was fond of comparing theme park design to moviemaking, and often used filmmaking techniques in the Disney parks, such as the technique of forced perspective.[25] Forced perspective is a design technique in which the designer plays with the scale of an object in order to affect the viewer's perception of the object's size. One of the most dramatic examples of forced perspective in the Disney Parks is Cinderella's Castle. The scale of architectural elements is much smaller in the upper reaches of the castle compared to the foundation, making it seem significantly taller than its actual height of 189 feet.[26]

---

22 Ibid., p. 90.
23 Marling.
24 Hench, John; Peggy Van Pelt (2003). *Designing Disney: Imagineering and the Art of the Show*. New York: Disney Editions.
25 Hench and van Pelt.
26 Wright.

The Power of Social Leadership

Project managers and team members like to be told that they have done a good job. It is a motivational force encouraging them to continue performing well. However, acknowledgment does not have to come with words; it can come from results. Project teams at Disney get to see the fruits of their efforts and the value they have created. At Disney's Imagineering Division, the fact that millions of visitors pass through the gates of Disney theme parks each year is probably the greatest form of motivation.

Walt Disney Company does acknowledge some Imagineers in other ways. Disney established a society called "Imagineering Legends." Three of their most prominent Imagineering Legends are John Hench (65 years with Disney), Claude Coats (54 years with Disney), and Martin Sklar (53 years with Disney). The contributions that these three Imagineers have made appear throughout the Disney theme park attractions worldwide. The goal of all Imagineers at Disney may very well be the acknowledgment of becoming an Imagineering Legend.

The Need for Additional Skills

All projects have special characteristics that may mandate a unique set of project management skills above and beyond what we teach using the ***PMBOK® Guide***. Some of the additional skills that Imagineers may need can be summarized as:

- The ability to envision a story
- The ability to brainstorm
- The ability to create a storyboard and build mockups in various stages of detail
- A willingness to work with a multitude of disciplines in a team environment
- An understanding of theme park design requirements
- Recognizing that the customers and stakeholders range from toddlers to senior citizens
- An ability to envision the attraction through the eyes and shoes of the guests
- An understanding of the importance of safety, quality, and aesthetic value as additional competing constraints
- A passion for aesthetic details
- An understanding of the importance of colors and the relationship between colors and emotions
- An understanding of how music, animatronics, architecture, and landscaping must support the story

Obviously, this list is not inclusive of all skills. But it does show that not everyone can fulfill their desire to be an Imagineer for Disney. But these skills do apply to many of the projects that most project managers are struggling with. Learning and applying these skills for future projects could very well make all of us better project managers.

References

Alcorn, Steve, and David Green. (2007). *Building a Better Mouse: The Story of the Electronic Imagineers Who Designed Epcot*. Themeperks Press ISBN 0-9729777-3-2.

Altier, William J. (1999). *The Thinking Manager's Toolbox*. New York: Oxford Press.

Altier, William J. (Sept 1981). Why managers fail to solve problems quickly. *Management Review* 70 (9): 36.

Ben Mahmoud-Jouini, S., Midler, C. & Silberzahn, P. (2016, Apr–May). Contributions of design thinking to project management in an innovation context. *Project Management Journal*, DOI: 10.1002/pmj.21577, 144–156.

Brown, T. (2008). Design thinking. *Harvard Business Review* 86 (6): 84.

Chamorro-Premuzic, Tomas (2015). Why group brainstorming is a waste of time, *Harvard Business Review*, March 25, digital article, pp. 2–4.

Chamorro-Premuzic, Tomas. (2015). Why brainstorming works better online. *Harvard Business Review*, April 2, digital article, pp. 2–4.

Doneva, Svetlana. (2011). Keep that box firmly closed. *Finweek, Mar* 24, p. 21.

Fainberg, Greg Z. (2009). How to solve any problem. Lexington, KY: Cexino.

Ghez, Didier; Littaye, Alain; (2002), Translated into English by Cohn, Danielle. *Disneyland Paris From Sketch to Reality*. Nouveau Millénaire Editions ISBN 2-9517883-1-2.

Gobble, MaryAnne M. (2014). The persistence of brainstorming. *Research Technology Management* 57 (1) (Jan/Feb): 64–66.

Hench, John. (2003). *Designing Disney*, Disney Editions, p. 2.

Hench, John, with Peggy Van Pelt. (2003). *Designing Disney: Imagineering and the Art of the Show*. Disney Editions ISBN 0-7868-5406-5.

Imagineers, The (as "The Disney Imagineers"). (2005). *The Imagineering Workout: Exercises to Shape Your Creative Muscles*. Disney Editions ISBN 0-7868-5554-1.

Imagineers, The. (1996). *Walt Disney Imagineering: A Behind the Dreams Look At Making the Magic Real*. Disney Editions. ISBN 0-7868-6246-7 (hardcover); 1998, ISBN 0-7868-8372-3 (paperback).

Imagineers, The. (2003). *The Imagineering Way: Ideas to Ignite Your Creativity*. Disney Editions ISBN 0-7868-5401-4.

Imagineers, The. (2005). *The Imagineering Field Guide to Magic Kingdom at Walt Disney World*. Disney Editions ISBN 0-7868-5553-3.

Imagineers, The. (2006). *The Imagineering Field Guide to Epcot at Walt Disney World*. Disney Editions ISBN 0-7868-4886-3.

Imagineers, The. (2007). *The Imagineering Field Guide to Animal Kingdom at Walt Disney World*. Disney Editions ISBN 1-4231-0320-3, ISBN 978-1-4231-0320-2.

Imagineers, The. (2008). *The Imagineering Field Guide to Disneyland*. Disney Editions ISBN 1-4231-0975-9, ISBN 978-1-4231-0975-4.

Imagineers, The. (2010). *Walt Disney Imagineering: A Behind the Dreams Look at Making More Magic Real*. Disney Editions. ISBN 1-4231-0766-7 (hardcover).

Kurtti, Jeff. (2006). *Walt Disney's Legends of Imagineering and the Genesis of the Disney Theme Park*. Disney Editions ISBN 0-7868-5559-2.

Markman, Art. (2015). The problem-solving process that prevents groupthink, *Harvard Business Review*, November 25: 2–4.

Marling, Karal (1997). *Designing Disney's Theme Parks*. Paris, New York: Flammarion.

McCafferty, Dennis. (2014). Tips for building successful virtual teams. *Baseline*, Apr 9: 1.

Moore, Laura B. (2012). Brainstorming your team to happiness. *PM World Today* 1 (5), Special section, pp. 1–4.

Mootee, I. (2013). *Design Thinking for Strategic Innovation*. Hoboken, NJ: John Wiley and Sons, 60.

Proctor, Tony (1999). *Creative Problem Solving for Manager*. 3rd Edition. Routledge, United Kingdom.

Scribner, George, and Jerry Rees (Directors) (2007). *Disneyland: Secrets, Stories, and Magic* (DVD). Walt Disney Video. Surrell, Jason. (2007). *The Disney Mountains: Imagineering at Its Peak*. Disney Editions ISBN 1-4231-0155-3

Surrell, Jason. *Pirates of the Caribbean: From the Magic Kingdom to the Movies.* Disney Editions (2007). ISBN 1-4176-9274-X, ISBN 978-1-4176-9274-3.

Surrell, Jason. *The Haunted Mansion: From the Magic Kingdom to the Movies.* Disney Editions (2003). ISBN 978-0-7868-5419-6.

Wright, Alex. (2005). *Imagineers. The Imagineering Field Guide to Magic Kingdom at Walt Disney World.* New York: Disney Editions.

6

Pillar 6: New Forms of Project Leadership

6.0 Introduction

Most people seem to agree that effective leadership in project management can contribute significantly to a successful outcome. Unfortunately, there have been limited empirical studies and research on project management leadership styles and how it impacts the performance of team members, on their present and future assignments, compared to the volumes of information related to project management processes, methodologies, tools, and techniques.

There has also been limited research until now on crisis leadership. The pandemic due to Covid-19 has taught us that all companies can be affected by crises, and that crises leadership processes do not necessarily use the same processes, tools, and techniques as with traditional projects. The behavioral expectations from crisis leadership may be significantly different than traditional project behaviors due to timing, risks, involvement by stakeholders, and the impact of the enterprise environmental factors.

In the past, leadership research focused on the identifying the causes of project failure and seeing if any of the causes could be attributed to the leadership style that was used. The intent was to identify leadership effectiveness models that might eliminate many of the issues related to project failures.

6.1 Issues with Leadership Studies

More recently, the focus on project management leadership styles research has shifted from causes of project failure to identifying the factors, other than financial indicators, that contributed heavily to project success. Researchers believed that there is a strong correlation between project success and project leadership. The problem with most of their findings was in the definition of project success.

Project success does not fit into any predefined criteria. There can be a different definition of success in each industry and even in the same industry, and the meaning of success can change from project to project and in each life cycle phase. Success is no longer measured by how well we adhered to the time, cost, and scope requirements. Success now includes other components such as the creation of tangible or intangible business benefits and business value, and their sustainability.

Most of the early leadership research focused on specific industries such as health care, finance, IT, pharmaceutical, and construction. Some researchers focused on large versus small projects, gender, culture, and the impact on leadership by the type of contract. A study by Thamhain (2004) failed to find any significant common relationship between leadership and project performance. However, some research results focused on specific types of projects. Strang (2010) identified a leadership style that seemed to work well specifically in new product development projects.

Until recently, determining project management team leadership effectiveness has not received the attention it deserved. Project management leadership was focused on managing traditional projects where the emphasis was on short-term profitability and problem-solving rather than motivating team members. Significantly more research is needed with an emphasis on the impact on team members on their current and future assignments rather than project deliverables.

6.2 Selecting the Leader

Unlike traditional organizations where management positions are considered as full-time permanent positions and often filled based on the leadership style of the individual selected, project manager assignments are more likely to be filled based on the length of the project, availability of personnel, technology and scope requirements, or relationships with certain clients and stakeholders. Because projects must come to an end, and usually the sooner the better, emphasis has been on the project manager's authority and power rather than his/her leadership capability. Fortunately, this is now changing with more research on the behavioral and social expectations of personnel resulting from effective project management leadership. Companies are no longer relying on rules and regulations dictating how people should act.

Companies are now recognizing that filling functional management positions may be impacted by the company's level of project management maturity. A Tier 1 auto supplier had a history of filling functional management positions based mainly upon the individual's level of technical expertise even though they also maintained a technical career path ladder. As the need for effective project management became apparent, the company decided that functional management positions should be staffed with people that openly, honestly, and ethically communicate and collaborate with project teams. The communication and collaboration skills became more important than the technical skills.

6.3 Introduction to Leadership Styles

Traditional leadership can be defined as a "social process" where a person must enlist the aid of others, such as individuals, teams, or even an entire organization, to accomplish a task or group of tasks. Even though the definition seems simple enough, there are several types of leadership that can be chosen when considering the environment and the tasks. Each leadership style comes with advantages and disadvantages based on the application even though there may be common leader attributes or behaviors such as intelligence, trustworthiness, and courage.

In the early years of project management, companies utilized a one-size-fits-all methodology accompanied by bureaucratic leadership. Project management was driven by rigid policies and procedures so that senior management would maintain command and control over all projects.

The goal of leadership was the profitability of the project with little concern for the workers. This limited the ability of the project manager to motivate workers and to be creative. Workers often did as little as possible to receive their paychecks. The leadership style to be used was often dictated by senior management or the project sponsors.

Some leadership styles often discussed in project management are relatively easy to understand. Three early leadership styles that were studied by Lewin et al. (1939) were authoritarian, participative, and laissez-faire.

- **Authoritarian or autocratic:** All decision-making authority is centralized in the hands of the leader. This allows for quick decision-making, and the leader decides when, if at all, to inform the team of his/her decision. Team members may not be asked for ideas or opinions.
- **Participative or democratic:** Often referred to as shared leadership, this style allows the leader to share decision-making with the team. Team members provide opinions and ideas, but this has the disadvantage of requiring more time than might be available.
- **Laissez-faire or free rein:** In this style, all decision-making may be passed on to the subordinates.

These three leadership styles often appear in project management textbooks. The authoritarian style may appear on R&D projects when the project manager position is staffed with the person who possesses a command of the technology. Authoritarian styles are often use by prima donnas. Participative leadership can occur when the project manager may possess just a cursory understanding of the technology and must rely upon the team for decision-making support. Laissez-faire leadership can occur when the project manager position is filled with someone from marketing or sales because of their relationship with customers and stakeholders, and the team makes most of the decisions.

A leadership style which is often discussed in project management coursework is situational leadership. This approach assumes that each project may have unique characteristics due to different business cases, stakeholders, time durations, costs, constraints, and assumptions. Because each project has unique characteristics, it may not be possible to identify an optimal leadership profile for a project manager. This has plagued project management educators for decades in their quest to identify an ideal leadership style. As stated by Hemphill (1949), "What an individual actually does when acting as a leader is in large part dependent upon characteristics of the situation in which he functions."

Sometimes, leadership styles can be decomposed and provide us with insight into other leadership models. As an example, Fiedler (1967) identified a situational contingency leadership style that focused on the leader's effectiveness. Fiedler defined two types of leaders: those who tend to accomplish the task by developing good relationships with the group (relationship-oriented), and those who have as their prime concern carrying out the task itself (task-oriented).

Relationship-oriented leadership is a contrasting style in which the leader is heavily focused on the relationships amongst the group and is generally more concerned with the overall well-being and satisfaction of group members. Relationship-oriented leaders emphasize communication within the group, show trust and confidence in group members, and show appreciation for work done. However, the emphasis on relationships may cause productivity to suffer.

Task-oriented leadership is a style in which the leader is focused on the tasks that need to be performed to meet a certain production goal. Task-oriented leaders are generally more concerned with producing a step-by-step solution for a given problem or goal, strictly making sure the deadlines are met, results are achieved, and reaching target outcomes. This is often accomplished with little regard for the well-being of the team members.

According to Fiedler, there is no ideal leader. Both task-oriented and relationship-oriented leaders can be effective if their leadership orientation fits the situation. When there is a good leader-member relationship, highly structured tasks, and high leader position power, the situation is considered a "favorable situation." Fiedler found that task-oriented leaders are more effective in extremely favorable or unfavorable situations, whereas relationship-oriented leaders perform best in situations with intermediate favorability. Most of the situational/contingency theories assume that leaders can change their behavior to meet differing circumstances or widen their behavioral range at will, when in practice many find it hard to do so because of unconscious beliefs, fears, or ingrained habits.

There are many other types of project management leadership, some of which are derivatives of the leadership styles mentioned previously. Some of these include:[1]

- **Transactional:** In this style, the leader promotes compliance by the followers or team using both rewards and punishment. This usually keeps workers motivated, but only for the short term. The project leader will establish procedures and set standards for efficiency usually for projects that must be carried out in a specific way such that everything will flow smoothly. The focus is on near-term results rather than forward-thinking ideas. Transactional leadership sets bonuses, merit awards, or other forms of recognition when certain goals are met.
- **Transformational:** Transformational leaders interact closely with followers, focus on higher order intrinsic needs, and raise consciousness about the significance of specific outcomes and new ways in which those outcomes might be achieved. Transformational leaders want followers to achieve intrinsic motivation and job fulfillment. Transformational leadership is interactive and inspiring, whereas transactional leadership is generally passive. Transformational leaders are pragmatic and think outside the box when solving problems whereas transactional leaders tend to think inside the box.
- Transformational leadership serves to enhance the motivation, morale, and job performance of followers through a variety of mechanisms; these include connecting the follower's sense of identity to a project and to the collective identity of the organization; being a role model for followers in order to inspire them and to raise their interest in the project; challenging followers to take greater ownership for their work, and understanding the strengths and weaknesses of followers, which allows the leader to align followers with tasks that enhance their performance. It is also important to understand the qualities transformational leadership can bring to a work organization. Transformational leaders are strong in their abilities to adapt to different situations, share a collective consciousness, self-manage, and be inspirational while leading a group of employees.
- **Authentic:** Authentic leadership is an approach to leadership that emphasizes building the leader's legitimacy through honest relationships with followers based on an ethical foundation. Generally, authentic leaders are positive people with truthful self-concepts who promote openness. By building trust and generating enthusiastic support from their subordinates, authentic leaders can improve individual and team performance. This approach has been fully embraced by many leaders and leadership coaches who view authentic leadership as an alternative to leaders who emphasize profit and share price over people and ethics. Authentic leadership is a growing area of study in academic research on leadership, which has recently grown from obscurity to the beginnings of a fully mature concept (Gardner et al. 2011).

1 Portions of this section have been adapted from Wikipedia, the Free Encyclopedia: leadership, transactional leadership, and transformational leadership

6.4 Project Management Challenges

Given the number of different leadership styles, why has it been so difficult to define a leadership style workable for a project management environment? Most leadership styles are created from observations and testing in a reasonably stable and controlled setting such as salespeople working with consumers so that we can understand the challenges that both the leader and workers must face. An example might be observing bank tellers or waiters/waitresses interacting with customers to see the effect that various leadership styles would have on their performance and then to select the best leadership approach to achieve the desired outcomes.

Project management environments are generally unstable and likely to change from project to project. Each company, even in the same industry and with similar projects, can operate in different settings based on a variety of factors. This is the reason why some educators believed that, although no definitive leadership style is recommended, project management was closely aligned to situational leadership practices.

The project management environment and setting contain many challenges that make the selection of an effective leadership style difficult. Some of the challenges include the following:

- In some organizations, becoming a project manager could be a part-time assignment that one must perform in addition to one's normal job. It is not uncommon for the "temporary" or acting project managers to then provide little interest in effective project leadership, knowing that their project will come to an end and their performance reviews may not be based on the success or failure of their part-time assignment as a project manager.
- Project managers may not have the authority to hire or fire people from a project. Project staffing is performed by functional managers, although project managers can make a request for specific resources.
- Project managers may possess little authority, if any, over the assigned resources and yet they must provide some form of leadership. Line managers may possess almost all the authority and quite often project managers need the line managers to provide frequent instructions to the workers on how tasks should be performed.
- Project team members come from diverse backgrounds and must work together quickly. Based on the length of the project, this could be an issue if there is not enough time for them to get to know one another and adjust appropriately.
- Project management is often regarded as an environment identified by the workers as "multiple boss reporting." The workers must report directly or indirectly to their functional manager and each project manager on whose project they are working. Workers may be assigned to multiple projects concurrently. The workers must then succumb to multiple leadership styles.
- Line managers often retain all rights for wage and salary administration for the people that the project manager is expected to provide leadership for. This may eliminate the project manager's ability to motivate people using predefined rewards, since line managers are responsible for motivation of the workers through performance reviews.
- If team members are given the choice as to where to provide their loyalty, the project managers or their line manager, the decision is almost always in favor of where their performance review would come from, which is of course the line organization. Even if the project managers are asked to provide input on worker performance to the functional managers during performance reviews, the final decision on rewards rests with the functional managers.
- Without some control over the workers, project managers may not treat people as a valued asset but, rather, as a project labor cost to be removed from the project as soon as possible.

- Based on the length of the project, many project managers do not consider how their leadership skills affect the team because, once the project is completed, the project managers would be reassigned to other projects and might never work with these same people again. This would be an example of transactional leadership, but without the power to administer rewards or punishment. The length of the project, whether the team members are assigned full-time or part-time, and whether the project manager expects to work again with these team members in the future, impacts how the project manager provides leadership. The result is that project managers may have little interest in helping workers improve their skills for future projects.
- Most team members may be assigned to several projects at the same time. Project managers cannot "force" the assigned functional employees to perform the work on their project in a timely manner without support from the line managers.

Given the existence of these challenges, and with little research data to link them to leadership styles, project management researchers focused on the role of the project manager and the power and authority he/she needed to be successful. As stated by Clarke (2012),

> Most project management studies focus on understanding the roles and power position of the project manager, with very little research being dedicated to understanding the effect of the leadership style on the project team.

For several decades, most of the research and changes that took place in project management were with the processes, tools, and techniques rather than project management leadership. As project management evolved, companies created enterprise project management methodologies that were composed of multiple tools displayed as forms, guidelines, templates, and checklists. The tools were designed to increase the chances of repeatable project success and used on multiple projects. Ideas for the additional tools often came from an analysis of best practices and lessons learned captured at the end of each project. Many of the new tools came from best practices learned from project mistakes, with the hope that the mistakes would not be repeated on future projects. Project teams now could have as many as 50 different tools to be used – but mostly quantitative rather than behavioral tools. Companies believed that the processes, tools, and techniques were more important than the leadership styles for repeatable project success.

6.5 Leadership and Cultures

Most project team members understand the complexity of working on multiple project teams concurrently where each project manager may adopt a different leadership style. Project managers may choose a leadership style based on the complexity and clarity of the scope, environmental risks, previous experience with certain stakeholders, management's expectations, emotional maturity level of the team, and the team's experience working in a project management environment. Team members are usually willing to accept, and possibly endure, any leadership style selected by the project manager and expect it to become the culture for the duration of the project. The problem occurs when PMs change their leadership style during the execution of the project without understanding the impact on the expected project outcome.

PMs generally face a dilemma of whether to be an authoritarian (i.e., directive) or social leader. Authoritarian leaders expect team members to comply with instructions and may provide team members limited opportunities to be creative and anticipate problems. Social leadership focuses on collaboration, trust, and empowerment. Changing leadership styles during a project from social to authoritarian can create confusion and alienate team members when they believe they have lost their empowerment and no longer trust the project manager. Going from authoritarian to social can be equally as bad if team members believe that this is temporary and can change back quickly. Team members may have to live with fear or uncertainty, which generates mistrust and may encourage team members to resist changes. The result can be a negative impact on project outcomes.

Effective project leadership styles are often hard to maintain unless the project and corporate cultures are aligned. While the PM may be able to create a cooperative project team culture, it must be supported by a similar corporate culture that encourages ideas to flow freely, understands the strengths and weaknesses of the personnel and has confidence in their abilities. Even when crises occur, team members seem to perform better when corporate and project cultures are similar.

6.6 Excellence in Action: Project Leadership for the Smart Mission[2]

Adaptive Capabilities for Dynamic Knowledge

On February 1, 2003, the space shuttle *Columbia* (STS-107) disintegrated upon reentering the Earth's atmosphere, killing all seven crew members.

The ensuing *Columbia* Accident Investigation Board indicated that in addition to the technical causes of the failure, there were underlying organization factors that contributed to the failure. The report identified the need for a culture committed to organization learning, where communication, collaboration, and openness of ideas would be the norm. Subsequent studies painfully pointed out the need for a culture of more effective communication and knowledge sharing.

Six years later, Ed was attending the Flight Readiness Review for Space Shuttle 119. He was asked to be in the decision room to watch for signs of how effectively the project team was working and collaborating. Very early in the morning, a problem was the focus of discussion. On the previous shuttle flight, it was detected that there was an unexpected hydrogen flow increase from one of the shuttle's main engines. The flow valve did not operate as predicted. The cause of the flow valve problem was not understood by engineering, and there was not clear agreement on the potential impact.

During the next 14 hours of review, he watched what would become one of his proudest moments working at NASA. The broad community of NASA and its industry partners engaged in active conversation, open sharing of ideas, transparent decision-making, and intensive participation. It was a most impressive example of collaborative teamwork by a vast team under performance pressure.

---

2 Material in this section was provided by Dr. Edward Hoffman, Laurence Prusak, and Matthew Kohut. Dr. Hoffman was the founding Director of the NASA Academy of Program, Project, & Engineering Leadership (APPEL), and the first NASA Chief Knowledge Officer. He is a senior lecturer at Columbia University and has co-authored the book, The Smart Mission: NASA's Lessons for Managing Knowledge, People, and Projects. 2022. Hoffman, E., Kohut, M., Prusak, L. to be published MIT Press. Matthew Kohut has served as a communications consultant for fifteen years. Laurence Prusak is a researcher, consultant, and teacher who has been studying knowledge and learning for the past thirty years.

The eventual decision to launch would involve a team of over 1000 people across the nation communicating, collaborating, arguing, and innovating to achieve a solution. Only when NASA's safety and engineering communities said, "Go!" would the launch move forward. The engineering team would eventually design and patent a new noninvasive inspection technique.

The situation provided an opportunity to watch an exceptional collaboration that leveraged data, information, technology, and social capability. Later that night, as Ed was driving to his hotel, it was hard to believe that these two events – one catastrophic, one successful – occurred in the same organization, almost exactly six years apart. And yet both involved the same organization, program, most of the same people, teams, technology, processes, and of course the same industry.

So, what can explain the extreme performance variability? How can we create conditions to better ensure outcomes of agility, coordination, and high performance? This is one of the critical questions of organizational life, and it has become more important with the increasing uncertainty and demand for leadership in project teams.

The three authors have spent well over have a century working this issue, and our conclusion is that project leadership represents a commitment to certain principles that create a smart mission. A smart mission is a high-performing team that using signature practices around knowledge, learning, teaming, collaboration, story, and culture.

Our search for what leads to both productive work and well-being goes beyond our combined experience working within NASA on efforts to increase the likelihood of mission success. It includes individual research that started in different domains, but led to a unified viewpoint of leadership, team performance, and organizational success.

Laurence Prusak (Larry) started researching the characteristics that lead to organizational growth and success through a seminal work *Working Knowledge: How Organizations Manage What They Know* (Davenport & Prusak, 1998, Harvard Business School Press). This influential work helped establish the importance of knowledge markets. The most successful projects are dependent on creating, finding, and accessing expertise and solutions. The challenge for project teams and organizations is creating a culture that values knowledge sharing and learning. This rubs against the traditional project management demand for low overhead and cost. Learning and knowledge come with costs, and resources will increasingly be required for dynamic knowledge capabilities. Projects are part of the most rapid expansion of workplace changes, demanding new ways of working, adapting, and responding. The Agile rebellion prioritized learning, collaboration, and adaptability over process, control, and linear methods. In such a context, knowledge cultures are a defining characteristic.

Edward Hoffman (Ed) started researching complex learning systems and project teams in the late 1980s. As a result of the *Challenger* space shuttle disaster, Ed was assigned to design, develop, and sustain a NASA Academy for Program, Project, and Engineering Leadership (APPEL) in order to establish a NASA culture of learning, knowledge, and project team performance. It was within the challenge of designing an Academy for project leader and team learning that he began to understand the importance of story as a required capability in the project world. Adaptive leadership requires experimentation and reflection, and stories represent the most basic unit of knowledge. They encourage a leadership mindset of tradeoffs, conversations, and approaching challenges from a perspective that values collaboration, meaning, and respect. Stories also require time – time to understand, communicate, adapt, and learn.

(Project Management Success Stories: Lessons from Leaders. Laufer, A. and Hoffman, E. 2000. Wiley)

Matthew Kohut (Matt) discovered his area of research through his interest in discovering the qualities that lead to leadership influence and appreciation. *(Compelling People: The Hidden Qualities That Make Us Influential.* (2013) Neffinger, J, Kohut, M. Hudson Street Press.) This work applies not just for individuals, but as a description of the capabilities for project leadership. The key insight is that in a world dependent on change and new ideas, leadership requires both strength and warmth. The concept behind strength is a capability that demands attention based on competence and assertiveness. Equally important for project leadership is the ability to project warmth, a sense of shared concerns, interests, and emotions that can manifest as empathy or inclusion and open the door to trust.

Project leadership requires the ability to create and lead teams that excel through adaptability and continuous learning. This calls for rethinking and undoing inflexible processes and controls, and exchanging those models for a world of dynamic knowledge and learning. The project world will benefit from smart missions composed of the following six principles.

(*The Smart Mission: Lessons from NASA on Knowledge and Projects.* 2022. Hoffman, E., Kohut, M., Prusak, L. MIT Press.)

- Knowledge – rapidly creating, accessing, and integrating capability
- Learning – responding to a dynamic world, through deep learning
- Story – building conversational teams that practice inclusion and respect through a sense of shared meaning
- Teaming – recognizing that performance happens at the team level, and building team capability
- Culture – designing cultures of sharing, psychological safety, and adaptability
- Global collaboration – understanding that projects are global and work only through collaborative leadership

Project leadership for the smart mission recognizes that the most important factor is people. We can't express this enough. Most projects verbalize and recognize the importance of people, but there is lack of clarity as to what this means. Team performance is high when people are focused on the future of their work. This includes sharing knowledge and cultivating a culture where learning and honesty is the norm. In the "tale of two shuttles," a significant difference was how each team handled learning. In the success case there was open exchange, communication, argument, dissenting opinions, and a commitment to learning through conversation. This is the "sound of success": the energy of engaged and interactive people. In the tragic *Columbia* disaster, there was limited exchange and learning.

Our most successful teams nurture an environment of respect and inclusion. All members feel that they are heard and have something to contribute. This is enabled by clarity around the mission, vision, and goals, as well as through deliberate project leadership that prioritizes the six principles of a smart mission.

6.7 Leadership and Stakeholder Relations Management

Stakeholders are, in one way or another, individuals, companies, or organizations that may be affected by the outcome of the project or the way in which the project is managed. Stakeholders can be affected throughout the project either directly or indirectly or may function simply as observers. Stakeholders can shift from a passive behavior to becoming an active member of the team and participate in critical decisions.

How stakeholders interface with the project is often predicated upon the leadership style that the project manager exhibits. Project managers do not manage stakeholders. Rather, they manage the relationship with the stakeholders by choosing an effective leadership style that can be different for each stakeholder. In the early years of project management, project managers were afraid that stakeholders could interfere with the execution of the project and interfacing with stakeholders was avoided even when their involvement could be beneficial. Today, with the growth of strategic projects, stakeholder relations management practices are essential.

On small or traditional projects, the project team generally interfaces with just the project sponsor as the primary stakeholder, and the sponsor usually is assigned from the organization that funds the project. This is true for both internal and external projects. But the larger the project, the greater the number of stakeholders you must interface with. The situation becomes even more potentially problematic if you have many stakeholders, geographically dispersed, all at different levels of management in their respective hierarchy, each with a different level of authority, and language and cultural differences. Trying to interface with all these people on a regular basis, especially on a large, complex project, is very time-consuming and requires special leadership skills.

Stakeholder Complexities

One of the complexities of stakeholder relationship management is to select an appropriate leadership style that appeases all the stakeholders without sacrificing your company's long-term mission or vision for the project. Also, your company may have long-term objectives in mind for this project and those objectives may not necessarily be aligned to the project's objectives or each stakeholder's objectives. Lining up all the stakeholders in a row and getting them to uniformly agree to all decisions is more wishful thinking than reality. You may discover that it is impossible to get all the stakeholders to agree and you must simply hope to placate as many as possible at a given point of time. Different leadership styles for each stakeholder may be necessary.

Stakeholder relationship management cannot work effectively without commitments and support from all the stakeholders. Obtaining these commitments and support can be difficult if the stakeholders cannot see what is in it for them at the completion of the project, namely the business value and benefits that they expect. The problem is that what one stakeholder perceives as value, another stakeholder may have a completely different perception or a desire for a different form of value. For example, one stakeholder could view the project as a symbol of prestige. Another stakeholder could perceive the value as simply keeping their people employed. A third stakeholder could see value in the final deliverables of the project and the inherent quality in it. And a fourth stakeholder could see the project as an opportunity for future work with clients.

Getting stakeholders to be committed to the project and see the ultimate value in supporting it is easy if you are willing to allow each stakeholder to have their own views on the project, its objectives, success criteria and ultimate value. But getting all the stakeholders to reach an agreement is difficult, if not impossible, especially if there are several stakeholders.

Another form of agreements involves how stakeholders will interact with each other. It may be necessary for certain stakeholders to interact with one another and support one another regarding sharing resources, providing financial support in a timely manner, and the sharing of intellectual property.

While all stakeholders recognize the necessity for these agreements, they can be impacted by politics, economic conditions, and other enterprise environmental factors that may be beyond the control of the project manager. Certain countries may not be willing to work with other countries because of culture, religion, views on human rights, and other such factors.

For the project manager, obtaining these agreements right at the beginning of the project is essential. Some project managers are fortunate in being able to do this while others are not. Leadership changes in certain governments may make it difficult to enforce these agreements on complex projects.

It is important for the project manager to fully understand the issues and challenges facing each of the stakeholders. Although it may seem unrealistic, some stakeholders can have different views on the time requirements of the project. In some emerging nations, the construction of a new hospital in a highly populated area drives the commitment for the project even though the project could be late by a year or longer. People just want to know that it will eventually be built.

In some cultures, workers cannot be fired. Because they believe they have job security, it may be impossible to get them to work faster or better. In some countries, there may be as many as 50 paid holidays for the workers, and this can impact the project manager's schedule.

Not all workers in each country have the same skill level even though they have the same title. For example, a senior engineer in one country may be perceived as having the same skills as a lower grade engineer in another country. In some locations that may have a shortage of labor and workers are assign to tasks based on availability rather than capability.

In some countries, power and authority, as well as belonging to the right political party, are symbols of prestige. People in these positions may not view the project manager as their equal and may direct all their communications to the project sponsor. In this case, it is possible that salary is less important than power and authority. All these factors can impact the project manager's leadership style.

It is important to realize that not all the stakeholders may want the project to be successful. This will happen if stakeholders believe that they may lose power, authority, hierarchical positions in their company, or in a worse case, even lose their job. Sometimes these stakeholders will either remain silent or even be supporters of the project until the end date approaches. If the project is regarded as unsuccessful, these stakeholders may respond by saying "I told you so." If it appears that the project may be a success, these stakeholders may suddenly transform from adversaries or the silent majority to supporters.

It is quite difficult to identify these people and their hidden agendas. These people can hide their true feelings and be reluctant to share information. There are often no tell-tale or early warnings signs that indicate their true belief in the project. However, if the stakeholders are reluctant to approve scope changes, provide additional investment or assign highly qualified resources, this could be an indication that they may have lost confidence in the project.

Poor Understanding of Project Management

Not all stakeholders understand project management. Not all stakeholders understand the role of a project sponsor. And not all stakeholders understand how to interface with a project or the project manager even though they readily accept and support the project and its mission. Simply stated, most stakeholders are never trained in how to properly function as a stakeholder. Unfortunately, this cannot be detected early on but will become apparent as the project progresses.

Some stakeholders may be under the impression that they are merely observers and need not participate in decision-making or authorization of scope changes. For some stakeholders that desire to be just observers, this could be a rude awakening. Some will accept the new role while others will not. Those that do not accept the new role usually are fearful that participating in a decision that turns out to be wrong can be the end of their political career.

Some stakeholders view their role as that of a micromanager often usurping the authority of the project manager by making decisions that they may not necessarily be authorized to make, at least not alone. Stakeholders that attempt to micromanage can do significantly more harm to the project than stakeholders that remain as observers.

It may be a good idea for the project manager to prepare a list of expectations that he/she has of the stakeholders. This is essential even though stakeholders support the existence of the project. Role clarification for stakeholders should be accomplished early on the same way that the project manager provides role clarification for the team members at the initial kickoff meeting for the project.

Simply because the deliverable is provided according to a predetermined business case and set of constraints is no guarantee that the client will perceive value in the deliverable. It is true that clients track budgets and schedules, but it is the value at the end that makes the project a success or failure. Leadership styles that the project manager uses must emphasize the business benefits and value rather than just deliverables.

The ultimate objective of all projects should be to produce a deliverable that meets expectations and achieves the desired value. This should be the goal of the project manager as well as the client. While we always seem to emphasize the importance of the triple constraints when defining the project, we spend little time in defining the value characteristics that we expect in the final deliverables.

The value component or definition must be a joint agreement between the customer and the contractor (buyer / seller) during the initiation stage of the project. Also, in the ideal situation, the definition of value is aligned with the strategic objectives of both the stakeholders and the project manager.

Warren Buffett emphasizes the difference between price and perceived value when he stated, "Price is what you pay. Value is what you get." Most people believe that customers pay for deliverables. This is not necessarily true. Customers pay for the value they expect to receive from the deliverable. If the deliverable has not achieved value or has limited value, the result may be a dissatisfied customer.

Some people believe that a customer's greatest interest is quality. In other words, "Quality comes first!" While that may seem to be true on the surface, the customer generally does not expect to pay an extraordinary amount of money just for high quality. Quality is just one component in the value equation. Value is significantly more than just quality.

There is a common belief in project management that whoever makes the decisions on the project has the ultimate responsibility for the project's success or failure. This is not true. While project managers may discover that a multitude of decisions are being made by the client and the key stakeholders, the ultimate responsibility for success or failure rest with the project manager. This is just like quality; project managers can delegate work to others, including some decision-making, but the project manager retains the ultimate responsibility for the quality of the project's outcome.

It is not uncommon on large, complex projects that the role of the project manager becomes more of a facilitator or coordination of decisions made by others, especially active stakeholders. This happens because the project manager most likely has just an understanding of the technology rather than a complete command of the technology. The larger the project, the greater the tendency for the project manager to possess an understanding rather than a command of the technology.

Components of Stakeholder Relations Management

On the micro level, we can define stakeholder relations management using the six processes shown in Figure 6-1.

- **Identify the stakeholders.** This step may require support from the project sponsor, sales, and the executive management team. Even then, there is no guarantee that all the stakeholders will be identified.
- **Conduct stakeholder analysis.** This requires an understanding of which stakeholders are key stakeholders that have influence, the ability, and authority to make decisions, and can make or break the project.
- **Perform stakeholder engagements.** This step is when the project manager and the project team get to know the stakeholders.
- **Identify stakeholder information flow.** This step is the identification of the information flow network and the preparation of the necessary reports for each stakeholder.
- **Abide by agreements.** This step enforces stakeholder agreements made during the initiation and planning stages of the project.
- **Debrief the stakeholders.** This step occurs after contract closure and is to capture lessons learned and best practices for improvements on the next project involving these stakeholders.

Stakeholder management begins with stakeholder identification. This is easier said than done, especially if the project is multinational. Stakeholders can exist at any level of management. Corporate stakeholders are often easier to identify than political or government stakeholders.

Each stakeholder is an essential piece of the project puzzle. Stakeholders must work together and usually interact with the project through governance activities. Therefore, it is essential to know which stakeholders will participate in governance and which will not.

As part of stakeholder identification, the project manager must know whether he/she has the authority or perceived status to interface with the stakeholders. Some stakeholders perceive themselves as higher stature that the project manager and, in this case, the project sponsor may be the person to maintain interactions.

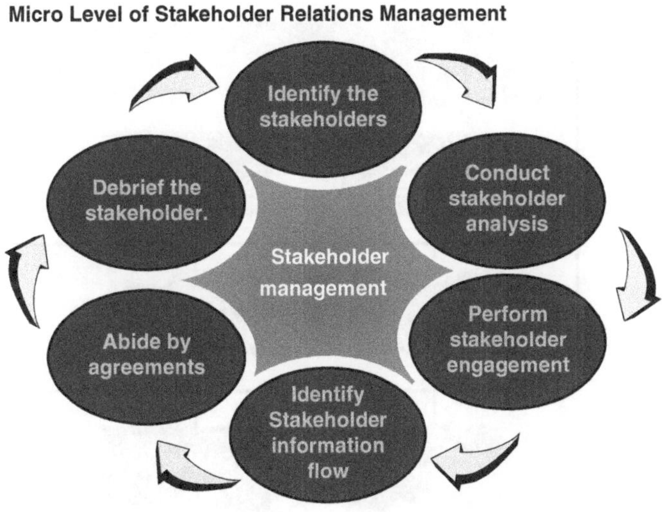

Figure 6-1 Micro Level of Stakeholder Relations Management

Classification of Stakeholders

There are several ways in which stakeholders can be identified. More than one way can be used on projects, and they can include:

- **Groups:** This could be financial institutions or creditors.
- **Individuals:** This could be by name or title, such as the CIO, COO, CEO, or just the name of the contact person in the stakeholder's organization.
- **Contribution:** This could be according to financial contributor, resource contributor, or technology contributor.
- **Other factors:** This could be according to the authority to make decisions, or other such factors.

Figure 6-2 shows another typical classification system for stakeholders. For simplicity's sake, the stakeholders can be classified as:

- Organizational stakeholders
- Product/market stakeholders
- Capital market stakeholders

The advantage of this system is that it appears as an organizational chart, and the names of the individuals can be place under each category.

Managing a project where stakeholders have different interests can be challenging. Consider a company that has a complex project to produce a new healthcare product. Consumers want to believe that the products developed will be safe and fit for use. Stockholders are more concerned with market share that can increase the stock's selling price and increase the dividend. Lending institutions may be less concerned about product safety and more concerned about the revenue stream of the products such that cash flow can repay the debt.

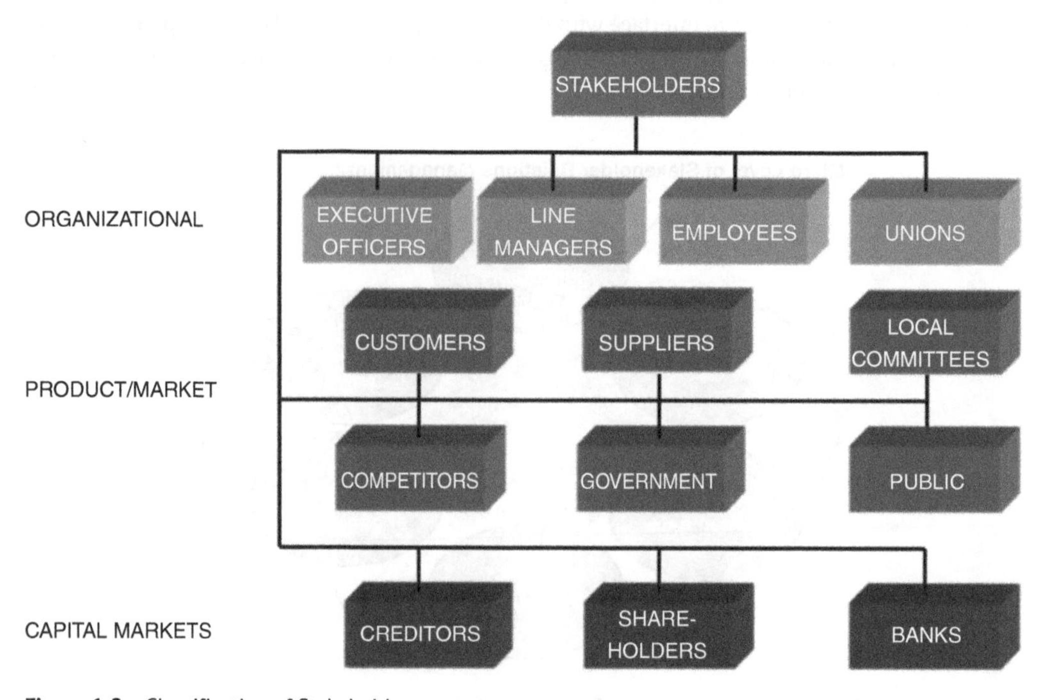

Figure 6-2 Classification of Stakeholders

Government agencies may have only one concern: protecting public health. Management must worry about health and product safety, such that the image and reputation of the company will not be damaged if any bad news appears. Employees may provide lip service to concerns of product safety, whereas their real concern might be employment in the firm.

On large, complex projects with a multitude of stakeholders, it may be impossible for the project manager to properly cater to all the stakeholders. Therefore, the project manager must know who are the most influential stakeholders that can provide the greatest support on the project.

Not all stakeholders are equal in influence, power, or the authority to make decisions in a timely manner. It is imperative for the project manager to know who sits on top on the list.

Finally, it is important to remember that stakeholders can change over the life of a project, especially if it is a long-term project. Also, the importance of certain stakeholders can change over the life of a project and in each life-cycle phase. The stakeholder list is therefore an organic document subject to change.

Stakeholder Mapping

Stakeholder mapping is most frequently displayed on a grid comparing their power and their level of interest. This appears in Figure 6-3.

- **Manage closely:** These are high power, interested people that can make or break your project. You must put forth the greatest leadership effort to satisfy them. Be aware that there are factors that can cause them to change quadrants rapidly.
- **Keep satisfied:** These are high-power, less-interested people who can also make or break your project. You must put forth some effort to satisfy them but not with excessive detail that can lead to boredom and total disinterest. They may not get involved until the end of the project approaches.
- **Keep informed:** These are people with limited power but keenly interested in the project. They can function as an early warning system of approaching problems and may be technically astute to assist with some technical issues. These are the stakeholders that often provide hidden opportunities.
- **Monitor only:** These are people with limited power and may not be interested in the project unless a disaster occurs. Provide them with some information but not with too much detail such that they will become disinterested or bored.

Stakeholder mapping can also take place with the names of the people placed in the appropriate quadrants in Figure 6-3. The names can be color-coded to identify the supporters or advocates, blockers, and those that appear neutral. It is important to note that supporters, blockers, and

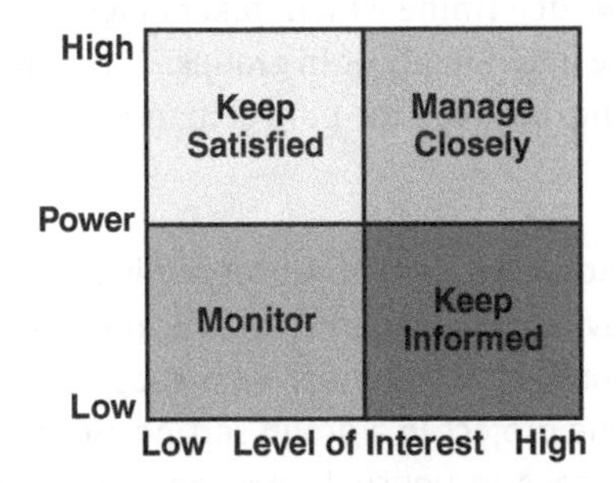

Figure 6-3 Stakeholder Mapping

neutral positions can appear in any quadrant, and that names can move from quadrant to quadrant based on changes that occur in each life-cycle phase.

The larger the project, the more important it becomes to know who is and is not an influential or key stakeholder. Although you must win the support of all stakeholders, or at least try to do so, the key stakeholders come first. Key stakeholders may be able to provide the project manager with assistance with the identification of enterprise environmental factors that can impact the project. This could include forecasting on the host country's political and economic conditions, possibility for additional funding and other such issues. In some cases, the stakeholders may have software tools that can support the project manager's organizational process assets.

Thus far, we have discussed the importance of winning over the key or influential stakeholders. There is also a valid argument for winning over the stakeholders that are unimportant. While some stakeholders may appear to be unimportant, that can change rapidly. For example, an unimportant stakeholder suddenly discovers that a scope change is about to be approved and that scope change can seriously impact the unimportant stakeholder, perhaps politically. Now, the unimportant stakeholder becomes a key stakeholder.

Another example occurs on longer-term projects where stakeholders may change over time perhaps because of politics, promotions, retirements, or reassignments. The new stakeholder may suddenly want to be an important stakeholder whereas his/her predecessor was more of an observer.

Finally, stakeholders may be relatively quiet in one life cycle because of limited involvement but become more active in other life cycle where they must participate. The same may hold true for people that are key stakeholders in early life-cycle phases and just observers in later phases. The project team must know who the stakeholders are. The team must also be able to determine which stakeholders are critical stakeholders at a specific point in time.

Stakeholder Engagement

Stakeholder engagement is when you physically meet with the stakeholders and determine their needs and expectations:

- Understand them and their expectations.
- Understand their needs.
- Value their opinions.
- Find ways to win their support on a continuous basis.
- Identify any stakeholder problems early-on that can influence the project.

Even though stakeholder engagement follows stakeholder identification, it is often through stakeholder engagement that we determine which stakeholders are supporters and which are not. Stakeholder engagement can be individually or in groups. This may also be viewed as the first step in building a trusting relationship between the project manager and the stakeholders. As stated by Tres Roeder (2013):

> The project manager should be vigilant to observe differences in stakeholder behavior when they are meeting individually compared to meeting in a group. The project manager should not assume that a project stakeholder who supports the project in an individual meeting will also support the project in a group setting. Group dynamics are complex and can cause individuals to behave in unusual and unexpected ways. It is well known that

peer pressure among teenagers can cause unwitting participants to take actions they otherwise would not take. This dynamic happens with adults, too. Group and organizational pressures can influence stakeholder perspectives.

The best way for the project managers to increase their awareness of how stakeholders will react in a group setting is simply to observe them in groups. The project manager should note any differences in perspectives and viewpoints between individual meetings with the stakeholder and group meetings with the stakeholder. When observing the stakeholder's group interactions, the savvy project manager will notice who else is in the meeting. It may be noted that certain stakeholders will adjust their perspectives if specific people are in attendance.

As part of stakeholder engagements, it is necessary for the project manager to understand each stakeholder's interests. One of the ways to accomplish this is to ask the stakeholders (usually the key stakeholders) what information they would like to see in performance reports. This information will help identify the key performance indicators (KPI) needed to service this stakeholder.

Each stakeholder may have a different set of KPI interests. This then becomes a costly endeavor for the project manager to maintain multiple KPI tracking and reporting flows, but it is a necessity for successful stakeholder relations management. Getting all the stakeholder to agree on a uniform set of KPI reports and dashboards will be almost impossible.

The list of the number of issues facing all the stakeholders could be quite long. There must be an agreement on what information is needed for each stakeholder, when the information is needed, and in what format the information will be presented. Some stakeholders may want a daily or weekly information flow whereas others may be happy with monthly data. For the most part, the information will be provided via dashboard reporting systems.

The more complex the project, the greater the need for virtual teams. Virtual teams thrive on effective communications. And if the information provided in the performance reports and the tracking of the KPI are accurate, trust will build up between the project team and the various stakeholders.

Since virtual teams may be remotely located from where the work is taking place, they must rely heavily upon communications and be made to feel that the information they receive is true. As part of stakeholder information flow for both virtual and non-virtual teams, the project manager must:

- Prepare a communication plan that identifies the reporting needs of each stakeholder (amount of information, level of detail, etc.).
- Identify stakeholder specific KPI.
- Identify communication protocols.
- Identify any proprietary information requirements or security needs.
- Continuously focus on the value and benefits at completion of the project.

The need for effective stakeholder communications is clear:

- Communicating with stakeholders on a regular basis is a necessity.
- By knowing the stakeholders, you may be able to anticipate their actions.
- Effective stakeholder communication builds trust.
- Virtual teams thrive on effective stakeholder communications.
- Although we classify stakeholders by groups or organizations, we still communicate with people.
- Ineffective stakeholder communications can cause a supporter to become a blocker.

There are two types of project review meetings: those with the project team and those with the stakeholders. Team meetings with stakeholders have their own characteristics. The items discussed in the stakeholder meetings include:

- Review of the stakeholder specific KPI information
- Discussion of how well project management is working
- Forecasts for the time, cost, benefits, and value expected at the end of the project.

These meetings also are used to resolve problems. Project managers must find solutions to problems such that multiple stakeholders are satisfied simultaneously or else the project manager will be overwhelmed in meetings. It may be necessary to have multiple stakeholders attending the same meeting if there are common issues to be resolved.

The opportunities for scope changes abound on every project. On large, complex project, there may be one or more individuals assigned to the project office simply for the management of scope changes. Scope changes can be approved and implemented incrementally as the project progresses, or all scope changes can be withheld until after the project is completed and then implemented as an enhancement project.

On complex projects with many stakeholders, the advantages of a scope change for one stakeholder could be seen as a disadvantage for other stakeholders. Stakeholders often recommend and approve scope changes for what is in their own personal interest, while neglecting those around them.

In the past, most companies had enterprise project management methodologies that focused upon linear thinking. All project work followed well-established life-cycle phases. The project manager also had forms, guidelines, templates, and checklists for each phase. This linear thinking may not be appropriate for many of today's complex projects.

Today's complex projects may require a fluid or flexible framework that can be custom designed to applied differently for each client and stakeholder. There may be different tools for each client. Therefore, project managers may need to use outside-of-the-box thinking to give each stakeholder the attention he/she expects.

Part of stakeholder identification and engagement is the establishment of agreements between the stakeholders and the project manager as well as the stakeholders themselves. These agreements must be enforced throughout the project. The project manager must:

- Identify all agreements between stakeholders (i.e., funding limitations, sharing of information, approval cycle for changes, etc.).
- Identify how politics may change stakeholder agreements.
- Identify which stakeholders may be replaced during the project (i.e., retirement, promotion, change of assignment, politics, etc.).

The project manager must be prepared for the fact that not all agreements will be honored.

These types of stakeholder debriefing sessions occur at the closure of the project, usually after contractual closure. Some companies have a life-cycle phase after contractual closure called Customer Satisfaction Management for the purpose of building a strategic partnership with you client and the stakeholders. The intent of these sessions is to determine:

- What did we do well on this project, performance-wise and regarding stakeholder management?
- What did we do poorly on this project performance-wise and regarding stakeholder management?
- What are the areas for improvement?

Attendance in the meeting is not restricted to just the project manager and the stakeholders. Attendance can include the sales team that sold the stakeholders on the project, the sponsors that interfaced with the stakeholders, and senior management hoping for additional work.

Satisfaction management surveys can occur throughout the project as well as at the end. There are three areas of interest in these surveys:

- Hard satisfaction management data such as performance data such as KPI data and milestone/deliverable measurements
- Soft satisfaction management data such as effective stakeholder interfacing and communications
- Ethical behavior data

Some companies use templates for this with boxes to be checked, ranging from completely unsatisfied to completely satisfied. When boxes are checked that are not in the completely satisfied or just satisfied categories, guidelines may exist on what to do to next to proceed to a higher level of satisfaction.

There are critical factors that must be considered for successful stakeholder relations management:

- Effective stakeholder management takes time. It may be necessary to share this responsibility with sponsor, executives, and members of the project team.
- Based on the number of stakeholders, it may not be possible to address their concerns face-to-face. You must maximize your ability to communicate via the internet. This is also important when managing virtual teams.
- Regardless of the number of stakeholders, documentation on the working relationships with the stakeholders must be archived. This is critical for success on future projects.

Effective stakeholder management can be the difference between an outstanding success and a terrible failure. Successful stakeholder relations management can result in agreements. The resulting benefits will be:

- Better decision-making and in a timely manner
- Better control of scope changes; prevention of unnecessary changes
- Follow-on work from stakeholders
- End user satisfaction and loyalty
- Minimizing the impact that politics can have on your project

Sometimes, regardless of how hard we try, we will fail at stakeholder relations management. Typical reasons include:

- Invite stakeholders to participate too early, thus encouraging scope changes and costly delays.
- Invite stakeholders to participate too late such that their views cannot be considered without costly delays.
- Invite the wrong stakeholders to participate in critical decisions, thus leading to unnecessary changes and criticism by key stakeholders.
- Key stakeholders become disinterested in the project.
- Key stakeholders are impatient with the lack of progress.
- Allow the key stakeholders to believe that their contributions are meaningless.
- Manage the project with an unethical leadership style or interfacing with the stakeholders in an unethical manner.

Project management social leadership practices, which are discussed later in this chapter, should significantly improve stakeholder relations management practices.

6.8 The Changing Leadership Landscape

Project management was initially used for traditional or operational projects that were well defined at the start. Most workers had a clear understanding of the scope of the projects and what tasks they were expected to do. Even if the workers disliked the assignment, they endured the pain knowing the project would eventually come to an end and they could move on to another hopefully more pleasing assignment.

Project management training programs focused mainly on educating people on the use of the earned value measurement system (EVMS), rather than behavioral or leadership issues. If leadership were discussed, the focus was on the importance of and decision-making necessary for maintaining the time and cost constraints and making sure that all work performed was in scope and aligned to the business case and statement of work. Communication between the leader and the team members was often done only at weekly or monthly team meetings unless there were problems.

Advances in the project management processes, tools, and techniques, along with more companies recognizing the benefits of good project management practices, convinced senior management that project management should be used throughout the company on other than just traditional projects. These included projects related to strategic planning, innovation, R&D, marketing, and so forth that were not well-defined, had a need for creativity, and required effective leadership and behavioral tools. As the management guru Tom Peters stated, "We are now managing our (entire) business by projects."

These new or strategic projects, which required a strategic project leadership model, focused on creating competitive advantages, change management, and thinking outside of the box. Meeting the strategic goals and objectives made it clear that traditional tools, techniques, and processes may be ineffective on these types of projects. Strategic leaderships models must focus on people skills and motivate the team by providing them with a vision. The vision may also require the team to build relationships both inside and outside the project team, possibly including heavy involvement with stakeholders.

As project management continued to evolve and spread throughout each business unit, companies moved away from collocated teams to distributed or virtual teams. Project durations were getting longer, and employees were being assigned full-time to many of the projects. Now, additional tools, more specifically behavioral tools, were needed to help support the new forms of project communications and behavioral expectations that would be required in dealing with a multitude of people for a longer time. This included an understanding and expectations of the way that the project manager would interact toward the project team, and vice versa. Project managers, as well as team members, were now expected to communicate with everyone, multi-nationally if necessary, including suppliers, distributors, stakeholders, and government agencies.

As projects became larger and required more staff, the project management organizational environment became more complex, demanding, and with a much higher degree of performance risk. Project managers recognized that they could not handle all the new challenges by themselves using traditional leadership practices based solely upon authority and power but had to rely on the assigned diversified workforce to assist in resolving critical issues.

Project teams were under higher levels of stress, having to deal with difficult clients and demanding stakeholders, and yet maintain a warm and welcoming expression when interfacing with all parties. Project management organizations were now facing many of the same issues that were identified previously in service-type organizations. As stated by Lu et al. (2019):

> Under pressure from life and work, many employees understandably experience negative emotions such as anxiety, anger, dissatisfaction, or depression (Jensen et al. 2013). However, most companies discourage employees from expressing those emotions at work (Hochschild 1979, 1983). Accordingly, employees must manage them to comport with workplace expectations (Scott and Barnes 2011). Self-managing those feelings is work – emotional labor.

Understanding Emotional Labor

Emotional labor is the way that workers manage their feelings to fulfill the emotional requirements of a job (Hochschild 1983; Grandey 2000). It is not just being courteous or respectful. Team members are expected to control their emotions, whether felt or not, when interfacing with customers, stakeholders, superiors, and co-workers. This includes the suppression of certain emotions that might cause customers and stakeholders to have negative feelings about the project's performance or generating unfelt feelings to disguise how they actually feel.

Hochschild (1983) divided emotional labor into two categories: surface acting and deep acting. In surface acting, workers are expected and often pressured to hide their true feelings and display emotions – perhaps using facial expressions that they do not actually feel. As an example, a waitress is expected to smile and act positively toward all customers. Deep acting is when the workers try to regulate their inner-most feelings by adjusting their thinking to feel the emotion that they are expressing.

Managing Emotional Labor

Emotional labor management in the project management workplace and the impact on project success is now drawing more attention from researchers than in the past. However, there is still little information about how different project management leadership styles affect employees' emotional labor. The failure to understand and manage emotional labor in a project management environment can incur significantly more strategic and business damage than in a service organization. Alienation and loss of just one important customer could result in a loss of more than a billion dollars in future business.

For many companies, the solution appeared to be adopting an approach used in some service-oriented companies where management's leadership emphasized "the customer is always right." Although this seemed like a good approach, it denied team members the right to express their negative feelings and often incurred harassment and abuse by the leader if not followed. Other issues, as stated by Sendjaya et al. (2008), included:

> Contemporary organizations are plagued by systemic problems such as bullying leadership (Einarsen 1999), abuse of power (Sankowsky 1995), unethical practices (Currall and Epstein 2003), toxic emotions (Frost 2003), social isolation and alienation in the workplace (Sarros et al. 2002), and the violation of employees' psychological well-being and work-life balance (De Cieri et al. 2005; Thornthwaite 2004; Wright and Cropanzano 2004).

Some companies believed that the way that an employee displays his/her feelings should be consistent with the company's view on business ethics. As such, companies conducted seminars on business ethics that dealt with ways to interface with clients and stakeholders, the control of emotional labor, the type of information and intellectual property that can and cannot be released, and promises that can and cannot be made.

Although companies had good intentions by conducting these seminars on ethics, there were some negative results, especially related to the use of surface acting to control emotional labor. When companies encourage surface acting, employees feel frustrated that their concerns will not be heard. They may feel unsafe to express their true feelings for fear of termination or upsetting management and the status quo. The result can lead to employee turnover, burnout, fatigue, frustration, absenteeism, poor performance, and possibly mental disorders.

The human costs of "service with a smile" requirements can be accompanied by more problems than benefits, especially in a project management environment where team members must manage their emotions around leaders and colleagues as well as the customers. According to Grandey et al. (2015):

> We articulate how formalized emotion display requirements limit self-determination by threatening the autonomy, competence, and belongingness needs of employees. Further, via an organizational justice lens, we argue that emotional labor is an unfair labor practice because employees in such circumstances are (1) undervalued by the organization (constituting distributive injustice); (2) disrespected by customers (constituting interactional injustice); and (3) self-undermined by organizational policies (constituting procedural injustice). We then argue for bringing light to the dark side of emotional labor with a "modest proposal": Organizations and customers should abandon formalized emotion display expectations and replace such efforts with more humanistic practices that support and value employees, engendering positive climates and an authentically positive workforce.

6.9 Servant Leadership

The importance of emotional labor attracted the attention of scholars and practitioners following Greenleaf's introduction to the concept of servant leadership in 1970. Over the past two decades, growing empirical research has shown that servant leadership enables companies to develop and maintain a competitive advantage. Servant leadership is a philosophy where the main goal of the leader is to serve the team rather than the team working to serve the leader. If this is done effectively, the employees acquire opportunities for personal and professional growth and the company grows as well through the workers' commitment and engagement. Many of the adverse effects of emotional labor mentioned previously can be reduced or eliminated.

Unlike most project managers who focus on and emphasize their personal or organizational interests, servant leaders recognize the importance of participation and interaction to get project team members to work together. This includes encouraging team members to network among themselves. Servant leaders must be empathic in their ability to create an environment where the team members feel safe to express their personal and professional feelings. In turn, the team members will feel empowered and have the freedom and autonomy to state their concerns, thoughts, opinions, and recommendations, as well as making decisions on how they will perform their tasks.

Servant leaders are expected to behave in an ethical manner and must possess some degree of visible devotion to fulfilling employee needs and development. There must be an acceptance of "imperfect" employees and visible compassion for troubled workers. Servant leaders must have a tolerance for mistakes without making judgments.

When servant leadership works well, bonds are built between leaders and workers such that misunderstandings are minimized. Workers have better attitudes about their job, demonstrate trust in their leaders, and feel safe at work. Workers devote themselves to the job and often perform extra work. Servant leadership helps develop a better working relationship between project teams and customers, which can lead to repeat business opportunities.

Since the 1970s, considerable research has been performed related to the link between controlling emotional labor and servant leadership. Spear (1995, 2002) elaborated on Greenleaf's work by identifying 10 characteristics of a servant leader:

- **Listening.** Listening is a willingness to openly accept the ideas, opinions, and suggestions of workers.
- **Empathy.** Empathy extends listening when leaders can put themselves in the situation that others say they are in and empathize with them and their feelings. This is accepting people for who they are.
- **Healing.** The ability of a leader to help workers endure the disappointment and emotional pain that comes from broken dreams, hopes, and other challenges.
- **Awareness.** The ability of the leader to identify cues and signs in the environment to help workers perform better.
- **Persuasion.** Persuasion or persuasion mapping enables the leader to identify the needs of the workers and focus on the importance of their work without the use of formal authority or legitimate power.
- **Conceptualization.** The ability of the leader to think about the future rather than just present-day needs and to encourage workers to use mental models to expand the creativity processes.
- **Foresight.** This includes using intuition to anticipate and predict the future for the benefit of the workers and the organization.
- **Stewardship.** Stewardship involves preparing the organization and its members for great contributions to society thereby willing to serve others.[3]
- **Growth.** Working with team members, possibly on a one-on-one basis, to get them motivated. This, in turn, can lead to employee satisfaction and the worker is encouraged to perform extra work.
- **Community building.** Encouraging the workers to view the organization and the team as a community where workers communicate with each other to address their issues.

The 10 characteristics opened the door for empirical studies and volumes of literature on servant leadership theory. Some papers discuss only a few of these characteristics. Barbuto and Wheeler (2006) addressed five characteristics, namely altruistic calling, emotional healing, wisdom, persuasive mapping, and organizational stewardship.

Most studies on servant leadership focus on traditional rather than project management organizational environments. A study by Krog et al. (2015) focused on servant leadership in a project

3 For an example of stewardship, see "Why Social Impact Matters," *Pulse of the Profession*® In-Depth Report, 2020, The Project Management Institute, Newtown Square, PA.

management environment and included the role of the project sponsor using the five characteristics discussed by Barbuto and Wheeler (2006):

> Most project leadership studies focus on understanding the role and power position of the project manager, with very little research on understanding the effect of leadership style on a project team and the project's success. Although there is no definitive leadership style that is preferred when leading projects, researchers have recognized servant leadership as being a model that may assist in overcoming many of the challenges a project leader may face. More specifically, this study aimed to understand the relationship between a project sponsor's servant leadership traits of altruistic calling, emotional healing, wisdom, persuasive mapping and organizational stewardship and a project team's empowerment, commitment, trust, and innovative behavior.

Despite the practical credibility of effective servant leadership, there are situations in project management that can make implementation difficult. If the leader is recognized as an expert in a certain field of study or has a reputation as a successful leader, workers may not feel empowered to offer their opinion. Another issue is the length of the project. The goal of servant leadership is for the workers to have trust in the leader's decisions and management style. When trust exists, team members are usually willing to take more risks. This may be difficult to do on short-term projects, whereas on long-term efforts there is sufficient time to cultivate interpersonal relationships, and for the dynamics of trust on team member interactions to materialize.

6.10 Social Project Management Leadership

Traditional project management leadership, based heavily on authority and power, is changing to social project management leadership that emphasizes interpersonal skills. There are four factors driving the change:

1. **Early research on servant leadership was restricted to service-oriented firms.** Today, the concepts behind servant leadership have expanded to other applications including project management.
2. **There has been growth in less traditional types of projects.** Many of these new types of projects are not well-defined initially, and they require significant interaction throughout the life of the project between the project manager and the team members. Project managers can no longer rely on project initiation documentation with a well-defined business case, statement of work and listing of requirements, to determine the leadership style needed. Scope changes will occur more frequently than in the past, and this will necessitate more interfacing activities.
3. **New tools and techniques have entered the project management landscape.** These include behavioral tools, many of which are needed to support the new types of projects. Some of these techniques are the need for brainstorming, creativity, design thinking, creative problem-solving, active listening, and emotional intelligence. These new techniques have a strong behavioral basis and require that project managers and team members interface and interact much more frequently than in the past. New flexible methodologies, such as Agile and Scrum, have components that emphasize many of the behavioral and interpersonal issues not addressed in the traditional Waterfall approach.

4. **The desire for companies to become multinational might be the most serious concern.** Most project personnel are poorly trained, if at all, in how religions, politics and cultures impact how project leadership should take place when projects are managed away from the country in which the parent company resides. The impact that religious, political, and cultural issues can instill on servant leadership and emotional labor is still unknown. These issues must be addressed considering that we are using virtual project management teams without fully understanding the impact of leadership on teams with diversified staffing.

The good news is that more studies are being conducted that focus on identifying effective leadership traits (Muller and Turner 2007, 2010) and Shao and Muller (2011). These studies identified leadership competencies categorized into emotional competencies, managerial competencies, and intellectual competencies. These types of studies open the door for a better understanding of effective social project management leadership and may eliminate or explain the inconsistent findings in earlier research.

Projects are managed by people, not tools. There must be a concerted effort by companies, as well as the project management community at large, to recognize the growing importance and need for effective social project management leadership supported by an understanding of servant leadership concepts and emotional labor.

With the growth in project management metrics and the ability to measure anything and everything, we believe today that we can establish tangible and/or intangible metrics that can measure the success of social project management leadership. The capturing of best practices and lessons learned in the future will also include the effectiveness of project management leadership and the impact on project teams.

6.11 The Growth in Importance of Crisis Leadership

Crises can occur in any company and at any time. The pandemic resulting from COVID-19 has shown virtually every company worldwide that they may not be immune from serious disruptions to their business. What companies fail to realize, often until it is too late, is that traditional project management practices and leadership styles may be ineffective during crises. Leadership with virtual teams is more complicated than with seeing the entire workforce in person at a weekly team meeting.

Crisis leadership requires an examination of the processes that are essential for an organization and its management when dealing with crises. Even though many of these processes, and the accompanying tools and techniques, are based on best practices and lessons learned from experience, they may not be applicable to crises-related projects without some modifications.

How companies respond to the crisis is critical. Thanks to usually excessive media coverage, the world watches how companies respond to a crisis. Based on the outcome, the public then categorizes the company as either a victim or villain in the way the crisis was managed. What is expected to be discussed in journal articles will be the project management processes that were used and the accompanying leadership styles.

Most companies today capture best practices and lessons learned from projects during execution and at closure. The best practices look at what the company may have done right and wrong. However, what has been lacking until recently, thanks largely to the pandemic, is a detailed look at the effectiveness of the leadership style that was used and how team members responded. The more in-depth look specifically at crisis leadership can give companies guidance on what type of individuals are best suited to manage crisis projects in the future.

Project managers have become accustomed to managing within a structured process such as an enterprise project management methodology. The statement of work may have gone through several iterations and is now clearly defined. A work breakdown structure exists, and everyone understands their roles and responsibilities as defined in the responsibility assignment matrix (RAM). All of this took time to do.

This is the environment we all take for granted. But now let us change the scenario a bit. The president of the company calls you into his office and informs you that several people have just died using one of your company's products. You are being placed in charge of this crisis project. The lobby of the building is swamped with the news media, all of whom want to talk to you to hear your plan for addressing the crisis. The president informs you that the media knows you have been assigned as the project manager, and that a news conference has been set up for one hour from now. The president also asserts that he wants to see your plan for managing the crisis no later than 10:00 p.m. Where do you begin? What should you do first? Time is now an extremely inflexible constraint rather than merely a constraint that may be able to be changed. Time does not exist to perform all of the activities you are accustomed to doing. You may need to make hundreds if not thousands of decisions quickly, and many of these are decisions you never thought that you would have to make. This is crisis project management. What leadership style is best for this type of environment?

Understanding Crisis Management

The field of crisis management is generally acknowledged to have started in 1982 when seven people died after ingesting extra strength Tylenol capsules that were laced with cyanide. Johnson & Johnson, the parent of Tylenol, handled the situation in such a manner that it became the standard for crisis management.

Today, crises are neither rare nor random. They are part of our every-day lives. Crises cannot always be foreseen or prevented, but when they occur, we must do everything possible to manage them effectively. We must also identify lessons learned and best practices so that mistakes are not repeated on future crises that will certainly occur.

Some crises are so well-entrenched in our minds that they are continuously referenced in a variety of courses in business schools. Some crises that have become icons in society include:

- Hurricane Katrina
- Mad cow disease
- Space shuttle *Challenger* explosion
- Space shuttle *Columbia* reentry disaster
- The Tylenol poisonings
- The Union Carbide chemical plant explosion in Bhopal, India
- The Exxon *Valdez* oil spill
- The Chernobyl nuclear disaster
- The Three Mile Island nuclear disaster
- The Russian submarine *Kursk* disaster
- Enron and WorldCom bankruptcies

Some crises are the result of acts of God or natural disasters. The public is generally forgiving when these occur. Crisis management, however, deals primarily with man-made crises such as product tampering, fraud, and environmental contamination. Unlike natural disasters, these man-made crises are not inevitable, and the general public knows this and is quite unforgiving.

When the Exxon *Valdez* oil spill occurred, Exxon refused to face the media for five days. Eventually, Exxon blamed the ship's captain for the accident, and verbally attacked the Alaska Department of the Environment for hampering its emergency efforts. Stonewalling the media and assuming a defensive posture created extensive negative publicity for Exxon.

Most companies neither have any processes in place to anticipate these crises, even though they perform risk management activities, nor do they know how to manage them effectively after they occur. When lives are lost because of man-made crises, the unforgiving public becomes extremely critical of the companies responsible for the crises. Corporate reputations are very fragile. Reputations that had taken years to develop can be destroyed in hours or days.

Some people contend that with effective risk management practices and a better understanding of crisis leadership skills, these crises can be prevented or minimize the damage. While it is true that looking at the risk triggers can prevent some crises, not all crises can be prevented. However, best practices in crisis management can be developed and implemented such that when a crisis occurs, we can prevent a bad situation from getting worse.

For some time, corporations in specific industries have found it necessary to simulate and analyze worst-case scenarios for their products and services. Product tampering would be an example. These worst-case scenarios have been referred to as contingency plans, emergency plans or disaster plans. These scenarios are designed around "known unknowns," where at least partial information exists on what events could happen.

Crisis management requires a heads-up approach with a very quick reaction time combined with a concerted effort on the part of possibly all employees. A nontraditional type of leadership style is necessary. In crisis management, decisions must be made quickly, often without even partial information and perhaps before the full extent of the damages are known. Events happen so quickly and so unpredictably that it may be impossible to perform any kind of planning. Roles and responsibilities of key individuals may change daily. There may be active involvement by a majority of the stakeholders many of which had previously been silent. Company survival could rest entirely on how well a company manages the crisis.

Crises can occur within any company, irrespective of the size. The larger the company involved in the crisis, the greater the media coverage. Also, crises can occur when things are going extremely well. The management guru Peter Drucker noted that companies that have been overwhelmingly successful for a long time tend to become complacent, even though the initial assumptions and environmental conditions have changed. Under these conditions, crises are more likely to occur. Drucker calls this "the failure of success."

Crises Lessons Learned

It is important to examine the lessons learned from previous crises. Some of the lessons learned from previous crises include:

- Failing to realize that a crisis exists can create catastrophic delays in addressing the crisis.
- Early warning signs are either marginally addressed or not taken seriously.
- Failing to take immediate responsibility for the crisis and develop a crisis management plan can make the situation worse.
- When more than one company is involved, each company blames the other, leaving the public with the belief that neither company could be trusted.
- Each company involved in the crisis takes a different approach to solving the crises.
- Actions must reinforce words; otherwise, the public will become nonbelievers.
- Disregarding public opinion or lying to the public can make the situation much worse.

- It is vital to accept social responsibility – appearing at the site of the crisis and showing compassion for the victims and their families.
- On crisis projects, the (executive) project sponsor may be more actively involved and may end up performing as the project manager as well.
- The project sponsor may function as the corporate spokesperson, responsible for all crisis communications. Strong communication skills are therefore mandatory.
- The company must be willing to seek help from all stakeholders and possibly also government agencies.
- Corporate social responsibility must be a much higher priority than corporate profitability.
- There is no cover in defending actions by arguing that they were legally correct when it is clear they were morally and ethically incorrect.
- Use the media to your advantage. Attacking the media can make the situation even worse.
- The longer the crisis remains in the public's eye, the greater the tendency for the company to be portrayed as a villain rather than as a victim.
- Some crises are created due to a poor organizational culture.
- The chain of command must not insulate managers and executives from hearing bad news.
- Management must not refuse to listen to workers who are pleading for help.
- When there exists a concern for human life, that trumps the pressure to maintain the schedule at all costs.

An interesting fact about the lessons learned and best practices listed above is that they lack information related to the leadership style that a project manager should consider using when managing crisis projects. Almost all of the management issues are related to senior management or governance personnel.

Victims versus Villains

The court of public opinion usually casts the deciding ballot as to whether the company involved in the crisis should be treated as a victim or a villain in the way they handled the crisis. The two determining factors are most often the company's demonstration of corporate social responsibility during the crisis and how well they dealt with the media.

During the Tylenol poisonings, Johnson & Johnson's openness with the media, willingness to accept full responsibility for their products, and rapid response to the crisis irrespective of the cost were certainly viewed favorably by the general public. Johnson & Johnson was viewed as a victim of the crisis. Other companies, on the other hand, have been viewed as a villain despite their belief that they were doing good for humanity.

Crises Life-Cycle Phases

Crises can be shown to go through the life cycles illustrated in Figure 6-4. Unlike traditional project management life-cycle phases, each of these phases can be measured in hours or days rather than months. Different leadership may be required in each life-cycle phase. Unsuccessful management of any of these phases could lead to a corporate disaster.

Most crises are preceded by early warning signs or risk triggers indicating that a crisis may occur. This is the early warning phase. Typical warning signals might include violations of safety protocols during technology development, warnings from government agencies, public discontent, complaints from customers and warnings/concerns from lower-level employees.

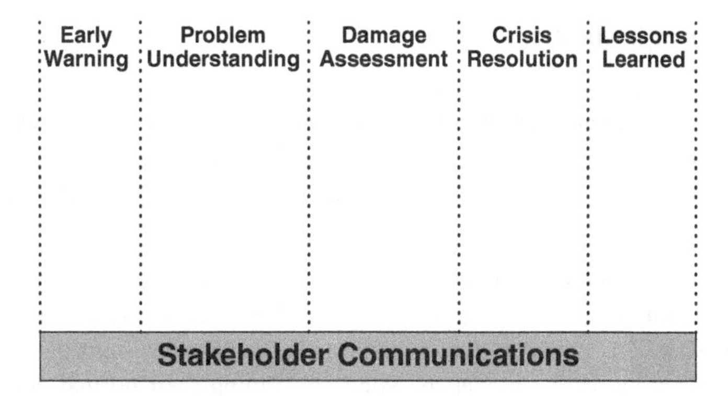

Figure 6-4 Crises Life-Cycle Phases

Historically, many companies were poor at understanding risk management, especially at evaluation of early warning signs. Today, project managers are trained in the concepts of risk management, but specifically related to the management of the project, or with the development of the product. Once the product is commercialized, the most serious early warning indicators can appear and, by that time, the project manager may be reassigned to another project. Someone else must then evaluate the early warning sings.

Early warning signs are indicators of potential risks. Time and money are a necessity for evaluation of these indicators, which preclude the ability to evaluate all risks. Therefore, companies must be selective in the risks they consider.

The next life-cycle phase is understanding the problem causing the crisis. For example, during the Tylenol poisonings, once the deaths were related to the Tylenol capsules, the first concern was to discover whether the capsules were contaminated during the manufacturing process (i.e., an inside job) or during distribution and sales (i.e., an outside job). Without a fact-based understanding of the crisis, the media can formulate their own cause of the problem and pressure the company to follow the wrong path.

The third life-cycle phase is the damage assessment phase. The magnitude of the damage will usually determine the method of resolution. Underestimating the magnitude of the damage and procrastination can cause the problem to escalate to a point where the cost of correcting the problem can grow by orders of magnitude. Intel found this out the hard way.

The crisis resolution stage is where the company announces their approach to resolve the crisis. The way the public views the company's handling of the crisis has the potential to make or break the company.

The final stage, lessons learned, mandates that companies learn not only from their own crises but also from how others handled their crises. Learning from the mistakes of others is better than learning from one's own mistakes.

Perhaps the most critical component in Figure 6-4 is stakeholder communications. When a crisis occurs, the assigned project manager may need to communicate with stakeholders that previously were of minor importance, such as the media and government agencies, and all of whom you have competing interests. These competing interests mandate that the project managers understand stakeholder needs and objectives, and also possess strong communication skills, conflict resolution skills, and negotiation skills. Therefore, what types of skills or behavior should project managers possess for these crisis situations? How do we determine or evaluate potential crisis leaders?

Project Management Implications

While it is true that every crisis has its unique characteristics, there is some commonality that can affect project management and selection of the appropriate crisis leadership style. Not all project managers, even those who have been successful managing traditional or strategic projects, have the necessary qualifications for managing crisis projects. Some implications for project managers include:

- **Know who will lead the crisis team.** It is important to understand who will be leading the crisis team. It is quite rare that a project manager will be given the responsibility to manage a crisis team, at least with our definition of a crisis. Many of the decisions that need to be made are not those made by project managers when performing their normal duties. The project sponsor will most likely assume a dual role and be the leader of the project team as well as acting as the sponsor. As in the Tylenol case, it is common for the CEO to assume primary responsibility for managing the crisis team. The leader of the crisis team must have complete authority to commit corporate resources to the project. The project manager, as we know it, will function in an assistant project manager capacity.
- **Form a crisis committee.** In time of a crisis, there will exist a crisis committee composed of the senior-most levels of management. The crisis committee will also have multifunctional membership. Project managers and assistant project managers will then report to the entire membership of the committee rather than to a single sponsor.
- **Communicate effectively.** The leader of the crisis team will be the primary spokesperson for the crisis and ultimately responsible for all media communications. The media cannot be ignored and has the power to portray the company as either a victim or a villain. The senior-most levels of management, especially those executives with professional communication skills, must do crisis communication with the media. It is essential that the corporation speak with one voice, accompanied by swiftness, honesty, openness, sincerity and compassion for the victims and their families. Information must not be withheld from the public. Withholding information from the media with the excuse that the information is incomplete may be viewed as stonewalling.
- **Manage stakeholder interests.** The crisis team must identify all the parties affected by the crisis. This includes bankers, stockholders, employees, suppliers, customers, top management, government agencies, and others. Each stakeholder can have a different interest in how the crisis is resolved, such as a financial, medical, environmental, political, or social interest. The crisis team must also be willing to ask for help from external agencies such as the FBI, Environmental Protection Agency, Federal Emergency Management Agency, and the Red Cross. The assistance of these external stakeholders can be invaluable.
- **Assume responsibility.** The company must accept responsibility for their actions or inactions immediately, and without being coerced into doing so. This will most likely fair well with the media.
- **Respond quickly.** In every crisis, there is usually a small window of opportunity where quick and decisive action can limit or even reduce the damages. Another reason for a quick response is because of the media. The longer the company takes to act, the greater the likelihood the media will look upon the company unfavorably.
- **Show compassion.** The respect for people is mandatory. It is essential that the company expresses and demonstrates compassion for all injured parties and their families, irrespective of who was actually at fault for the crisis. The emotions of the victims and their families can be

expected to run high. The public expects the company to demonstrate compassion. This also includes being on the scene of the disaster as quickly as possible. Delaying a visit to the crisis scene may be viewed as a lack of compassion or, even worse, that the company is hiding something.

- **Document everything.** Because of the multitude of legal issues that may be encountered during a crisis, most of the decisions made will need to be clearly documented. The project manager and the associated team members should possess strong writing skills.

6.12 The Growth in Competency Models

One of the expected changes over the next several years will be the conversion from project management job descriptions to project management competency models, largely due to the need for advanced leadership skills as discussed in the previous sections. Traditional project management job descriptions were based on three words: *role, authority,* and *power*. Job descriptions were easy to prepare, since they were aligned with the firm's one-size-fits-all methodology and applied only to traditional projects. As identified in Chapter 1, project managers are now expected to manage strategic as well as traditional projects. There will be a significant variability in roles and responsibilities incurred with the new projects. Standardizing roles may no longer be possible.

Job Descriptions versus Competency Models

Some experts believe that job descriptions and competency models are the same, but there are differences. **Job descriptions** tend to focus on a particular job (i.e., traditional projects) and what might be in the best interest of a client or functional department. **Competency models** focus on the skillset needed for what is in the best interest of the entire company (i.e., strategic projects) over the long term. Skills needed to support job descriptions may be able to be developed over a short horizon, whereas skills needed for competencies may take years to develop, and the required skills are everchanging. This becomes extremely important when considering the leadership styles and behaviors discussed in this chapter.

Another difference is that job descriptions often provide employees with career path opportunities – but those opportunities are restricted to their functional organization. Competency models provide both vertical and horizontal career path opportunities over the entire company. This becomes a critical issue for both the company and the workforce when project management capabilities are treated as a strategic competency rather than just another career path position, as discussed in earlier chapters. Advancement opportunities based on job descriptions in some companies are junior project managers, project managers, and senior project managers, and may be predicated on years of experience, whereas competency models focus on demonstration of the required skillsets. Levels of competencies may be defined simply as basic, advanced, and expert.

Determining the components of a job description or competency model can be highly subjective. Some companies took the easy way out in the past when preparing project management job descriptions by taking information from PMI's *PMBOK® Guide* and the *Standard for Project Management*. If the job descriptions are designed for traditional projects, then they may be transferable to other companies. Competency models may be company-specific and subject to continuous updates as the business base changes. Unlike job descriptions, competency models emanate from workforce capabilities and can create a sustainable competitive advantage.

Categories of Executive Leadership Competencies[4]

Companies have been using competencies for decades, but for specific jobs such as succession planning for future executive and managerial positions. The drivers at 3M for a leadership competency model were the strategic changes taking place in their business – namely, shrinking profit margins, narrower product differentiation, and pressure to reduce prices. The leadership competencies were divided into three categories:

FUNDAMENTAL:

- **ETHICS AND INTEGRITY**
  Exhibits uncompromising integrity and commitment to 3M's corporate values, human resource principles, and business conduct policies. Builds trust and instills self-confidence through mutually respectful, ongoing communication.
- **INTELLECTUAL CAPACITY**
  Assimilates and synthesizes information rapidly, recognizes the complexity in issues, challenges assumptions, and faces up to reality. Capable of handling multiple, complex, and paradoxical situations. Communicates clearly, concisely, and with appropriate simplicity.
- **MATURITY AND JUDGMENT**
  Demonstrates resiliency and sound judgment in dealing with business and corporate challenges. Recognizes when a decision must be made and acts in a considered and timely manner. Deals effectively with ambiguity and learns from success and failure.

ESSENTIAL:

- **CUSTOMER ORIENTATION**
  Works constantly to provide superior value to the 3M customer, making each interaction a positive one.
- **DEVELOPING PEOPLE**
  Selects and retains an excellent workforce within an environment that values diversity and respects individuality. Promotes continuous learning and the development of self and others to achieve maximum potential. Gives and seeks open and authentic feedback.
- **INSPIRING OTHERS**
  Positively affects the behavior of others, motivating them to achieve personal satisfaction and high performance through a sense of purpose and spirit of cooperation. Leads by example.
- **BUSINESS HEALTH AND RESULTS**
  Identifies and successfully generates product, market, and geographic growth opportunities, while consistently delivering positive short-term business results. Continually searches for ways to add value and position the organization for future success.

VISIONARY:

- **GLOBAL PERSPECTIVE**
  Operates from an awareness of 3M's global markets, capabilities, and resources. Exerts global leadership and works respectfully in multicultural environments to 3M's advantage.
- **VISION AND STRATEGY**
  Creates and communicates a customer-focused vision, corporately aligned and engaging all employees in pursuit of a common goal.

---

4 Adapted from M. E. Alldredge and K. J. Nilan (2000), 3M's Leadership Competency Model: An internally developed solution, *Human Resource Management*, 39 (2, 3), (Summer/Fall): 133–145. ©2000 John Wiley & Sons, Inc.

- **NURTURING INNOVATION**
  Creates and sustains an environment that supports experimentation, rewards risk taking, reinforces curiosity, and challenges the status quo through freedom and openness without judgment. Influences the future to 3M's advantage.
- **BUILDING ALLIANCES**
  Builds and leverages mutually beneficial relationships and networks, both internal and external, which generate multiple opportunities for 3M.
- **ORGANIZATIONAL AGILITY**
  Knows, respects, and leverages 3M culture and assets. Leads integrated change within a business unit to achieve sustainable competitive advantage. Utilizes teams intentionally and appropriately.

The uniqueness and targeting of competency models make it difficult to transfer them to other companies. As an example, the Ethics and Integrity component in the Fundamentals category includes values, which 3M is committed to as a means of building trust. The following are the 3M values:

- We satisfy customers with superior quality, value, and service.
- We provide our investors with a fair rate of return through sustained quality growth.
- We respect our social and physical environment.
- We work to make 3M a company that employees are proud to be part of.

While some of the categories may be common among several management positions and industries, others are unique. In health care and clinical work, crisis management and crisis leadership components are often included as subcategories under leadership and/or management.

6.13 Project Management Core Competency Models

Several factors other than strategic importance must be considered when creating project management competency models. Many of the factors are related to the types of projects that are part of the firm's core business and may include:

- Dollar value of the projects
- Length of the projects
- Complexity of the projects
- Risks associated with the projects
- Targeted customer base

Components of the project management core competency model based on information presented in Chapters 5 and 6 could be the components shown in Figure 6-5.

One of the ways that company can prepare workers for the future is through training. Each of the components in a competency model may be accompanied by training courses. Let us assume that the components shown in Figure 6-5 can be listed under three major categories: basic, intermediate, and advanced project management skills. The three major categories can be plotted as shown in Figure 6-6 and, on each axis, we can list the courses offered to the workers.

The training is not restricted just to the skills needed on each axis. Competency models focus on company core competencies, and there will still be courses related to the use of company policies and procedures required for the execution of the company's business models, and how they now interact with project management practices on strategic projects.

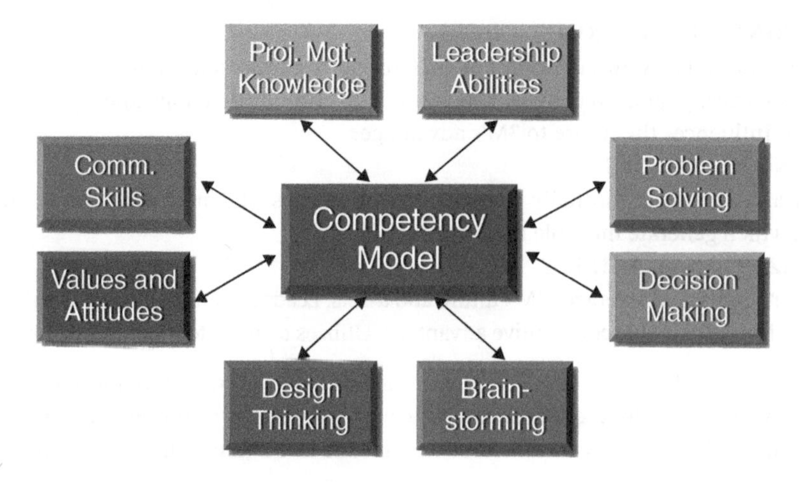

Figure 6-5 Typical Components in a Project Management Core Competency Model

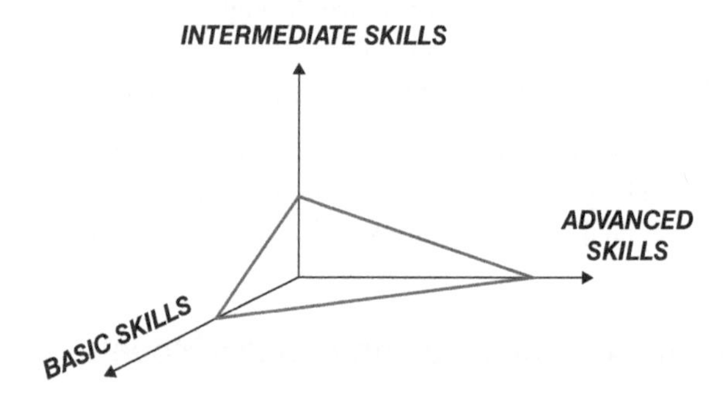

Figure 6-6 Major Categories for Identifying Training Courses

6.14 Excellence in Action: Eli Lilly[5]

Eli Lilly has perhaps one of the most comprehensive and effective competency models in industry today. The model was developed more than 15 years ago and remains today as one of the best competency models developed for innovation and R&D project management. The model encompasses many of the core competencies that have been discussed in this book.

Martin D. Hynes III, formerly director, pharmaceutical projects management (PPM), was the key sponsor of the initiative to develop the competency model. Thomas J. Konechnik, formerly operations manager, pharmaceutical projects management, was responsible for the implementation and integration of the competency model with other processes within the PPM group. The basis for the competency model is described here.

Lilly Research Laboratories project management competencies are classified under three major areas:

Scientific/Technical Expertise

- Knows the Business: Brings an understanding of the drug development process and organizational realities to bear on decisions.

5 ©2021 The Eli Lilly Company. All Rights Reserved. Reproduced with Permission.

- Initiates Action: Takes proactive steps to address needs or problems before the situation requires it.
- Thinks Critically: Seeks facts, data, or expert opinion to guide a decision or course of action.
- Manages Risks: Anticipates and allows for changes in priorities, schedules, and resources and changes due to scientific/technical issues.

Process Skills

- Communicates Clearly: Listens well and provides information that is easily understood and useful to others.
- Attention to Details: Maintains complete and detailed records of plans, meeting minutes, agreements.
- Structures the Process: Constructs, adapts, or follows a logical process to ensure achievement of objectives and goals.

Leadership

- Focuses on Results: Continually focuses own and others' attention on realistic milestones and deliverables.
- Builds a Team: Creates an environment of cooperation and mutual accountability within and across functions to achieve common objectives.
- Manages Complexity: Organizes, plans, and monitors multiple activities, people, and resources.
- Makes Tough Decisions: Demonstrates assurance in own abilities, judgments, and capabilities; assumes accountability for actions.
- Builds Strategic Support: Gets the support and level of effort needed from senior management and others to keep project on track.

We examine each of these competencies in more detail below.

1. **Knows the Business:** Brings an understanding of the drug development process and organizational realities to bear on decisions.

   Project managers/associates who demonstrate this competency will:
   - Recognize how other functions in Eli Lilly impact the success of a development effort.
   - Use knowledge of what activities are taking place in the project as a whole to establish credibility.
   - Know when team members in own and other functions will need additional support to complete an assignment/activity.
   - Generate questions based on understanding of nonobvious interactions of different parts of the project.
   - Focus attention on the issues and assumptions that have the greatest impact on the success of a particular project activity or task.
   - Understand/recognize political issues/structures of the organization.
   - Use understanding of competing functional and business priorities to reality test project plans, assumptions, time estimates, and commitments from the functions.
   - Pinpoint consequences to the project of decisions and events in other parts of the organization.
   - Recognize and respond to the different perspectives and operating realities of different parts of the organization.
   - Consider the long-term implications (pro and con) of decisions.
   - Understand the financial implications of different choices.

   Project managers/associates who do not demonstrate this competency will:
   - Rely on resource and time estimates from those responsible for an activity or task.
   - Make decisions based on what ideally should happen.
   - Build plans and timelines by rolling up individual timelines and so on.

- Perceive delays as conscious acts on the part of other parts of the organization.
- Assume that team members understand how their activities impact other parts of the project.
- Focus attention on providing accurate accounts of what has happened.
- Avoid changing plans until forced to do so.
- Wait for team members to ask for assistance.

Selected consequences for projects/business of not demonstrating this competency are:
- Project manager or associate may rely on senior management to resolve issues and obtain resources.
- Proposed project timelines may be significantly reworked to meet current guidelines.
- Attention may be focused on secondary issues rather than central business or technical issues.
- Current commitments, suppliers, and so on, may be continued regardless of reliability and value.
- Project deliverables may be compromised by changes in other parts of Lilly.
- Project plans may have adverse impact on other parts of the organization.

2. **Initiates Action:** Takes proactive steps to address needs or problems before the situation requires it.
   Project managers/associates who demonstrate this competency will:
- Follow up immediately when unanticipated events occur.
- Push for immediate action to resolve issues and make choices.
- Frame decisions and options for project team, not simply facilitate discussions.
- Take on responsibility for dealing with issues for which no one else is taking responsibility.
- Formulate proposals and action plans when a need or gap is identified.
- Quickly surface and raise issues with project team and others.
- Let others know early on when issues have major implications for project.
- Take action to ensure that relevant players are included by others in critical processes or discussions.

Project managers/associates who do not demonstrate this competency will:
- Focus efforts on ensuring that all sides of issues are explored.
- Ask others to formulate initial responses or plans to issues or emerging events.
- Let functional areas resolve resource issues on their own.
- Raise difficult issues or potential problems after their impact is fully understood.
- Avoid interfering or intervening in areas outside own area of expertise.
- Assume team members and others will respond as soon as they can.
- Defer to more experienced team members on how to handle an issue.

Selected consequences for projects/business of not demonstrating this competency are:
- Senior management may be surprised by project-related events.
- Project activities may be delayed due to "miscommunications" or to waiting for functions to respond.
- Effort and resources may be wasted or underutilized.
- Multiple approaches may be pursued in parallel.
- Difficult issues may be left unresolved.

3. **Thinks Critically:** Seeks facts, data, or expert opinion to guide a decision or course of action.
   Project managers/associates who demonstrate this competency will:
- Seek input from people with expertise or first-hand knowledge of issues and so on.

- Ask tough, incisive questions to clarify time estimates or to challenge assumptions and be able to understand the answers.
- Immerse self in project information to quickly gain a thorough understanding of a project's status and key issues.
- Focus attention on key assumptions and root causes when problems or issues arise.
- Quickly and succinctly summarize lengthy discussions.
- Gather data on past projects, and so on, to help determine best future options for a project.
- Push to get sufficient facts and data in order to make a sound judgment.
- Assimilate large volumes of information from many different sources.
- Use formal decision tools when appropriate to evaluate alternatives and identify risks and issues.

Project managers/associates who do not demonstrate this competency will:
- Accept traditional assumptions regarding resource requirements and time estimates.
- Rely on team members to provide information needed.
- Push for a new milestone without determining the reason previous milestone was missed.
- Summarize details of discussions and arguments without drawing conclusions.
- Limit inquiries to standard sources of information.
- Use procedures and tools that are readily available.
- Define role narrowly as facilitating and documenting team members' discussions.

Selected consequences for projects/business of not demonstrating this competency are:
- Commitments may be made to unrealistic or untested dates.
- High-risk approaches may be adopted without explicit acknowledgment.
- Projects may take longer to complete than necessary.
- New findings and results may be incorporated slowly only into current Lilly practices.
- Major problems may arise unexpectedly.
- Same issues may be revisited.
- Project plan may remain unchanged despite major shifts in resources, people, and priorities.

4. **Manages Risks:** Anticipates and allows for changes in priorities, schedules, resources, and changes due to scientific/technical issues.

Project managers/associates who demonstrate this competency will:
- Double-check validity of key data and assumptions before making controversial or potentially risky decisions.
- Create a contingency plan when pursuing options that have clear risks associated with them.
- Maintain ongoing direct contact with "risky" or critical path activities to understand progress.
- Push team members to identify all the assumptions implicit in their estimates and commitments.
- Stay in regular contact with those whose decisions impact the project.
- Let management and others know early on the risks associated with a particular plan of action.
- Argue for level of resources and time estimates that allow for predictable "unexpected" events.
- Pinpoint major sources of scientific risks.

Project managers/associates who do not demonstrate this competency will:
- Remain optimistic regardless of progress.
- Agree to project timelines despite serious reservations.

- Value innovation and new ideas despite attendant risks.
- Accept less experienced team members in key areas.
- Give individuals freedom to explore different options.
- Accept estimates and assessments with minimal discussion.

Selected consequences for projects/business of not demonstrating this competency are:

- Projects may take longer to complete than necessary.
- Project may have difficulty responding to shifts in organizational priorities.
- Major delays could occur if proposed innovative approach proves inappropriate.
- Known problem areas may remain sources of difficulties.
- Project plans may be subject to dramatic revisions.

5. **Communicates Clearly:** Listens well and provides information that is easily understood and useful to others.

Project managers/associates who demonstrate this competency will:

- Present technical and other complex issues in a concise, clear, and compelling manner.
- Target or position communication to address needs or level of understanding of recipient(s) (e.g., medical, senior management).
- Filter data to provide the most relevant information (e.g., does not go over all details but knows when and how to provide an overall view).
- Keep others informed in a timely manner about decision or issues that may impact them.
- Facilitate and encourage open communication among team members.
- Set up mechanisms for regular communications with team members in remote locations.
- Accurately capture key points of complex or extended discussions.
- Spend the time necessary to prepare presentations for management.
- Effectively communicate and represent technical arguments outside own area of expertise.

Project managers/associates who do not demonstrate this competency will:

- Provide all the available details.
- See multiple reminders or messages as inefficient.
- Expect team members to understand technical terms of each other's specialties.
- Reuse communication and briefing materials with different audiences.
- Limit communications to periodic updates.
- Invite to meetings only those who (are presumed to) need to be there or who have something to contribute.
- Rely on technical experts to provide briefings in specialized, technical areas.

Selected consequences for projects/business of not demonstrating this competency are:

- Individuals outside of the immediate team may have little understanding of the project.
- Other projects may be disrupted by "fire drills" or last-minute changes in plan.
- Key decisions and discussions may be inadequately documented.
- Management briefings may be experienced as ordeals by team and management.
- Resources/effort may be wasted or misapplied.

6. **Pays Attention to Details:** Systematically documents, tracks, and organizes project details.

Project managers/associates who demonstrate this competency will:

- Remind individuals of due dates and other requirements.
- Ensure that all relevant parties are informed of meetings and decisions.
- Prepare timely, accurate, and complete minutes of meetings.

- Continually update or adjust project documents to reflect decisions and changes.
- Check the validity of key assumptions in building the plan.
- Follow up to ensure that commitments are understood.

Project managers/associates who do not demonstrate this competency will:

- Assume that others are tracking the details.
- See formal reviews as intrusions and waste of time.
- Choose procedures that are least demanding in terms of tracking details.
- Only sporadically review and update or adjust project documents to reflect decisions and other changes.
- Limit project documentation to those formally required.
- Rely on meeting notes as adequate documentation of meetings.

Selected consequences for projects/business of not demonstrating this competency are:

- Coordination with other parts of the organization may be lacking.
- Documentation may be incomplete or difficult to use to review project issues.
- Disagreements may arise as to what was committed to.
- Project may be excessively dependent on the physical presence of manager or associate.

7. **Structures the Process:** Constructs, adapts, or follows a logical process to ensure achievement of objectives and goals.

Project managers/associates who demonstrate this competency will:

- Choose milestones that the team can use for assessing progress.
- Structure meetings to ensure agenda items are covered.
- Identify sequence of steps needed to execute project management process.
- Maintain up-to-date documentation that maps expectations for individual team members.
- Use available planning tools to standardize procedures and structure activities.
- Create simple tools to help team members track, organize, and communicate information.
- Build a process that efficiently uses team members' time, while allowing them to participate in project decision; all team members should not attend all meetings.
- Review implications of discussion or decisions for the project plan as mechanism for summarizing and clarifying discussions.
- Keep discussions moving by noting disagreements rather than trying to resolve them there and then.
- Create and use a process to ensure priorities are established and project strategy is defined.

Project managers/associates who do not demonstrate this competency will:

- Trust that experienced team members know what they are doing.
- Treat complex sequences of activities as a whole.
- Share responsibility for running meetings, formulating agendas, and so on.
- Create plans and documents that are as complete and detailed as possible.
- Provide written documentation only when asked for.
- Allow team members to have their say.

Selected consequences for projects/business of not demonstrating this competency are:

- Projects may receive significantly different levels of attention.
- Project may lack a single direction or focus.
- Planning documents may be incomplete or out of date.
- Presentations and briefings may require large amounts of additional work.

- Meetings may be seen as unproductive.
- Key issues may be left unresolved.
- Other parts of the organization may be unclear about what is expected and when.

8. **Focuses on Results:** Continually focuses own and others' attention on realistic project milestones and deliverables.

   Project managers/associates who demonstrate this competency will:
- Stress need to keep project-related activities moving forward.
- Continually focus on ultimate deliverables (e.g., product to market, affirm/disconfirm merits of compound, value of product/program to Lilly) (manager).
- Choose actions in terms of what needs to be accomplished rather than seeking optimal solutions or answers.
- Remind project team members of key project milestones and schedules.
- Keep key milestones visible to the team.
- Use fundamental objective of project as means of evaluating option driving decisions in a timely fashion.
- Push team members to make explicit and public commitments to deliverables.
- Terminate projects or low-value activities in timely fashion.

   Project managers/associates who do not demonstrate this competency will:
- Assume that team members have a clear understanding of project deliverables and milestones.
- Approach tasks and issues only when they become absolutely critical.
- Downplay or overlook negative results or outcomes.
- Keep pushing to meet original objectives in spite of new data/major changes.
- Pursue activities unrelated to original project requirements.
- Trust that definite plans will be agreed to once team members are involved in the project.
- Allow unqualified individuals to remain on tasks.
- Make attendance at project planning meetings discretionary.

   Selected consequences for projects/business of not demonstrating this competency are:
- Milestones may be missed without adequate explanation.
- Functional areas may be surprised at demand for key resources.
- Commitments may be made to unreasonable or unrealistic goals or schedules.
- Projects may take longer to complete than necessary.
- Objectives and priorities may differ significantly from one team member to another.

9. **Builds a Team:** Creates an environment of cooperation and mutual accountability within and across functions to achieve common objectives.

   Project managers/associates who demonstrate this competency will:
- Openly acknowledge different viewpoints and disagreements.
- Actively encourage all team members to participate regardless of their functional background or level in the organization.
- Devote time and resources explicitly to building a team identity and a set of shared objectives.
- Maintain objectivity; avoid personalizing issues and disagreements.
- Establish one-on-one relationship with team members.
- Encourage team members to contribute input in areas outside functional areas.
- Involve team members in the planning process from beginning to end.
- Recognize and tap into the experience and expertise that each team member possesses.

- Solicit input and involvement from different functions prior to their major involvement.
- Once a decision is made, insist that team accept it until additional data become available.
- Push for explicit commitment from team members when resolving controversial issues.

Project managers/associates who do not demonstrate this competency will:

- State what can and cannot be done.
- Assume that mature professionals need little support or team recognition.
- Limit contacts with team members to formal meetings and discussions.
- Treat issues that impact a team member's performance as the responsibility of functional line management.
- Help others only when explicitly asked to do so.
- Be openly critical about other team members' contributions or attitudes.
- Revisit decisions when team members resurface issues.

Selected consequences for projects/business of not demonstrating this competency are:

- Team members may be unclear as to their responsibilities.
- Key individuals may move onto other projects.
- Obstacles and setbacks may undermine overall effort.
- Conflicts over priorities within project team may get escalated to senior management.
- Responsibility for project may get diffused.
- Team members may be reluctant to provide each other with support or accommodate special requests.

10. **Manages Complexity:** Organizes, plans, and monitors multiple activities, people, and resources.

Project managers/associates who demonstrate this competency will:

- Remain calm when under personal attack or extreme pressure.
- Monitor progress on frequent and consistent basis.
- Focus personal efforts on most critical tasks: apply 80–20 rule.
- Carefully document commitments and responsibilities.
- Define tasks and activities to all for monitoring and a sense of progress.
- Break activities and assignments into components that appear doable.
- Balance and optimize workloads among different groups and individuals.
- Quickly pull together special teams or use outside experts in order to address emergencies or unusual circumstances.
- Debrief to capture "best practices" and "lessons learned."

Project managers/associates who do not demonstrate this competency will:

- Limit the number of reviews to maximize time available to team members.
- Stay on top of all the details.
- Depend on team members to keep track of their own progress.
- Let others know how they feel about an issue or individual.
- Rely on the team to address issues.
- Assume individuals recognize and learn from their own mistakes.

Selected consequences for projects/business of not demonstrating this competency are:

- Projects may receive significantly different levels of attention.
- Projects may take on a life of their own with no clear direction or attainable outcome.
- Responsibility for decisions may be diffused among team members.

- Exact status of projects may be difficult to determine.
- Major issues can become unmanageable.
- Activities of different parts of the business may be uncoordinated.
- Conflicts may continually surface between project leadership and other parts of Lilly.

11. **Makes Tough Decisions:** Demonstrates assurance in own abilities, judgments, and capabilities; assumes accountability for actions.

    Project managers/associates who demonstrate this competency will:
    - Challenge the way things are done and make decisions about how things will get done.
    - Force others to deal with the unpleasant realities of a situation.
    - Push for reassessment of controversial decisions by management when new information/data become available.
    - Bring issues with significant impact to the attention of others.
    - Consciously use past experience and historical data to persuade others.
    - Confront individuals who are not meeting their commitments.
    - Push line management to replace individuals who fail to meet expectations.
    - Challenge continued investment in a project if data suggest it will not succeed.
    - Pursue or adopt innovative procedures that offer significant potential benefits even where limited prior experience is available.

    Project managers/associates who do not demonstrate this competency will:
    - Defer to the ideas of more experienced team members.
    - Give others the benefit of the doubt around missed commitments.
    - Hold off making decisions until the last possible moment.
    - Pursue multiple options rather than halt work on alternative approaches.
    - Wait for explicit support from others before raising difficult issues.
    - Accept senior managers' decisions as "nonnegotiable."
    - Rely on the team to make controversial decisions.
    - Provide problematic performers with additional resources and time.

    Selected consequences for projects/business of not demonstrating this competency are:
    - Projects may take longer to complete than necessary.
    - Failing projects may be allowed to linger.
    - Decisions may be delegated upward.
    - Morale of team may be undermined by nonperformance of certain team members.
    - "Bad news" may not be communicated until the last minute.
    - Key individuals may "bum out" in effort to play catch-up.

12. **Builds Strategic Support:** Gets the support and level of effort needed from senior management and others to keep projects on track.

    Project managers/associates who demonstrate this competency will:
    - Assume responsibility for championing the projects while demonstrating a balance between passion and objectivity.
    - Tailor arguments and presentations to address key concerns of influential decision makers.
    - Familiarize self with operational and business concerns of major functions within Lilly.
    - Use network of contacts to determine best way to surface an issue or make a proposal.
    - Push for active involvement of individuals with the experience and influence needed to make things happen.
    - Pinpoint the distribution of influence in conflict situations.
    - Presell controversial ideas or information.

- Select presenter to ensure appropriate message is sent.
- Ask senior management to help position issues with other senior managers.

Project managers/associates who do not demonstrate this competency will:

- Meet senior management and project sponsors only in formal settings.
- Propose major shifts in direction in group meetings.
- Make contact with key decision makers when faced with obstacles or problems.
- Limit number of face-to-face contacts with "global" partners.
- Treat individuals as equally important.
- Avoid the appearance of politicking.
- Depend on other team members to communicate to senior managers in unfamiliar parts of Lilly.

Selected consequences for projects/business of not demonstrating this competency are:

- Viable projects may be killed without clear articulation of benefits.
- "Cultural differences" may limit success of global projects.
- Decisions may be made without the input of key individuals.
- Resistance to changes in project scope or direction may become entrenched before merits of proposals are understood.
- Key individuals/organizations may never buy in to a project's direction or scope.

Minor conflicts may escalate and drag on.

6.15 Conclusions

The growth in project management competency models is inevitable and emphasis will most likely be in leadership processes and leadership behavior. With the expansion of project management into strategic projects, crises leadership practices will also be included in competency models.

References

Barbuto J. E. and Wheeler, D. W. (2006). Scale development and construct clarification of servant leadership. *Group & Organization Management* 31(3): 300–326.

Clarke, N. (2012). Leadership in projects: what we know from the literature and new insights. *Team Performance Management* 18(3/4): 128–148.

Currall, S. C., and Epstein, M. J. (2003). The fragility of organizational trust: lessons from the rise and fall of Enron. *Organizational Dynamics* 32: 193–206.

De Cieri, H., Holmes, B., Abbott, J., and Pettit, T. (2005). Achievements and challenges for work/life balance strategies in Australian organizations. *International Journal of Human Resource Management* 16: 90–103.

Einarsen, S. (1999). The nature and causes of bullying at work. *International Journal of Manpower* 20: 16–27.

Fiedler, F. E. (1967). *A Theory of Leadership Effectiveness.* McGraw-Hill: Harper and Row Publishers Inc.

Frost, P. J. (2003). *Toxic Emotions at Work: How Compassionate Managers Handle Pain and Conflict.* Boston: Harvard Business School Publications.

Gardner, W. L., Cogliser, C.C., Davis, K.M., and Dickens, M.P. (2011). Authentic leadership: A review of the literature and research agenda. *Leadership Quarterly* 22: 1120–1145.

Grandey, A. A. (2000). Emotion regulation in the workplace: A new way to conceptualize emotional labor. *Journal of Occupational Health Psychology* 5 (1): 59–100.

Grandey, A. A., Rupp, D. and Brice W. N. (2015). Emotional labor threatens decent work: A proposal to eradicate emotional display rules. *Journal of Organizational Behavior* 36: 770–785. Published online 21 May 2015 in Wiley Online Library (wileyonlinelibrary.com)

Greenleaf, R. K. (1970). *The Servant as a Leader*. Indianapolis, IN: Greenleaf Center.

Hemphill, J. K. (1949). *Situational Factors in Leadership*. Columbus: Ohio State University Bureau of Educational Research.

Hochschild, A. R. (1979). Emotion work, feeling rules, and social structure. *American Journal of Sociology* 85(3): 551–575.

Hochschild, A. R. (1983). *The Managed Heart*. Berkeley: University of California Press.

Jensen, J. M., Patel, P. C., and Messersmith, J. G. (2013). High-performance work systems and job control consequences for anxiety, role overload, and turnover intentions. *Journal of Management* 39 (6): 1699–1724.

Krog, C., and Govender, K. (2015). Servant leadership and project management: examining the effects of leadership style on project success. *Proceedings of the European Conference on Management, Leadership & Governance*, 201–210.

Lewin, K., Lippitt, R., and White, R. (1939). Patterns of aggressive behavior in experimentally created social climates. *The Journal of Social Psychology:* 271–301.

Lu, J., Zhang, Z. and Jia, M. (2019). Does Servant Leadership Affect Employees' Emotional Labor? A Social Information-Processing Perspective. *Journal of Business Ethics* 159(2): 507–518.

Muller, R., and Turner, J. R. (2007). Matching the project manager's leadership style to project type. *International Journal of Project Management* 25(1): 21–32.

Muller, R., and Turner, J. R. (2010). Leadership competency profiles of successful project managers. *International Journal of Project Management* 28: 437–448.

Roeder, T. (2013). *Managing Project Stakeholders*. Hoboken, NJ: John Wiley and Sons, 158.

Sankowsky, D. (1995). The charismatic leader as narcissist: understanding the of power. *Organizational Dynamics* 23: 57–71.

Sarros, J. C., Tanewski, G. A., Winter, R. P., Santora, J. C. and Densten, I. L. (2002). Work alienation and organizational leadership. *British Journal of Management* 13: 285–304.

Scott, B. A., and Barnes, C. M. (2011). A multilevel field investigation of emotional labor affect, work withdrawal, and gender. *Academy of Management Journal* 54(1): 116–136.

Sendjaya, S., Sarros, J. C., and Santora, J. C. (2008). Defining and measuring servant leadership behavior in organizations. *Journal of Management Studies (Wiley-Blackwell)* 45 (2): 402–424.

Shao, S. and Muller, R. (2011). The development of constructs of program context and program success: a qualitative study. *International Journal of Project Management* 29: 947–959.

Spears, L. C. (1995). *Reflections on leadership: How Robert K. Greenleaf's Theory of Servant Leadership Influenced Today's Top Management Thinkers*. New York: John Wiley.

Spears, L. C. (2002). Introduction: Tracing the past, present and future of servant-leadership. In L. C. Spears (Ed.), *Focus on Leadership* (pp. 1–18). New York: John Wiley.

Strang, K. D. (2010). Leadership substitutes and personality impact on time and quality in virtual new product development projects. *Project Management Journal* 42(1): 73–90.

Thamhain, H. J. (2004). Linkages of project environment to performance lessons for team leadership. *International Journal of Project Management* 22: 533–544.

Thornthwaite, L. (2004). Working time and work-family balance: a review of employees' preferences. *Asia Pacific Journal of Human Resources* 42: 166–184.

Wright, T. A., and Cropanzano, R. (2004). The role of psychological well-being in job performance: a fresh look at an age-old quest. *Organizational Dynamics* 33: 338–351.

7

Pillar 7: Organizational Cultural Shift to the Project Way of Working

7.0 Introduction

Projects are becoming more and more the "modus operandi" of organizations. There is no difference anymore between what a normal process and a project is.

This kind of approach requires a relevant change in the way organizations approach their projects and, consequently, shape the work culture.

Adapting from Beswick, Toma, and Vargas (2020), the ecosystem where projects evolve consists of many elements that need to work in sync for growth to happen. This ecosystem is composed of five core pillars: **strategy**, **leadership**, **management**, **culture,** and **processes**.

It is impossible to make "ideas a reality" without changes in how the work, the decision-making process, diversity, and several other aspects need to be tailored and adapted to the current environment.

7.1. The Need for Cultural Shift

Organizational culture can be defined as a set of features, assumptions, and behaviors identified in the organization's approach to solving problems, defining policies and processes, developing new products, and handling relationships with employees, among other topics.

A key reason for the challenges that organizations face today is related to a "way of working" that does not respond to the current needs of the market and different stakeholders.

Four factors have an intrinsic relationship with the needed cultural change: the new workforce behavior; streamlining processes; flexibility, adaptability, and reaction; and diversity.

New Workforce Behavior

The project manager or team member today is different from the past – and not only because they are younger or with another college degree or background.

They are different because they have different expectations and aren't necessarily motivated by the same incentives people were 10 or 20 years ago.

The new workforce wants to connect their own interests with the organization. Their journey is associated with the journey of the organization (Tabrizi & Terrell, 2013).

Another important aspect is the concept of *workforce*. In the past, the workforce was restricted to the organization's employees performing internal work and being paid by the organization.

Research from MIT Sloan School of Management in partnership with Delloite (MIT, 2021) pointed out that 87 percent of the interviewees (almost 5,000 professionals) have a broader definition of the workforce that includes suppliers and contractors, and this wider group of stakeholders has their own expectations and transformative journey.

This new behavior has led to the concept of **human-centric transformation** that the Brightline Initiative[1] explored in its Transformation Compass, where the connection between the interests and expectations of the workforce is aligned with the interests and expectations of the organization to maximize results and drive changing behaviors (Figure 7-1).

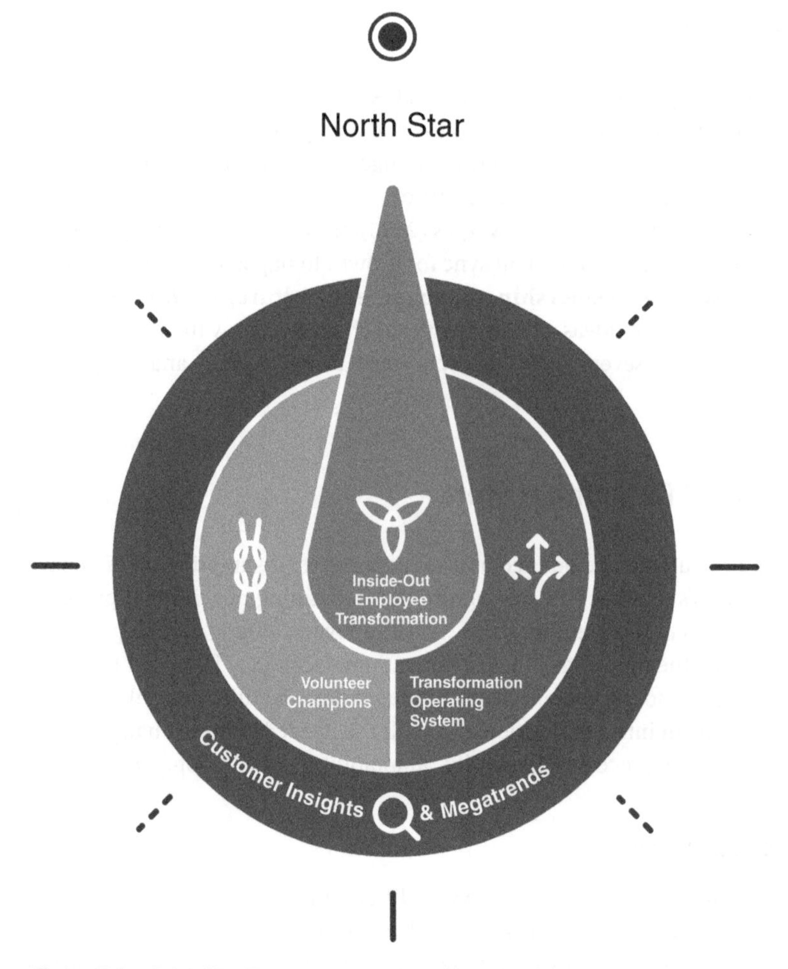

Figure 7-1 Brightline Transformation Compass (PMI, 2019)

1 Brightline Initiative is a Project Management Institute think tank aiming to support C-level executives on understanding the challenges to connect ideas with reality.

Streamlining Processes

The current challenges require processes that are simple and effective. When startup companies are compared with incumbent enterprises, it is perceived that they have far simpler methods due to their shape and constitution. It is a clear advantage to promote change.

The challenge is to replicate in some ways the streamlined processes in large incumbent organizations. Together with the challenge on the business itself, there is a visible challenge with the workforce culture. Changing processes is always easier than changing people's attitudes.

A solution that companies like Volkswagen did was to create something called "One strategy, two approaches." Instead of disrupting the entire value chain, they split the strategy execution and the projects into two teams: one taking care of the improvements of the traditional car industry and another operating like a startup to rethink mobility (EIU & Brightline, 2017).

Flexibility, Adaptability, and Reaction

Combined with the simplification of processes, organizations, must foster the ability of their teams to be flexible and adaptable to react to the changing environment.

The "control freak" approach is over. There is no possibility to behave like if a project was a closed system. Today, aspects of volatility, uncertainty, complexity, and ambiguity (VUCA) are present in all types of projects.

However, the organization must create ways to boost psychological safety at work, allowing people to freely exchange ideas, concerns, and questions that would not be asked otherwise (Edmondson, A. C. & Hugander, P., 2021).

This psychological safety improves people's ability to be adaptable and flexible.

The Standards for Project Management and the PMBOK® Guide 7th Edition published by PMI (PMI, 2021) reinforce these capabilities to react and adapt to all the 12 principles of project management. However, two of them are intrinsically related to this behavior.

1. Embrace adaptability and resiliency.
2. Navigate complexity.

Diversity

Improving diversity in the workforce is one of the key factors for delivering transformative change and projects.

Diversity goes above and beyond ethnicity. It includes involving people from different social and ethnic backgrounds, gender, sexual orientations, religious beliefs, etc.

Diversity fosters adaptability and a changing mindset due to the following factors:

- People are accepted and valued even when they think differently.
- When associated with psychological safety, people feel safe to share risks, concerns, and new disruptive ideas.
- Diversity brings different perspectives to the analysis, improving creative responses and actions.
- Diversity improves innovation and problem-solving.
- Improves capability to understand client's needs.
- People with different backgrounds bring a different set of skills.
- Diversity reduces group thinking.

In this chapter, we will share two case studies. The first one is the GEA's approach to project management. GEA Group is a leading engineering company that implements industrial and manufacturing plants to sectors like beverage, food, dairy, pharma, among several others. Their approach is entirely focused on the human perspective of managing engineering projects.

The second one is a testimonial of the director responsible for building the fourth largest Hydroelectric Power Plant in the world: the Belo Monte HPP. He shares the challenges of managing a large, complex, and divergent group of stakeholders working together in one of the most challenging environments to operate: the Amazon Forest. From handling the "Piracema," the reproduction cycle of fishes in the Xingu River basin, to replacing complex turbine parts to deliver the project as expected.

7.2 Excellence in Action: GEA[2] Project Management in GEA Process Engineering: Our Vision for the Future

Introduction

During the last two decades, changes in project management philosophy have been very significant. Not only has the discipline changed from the methodological perspective (sequential, Agile, hybrid) but also from the strategical point of view. The project management approach has moved from the individual project perspective to the management of portfolio of projects, passing throughout the establishment of Project Management Offices as the best tool for enhancing the decision-making within the company for doing the business more sustainable as well as more connected to the priorities of the company and the customers.

Those changes had paid an effect in the project managers, who have moved from being "egoist" people thinking about fulfilling the objectives of the project, mainly dominated by the project triple constrain (cost, time, and scope) to, along with it, be more conscious about the other aspects surrounding the project environment and, what is more important, the effects that their decisions have in the company as a whole. Undoubtedly, the evolution of the relationship within the companies, their customers, and their contractors must be taken into account. In the twenty-first century, nothing is isolated, and the outdated silo thinking has moved aside to give pass to a multicultural, multidisciplinary, and collaborative approach. Therefore, project management must evolve, like many other disciplines in the company, to follow that pattern. Not taking it into consideration could result in a significant deterioration of the business, paying a heavy toll not only in the finances of the companies but also in their own chances of surviving.

Moreover, the project management discipline has been considered a very important asset for the overall economy, and there are studies that have already evaluated the contribution of the discipline to the GDP. This has given project management a very significant relevance from the macroeconomic perspective.

That being said, let us move further to draft what, from our perspective, could be the future of the project management and its environment during the next decades.

2 Material in this section has been provided by Miguel Antonio Martínez Carrizo, Head of the Project Management Office, Liquid & Food Technologies EMEA South, Liquid & Powder Technologies Division, GEA Process Engineering S.A.

New Challenges, New Methods

Yes, our society, our economy, and our culture have evolved from the linear/sequential approach, common to previous generations, to an entangled way of living that marks all aspects of our life. Not only the continuous change, but the speed of that change is permanently sculpting our lives and therefore, the way we work too. Following it, project management has evolved, too. The sequential way of developing projects has changed in some projects by using different techniques that were named in the 1980s and 1990s as Agile or adaptive. Instead of trying to describe these techniques (there is an extensive amount of literature about it), we are going to explain how our company has adapted to them.

Basically, our traditional projects (mainly dedicated to design liquid food and beverage installations) have been adapted to the changing environment that the new technologies are imposing. Our customers are not only worried about having an efficient installation that can process correctly, under sanitary and food conditions, their recipes. They need to know in real time how the process is doing as well as monitor and control the main business KPIs (production rates, energy consumptions, fulfilling of customers' orders, etc.). Then the scope, mostly focused on the process part, is now including the management side of it that allows our customers to make decisions in a truly real time environment.

Therefore, the control system, considered in the past a secondary aspect of the installations, always working behind the scenes, has truly become the brain of the installation we built, taking care not only about the control of it but also about its connectivity with other entities inside and outside the company that provides valuable business information to the highest levels of the management of our customers.

In that regard, the changing needs of the business have moved us to adapt ourselves and manage at least part of our projects throughout an intensive interaction with customers, contractors, and other stakeholders by using some of the Agile techniques. In those projects we have passed from the linear/sequential (Waterfall) approach to a hybrid one where at least the control system and its surroundings are subjected to a high adaptive environment.

Furthermore, the continuous update of the customers' control system has created a new category of projects in the food industry called migrations dedicated mainly to upgrade the control system and provide new functionalities to the customers. They are projects purely adaptive that are heavily subjected to the customer and market needs.

Definitely, and although the "traditional" way of doing projects is going to remain, especially for large and very large projects with a strong construction part, the hybrid and adaptive approach are gaining ground, are here to stay and will be more frequent in the future.

Independent of the methodology used, let us focus on the support mechanisms that our project management people are putting in place for making this new reality and upcoming future associated to it more solid and sustainable.

The Project Sponsor Role

There is an extensive amount of literature about the functions and relevance of the project sponsor in the success of projects development. Here we are going to describe how we understand this role within our company and instead of giving a list of topics, let us group them according to the following functional criteria:

Strategical

- The project sponsor guarantees that the company policies and strategies are followed during the execution of the projects by being a guardrail for the project managers. Also, to prioritize the projects according with the strategy of the company.

- But not only does the project sponsor take care of achieving the targets set up in the highest levels of the company, he/she also takes care of the way of achieving them. Ethic and professional considerations are a very important part of the business. Not everything is allowed, and the project sponsor guarantees that the project managers and the project teams, as an important part of the image of the company for being in contact with internal and external stakeholders (and sometimes with public audiences), behave following the company directives and good practices in project management.
- The project sponsor plays an important role in the strategic developments of the company, particularly on the continuous improvement and innovation initiatives. The project sponsor acts as the driver of the learnings coming from the projects into effective policies and practices within the company. From that perspective, the inclusion of the project sponsor (usually the head of project management) in the C group is now a common practice in many companies, especially when executing projects is one of their core businesses.

Operational

Along with the regular management (plan, organize, coordinate, and monitor), there are other functions that deserve to be mentioned here:

- Taking care of the right execution of the projects and, in case of conflict, by assigning priorities between the different projects.
- Guarding the project management governance determined by the company, being the person accountable of the project management quality of the projects.
- Supporting the project manager when the project is stressed and subjected to issues or crises that the project team is not capable of managing. The intervention of the project sponsor is sometimes key for releasing resources from other departments (or from external sources) or for managing the relationship with stakeholders, the project team members, or the "creep" of the project – in some cases, representing GEA and its interests, being in those cases the speaker within the company.

Reporting

The project sponsor reviews with the project managers the status of the project in regular basis and, according to their feedback, the project sponsor reports to higher levels of management, usually following the overall portfolio perspective and also paying special attention to strategical, very large and/or complex projects. Reporting is focused mainly on finance indicators, but it should consider others that could affect the business development and sustainability of the company.

Interfaces

Interfaces occur mainly with other departments of the company, and particularly with their heads. The matrix structure predominant in the company create conflicts, especially when assigning resources to the projects. The project sponsor has the duty to support the project manager in his/her interaction with the different departments for providing the needed resources to develop the project in the best possible way and prioritizing according to the company strategy. Also, as mentioned above, the interaction with intermediate and high levels of management of the customers is very frequent and it is the duty of the project sponsor to promote this interaction for being ready when the projects have issues that the project managers cannot solve.

Needless to say, the project sponsor, along with being knowledgeable in the project management discipline, must have long experience managing projects.

To summarize, today the project sponsor is the main factor of influence during the project development as well as the ultimate accountable of the success of the project, but will gain even more relevance in the future due to the increase complexity present in the projects and in the relationships inside the company and external stakeholders. He/she has to guide the project managers, making sure that they are capable of seeing the woods while managing the trees.

Project Steering Committees

As we have mentioned, there is an increased importance of the company strategy in the development of the projects, especially when they are very large, strategical, or complex projects.

In those cases, GEA promotes the creation of the Project Steering Committee as a mechanism of advice and counseling to the project manager in his/her way of managing the project. It is a way of looking for the involvement of the most relevant internal stakeholders, so they would be able to be informed firsthand about the project development, give the project manager the chance to pulse their opinion in a direct way and obtain the benefits not only from their expertise but also from their view of the project, usually from a very different perspective. It is also a way for avoiding misunderstandings and confusions and create trust and confidence to the project manager and by extension to the Project Team.

In special cases, there is also a Project Steering Committee with the involvement of the customer stakeholders. On doing it, the project has another forum where to discuss the project issues, prioritizing the business perspective more than the operating one and, whenever the conflicts arise, without the typical frictions between the project team members.

From our company perspective, the main functions of the Project Steering Committee could be summarized as follows (not necessarily listed in order of importance):

- Support and provide guidance to the project manager in all the strategic and major issues of the project as well as in the businesses related to it.
- Provide high-level monitoring and control of the project, keeping it focused on its contractual and noncontractual objectives (such as the management of cost, quality, resources, stakeholders, finance, etc.) that are key for the project development.
- Ensure the compliance and adherence to the GEA policies and rules.
- Safeguard that effort and expenditure are appropriated to accomplishing project objectives and project completion.
- Authorize the change of phase and ultimately the closure of the project.
- Represent the legitimate interests of the different stakeholders (people or groups) involved in the project, including customer and contractors, and, in case of conflict of interests, to reconcile them in the best way for the project and for GEA.

In order to accomplish the above functions, it is expected that the Steering Committee members (again, not in order of importance):

- Have the availability not only to attend to the Steering Committee meetings but also to do it actively by reviewing the documentation provided in advance by the project manager. The periodicity would be determined by its members in the first meeting, but usually it is approximately every six or eight weeks.
- Understand the strategic implications and outcomes of the project and be genuinely interested on them.

- Be willing to be not only an advocate for the project but also to question constructively the feedback provided by the project manager.
- In the case of conflicts of interest with other members of the Steering Committee, be flexible for reaching an agreed resolution.

Finally, to remark the importance of the involvement of the top stakeholders of the very large, strategical, and complex projects for promoting the change of culture that the new times require. The interaction between both groups (top management and project managers) acts as a catalyzer for, from top to down and vice versa, creating the new setup of the company relationships that is very beneficial not just to create confidence and trust but also to know better the business from its different perspectives, facilitating the decision-making process, the strategic thinking and, at the end, the sustainability of the company.

The Ultimate Tool, The Project Management Office (PMO)

The new set up required in project management for the twenty-first century would not be completed without the truly tool that moves the profession in the private and public companies. It is not enough for project managers to follow what the main institutions in project management are dictating. Even the usual practices and bodies of knowledge have already reached the category of governmental standards, to have departments that assure the good practices in all aspects of Project Management are required.

In that regard, The Project Management Offices (PMOs) have evolved too. During the last decades of the twentieth century, the PMOs were mere occasional spaces created for particular projects (in most cases, internal ones), with very limited life and scope. But in the last years, they have become the cornerstone of well-settled project management practices. Their responsibility and functions are key for developing strategic competencies in project management, give more leverage within the organizations to the profession and be the reference to follow in these complicated times where the VUCA (volatility, uncertainty, complexity, and ambiguity) philosophy are present.

Several sides of the PMOs have to be considered:

- Project managers have to feel they have a home where they stay comfortable. Old times where they belong to one of the departments of the company are gone. The profession has its own characteristics, mechanisms, standards, and practices and even when the projects are loaded with heavy technical scope, a professional project manager would make the difference in the success of the project.
- In addition to the sense of belonging, the PMOs are a key part for providing to the project managers the traditional and new competences required by the profession as well as the ones part of the DNA of the company. On doing it the PMO is going to achieve the perfect match for the project managers as well as creating and nurturing talent within the company ready to fill the leadership pipeline.
- The PMOs are the guardians of the good practices of the project managers by implementing the governance on the projects. Since the definition of project success is mutating constantly, someone has to take care of securing the development of the projects in the proper way. In that regard, the head of the PMO, usually the project sponsor of the projects, is the one accountable that the quality on their execution will be planned, assured, and controlled.
- It has to provide the proper tools for facilitating the work to the project managers. Now and more intensively in the future, the project managers have to dedicate themselves to project tasks adding value to the project as well as to the organizations, being the responsibility of the PMO

to be up to date in the latest technologies and tools supporting the project management tasks. Having a good set of tools that make data analysis, reporting, project controlling, etc. is going to be a challenge in the next years. Artificial intelligence, data management, visualization and communication techniques, decision trees and, in general, the digitalization tools, are going to play a significant role in the day to day of the project managers for allowing them to make decisions more solidly and to do truly added value tasks (instead of those that are pure repetition).
- Increase the project management maturity of the company, passing from recognizing value and manage requirements to meet the needs strategically executed and in an optimized way. That would allow the company to pass from executing just the regular projects to the complex and innovative ones. At the end, the PMO is responsible for transforming the project management practice in a company from being another capability to a core skill, able to work with long-term vision, in a collaborative, holistic, and multidisciplinary way.
- The traditional function based on the data storage and financial reporting has to move to a value driven office, where management and leadership find their place, where the strategic business objectives of the company are achieved throughout the planning, prioritization, facilitation, collaboration, communication, training, coaching, continuous improvement. . .

Therefore, the PMO, as the house of the project managers and as an important department for maintaining the pulse of the company, has to be the frame, a solid one, where the excellence in the project practice spread to the different departments of the company for achieving the brilliance expected in the new times.

Finally, we cannot conclude with the PMO without remarking its relevance on one of the most important tendencies, particularly in the companies dedicated to developing projects, which is to move from project management toward the organizational project management (OPM). The holistic approach of the OPMs allows the "projectified" companies to tackle its different corporative structures, moving its organizational set up from the current hierarchical perspective toward a set of organizational layers able to promote a new set or relationships based on the interaction, adhesion, and cohesion. It would allow PMOs to address in a constant change environment the challenges present in the projects as well as to significantly increase the maturity of the company.

Conclusions

In our opinion, the best recipe for facing the new challenges in project management is to provide the project managers with the right education and the right supporting tools and mechanisms, particularly from the human perspective (in terms on counselling and advice).

If that is complemented with a place/department that serves as the guardian of the fundamentals of the business and company culture (governance, competences, motivation, ethics, vision, and mission, etc.), including the provision of the right tools for making the work of the project managers more value added, then we will have the perfect match for assuring now and in the future the success of the projects and, by extension, the success of the company.

Besides, if we add the benefits in terms of growing the maturity of the company and its organizational setup, then we would be able to leave the company ready for facing the changing future in a robust and reliable way.

Summarizing, if we want to be driven by passion and enjoyment (instead of fear and anxiety) and promote strategic thinking, innovation, and holistic behaviors (instead of resistance to change and silo thinking habits), to combine successful management and leadership (instead of choosing between management or leadership), this is definitely the path to follow.

7.3 Excellence in Action: Norte Energia Belo Monte Hydroelectric Power Plant[3]

The Mission

Faced with certain challenges, it is natural even for an experienced project manager to hesitate. So, imagine hearing the offer to take over the implementation of the Belo Monte Hydroelectric Powerplant (HPP), where nine turbine-Generator join Units (GUs) remained to be implemented, within a period available for completion of only 18 months (Figure 7-2). I felt like that when I received this invitation in June 2018.

Even with the knowledge of HPPs implementation and the project itself I already had, the decision to say yes would have to be taken in good foundations. Even the fact I had been playing the role of director of Regulation and Commercialization (DRC) during all the year before at 'Norte Energia Sociedade Anônima' (NESA), a special purpose company responsible for the concession of Belo Monte HPP, and having been before member of the company's board of directors and technical committee since 2012, wasn't helping me.

On the contrary, this was making the expectation around my name even bigger, and there was not the slightest margin for mistakes.

The Belo Monte HPP is the fourth-largest HPP in the world. It has an installed capacity of 11,233MW, with 18 GUs in the main powerhouse with a nominal capacity of 611.11MW each, plus 6 GUs in the auxiliary powerhouse with 38.38MW each. Just for comparison, one single of these GUs has the capacity to generate almost the same amount of energy as the Angra I do, the Brazilian Nuclear Power Plant located in Angra dos Reis City, State of Rio de Janeiro, with 640MW.

This HPP has been under operation since January 2016, at a full capacity of its two reservoirs, main and secondary, its interconnection channel between the reservoirs, its 27 dikes and 6 dams, including the main dam and its spillway, the system of transposition of boats over the dam, the auxiliary powerhouse, the 238kV substation and the fish transposition system (these last six parts are located on the Pimental site, 50 kilometers far away of Belo Monte site) (see Figures 7-3 to 7-10).

In June 2018, there were only a few contractual pending issues for the commissioning of these structures and devices to be finalized, highlighting the safety conditions of dams and dikes.[4] The main powerhouse went in operation in February 2016 with one GU. In June 2018, eight GUs were under operation, and the ninth was being commissioned, with remaining the tenth to eighteenth GUs to be implemented. Although more than 80 percent of the US$12.8 billion total Capex for the project had already been carried out, only 49 percent of the nominal capacity of the HPP were available for operation.

Belo Monte HPP is a world-renowned project, particularly because of its political and socio-environmental controversies since is located in the heart of the Brazilian Amazon, on the Xingu River, and has aroused several feelings, due to indigenous concerns (Figure 7-4). However, few

3 Material in this section has been provided by Flávio Dutra Doehler, Former Director of Norte Energia SA. Flavio has more than 30 years of experience in the Hydroelectric Power Plants (HPP) and worked in Belo Monte from 2017 to 2020.

4 Following the relevant Brazilian Dam Safety Law, Nº 1334/10 under the new management of the DPI, the 33 Emergency Assistance Plans - PAEs were redone, as one of the requirements for the commissioning and reception by NESA of 33 dikes and dams, which until then were under the responsibility of the Main Contractor for Civil Works, being carried out until, this time by its control, monitoring and reporting, under inspection by The Electricity Government Agency, always in accordance with legal and regulatory precepts.

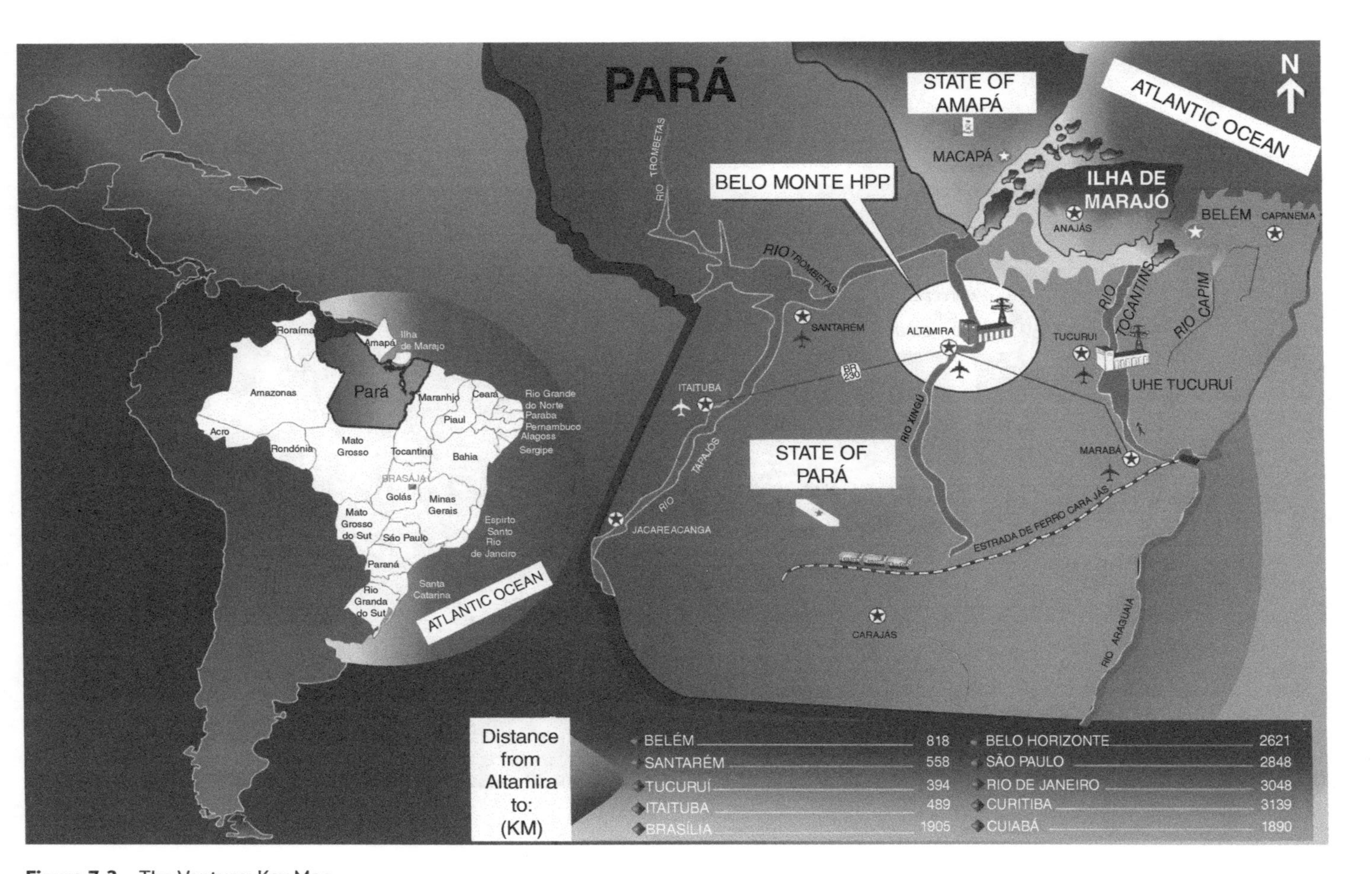

Figure 7-2 The Venture: Key Map

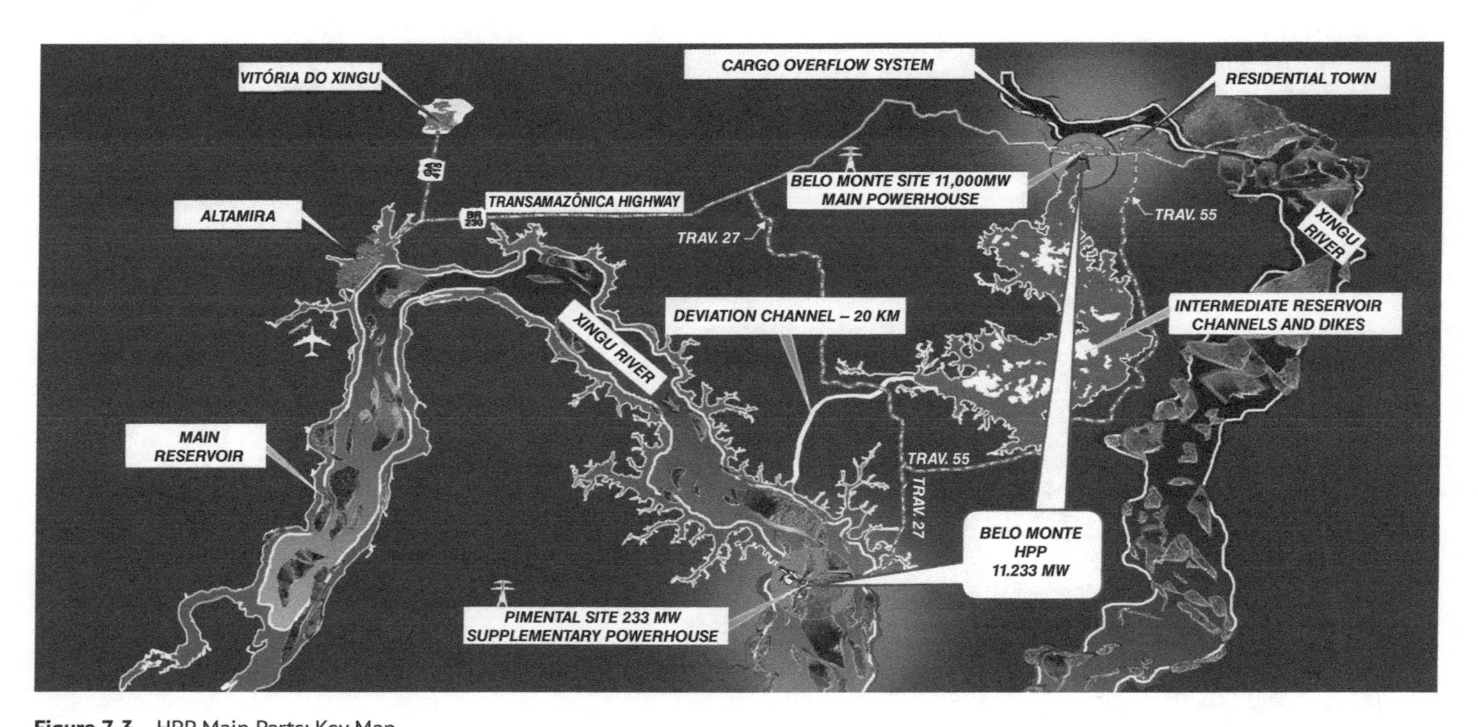

Figure 7-3 HPP Main Parts: Key Map

Figure 7-4 Pimental Site. Aerial partial view since downstream in May 2019 (from the left: the spillway, the auxiliary powerhouse, the fish transposition system, and the 230kV substation).

people have an idea of its local benefits thanks to the human development provided, reversing the condition of social and local biome deterioration that had been taking place for decades.

Even lesser is the recognition of the project as strategic, from the state's point of view, for providing human dignity and strengthening the country's sovereignty, not only by increasing the number of energy consumers, highlighting its importance for the Brazilian Electric Sector –(SEB)[5] for preserving the clean electric matrix, with less costly energy, but also because of the engineering and work capacity of Brazilians, with almost 100 percent of nationalization.

Returning to the point where I was reflecting on the invitation to take on the challenge, the time factor brought an extra weight. The answer had to be immediate. In general cases, the project manager responsibility in the implementation phases of HPPs is very high. In Belo Monte HPP case, due to its magnitude and the controversies around it spread around the world, it became much higher. However, the priority in my considerations at that moment was only trying to figure out in a very short period of time how to present objective solutions to deal with the challenge.

On the other hand, I knew that by taking over the implementation, as statutory holder of the Director of Production and Implementation (DPI)[6] I would directly assume the risk of liability for possible damages caused to the company and or to affected third parties and may be liable for them with my personal assets.[7] Thus, the subjective factor had to be restricted to the awareness of

---

5 SEB is the acronym of Setor Elétrico Brasileiro. Belo Monte has 6.6 percent of Brazil's installed capacity. Even though it does not have a regularization reservoir, as originally provided for in the predecessor project, which was called Kararô-Babaquara, operating on a "run-of-river", in the month of peak flow in the Xingu River, this percentage grows to up to 36 percent of energy consumed today in the country. Most of this energy is consumed in the Southeast and Northeast, by transporting energy through a transmission system composed of two TLs in direct current and several connections to the Northeast and Southeast transmission subsystems, enabling water storage in hydroelectric plants of these regions.

6 Who held at the same time the functions of the chief operating officer (COO) and of project manager (PM).

7 As established by the Brazilian Civil Code, Law 10,406/02, and Law of Corporations, 6,404/73.

Figure 7-5 Diversion Chanel: Aerial View, South to North

the opportunity to contribute so that all the investment made in the project up to that moment could return for shareholders and society at the planned results, through my direct participation in the implementation.

I had a weekend to analyze the contracts documents and progress reports to prepare myself for the decision. Although a more accurate diagnosis was required, as well as an assessment of the complementary resources that would be needed, I felt that there would not be another similar opportunity in my career, aware that I would not be embarking on an adventure, but facing my mission. As the deadline for deliveries required a productivity never achieved before, strong measures would have to be taken. However, the conditions for such measures to be received should first be established.

Thus, with updated information on the field implementation situation, which was restricted to records, as there was no way to gather verbal information with the team, therefore aggravating the existing management crisis, I listed the new main goals necessary for the implementation as well as the critical points to reach them and I introduced them to the CEO.

The only condition I set to accept the mission was that the entire company should be engaged in the same implementation objectives, having them as business commitments. After the necessary prior alignments with the executive board and board of directors, a meeting involving the members of these bodies was held with the entire DPI team.

I concisely presented my diagnosis of the situation, the main goals, and critical points that I had listed the weekend before. This meeting functioned more to launch a supposed new project, broadly affirming my commitment to management and staff, than as an inauguration of the new director. Therefore, each time it is mentioned hereinafter, it will be designated as the Kick-off Meeting.

In fact, remarkable changes had already been taking place at NESA for over a year. They were initiated from the new executive board and inauguration of its respective members, defined in May of the previous year, when that board became composed of four seats instead of the previous seven.

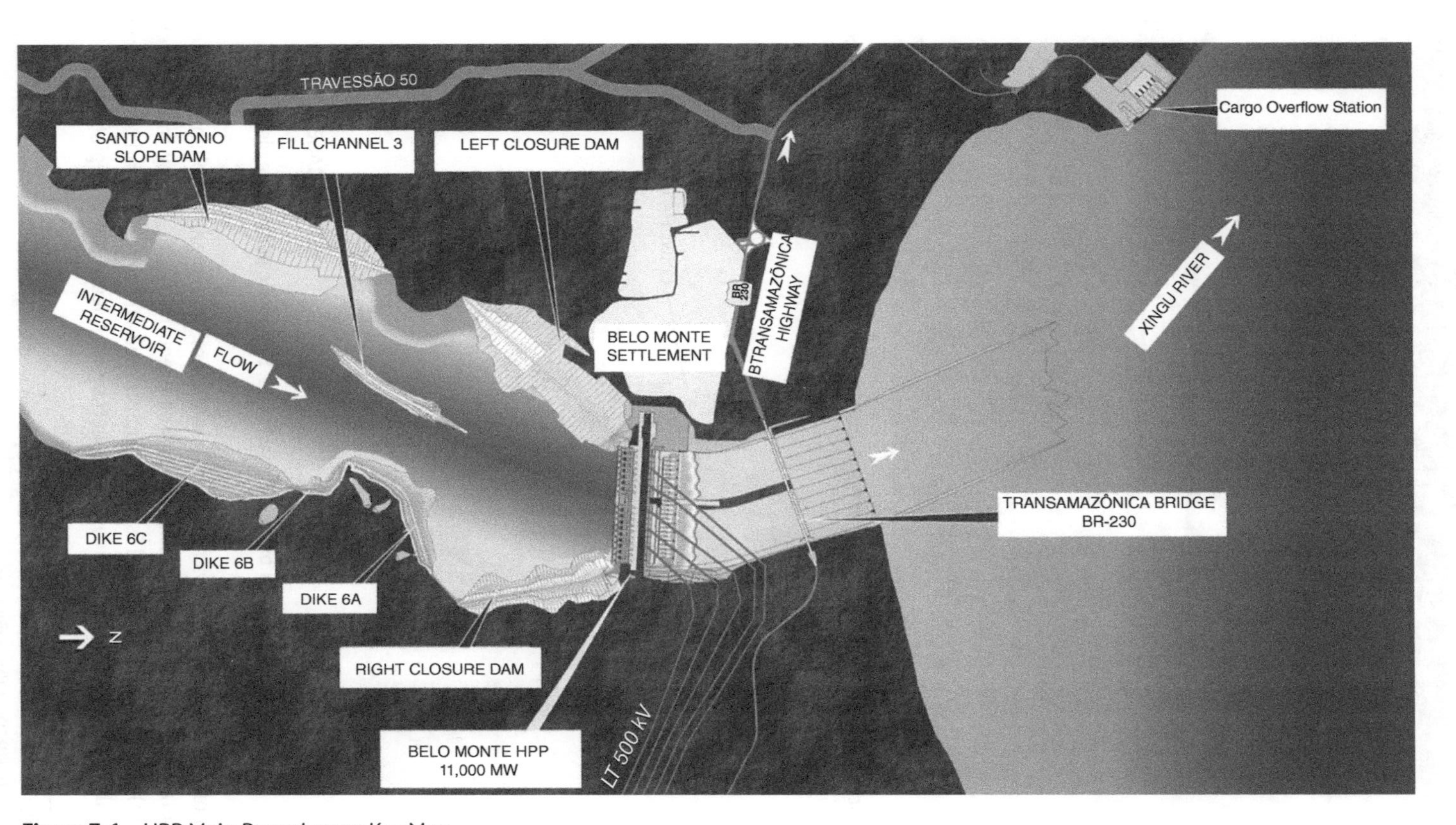

Figure 7-6 HPP Main Powerhouse: Key Map

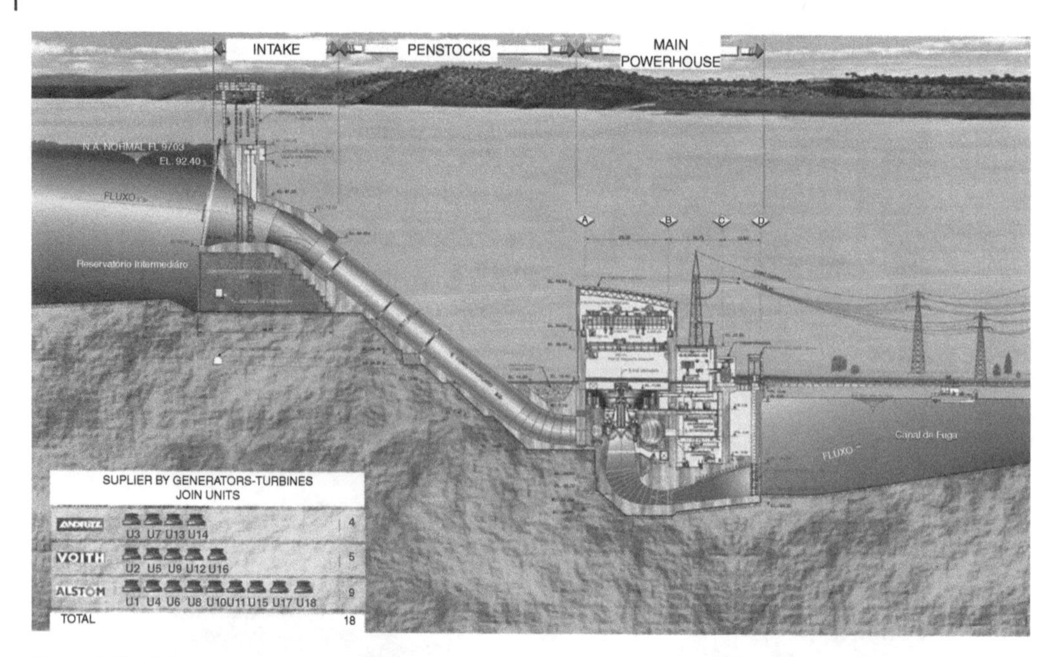

Figure 7-7 Schematic cross-section by one of Generator-Turbine Join Unit (GU): Main Powerhouse

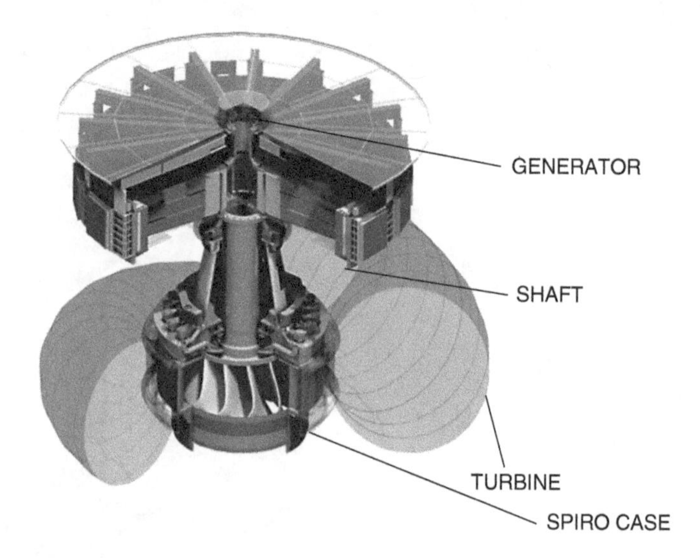

Figure 7-8 Parts of GU of Main Powerhouse: Schematic perspective view

A reform of the company's bylaws and the shareholders' agreement established the new structure, its operating conditions, and attributions. The executive board, led by the chief executive officer (CEO) was composed, and still is, of the mentioned DPI and DRC and the administrative and financial director (DAF)[8] also, working together in a lean and collegiate way.[9] The new

8 Who held the functions of chief financial officer (CFO), chief technology officer (CTO) and Chief human resources officer (CHRO), at the same time.

9 The DPI resulted from the merger of the former Civil Works and Supply and Assembly staffs, whose teams worked under different heads and now incorporating an additional function, to maintain and operate the HPP, to integrate one solely functional head. The presidency absorbed the former Social and Environmental Directorate (DS), and the Financial Directorate (DF) absorbed the old Management Directorate (DG).

directors had been hired through an evaluation and selection process among candidates from the market, under the coordination of an independent recruitment company, also selected by the board of directors among several well-known companies.[10]

The board's mission was to reverse the threats and impacts on the Belo Monte Business Plan, establishing and composing commitments, and means for their loyalty, above all to finalize the implementation, but also respecting future parameters, until the end of the concession.

Since mid-2017, important essential issues had already been solved, highlighting the financial equation of the project, based on the contributions of additional payment by the shareholders[11]; the settlement of outstanding financial issues with the Electric Energy Trading Chamber – CCEE, in the amount of approximately US$212 million, caused by the deficit in energy delivery in relation to the contracted amount, due to the hydrological shortage in 2017[12]; and the revision of the Implementation Schedule, of the Concession Contract, admitting one year of delay in the start-up of the last GU of the main powerhouse, without penalties.[13]

The solving of the schedule involved me in a particular way, as I was responsible for the DRC. In this condition, although the decisions of the board were collegiate, the burden of commitment to the implementation became even more potentiated, as the same agents assumed the objective responsibility to comply.

Identifying Critical Points

Thus, having established the mentioned important initial commitments, the concentration of efforts on the DPI should be even greater to ensure implementation within the terms and conditions recently agreed, justifying all the effort undertaken. However, the fact that the change of ownership of the DPI only took place at that time did not mean that over the last year there were no specific actions in this area.

---

10 I had the opportunity to be hired for this process, after leaving Companhia Energética de Minas Gerais – Cemig, where I worked for over 31 years, always as project manager and administrator of electricity generation and transmission projects.

11 The mismatch between projected financial sources and uses was mainly motivated by increases in direct and indirect costs due to new, unforeseen environmental constraints and restrictions, and judicial stoppages and social movements, but also by the impasse between shareholders regarding the definition of responsibility among them on the allocation of an average 914MW of energy to the free market, representing 20 percent of the energy produced by the HPP. The drop in the expected value of this energy in the market did not allow for the alternative of sale, which, as a result, provided for in the Financing Agreement with the National Bank for Economic Development - BNDES, reduced the degree of leverage and determined additional contributions of equity.

12 All energy traded in the interconnected electrical system (does not include plants apart of the National Grid, Sistema Integrado Nacional - SIN) is accounted for by the CCEE. In calculating the results, if a hydroelectric generator had a negative balance due to lack of rain in the basin in which it is located, the energy that was lacking is supplied by hydroelectric generators that had a positive balance in other basins, against payment of a tariff of optimization. This is called the Energy Reallocation Mechanism – MRE. Dispatches from the plants are made centrally and autonomously by the National System Operator. If, after this meeting of accounts, the hydroelectric generation of the entire system remains in deficit, resulting in a Generation Scaling Factor (GSF) index (fruit of the ratio between the amounts of generated and contracted energy) lower than 1, each electric energy provider from HPP must make up for its respective lack of energy, or replace it by purchase of energy in the market, or accept the liquidation by payment at the Settlement Price of Differences – PLD, stablished by National Grid Operator – ONS, plus charges, at the CCEE. In the case of Belo Monte, even after the company match the gap with the not contracted free energy to reduce the total deficit of US$515 (five hundred and fifteen) million, it was remaining close to US$212 (two hundred and twelve) million to pay.

13 Aneel Order number 4303 of 12/19/17 was issued after an oral defense by the DRC of NESA and higher "concatenate the schedule for the implementation of the Generating Units from 13 to 18 with data for the entry into commercial operation of the second bipole of the Xingu Substation ...", and the application of the penalty provided for in clause 10"... must be evaluated in the event of delay in the implementation of GU, ..."

Figure 7-9 Status of Main Powerhouse in April 2018: Aerial view from upstream

Figure 7-10 Status of Main Powerhouse: Inside view since right to left

The consolidation of the new functional structure itself was an innovation. The new structure resulted from the merger of teams from two previous departments, one dedicated to civil works and the other dedicated to the supply which now incorporated a new area for operation and maintenance, that had not existed until then.

The replacement of the electromechanical assembler consortium, hired for the main powerhouse (Figures 7-9 and 7-10), should be highlighted, due to its inability to fulfill the contract[14] The replacement of the assembler represented a special and difficult transition process, involving the entire management of the company, in a great effort, including the selection and hiring of the new assembler, COMGEV.[15]

The logic used was to seek to do "the most important first." In other words, a ranking of more sensitive points would give a clue as to what the priorities were. This logic was evident, since the sum of the value of only three contracts represented around 88 percent of the aforementioned Capex value.

---

14 The Belo Monte Assembler Consortium - CMBM, composed of Toyo Setal Empreendimentos Ltda, with 40 percent; Engevix Construções S.A., with 10 percent and Engevix Engenharia S.A., with 40 percent, leader of the consortium, had its contract for the assembly of the Main Powerhouse, dated April 2013, denounced by NESA in private trial because it had been irretrievably losing productivity.

15 The new Consortium COMGEV was composed by GE Energias Renováveis, 29 percent; by Enesa Engenharia Ltda, 42 percent and by Voith Hydro Ltda, 29 percent – this was the leader of the Consortium, which contract took over on January 18, 2018.

However, the transformation actions on assembly were defined, those related to civil works and the supply of electromechanical equipment should also be established. But that was not what happened. Only after a long elapse of time after the related negotiations to the assembly contractor, the Civil Works with the CCBM[16] and several pieces and parts of the Electromechanical supply with the ELM Consortium[17] had not arrived yet. In July 2018, the new assembly contract completed six months in force.

However, its performance so far was not sufficient to meet the recovery of the assembly schedule agreed with Aneel. In the follow-up and control of the implementation progress, the production and delay risk level indicators presented evolutions in the opposite direction to the expectation. In other words, the pace had been decreasing instead of increasing to reach the required goals. With this, there were already informal understandings between the NESA team and representatives of the new assembly consortium for adjustments to the schedule, foreseeing postponements[18].

There were also other issues, such as pending matters arising from the transition with the old assembler, contradictions between provisions that determined scope increases due to unforeseen events and merely referential surveys carried out during the proposal, but which should be confirmed by measurements throughout the contract, going in addition to the overall amount to be paid, among others.

Finally, strong interface issues between contracts directly affected their performance as well as contracts for civil works and electromechanical supplies. The contractual terms of these contracts deteriorated due to delays due to socio-environmental reasons and work stoppages by social movements, in addition to variations in scope. All these items were characterized by concurrent liability between contractors and the company.

The main contracts for civil works, assembly and supply functioned independently. In common, they had a commitment clause for promoting the integrated general schedule, which means the obligation for the contractor to have an active posture, to avoid problems that could restrict or impede the proper progress of the whole deployment, but they also had another clause that provided for excluding responsibility in case of issues arising by third parties, outside its own contracted scope.

The interfaces between the contracted scopes were not NESA's responsibility, but subject of its coordination. Since the financial problems faced by NESA, with the increasing deterioration of the contractual conditions, as mentioned above, the exclusions of liability of contractors had prevailed over activism in the obligation to do.

The DPI team consisted of fewer than a hundred professionals. Although considered a small number taking the size and complexity of the project into account, its dimensioning met the

---

16 The main civil works contract was firmed with the Construtor Belo Monte Consortium – CCBM, composed of Andrade Gutierrez Engenharia (leader), Construções e Comércio Camargo Correia, Construtora Norberto Odebrecht, Construtora OAS, Construtora Queiroz Galvão, Contern Construções e Comércio, Galvão Engenharia, Serveng – Civilsan Engenharia Associados de Engenharia, Cetenco Engenharia and J Malucelli Construtora de Obras, with the approximate amount of US$9.5 billion had a balance of only 2 percent remaining to be realized, however, only in the Main Powerhouse and its surroundings, about 200,000 m^3 of second-stage concrete finishing works were still to be carried out.

17 Consortium for the electromechanical equipment parts and systems supply for the Belo Monte HPP, comprising Alstom Energias Renováveis (which was incorporated into GE Energias Renováveis), Voith Hydro and Andritz Hydro with a contract worth approximately US$3.8 billion.

18 This finding was brought to the Management Team and the Board of Directors by a specific consultancy from Alvarez & Marsal, which evaluated the production and recovery capacity, revealing by projection that the delivery of the last GU would be delayed from eight months to a year, in terms of productivity and of risk from that time.

expected requirements thanks to the high degree of experience and specialization of its people and the high level of outsourcing of services and supplies, most hired by lump-sum contracts.[19]

The structure had 6 superintendencies and 15 managements. Two of them were not intended for implementation, the Superintendence of Operation and Maintenance and the Superintendence of Works in the Surrounding Area.[20] Two other superintendencies were designated as commercial, one dedicated to contractual measurements of civil works and the other to electromechanical contractual measurements.

Finally, there were two technical superintendencies that coordinated the implementation, respectively, of the civil works and the assembly and electromechanical supply. The scheduling and control of materials were subordinated to these superintendencies. In fact, this structure basically represented the sum of the existing superintendencies in the previous departments, the former Civil Works and the Supply and Assembly Directorate teams.

This same whole team, which were assembled from mid-2018, succeed until the beginning of 2016, enabling the start-up of the most parts of the HPP, as reported above, in spite of any delay caused by factors that were not under the control of the project management team, but impending social and environmental issues, as Aneel later accepted.

The small adjustments were made to optimize the resources, concerning more to bringing together and reducing the number of people who worked in the implementation than altering the essence or function of its members. Following the best practices in UHE implementations processes, here its focus was on the engineering knowledge of the professionals.

The priority was not to focus processes or project management, but HPP construction instead. The vertical management way was made more flexible only in special cases taking place *ad hoc,* after authorization and under the rigid leading of the bosses of each area.

In short, technical certification was more required than process or management certification. However, after the startup of the HPP under operation, marking the end of the major civil works, there was an inversion of roles in project management between the contractors and the NESA team.

There was a growing loss of pace of implementation caused mainly by NESA's financial problems, made worse with the need to recompose gaps of the delays already in place. The main issues to be answered were the declining assembling capacity, even within the new contract, and the lack of contractual coverage for new terms and additional scopes arising from civil works and supplies.[21]

These last needs were becoming hostage to specific assembly demands, which the contracted managers ignored, always claiming an exception and often demanding from NESA postures that should have been the initiative of the contractors themselves.

The NESA team then began to act more to remove impediments or restrictions than to actively manage the harmonic production and quality of processes and contracted deliveries. The instability caused by the merger of the Works Directorate and the Supply and Assembly teams into a

---

19 To get an idea the number of contracts existing in the DPI at the end of 2019 were 86, being third eight implementation contracts, 20 contracts for operation and maintenance scope and 28 consulting and advisory contracts.

20 At the same time that the replacement of the holder of the DPI took place, part of its team, about 20 people, related at Surroundings works, with construction of buildings and infrastructure for the affected municipalities as environmental compensation, leaved to be ruled by the CEO.

21 Due to delays, stoppages of works by social movements and changes in scope, as mentioned above, contractual adjustments through additive terms were necessary, not only in the new assembly contract, but also, simultaneously and interactively, in the contracts of electromechanical supply and civil works, as there were delays, losses from work stoppages by social movements and changes in scope, of concurrent responsibility between contractors and the company.

single Production and Implementation Directorate, soon followed by the replacement of the assembly consortium, only aggravated this condition.

The need to break with the past, which so many professionals from both NESA and contractors had been adopting, now assumed a kind of institutional condition. It didn't matter whether it was for those who played a leading role in the previous phase, and adopted this posture to protect themselves from responsibility of the liability of negative results, or for those who started to assume a relevant role in the new administration, who also went in the negative, disapproving the past line, exempting them, in the majority cases, from the obligation to recover and adopt a new conduct, focused on results – the attempt to deny the past had become a constant among the team.

Finally, denying the way the implementation process was carried out in the past constituted a strong argument to create the feeling of "scorched earth," justifying the replacement of past practices by entirely new ones, without the commitment to reconstitution.

There were also issues for the team to integrate as needed. As it was composed of professionals linked to civil engineering, coming from the works board, and of electrical and mechanical engineering professionals, coming from the supply and assembly board, the different cultures and attitudes, which often led to disputes and exchanges of reciprocal accusations for the interface issues that had occurred, became now untenable because they would certainly compromise the expected result. All this, plus the proximity of the end of the implementation, made noticeable the high lack of motivation of the team, inversely to the low lack of commitment to the practices for the results sought.

In addition to the related issues on the DPI team, the break in the continuity of processes, the loss of project references and the resulting threat to the implementation history, the assembly and electromechanical supply contracts had already been affected. In the first case, there were gaps in the availability of tools, materials, utensils, and equipment at the construction site alleged by the new assembler, comparing the list found in the sheds and yards with those that should have been left by the old assembler when it was demobilized.

This listing represented an amount that was paid in the previous contract and considered as a deduction in the new contract. Its availability would then be essential to the work of the new assembler. However, the joint inventory of these items that should be carried out by CMBM, COMGEVE, and NESA, as provided for the termination contract term for the old assembler demobilization, as well as to the new contractor inauguration, never had its report found. Now there was a claim by COMGEV for the immediate replacement of the missing items, or fixing the damaged ones, by NESA, so as not to jeopardize the progress of the assembly. That was since a clause excluding specific liability for these resources exempted the new assembly consortium from contractual sanctions for delays.

But the issues did not stop there.

The equipment and parts to be assembled, stored at the construction site, also had deficiencies, with lacs or physical deterioration of items that were listed as delivered in the packing slips. The recovery and replacement of these items could not await the conclusion of administrative or arbitration proceedings to be instituted.

The tight disposal of time for reach the key milestones required the immediate availability of components, in this case with more risks of jeopardizing the new schedule since its infancy. There were several critical issues, but, as an example, one of the most critical deserves to be highlighted. Two turbine shafts[22] had oxidation due to improper handling and storage that had affected the packaging causing damages its regions counterparts of the bearings and the brake track. Its

22 Parts with 3 meters in diameter, 9 in height and 112 tons.

due rectification by machining could mean the return to the factory, located more than three thousand kilometers away from the Belo Monte HPP, to be overcome by poor roads, fluvial and maritime transport, representing a great risk that could strongly threaten the delivery times.

The absence of the former assembler and the supplier's refusal to assume full responsibility for such deficiencies became an impasse. This because there was the legal responsibility of the CMBM, under the figure of the "trustee," who was contracted with the former assembler, responsible for all materials to be assembled since they arrive in the site.

In relationship with other issues intrinsic to electromechanical supply and civil works contracts, although at first not presented as an emergency issue, the postponement of their solution could bring repercussions from now on and growing in the interfaces. Not to mention that such untreated or mistreated issues could represent a "snowball," with major impacts still within the contractual terms, in addition to contractual liabilities even after the end of deliveries.

Thus, let us first deal with the Civil Works Contract. It had three scheduled dates for completion of parts of the works, compatible with the concession's original implementation schedule. The first one, 04/30/2018, marked the end of the operation and maintenance of the construction site and camp. The related services, which would be terminated, were still necessary after that date, as they focused on structures and products that served both NESA and its other contractors.

However, even after two months without contractual coverage, the CCBM continued to provide the services of electricity supply, meals, transport, water treatment and supply, sewage and garbage collection and treatment and property surveillance, always preceded by correspondence with reservations of that the costs after the contractual termination date for that item constituted NESA's debt. More than 3,800 people worked, and a good part of them lived in the construction site and camp in mid-2018. According to the updated schedule, the expectation was that the mobilization of resources for civil works would be maintained until the end of 2019.

The second milestone for the contractual completion was linked to the last concreting in the main powerhouse, or its surroundings, on 09/20/2018. The new schedule provided for the last concreting in January 2020.

The third contractual completion milestone established the closure of the activities of the residential village on 01/31/2019. This date did not need to be changed, as there would be time for NESA to arrange for the hiring of a company specialized in administration, security, and conservation of condominiums for the period in which the village was needed from then on, until its demobilization, scheduled for occur in mid-2020.

There were other allegations of contractual issues by the CCBM, such as the impediment, restrictions, and interference in the release of fronts by NESA and its other contractors, for second-stage civil works in the main powerhouse, causing delays that could not be imputed to him. Furthermore, there was a real competition for the powerhouse's own load-lifting devices, made available by NESA, causing immediate interfaces issues between CCBM and the new assembler, in addition to repercussions regarding its maintenance.

There was also a pending issue regarding the fulfillment of the contract, this time favorable to NESA, which was the compensation by CCBM to NESA of the differences between the quantities of construction steel and cement delivered by the contracting party to the contractor, and the amount of these materials actually used in structures, whose balance and settlement of accounts should be done after concreting is completed.

A last, but important, pending issue in the CCBM contract was the completion of the commissioning of dikes and dams, in accordance with the projects and standards, but also with the instrumentation and verifications of compliance with the proper dam safety procedures, in order to allow its final transfer to the responsibility of NESA. Until July 2018, these issues and demands had not been formally presented to the executive board and board of directors.

Regarding the supplies contract, the expectation seems not to present any major issue, since the implementation delays that occurred would result in greater margins of time for tasks and deliveries than those foreseen in the original schedule, benefiting the contractor. However, the ELM claimed the retention of physical deliveries, with increasing of storage and recharge costs and beyond representing more transport distances within the site. In addition, it also claimed fixed costs of extending mobilization for assembly supervision, including presenting values based on unit prices of supervisors applicable to the situation that occurred, as provided for in the contract proposal. The lack of supervision by the manufacturer was also an exclusion to guarantee the equipment after it started operating, indispensable not only for the safety of NESA, but also for financing protocols from the BNDES.

The shortages and deficiencies of parts and equipment stored at the construction site did not have a single cause. They also originated in the management of ELM contracts. Here, prepayment for factory-ready parts ended up losing the connection with controls for on-site delivery due to three factors.

First, shared responsibility. While the verification of the definitive tax documentation, which contained the list of materials delivered and their values, was the responsibility of DAF, the receipt of material for stock was responsibility of DFM. Communication failures between the two areas exacerbated the issue.

Second, the delay in implementation. Changes in the delivery schedule could no longer be adequately addressed, causing a gap in the connection between actual and formal deliveries.

And third, the end of the contract with the first automaker. Usually the parts that arrive at the construction site, delivered by the suppliers, are placed in the hands of the assembler for storage under strong legal guarantees to protect the contractor in the figure of the trustee.

This condition requires equal rigor in counterpart deliveries. The control issues in the delivery ended up exempting the former assembler from giving concrete answers regarding the completeness and integrity of the parts delivered when leaving the construction site.

If the above problems were not enough, there were risks of environmental accidents. In March 2018, at the time of the piracema,[23] during commissioning tests with water from GU No. 09, there was a death of fish and a consequent notification of a penalty by the Brazilian Environment Institute – Ibama, responsible for the licensing and environmental inspection of the project.

From then on, after several negotiations the NESA socio-environmental team with the institute, it determined the prohibition of GU matches without the protection of anti-fish devices.[24]

23 Piracema is the time for fish migration in Brazil, which go up rivers to spawn. Fishing or killing fish during this period can be considered an environmental crime. In the Amazon, it is legally defined between the 15th of November and the 30th of March of the following year. As the flow variations in the Xingu River are very high, the best period for tests, when there are outflows for the referred tests, largely coincides with Piracema. Although Belo Monte is equipped with a fish transposition system - STP, this does not prevent the risk of fish death during operation maneuvers of the main powerhouse of the HPP. The STP is located on the Pimental site, therefore above the stretch with waterfalls known as "Volta Grande do Xingu", something like "the great loop of Xingu River," with more than 50 kilometers long and about 90 meters on the top head above the level of the tailrace water of the main powerhouse, oriented against the water flow. Thus, at the confluence of the main powerhouse tailrace with the river, the fish tend to go in the direction of the larger flow of water coming from the main powerhouse, and only stop when the water velocity is greater than its strength to swim. They cannot enter the draft tube when the turbines are operating, but rather at stopping, when the speed drops, exposed to the risk of impacts when the turbines start again.

24 "Instituto Brasileiro do Meio Ambiente – Ibama" established this condition by letter that had force of changing the environmental licensing, without giving up other procedures for the operation to drive away fish, such as the use of sonar to identify shoals near the GUs, maneuvers with alternation in the operation of contiguous GUs and even from boats that made noise and additional movement in the waters, in order to keep the fish away to a safe distance that would allow new GU departures.

After approval by the board of directors, sets of grids were contracted for four interchangeable GUs, which should be installed in time to allow the commissioning of seven of the nine GUs that still remained to be implemented, considering that the other two of the missing GUs would be commissioned outside the 'piracema' season.[25]

The Main Transformations

Once the challenge was accepted, before adopting changes, it was necessary to analyze what had already happened in Belo Monte and its consequences. Over the course of more than a year, it was understandable, despite the classic recommendation of not promoting agent changes during the implementation of projects, why in this case it became inevitable. Thus, decisions to manage the crisis that could result from the restructuring of the executive board and the change of the assembler were adopted.

They may occur from the definition of the executive board and the replacement of the assembler that were adopted. However, the risk of delays and extra costs since 2018 had been increasing, which could bring a negative expectation of new more radical changes coming with the replacement of the DPI director. In fact, there were neither alternative resources nor time for significant changes, as the reversal of results needed to be immediate.

The only focus the leadership had was to make the structure already changed to work. The resources would be almost the same. Thus, the first step must be calibrating expectations – everyone in the same ship. Everyone would have an important role to play unless they did not align themselves with the emergency the condition imposed and not properly with leadership.

Therefore, the best metaphor to represent the situation we found ourselves in would be that of a plane crashing because of a breakdown in some devices and that could not land without prior repairs. One turbine had been replaced; mechanics were making adjustments to return the required performance, but there were still several repairs to be done, including on the others existing turbines.

In between, the commander is changed. The focus needs to be on reconciling outstanding repairs with the operation of the plane. Not only must the fall be reversed but the speed and safety of the flight must be increased. Naturally, the plane here meant the implementation process and the turbines meant contractors.

Considering that the DPI team had the necessary requirements, experienced and specialized, it should not be the object of the required transformation, but its agent. There was no place for reengineering, in the sense that this word is accepted in the corporate world, but for the rescue of good engineering. Anyway, it would not happen spontaneously.

Bad practices and their consequences needed to be removed to align and make key competencies of the team as a whole work. The expression "don't throw the baby out with the bath water" was a good example of what needed to be done. Once the objectives and conducting lines were defined, so that the consequent transformations could come by the hands of the

25 As an alternative to the above prohibition, GUs with grids could be operated, after the proof of their effectiveness. As they were not foreseen in the original power plant project, they would work fitted into the same slots used for the stop-log gates, at the exit structures of the enclosures that house the turbines, known as draft tube. So, they needed to be dismountable to allow the use of the stoplogs floodgates during drying maintenance time and have to be equipped with an automation device that make them able go down immediately if any stoppage in the operation of the GUs occurred, being suspended again, gradually opening the draft tube outlet, as the UG had its operation restarted, slowly, in order to provide a flow of water that would expel the fish, without hurting them.

team itself, feeding back the integration and alignment with the results, there was no time or space for trial and error. Without consolidating the definition phase, even the necessary demand for results, though based on good management and engineering practices, could be received with suspicion.

The fact is that, as the end of the implementation of the HPP approaching, most professionals, who brought their families to literally live around the HPP construction, started to feel not only the threat of the end of their job contracts, but also an uncertain change in the lives of their families.

Leadership cannot deliver the message of a "reverse reward." While it asks for engagement, with involvement and emotion, and cannot have as the counterpart old management recipes, with protocols and coldness. A new commitment to the team needed to be made. There must be convergent aiming the same results. The common focus translates into results for the company and professional assets resulting from the successful implementation.

The solution for this challenge should be expansion of the employment contract, even after the last delivery and preparation for a job transition, in addition to the classic variable remuneration based on results.

Of course, demands will occur, but in an environment structured for objectivity, with reciprocity and without polarization. Changes between leaders and professionals, of all levels, alternating positions based on objectives that are continually renewed, with clarity and method.

With this, the continuous renewal of reciprocal commitments between the leadership and the team should be cultivated in constant means of communication and integration, with care so that the relationship is intense throughout the entire implementation process.

A "transformation by successive self-organization" will take place. This idea is based on the principle that large potential differences are found in not organized environments and bring proportional gains. As in a closed energy system, it is enough to align the appropriate means of transformation to start it. So, from an activation energy, the process will occur, in a continuous way, with intermediate results functioning not only as a platform for new achievements, but incorporating themselves into the means of transformation.

In the case of Belo Monte, after the kick-off meeting, intense and continuous work to develop the so-called "team spirit" began. The line of conduct adopted was not new, as it is recommended in several personnel management manuals. However, its practical implementation required a nonexhaustive, persevering integration work, with a well-defined focus on deliverables.

In management meetings, the expected behavior of the team was never declared, but always the respect and consideration of the company to the team. Interactions programmed for daily life, in routines or in episodes, strengthened the procedures necessary for implementation.

Specific changes in the functional structure, adapting designations, exchanging positions were communicated and justified, with care, involving those directly interested, but also the entire team. They started to be well accepted and even required and incorporated as being the initiative of the professionals themselves. Tools and expedients started to be structured for the adequate monitoring, communication, interaction, integration, and participation of all. Although it is not the case to detail them here, some examples deserve the report below.

On the production line, there were impasses of responsibility that impeded attitudes of both the contractors and the DPI team. The reported passive posture of the contractors, in view of the deterioration of the contract interface problems and the professionals' rupture with the past, needed to be reversed.

Surveys and investigations as to the facts and responsibilities of the problems mentioned with items to be used in the assembly and components to be assembled had been started, but not aiming to meet deliveries, but the recovery, at any future time, of the damage incurred.

In order to constitute an effective procedure, in a "buffer" way, the recommendation of the "lessons learned booklet," in a similar situation would be to elect a permanent advisor who had participated intensively in the contracts management and in inventory control. He had to be chosen from among the team managers, providing reliable information and proposing actions to make up for absences in time, without prejudice to the continuity of commitments made in the past. However, this decision would pose the risk of increasing the lack of team integration. So, an "out of the box" decision was needed, which at the same time fulfilled the requirement of recomposing the process of supplying and integrating the team.

At that time, the former director of DFM was serving as an advisor to the CEO for surrounding socio-environmental works. I could see that he had the respect of most of the team. His coming to the role of DPI advisor would have a much broader scope than the rescue of contractual conditions and to break the negative to the past, already reported.

Since the kick-off meeting he was already present. Hence constituting him as an advisor occurred naturally, more because of his actions, to be incorporated into the process of resuming the implementation, and because of recognition of the team, than because of formalities, which became consequential[26].

DPI's organizational restructuring could not take place differently. Since the kick-off, what was being conceived was a functional base to promote the integration between civil and electromechanical engineering, and between the DPI team and the contractors, focused more on processes than on activities. So, the assumptions for its effectiveness would be to act in a matrix and multidisciplinary character, allowing its components to play the role of contractual managers, of owner's engineering inspectors and act as arbitrators, in cases of differences in positions or deadlocks between the contracted on account of interfaces.

Three superintendencies would suffice. The existing Operation and Maintenance Superintendence would be maintained to take care of and answer for all structures, devices and generating units in operation, in addition to the connections by networks and electrical protection systems of the Belo Monte HPP. It had an important interface with the other superintendencies, as it would also continue to be responsible for actively commissioning the GUs and devices of the main powerhouse.

The Contract Planning and Management Superintendence would coordinate the measurements and release of contractual payments, both for civil works contracts and for electromechanical contracts, which in the previous format were responsibility of the two respective commercial superintendencies. This superintendence also by the team that took care of programming in the technical areas of civil and electromechanical engineering.

The focus for planning would be obtained from the integration of scheduled checks, including interfaces, which would continuously feedback, through monitoring and control, the checks as to what had been planned. And the Implementation Superintendence directly monitored the

---

26 Curiously, at the time I appointed advisor I received some advice to reverse the situation, as I would be constituting a "parallel chief" that would soon kick me out of the process. With the phrase attributed to Peter Schulz, former CEO of Porsche in mind, "Hire character, train skills," I had no doubts. In that case, the skills to be trained between the former director and myself were in the realm of relationship. Regardless of the differences that occurred, the immediate and final results in the project spoke for themselves.

implementation in the field, through inspection. The logic employed to compose the existing functional structure was conservative and classic[27].

To support the intense interaction proposed in the development of the Structure aforementioned, the process of monitoring and controlling the implementation deserved to be reinforced with the adoption of its own project management tools and procedures. The Planning and Contract Management Superintendence could not fail to detain its coordination. However, as there was no way to train or incorporate into the existing team professionals trained in project management tools, to complement the function of this Superintendence, which, in short, would operate a virtual panel, a consultancy specialized in project management was immediately hired, which would act in an interactive and continuous way with the entire NESA team.[28]

In view of the high degree of outsourcing and the large contractual responsibilities involved in the implementation process, a law firm specialized in contract management was also simultaneously hired, aiming at the proper formalization of communication between NESA and the contractors COMGEV, CCBM, and ELM.[29]

To maintain continuous interaction between the DPI team and contractors, several weekly meetings were refocused on outputs and inputs, activity interdependencies, and intermediate and final delivery targets. In the weekly agenda, new specific field meetings were added.

One such modality gained prominence, the Daily Monitoring Meeting – RAD. The acronym became synonymous with the official daily reference, guiding the entire team to solve problems or interferences with priority, daily goals, becoming a symbol of the new moment of the enterprise. For half an hour, at the beginning of the day, the entire team involved with the work fronts would meet, standing in the lobby of the main powerhouse, to discuss what had happened the day before and what would merit adjustments, who would be responsible for them, in addition to verifying the fulfillment of contractual routines, such as records of works journals and files for nonconformities.

The schedule became interactive and per day, with the combination of how the previous measures would be taken, as well as the resources for course corrections that were shown to be necessary.

Management meetings with the entire board took place on a monthly basis, in them I have always recognized the CEO's response to the request made to him that the entire company should be committed to the implementation, as the main problems with other areas were exposed and resolved openly and immediately.

There were also monthly meetings on site between all the directors of the three main consortia of assembly, civil works, and supply, led by the NESA's CEO, which we called "P to P Meetings." Accountability charges were common. Nevertheless, alignments and issue referrals that had not been exhausted at the level of the DPI were no longer subjective and became corporate.

In parallel with the definition of the new structure and functions, including the one that would supply the role of the Project Office, the project communication and integration system was

27 Among the remaining five superintendencies, two of their incumbents were made specialists advisers, and another one resigned. So, to take the responsibility for the new Superintendence of Planning and Contract Management, a selection process was carried out based on a triple list among three managers from the former commercial and technical superintendencies of civil works.

28 Contract with Alvarez & Marsal to analyze, quantify and qualify the evolution of implementation and risk exposure, expanding the existing tools, also allowing an ex-ante view.

29 Contract with Demarest Lawyers to advise on claims and counterclaims and organize a database and letters exchanged with the CCBM relating to this matter, as well as the analysis of new letters and participation in the contract to manage the supply and assembly contracts.

established, changes in procedures and postures, aiming at compliance with contractual rules, technical standards and better UHE implementation and project management practices needed to be established. They literally started the day after the kick-off, when I had the opportunity to reverse three noncompliant processes.

In the first of them, the Technical Superintendent of Supply and Assembly sought out a completed protocol form, called purchase order, asking me to urgently approve it, in a considerable amount, as it was slightly below the maximum limit of competence from the principal. Budget relocation should be done, as there was no forecast for that contract. It involved the rental of a special trailer, with 23 axles and a capacity of 250 ton, and a crane with a capacity of 150 ton. The trailer was intended to transport one of the four turbines that had been stored outside the main powerhouse, making an exception to a condition of the supply contract that determined the discharge in the assembly area, inside the powerhouse.

The exception to the contractual condition was admitted by NESA. The crane would solve a critical assembly point, sometimes serving the assembler, sometimes the contractor for civil works, which would resolve interface disputes that would also be NESA's responsibility.

With regard to the transportation of the turbine, in a quick analysis of the respective contracts, I understood that the exceptional delivery alleged by the supplier was not characterized. I also noticed that the rent price was above the market, which would be justified by the fact that the equipment was already deployed in the work, given the alleged urgency. In addition to not authorizing the proposed contract, I notified the supplier to transport all four turbines to the powerhouse at their own initiative and cost, as the contract clearly provided for their obligation as to the discharge location. If it did not do so within 15 days, NESA would contract and discount and gloss an amount equivalent to this service, in invoices to be paid in the future, using its contractual prerogative.[30]

In another contractual analysis regarding the apparently noble attitude of NESA in contracting the crane, in the list of equipment included in the assembler's proposal, a crane with a capacity equivalent to that intended was stated as being its responsibility, but it had not been mobilized.

In a meeting with the assembler, I registered the determination for the aforementioned crane to be mobilized and glossed over the amount equivalent to the months in which it should already be on the job. It is important to emphasize the symbolism of these attitudes, encouraging the team to act actively, merely using contractual expedients. In face of his amazement at my unfriendly attitude, I never tired of explaining the decision-making sequence. Indispensable activities, even if apparently not contained in the contracts, should first be requested to the contractors, seeking a non-onerous negotiation, before being compulsorily interpreted as an obligation of NESA.

This change in posture alone would prevent the perpetuation of undue attribution to NESA, taking the initiative as to the missing and necessary resources, which distorted its purpose, serving as a gateway to a series of allegations of other exclusions.[31]

On the third day I was in my new duties, it was the turn of one of the managers of the same Superintendence from the previous episode to ask me to approve the shipment of two special welding machines, used to join the ends of the rotor bars of the generator and the stator, for maintenance in São Paulo, under allegation of risk of irreparable damage to the implementation schedule.

---

30 Several procedures of this nature were adopted, supported by the opinions of the lawyers hired for contract management.

31 Fulfilling another principle of integration between the team and the company, the Superintendent's insistence on the anti-contractual procedures that had been adopted resulted in his removal.

After the same usual questions, I found that, although there was no budget forecast for such activity, it would be recurring and was provided for in the contract. However, I did not agree that this activity should be NESA. In addition to diverting the time to be dedicated to contract management, it would continue as a "safe-conduct" for any attitude of the assembler that represented a loss of productivity but depended on the availability of tools by NESA.

I called the responsible COMGEV director and informed him that from that date on, any recovery or replacement of a tool or support device would be responsibility of the assembler consortium and that we would negotiate the commercial conditions afterward. After some meetings with the assembler, the procedure was formalized, becoming one of the items in a meeting of accounts between the parties, according to understandings that would result in the future addendum.

A third episode had an even greater symbology. Even before I arrived at work, I had received an email from the director of the civil works consortium congratulating me for the new functions and asking me to talk about a proposal for the meeting minutes that would regularize activities that were without contractual coverage, by account of the end of contractual terms, as reported above.

I remember that I met, by chance, with the director in the office of the Civil Works Technician Superintendent and I informed him that such meeting minutes would not be signed, as long as some delivery commitments were not assumed by CCBM at a sufficient level to lead to the approval of the board of directors of NESA.

Although his immediate reaction was not friendly, the attitude opened space for several negotiations that would allow the continuity of services without contractual coverage, without this becoming a liability, in addition to a greater balance of forces with NESA so that the respective contractual addendum, necessary for the completion of the civil works, could be negotiated and approved.

In addition to these isolated episodes, which became symbolic, the implementation management routine itself would bring opportunities for the necessary adjustments, with additional procedures agreed between the parties and validated with the company's senior management in real time.

We needed to be careful that the positions adopted did not signal, on the one hand, rigor in the face of unforeseen events and lack of contractual coverage, as an acceptance of the perpetuation of the inertia of the contractors, nor, on the other, a tacit renegotiation, whose changes without the coverage of a competent approval would bring indefinite additional costs in the future, in relation to the three priority contracts. In addition, managing the interfaces of these contracts required that all instruments be revisited simultaneously and continuously, with the involvement of management and the necessary stakeholders.

In individual contact relationships, the focus should always be on ways to reconstitute the original references, in terms of principles, scopes, and procedures. In the limit of the insufficiency of the contractual scope, rules for a future recomposition would be validated in higher instances.[32]

---

32 This type of procedure was inspired by the contractual composition mechanism for solving small disputes so as not to impede the progress of its execution, as is the case of the Dispute Adjudication Board (DRB), adopted by the International Chamber of Commerce, mentioned in the text of the civil works contract with CCBM, as arbitration agreement, or arbitration clause. However, it was a kind of compositional expedient even before the controversy was established, aiming at an adequate referral for higher validation, with the same purpose that it would not constitute an impasse, not even threatening the performance of the role of each party.

At the time I took over the DPI, it was time for measurements. Thus, the bulletins of the services performed by the assembler reached my hands for approval for payment. I had the opportunity to understand why financial and physical evolution diverged from each other over time. A unanimous explanation from the DPI team was that the high contractual down-payment amount, of 20 percent of the total contract, had taken the contractor's spirits off in the pursuit of production.

In fact, large portions of advances do not have a correspondent beneficial effect on contractual management. However, the behavior of the financial curve should follow the reduction of the physical evolution curve, but not the distance between them.[33]

New checks on contractual documents, and with simulations from March letters from the then director of DPI communicating adjustments to the contractual schedule and milestones schedule, to accommodate problems of lack of release of assembly fronts in the powerhouse, which would be the responsibility of NESA, showed that such letters represented new advances, of contractual payments, which would be the responsibility of the board of directors. The logic employed was to exchange assembly events at their final location for pre-assembly events in the yard, allowing the flow of payments to be maintained.[34]

After the main cause of low productivity has been clarified, I communicate the CEO and the cancellation of the revised milestones schedule formalized that with COMGEV and its consequences and, more suspending payments for events that were not compatible with the original contracted milestones schedule. The formal procedures required also notifying the contractor to maintain the contractual goals and schedules. The retained value represented tens of millions of US dollars. As expected, the assembler made a large mobilization to release the payment, which created an opportunity to seek a joint solution for releasing work fronts and other resources provided for the resumption of production.

On the other hand, the weights of payment events would be redefined, giving more value to this critical path and those that were essential prerequisites for the beginning of other critical activities. As a result, the original premises of the contract were resumed with the recovery by NESA of negotiating power, allowing the parties to jointly seek solutions beyond the original contracted scope.

The way found so that the controversies did not prevent the essential increase in production for the constitution of a kind of informal checking account, on the balance sheet for future confrontation of accounts between credit and debit items,[35] rules were immediately agreed between the

33 This point also did not escape the controversies between the parties. COMGEV claimed that the Service Start Order – OIS, which opened the deadline for the assembly schedule, had been delayed by almost a month, while the NESA team claimed that such delay was caused by COMGEV, due to the lack of presentation of the contractual guarantees.

34 In the so-called "Revisão 1 do Eventograma" document, milestones related to final events had been suppressed, such as demobilization and entry into commercial operation of GUs, and assembly delivery milestones "in the well" had reduced 187 percent, while milestones related to pre-assembly had been increased by 56 percent.

35 The credits column considered the values of the items removed from the contracted scope and that would deserve a reduction in the overall value for the transport of a test hub and a service bridge that would no longer come from the Tucuruí HPP, assembly of metal parts embedded in the concrete of second stage around the spiral box of the GUs, which would now be contracted with the civil works contractor itself, avoiding important interface restrictions. In the column of NESA debts, the amounts for obtaining or recovering missing or defective items in the funds transferred by CMBM, and for additional services and supplies not provided for in the original contractual scope, which are now considered indispensable, after prior understandings between the parties.

parties, procedures that provided the elements for a future formalization of the contractual addendum, with a Final Schedule agreement.[36]

The principles and main contractual rules were thus preserved, without preventing small adjustments due to unforeseen circumstances or contractual deficiencies being immediately resolved, allowing the contracting party to have information that all the effort, beyond what was foreseen, would be reimbursed. The executive board, the technical committee and the board of directors were gradually informed of their evolution, until the signing of the minutes note, which started to function as an integral part of the contractual documents. Thus, all the procedures adopted had the support of a competent approval body.

In the case of the civil works of the main powerhouse (more specifically second-stage concrete, concerning the remaining structural concrete and missing roof, NESA's interest was only its completion, and it did not matter whether it would be carried out by CCBM or another contractor.[37] Thus, three alternatives were analyzed (i) to maintain CCBM as contractor for the completion of the works, by means of an addendum (with no increase in overall costs); (ii) terminate the CCBM contract at that moment on and hire a third party to carry out the other works; or (iii) terminate the CCBM contract and NESA itself carry out the other works.

After several studies carried out, including with participation of independent consultants, it pointed out that the first alternative was the best. The big difference in costs between this alternative and the other two was found, in addition the risks of a different choice could be enormous, due to the existing contractual links, and delays have already affected the balance of the existing contract.[38] Even so, in this uncertain environment, several interaction meetings with the CCBM were necessary before the definitive renegotiation of the new Final Schedule with COMGEV, on 13/09/18.

Concerning to most construction site and encampment services, NESA would need to take over some of them, even retroactively, whatever the alternative adopted for the completion of the civil works, since the CCBM's responsibility for them had ended in May 2018. So, DPI assumed the direct payment to third parties of 80 percent of the total value of these services, or the reimbursement to CCBM of the equivalent that it had already paid.[39] The other 20 percent would be CCBM's

36 Step by step after understandings, records in meeting minutes were established, and validated in committees and higher instances, in order not to hinder the resumption of the pace necessary for assembly, resulting in a final meeting minutes, on September 13, 2018, in which a new schedule was established, compatible with what was agreed with the board of directors, a new milestones schedule, charging more heavy parcels in the final assembling parties and only the intermediaries ones whose delivery released critical work fronts, as was the case of the penstocks. Beyond that there were previous procedures just in case of necessity of services out of scope, working as a referral for the execution of a future amendment with the agreed conditions.

37 In July 2018, the need to resume civil works was brought to the NESA board of directors for the first time, aiming at the Final Schedule by the new holder of the DPI. The representatives of Eletrobrás, Eletronorte and Chesf, held 49.97 percent of the votes, stated on that occasion that they would not deliberate on the proposal for an amendment with the CCBM, as a judicial investigation was underway on account of criminal complaints against some consortium members, over-invoicing intended to illicit payments to public agents, having been the subject of a decision of the Federal Court of Accounts, which, after a detailed report, designated the contracted price items that would be under suspicion, determining that those shareholders would take appropriate measures for the corresponding reimbursement.

38 According to one of the lawyers' opinions, given the fact that the contract is "Contract at Global Price," the mere expiration of the term would not eliminate the reciprocal obligation. The Contract would only be exhausted with the effective exhaustion of its object.

39 Through this agreement, the DPI committed to pay by reimbursement in the future, against the presentation of debit notes, as was the consumption of electricity until March/2020, and surveillance until October/2018; collection, treatment, and disposal of solid waste until October/18 and purchase of chemical products for water and sewage treatment until September/2018.

own consumption linked to the activities contained in the global contract value, regardless of the date of completion of the operation and maintenance of the worksite and camp.

The understanding was that until there was a contractual breach, the production of concrete or construction of the powerhouse roof remained under the responsibility of CCBM, regardless of the amount of maintenance for the site and camp.

Once the CCBM proposal for the amendment to be signed with NESA was received, the amounts corresponding to the compensation for the performance of services contained in the operation of the construction site and camp, together with the remaining items of concrete and the powerhouse roof were distributed in payment events compatible with the Final Schedule as clauses of the new amendment to be signed.

The final value of the contract would not change, and the sum of the items due to fixed costs, and the extension of terms, as described above, was met by the reallocation of the balance of Worksheet P7 – "Support Services" of the contract, containing items of unit prices that would no longer be necessary for the project. NESA proposed, as a condition for its signature, the broad, flat, general, irrevocable, and total reciprocal discharge of all obligations performed up to the date of signature of the present, except with some exclusions which had been the subject of controversy between the parties to be discussed in the future, so as not to hinder the resumption of the works.[40]

The amendment to the civil works contract was signed only in December 2018. However, due to the form of management described above, informal measures were gradually adopted, immediately as needed, out of strict trust between the parties, removing interfacing impediments, using the margins and mechanisms of the contract in force.

The strategy's recomposition for the ELM contract was different. First, two components of this consortium were also members of the assembly consortium. So, as soon as the conditions to meet the Final Schedule with COMGEV were defined, this same schedule was already informally internalized in ELM. Second, the main problem to be faced did not depend on a contractual addendum, but on investigations regarding the critical problem of the lack of parts and components to be assembled.

As the search for missing components and parts was sought, the concern increased. For the initial assemblies of the COMGEV contract (9th to 12th GUs), "loans" of equal parts, to be used in future GUs, were admitted, with great caution, as this was an exceptionality, just to save time with the surveys and investigations. Specialized consultants were hired for new inventories and verification of materials control processes.[41] The solution found was the physical separation of all components and parts for each UG.

At the end of a work of more than six months, the missing items were listed in classification by supplier and their replacement was agreed in the minutes of a meeting between the parties that established the future calculation of responsibilities. If the responsibility were confirmed as belonging to ELM's consortium members,[42] there would be disallowances of future payments or use of retentions in guarantee for reimbursements in the corresponding amount.[43] If they were characterized as belonging to NESA (understood that this included the previous assembler consortium), the suppliers would be exempt from liability.

40 The sum of this reallocation corresponded to approximately US$42 million.

41 The investigations ended up resulting in the dismissal of a manager at NESA because he continued to defend non-contractual positions that prevented the problem from being properly addressed.

42 GE, Voith or Andritz.

43 Replacement would cost around US$1.84 million.

Several of these issues, which were not limited to the competence of the DPI, were brought up to the 'P to P' Meeting. At the end of the assembly, there were no parts or components missing, preserving the spare parts provided for in contracts.

The addendum with the ELM was the last one to be signed. This delay was a way for NESA to counteract the risks of lack of components that should be managed by the suppliers to complete the assembly parts. Thus, when negotiations were made for the amendment to the ELM contract, mainly aiming to adjust the contractual term to the Final Schedule, the main interest of the contractor was the provision of assembly supervisors for the period additional to the original, as a condition for the extension of guarantees from GUs 13 to 18. The conditions for replacement of missing items had already been agreed in the minutes, whose content was incorporated into the final addendum.[44]

Regarding the anti-fishing grids, in December 2018 the first four sets were already operating in tests, foreseeing an interchangeable use, when there was a stoppage of some GUs already in operation, due to interruption of the energy flow in the north-south transmission system of the SIN, resulting in new fish kills.

As a result, Ibama notified NESA in the following January, at the peak of flows that would allow greater energy production by the HPP, prohibiting it from carrying out departures and GUs that were not equipped with anti-fishing grids. This restriction occurred at a critical moment for the SIN, coinciding with the shutdown of UTN Angra II for maintenance and with restrictions on energy dispatch in the Rondônia-São Paulo system. Thus, because the demand on the SIN was not unmet at that time, new and intense negotiations with Ibama were necessary at the request of the ONS.

Temporarily, even without enough anti-fish grids, paralyzed GUs startups were allowed, since NESA promised to implement sets of anti-fish grids before commissioning tests related to each UG.[45]

Due to the great additional effort of the DPI team, all sets were contracted with a single company, and implemented on time, in a manner compatible with the Final Schedule.

Innovations

Once the management procedures were established as described above, the demands for solutions to recover deadlines started to occupy most of the agenda at the various ordinary and extraordinary meetings between the parties. The first of these was the development of a structure with a rotating mechanism to machine the two axes that had been oxidized.

The functioning would be reversed. Instead of the axis turning around itself, the cutting tools would rotate around the axis, parked vertically, promoting the machining. After understandings between ELM and NESA, considering the concurrent fault between the former assembler

---

44 The agreement recorded in the meeting minutes was that (i) a joint investigation is carried out on the initiative of NESA in time to retain the amount that is determined to be the responsibility of the ELM, before the contractual balance payable is exhausted; (ii) only after the investigation report, if the responsibility of the CMBM as custodian is proven, can the quantification of this loss to NESA be sought, if the sentence settlement phase is still possible in the ongoing arbitration process.

45 Two contracts were signed with a small company called Hydrostec for the elaboration of the executive project, manufacture, supply, assembly, and commissioning of 36 sets of anti-fish grilles to be installed at the draft tube exit of the main Powerhouse in Belo Monte.

and the supplier, it was established that the costs of such operation would be shared, in equal proportions.[46]

The supplier should take all measures related to the design and management to perform the machining services. NESA would have specialized consultancy for project approval and execution monitoring.

Another innovation that deserves to be highlighted was the inversion of the assembly sequence of the stator-rotor assembly in the 3-UG well. The normal sequence dictates that the generator stator has to be mounted in place before the rotor. It so happens that the descent of the rotor and its coupling to the turbine shaft is a requirement for a series of complementary assembly services in the enclosure between the turbine and the rotor. Lowering the rotor before the stator could mean a gain of about 20 days in the assembly, as these services could be performed in parallel with the assembly of the stator.[47]

The innovative way to assemble depended not only on the engagement of the assembler and supplier, but on their endorsement to maintain the guarantees. This was only possible due to the maturing of the work environment built over since June 2018.

Accomplished Mission

After one year, four months and 20 days, nine generating units were delivered into operation, commissioned, and under full warranty (Figure 7-11).

Figure 7-11 Main Powerhouse: Aerial view in October 2019

46 Since there was an arbitration process between NESA and CMBM to define amounts resulting from the contract termination, the costs arising from breach of contract would become part of the list of items subject to reimbursement to the contracting party.
47 The large dimensions of the Belo Monte GUs determined a big risk for this operation. The stator diameter is over 22 meters. However, the clearance between the rotor and the stator is only 2.9 centimeters, with a margin of only 0.5 centimeters.
48 These are approximate percentages because a good part of the additional amounts were already foreseen as contract scope or not due to deviations during the implementation, but to mistakes in the contract, loss of parts and components to be assembled, and delays happened prior to the period considered here.

An approximate average of 70 days per GU. The assembly forecast based on this kind of experience in Hydro Powerplant implementation worldwide set more than double the time for assembly each GU, or three months per UG.

There were no significant deviations of realized values from the forecast. Adding up the beyond scope services and the variations provided for payment in the assembly contract and the price variations for the permanence of technical supervisors as well as damaged assembly components represented an additional value of less than half a percent in relation to the contracted value, and less than ten percent of the amount disbursed for the scope corresponding to the 9th GUs assembly.[48]

7.4 Conclusions

The two cases studies are relevant examples of how it is essential to break with the past to adapt to the future and how agility, self-organization, and fast decision-making shape the new organizational culture that allows the delivery of successful projects.

References

Beswick, C., Toma, D., and Vargas, R (2020). *To Drive Innovation You Must Understand Your Ecosystems.* London: London School of Economics Business Review (https://blogs.lse.ac.uk/businessreview/2020/10/29/to-drive-innovation-you-must-understand-your-ecosystem/)

Edmondson, A. C. and Hugander, P. (2021). *4 Steps to Boost Psychological Safety at Your Workplace.* Cambridge: Harvard Business Review.

EIU and PMI (2017). *Brightline Initiative Closing the Gap: Designing and Delivering a Strategy that Works.* Newtown Square: Project Management Institute.

MIT (2021). *Workforce Ecosystems: A New Strategic Approach to the Future of Work.* Cambridge: Massachusetts Institute of Technology Sloan Management Review.

PMI (2019). *Brightline Transformation Compass.* Newtown Square: Project Management Institute. (https://www.brightline.org/resources/transformation-compass/)

PMI (2021). *The Standards for Project Management and the PMBOK Guide,* 7th Edition. Newtown Square: Project Management Institute.

Tabrizi, B., and Terrel, M. (2013). *Inside Out Transformation.* Ashland: Evolve Publishing.

8

Pillar 8: Adaptive Frameworks and Life Cycles

8.0 Background

In the early years of project management, executives were fearful that project managers might begin making decisions that were reserved for the senior levels of management. Executives realized quickly that effective command and control over all project decisions could be obtained using a singular methodology accompanied by forms, guidelines, templates, and checklists that could be continuously updated.

There was also a belief that achieving project management excellence, or maturity, was more likely with a repetitive process that could be used on every project. This repetitive process was referred to as the "one-size-fits-all" project management methodology. Companies maintained and supported a single methodology for project management. Several companies established internal organizations, such as a project management office (PMO), for continuous improvements to the singular methodology.

After years of implementation and use, the characteristics of a good methodology were defined as:

- A recommended level of detail acceptable to management
- Use of templates
- Standardized planning, scheduling, and cost control techniques
- Standardized reporting format for both in-house and customer use
- Flexibility for application to all projects, within limitations
- Flexibility for rapid improvements
- Easy for the customer to understand and follow
- Readily accepted and used throughout the company
- Use of standardized life-cycle phases (which can overlap) and end of phase gate review meetings
- Based on guidelines rather than policies and procedures
- Based on a good work ethic supported by a cooperative culture

The benefits of using a singular methodology were being published in project management texts and taught in training courses. Some of the benefits included:

- Faster "time to market" through better control of the project's scope
- Lower overall project risk

- Better decision-making process
- Greater customer satisfaction, which could lead to increased business
- More time available for value-added efforts, rather than internal politics and internal competition

8.1 The Risks of Using a Singular Methodology

For most companies, the singular methodology was treated as a set of principles that a company believed could be tailored and then applied to a specific situation or group of activities that have some degree of commonality. The project management methodology, often referred to as the Waterfall approach where everything is done sequentially, became the primary tool for the "command and control" of projects providing some degree of *standardization* in the execution of the work and *control* over the decision-making process. Standardization and control came at a price and provided some degree of limitation as to when the methodology could be used effectively. Typical limitations included:

- **Type of project:** Most methodologies assumed that the requirements of the project were reasonably well-defined at the onset of the project. Tradeoffs were primarily on time and cost rather than scope. This limited the use of the methodology to traditional or operational projects that were reasonably well-understood at the project approval stage and had a limited number of unknowns. Strategic projects, such as those involving innovation and had to be aligned to strategic business objectives rather than a clear statement of work, could not be easily managed using the waterfall methodology because of the large number of unknowns and the fact that they could change frequently.
- **Performance tracking:** With reasonable knowledge about the project's requirements, performance tracking was accomplished mainly using the triple constraints of time, cost, and scope. Nontraditional or strategic projects had significantly more constraints that required monitoring and therefore used other tracking systems than the project management methodology. Simply stated, the traditional methodology had extremely limited flexibility when applied to projects that were not operational.
- **Risk management:** Risk management is important on all types of projects. But on nontraditional or strategic projects, with the high number of unknowns that can change frequently over the life of the project, standard risk management practices that are included in traditional methodologies may be insufficient for risk assessment and mitigation practices.
- **Governance:** For traditional projects, governance was provided by a single person acting as the sponsor for the project. The methodology became the sponsor's primary vehicle for command and control and used with the mistaken belief that all decisions could be made by monitoring just the time, cost, and scope constraints.

Recognizing the risks and limitations, companies decided that the singular methodology would be used to satisfy almost all their traditional or operational projects, but other types of projects, especially related to strategic issues, would be managed by functional rather than project managers. Strategic projects that included innovation, R&D and entrepreneurship were now being managed by functional managers who were often allowed to use their own approach for managing these projects rather than follow the one-size-fits-all methodology. Part of the problem was that teams working on strategic projects wanted the "freedom to be creative as they see fit" and therefore did not want to be handcuffed by having to follow any form of rigid methodology.

8.2 Project Management Landscape Changes

Companies began to realize the benefits of using project management from their own successes, the capturing of lessons learned and best practices, and published research data. Furthermore, companies were convinced that almost all activities within the firm could now be regarded as a project and they were therefore managing their business by projects.

Initially, as the one-size-fits-all methodology began to be applied to nontraditional or strategic projects, the weaknesses in the singular methodology became apparent. Strategic projects, especially those that involve innovation, may not be completely definable at project initiation, the scope of work can change frequently during project execution, governance may appear in the form of committee governance with significantly more involvement by the customer or business owner, and a different form of project leadership is often needed.

The traditional risk management approach used on operational projects appeared to be insufficient for strategic projects. As an example, strategic projects require a risk management approach that emphasizes VUCA analyses:

- **V**olatility
- **U**ncertainty
- **C**omplexity
- **A**mbiguity

Significantly more risks appear on strategic projects where the requirements can change rapidly to satisfy turbulent business needs. This became quite apparent on IT projects that focused heavily on the traditional waterfall methodology that offered little flexibility. The introduction of an agile or flexible methodology solved some of the problems but created others. Agile was a flexible methodology or framework that focused heavily on better risk management activities but required a great deal of collaboration. Every methodology or framework comes with advantages and disadvantages.

The introduction of an agile methodology gave companies a choice between a rigid one-size-fits-all approach or a very flexible agile framework, both of which were beneficial for specific projects. Unfortunately, not all projects are perfect fits for either an extremely rigid or flexible approach. Some projects are middle-of-the-road projects that may fall in between rigid waterfall approaches and flexible agile frameworks.

8.3 The Need for Multiple Flexible Methodologies

Companies are recognizing the need for multiple frameworks from which to select the best approach that can be customized to a given project. The first major hurdle in the selection process of a framework is to determine which constraints on a project are flexible and which are inflexible. This is illustrated in Figure 8-1 where time, cost, and scope are represented by the tips of the triangle rather than the sides.

With traditional project management, scope is considered as an inflexible constraint. Project managers are trained to ensure that the scope of the project is well-defined before establishing the budget and schedule which can change over the duration of the project. With techniques such as Agile and Scrum, time and cost are treated as inflexible constraints and the project team must then determine how much scope can be created within the time and cost constraints.

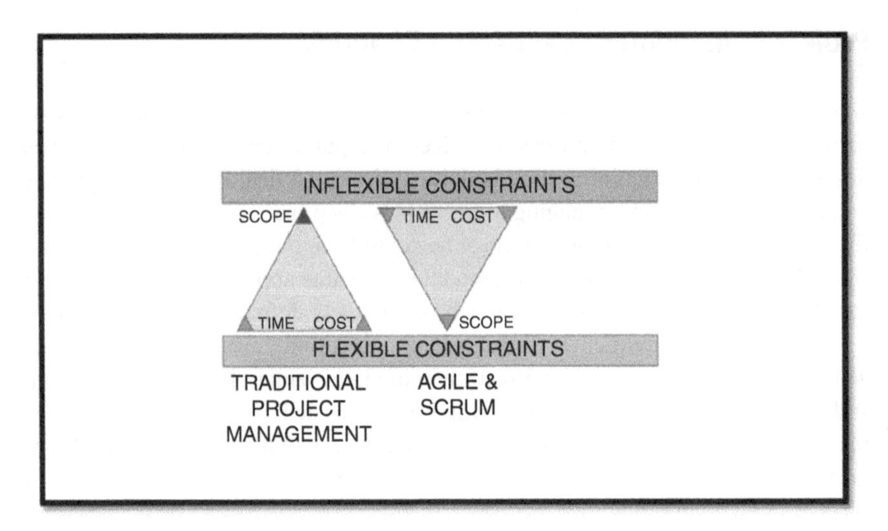

Figure 8-1 Flexible Versus Inflexible Constraints

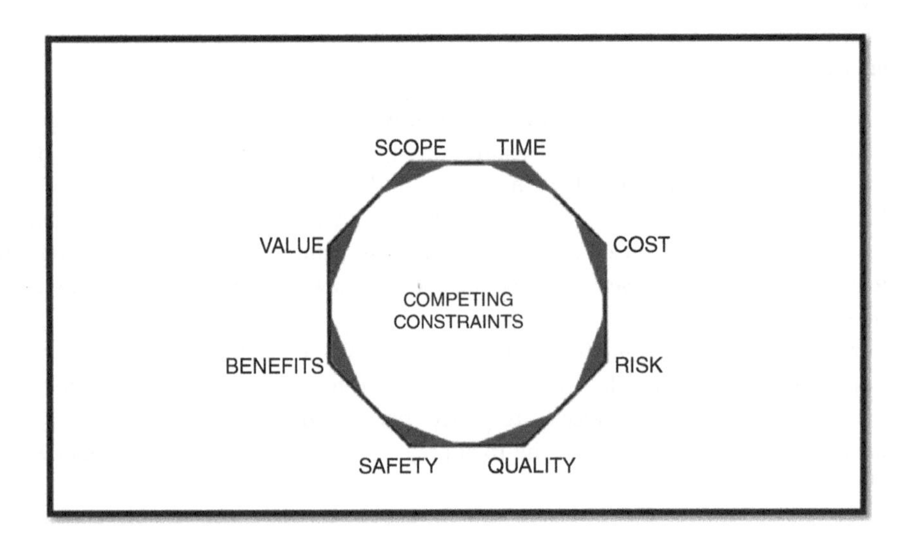

Figure 8-2 Possible Competing Constraints

Several years ago, with the introduction of the 4th edition of the ***PMBOK® Guide***, PMI began promoting the concept of competing constraints rather than the traditional view of just the triple constraints. As seen in Figure 8-2, this allows for project managers to consider more constraints when evaluating project management performance.

Figure 8-2 identifies eight possible competing constraints. The relative importance of each can change from project to project and can also impact how project success is defined. Some projects can have significantly more than eight competing constraints based on the impact of the enterprise environmental factors and the VUCA environment.

If the hexagon of competing constraints in Figure 8-2 is transposed onto Figure 8-1, as shown in Figure 8-3, one of the first steps that a project team must do is to determine which of the

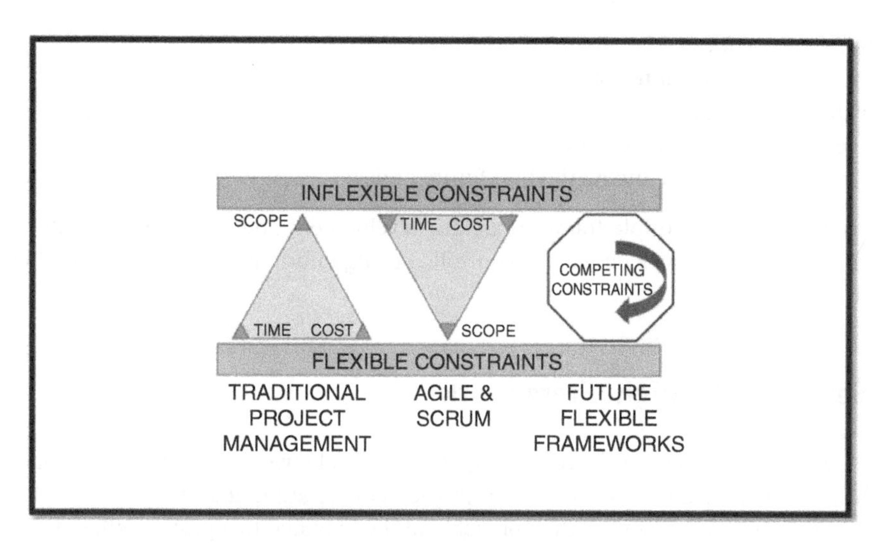

Figure 8-3 Flexibility and Inflexibility of Competing Constraints

competing constraints are inflexible and which are flexible. With many competing constraints, and based on the flexibility of the constraints, it is easy to see that companies may require many different types of frameworks in the future for project execution based on the project's characteristics.

Each framework could require a different set of life-cycle phases, different types of end-of-phase gate review meetings, and different types of procedural documentation based on the flexibility or inflexibility of the constraints.

One of the most important lessons learned over the past four decades from companies that used a singular methodology was the importance of procedural documentation for effective command and control of project management practices. Good procedural documentation will accelerate the project management maturity process, foster support at all levels of management, and greatly improve project communications.

There are many ways to create procedural documentation, and the type of procedural documentation selected is heavily biased on the type of framework and whether we wish to manage formally or informally. The project management policies, procedures, forms, and guidelines can provide some of these tools for delineating the processes, as well as a format for collecting, processing, and communicating project-related data in an orderly, standardized format. Project planning and tracking, however, involve more than just the generation of paperwork. They require the participation of the entire project team, including support departments, subcontractors, and top management, and this involvement fosters unity. Procedural documents help to:

- Provide guidelines and uniformity.
- Encourage useful, but minimum, documentation.
- Communicate information clearly and effectively.
- Standardize data formats.
- Unify project teams.
- Provide a basis for analysis.
- Ensure document agreements for future reference.
- Refuel commitments.

- Minimize paperwork.
- Minimize conflict and confusion.
- Delineate work packages.
- Bring new team members on board.
- Build an experience track and method for future projects.

If a company adopts multiple frameworks, the development and control of procedural documentation could create issues but can be controlled using information warehouses and business intelligence systems.

8.4 Selecting the Right Framework

Some practitioners envision the future as simply a decision between Waterfall, Agile, and Scrum as to which one will be a best fit for a given project. Others argue that new frameworks can be created from the best features of each approach and then applied to a project. What we do know with a reasonable degree of confidence is that new frameworks, with a great deal of flexibility and the ability to be customized, will certainly appear in the future. Deciding which framework is best suited to a given project will be the challenge, and project teams will be given the choice of which one to use.

Project teams of the future will begin each project by determining which approach will best suit their needs. This can be accomplished with checklists and questions that address characteristics of the project such as flexibility requirements, type of leadership needed, team skill levels needed, and the culture of the organization. The answers to the questions will then be pieced together to form a framework, which may be unique to a given project. Typical questions might include:

1. **How clear are the requirements and the linkage to the strategic business objectives?** On some projects, especially when innovation and/or R&D are required, it may be difficult to come up with well-defined objectives for the project even though the line-of-sight to the strategic business objectives is well known. These projects may focus more so on big, hairy, audacious goals (BHAGs) rather than on more well-defined objectives.

   When the requirements are unclear, then the project may be tentative in nature and subject to cancellation. You must also expect that changes will occur throughout the life of the project. These types of projects require highly flexible frameworks and a high degree of customer involvement.
2. **How likely is it that changes in the requirements will take place over the life of the project?** The greater the expectation of changes, the greater the need for a highly flexible approach. Changes may occur because of changing consumer tastes, needs or expectations. Allowing for too many changes to take place may get the project off track and result in a failed project that produces no benefits or business value. The size of the project is also important because larger projects are more susceptible to scope changes.

   In addition to the number of changes that may be needed, it is also important to know how much time will be allowed for the changes to take place. In critical situations where the changes may have to take place in days or weeks, a fast paced, flexible approach may be necessary with continuous involvement by stakeholders and decision makers.
3. **Will the customer expect all the features and functionality at the end of the project, or will the customer allow for incremental scope changes?** Incremental scope changes

allow the project to be broken down and completed in small increments that may increase the overall quality and tangible business value of the outcome. This may also provide less pressure on decision-making.
4. **Is the team collocated or virtual?** Projects that require a great deal of collaboration for decision-making may be more easily managed with a collocated team, especially when a large amount of scope changes are expected.
5. **If the project requires the creation of features to a product, where will the information come from for determining which features are necessary?** The answer to this question may require that the project team interface frequently with marketing and end users to make sure that the features are what the users desire. The ease by which the team can interface with the end users may be of critical importance.
6. **Is there success (and/or failure) criteria that will help us determine when the project is over?** With a poor or lack of success criteria, the project may require a great deal of flexibility, testing, and prototype development.
7. **How knowledgeable will the stakeholders be with the framework selected?** If the stakeholders are unfamiliar with the framework, a great deal of time may be wasted educating the customer on the framework selected and their expected role and responsibility in the framework. This may create a problem for stakeholders that exhibit a resistance to change.
8. **What metrics will the stakeholders and business owner require?** Waterfall methodologies focus on time, cost, and scope metrics. Flexible methodologies allow for other metrics such as business benefits and value achieved.

When a contractor uses a singular methodology, the contractor markets their approach in their proposal as part of competitive bidding. The singular methodology is usually aligned to the contractor's business model. However, when the contractor has a choice of several frameworks, selecting an approach that is closely aligned to the customer's business model may be received more favorably, improve customer and stakeholder relations, and provide the opportunity for follow-on business.

8.5 Be Careful What You Wish For

Selecting the right framework may seem like a relatively easy thing to do. However, as stated previously, all methodologies and frameworks come with disadvantages as well as advantages. Project teams must then hope for the best but plan for the worst. They must understand what can go wrong and select an approach where execution issues can be readily resolved in a timely manner. As companies become more knowledgeable in the use of frameworks, changing methodologies in midstream may become an option. Some of the questions focusing on "What can go wrong?" that should be addressed before finalizing the approach to be taken include:

- Are the customer's expectations realistic?
- Will the needs of the project be evolving or known at the onset?
- Can the required work be broken down and managed using small work packages and sprints or is it an all-or-nothing approach?
- Will the customer and stakeholders provide the necessary support and in a timely manner?
- Will the customer and/or stakeholder be overbearing and try to manage the project themselves?
- How much documentation will be required?

- Will the project team possess the necessary communications, teamwork, and innovation/technical skills?
- Will the team members be able to commit the necessary time to the project?
- Is the type of contract (i.e., fixed price, cost reimbursable, cost sharing, etc.) well-suited for the framework selected?

Selecting a highly flexible approach may seem on the surface to be the best way to go since mistakes and potential risks can be identified early, which then allows for faster corrective action to take place and prevents disasters from occurring. But what people seem to fail to realize is that the greater the level of flexibility, the more layers of management and supervision may need to be in place.

Today, there are several methodologies and frameworks available for project teams, such as Agile, Waterfall, Scrum, Prince2, and Rapid Application Development. In the future, we can expect the number of methodologies and frameworks available to increase significantly. Some type of criteria must be established to select the best approach for a given project.

8.6 Strategic Selection Implications

All companies strive for growth. Strategic plans are prepared identifying new products and services to be developed and new markets to be penetrated. Problems can occur when the selection or changeover to new processes and tools impacts the way that suppliers, distributors, or other strategic partners must perform their work to align to your needs.

Strategic planning often requires mergers and acquisitions with companies that have their own way of executing projects. Even the best-prepared strategic plans often fail when project operating systems are not compatible. Too many executives view strategic planning as planning only, with little consideration given to the implementation phase where project management processes, tools, and techniques are critical. Implementation success is vital during the merger and acquisition process because it impacts the decisions that need to be made concerning the selection of appropriate methodologies, frameworks, life-cycle phases, and gates discussed in this chapter. Some companies view mergers and acquisitions as a landlord–tenant relationship where the landlord makes all of the of the project management implementation decisions and the tenant must accept the results. The better approach is when they both see each other as equal partners and make decisions that satisfy everyone's needs.

Mergers and acquisitions focus on two components: preacquisition decision-making and post-acquisition integration of processes. Wall Street and financial institutions appear to be interested more in the near-term, financial impact of the acquisition rather than the long-term value that can be achieved through better project management and integrated processes. During the mid-1990s, companies rushed into acquisitions in less time than the company required for a capital expenditure approval. Virtually no consideration was given to the impact on project management and whether the expected best practices would be transferable to all parties. The result appeared to have been more failures than successes.

When a firm rushes into an acquisition, it typically allocates little time and effort on postacquisition integration. Yet, this is where the real impact of project management practices is felt, especially if people are removed from their comfort zone. When sufficient time is spent on preacquisition decision-making, both firms look at combining project management processes, sharing resources, transferring intellectual property, and the overall management of combined operations. If these issues are not addressed in the preacquisition phase, then a combative culture may occur during the postacquisition integration phase, resulting in resistance and undesirable results.

8.7 Excellence in Action: ServiceNow[1]

Introduction

Adaptive project frameworks have emerged in response to the increasing speed of change in business strategies and markets. To ensure continued alignment to strategic goals, a PMO must have the flexibility to continually adjust to changes not only in schedules and budgets, but also in project and program requirements and risks.

Adaptive can be defined as a PMO's ability to manage multiple methodologies and timescales to deliver on a portfolio of business outcomes.

The Rise of Agile

In recent years, many companies have made a clear pivot from traditional Waterfall methods to Agile for managing software development. Agile offers many benefits – chiefly, the ability to deliver value from projects or programs in shorter, faster timeframes.

Because of this benefit, Agile methodology has spread to other parts of the enterprise – from human-resources teams exploring strategies to redesign employee experience, to senior executives looking to accelerate strategic planning. Many corporate initiatives can benefit from following Agile methods, which rely on small teams, continuous delivery, frequent iterations (or sprints), and a focus on business value, user satisfaction, and velocity.

This approach accelerates time to market and allows for earlier detection of defects, which in turn improves product quality. Through Agile, delivery focuses on the value being delivered to the customer, rather than traditional constraints of time, scope, and cost.

This is especially valuable in environments when business requirements are ambiguous yet require delivery in a faster time frame. By splitting projects up and assigning them to autonomous teams, companies can more quickly test new ideas, measure success or failure, and pivot quickly as market conditions change or unexpected events – such as a global health crisis – radically alter strategic priorities.

This marks a break from traditional project management, where businesses may invest millions into ambitious multiyear initiatives without knowing whether, or how, they will deliver value upon their completion.

For example, in just over a year, a Fortune 500 housing and lifestyle solutions company that shifted to Agile enabled faster delivery of products and services, by assessing progress using portfolios focused on continuous product delivery instead of projects funded by annual budgets.

Since its Agile adoption, the company has seen:

- Time to market has improved by 30 percent in the first year
- Predictability, defined as the reliability of commitments such as costs and release dates to the business and stakeholders, has improved by 50 percent for the same period

1 Material in this section has been provided by Simon Grice (Senior Director, Innovation), Doug Page (Senior Manager, Product Management), Rani Pangam (Senior Director, IT Project Management), and Tony Pantaleo (Director, Product Success).

By following the Agile methodology, companies can clearly see the value of what it delivers and is able to focus on the future: mastering iteration over perfection, empowering change agents more quickly, and using a flexible model to better embrace change.

The Need for a Hybrid Approach

While Agile can deliver the nimbleness needed in the current business climate, in practice companies will face both short-term, fast-changing demands as well as investment needs with longer time horizons. Most companies will continue to have a mix of projects, which differ on needs, timescales, and desired outcomes.

Agile methodologies deliver tactical agility – the ability to iterate quickly in response to changing requirements – but do not necessarily provide strategic agility. At a business unit or organizational level, strategic agility equips companies with insight and discipline to remove systemic bottlenecks and quickly allocate resources to projects that promise to deliver the highest return.

The key to strategic agility is to follow a hybrid approach: one that applies the right framework, whether Waterfall or Agile, to the right need. The mix of Waterfall and Agile will vary across businesses – some may have a 50/50 split, while others are 20/80 or 80/20 – but few will operate with just one methodology. See Figure 8-4.

The Hybrid Approach Requires Two Competencies from PMOs

Aligning the Right Frameworks to Deliver Value

While Agile plays a leading a role, it does not fully displace projects requiring the structure of a traditional Waterfall approach. Management and oversight become less about governance, and more about ensuring that the right frameworks are being used to deliver value.

Take the example of an eye care company that uses a hybrid approach to drive value across the enterprise. Some of its products are governed by regulations in multiple countries. The company uses Waterfall to manage the highly complex, rigorous processes for products that require regulatory compliance. Other groups within the company use an Agile methodology to drive product and service innovation and accelerate time to market.

Another Fortune 500 company uses a 50/50 approach in applying both Waterfall and Agile. Waterfall provides the formal structure demanded for complex, large-scale projects while Agile enables IT to produce a continuous stream of deliverables in short timeframes. Less emphasis is placed on control and more on transparency and speed. A multiyear funding cycle improves both planning and throughput by eliminating the inevitable delays that come with annual budgeting.

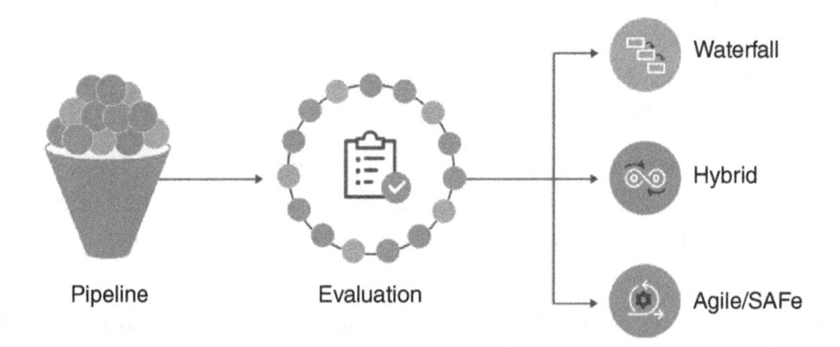

Figure 8-4 An Approach to Hybrid Project Management: Once projects enter the pipeline, project managers evaluate individual projects by requirements, schedules, resources, and outcomes, then select the most appropriate methodology for execution against ROI. A hybrid approach enables companies to easily consider factors unique to their business.

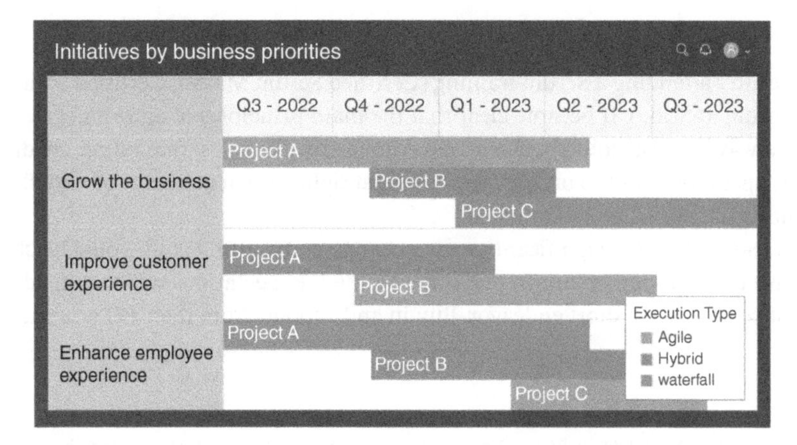

Figure 8-5 Project View by Business Priority: A platform approach gives project managers a central view of projects and programs by methodology – Agile, Hybrid, or Waterfall – and by progress against delivery date.

Providing Visibility across Different Frameworks

The second competency needed to manage hybrid frameworks is **visibility.** PMOs need visibility across different frameworks to advise business leaders on their portfolio investments. The PMO also needs to see bottlenecks across the enterprise, indicating where projects are siloed or aren't aligned to strategic goals. Typically, this visibility is inhibited by systems that are tied to specific methodologies.

Some companies choose to invest in a range of best-in-class systems and build integrations between these systems. However, with project data in different systems, real-time visibility into portfolio-level metrics can be limited. Merging data from multiple systems for reporting can be time-consuming and error-prone and managing multiple systems can be complex and costly. See Figure 8-5.

What's needed is a single digital platform that can frame the impact, return, and progress of the entire portfolio of projects, both Agile and Waterfall, across the company, for visibility and action. A single platform uses a single data repository with a single, unified workflow engine or platform. This approach generates real-time insights into what is happening across portfolios so everyone, from project managers to functional leads to senior executives, can understand how investments are performing and their alignment to strategic goals.

8.8 Excellence in Action: The International Institute for Learning[2]

The IIL Transformation Journey

In late 2016, the International Institute for Learning (IIL) decided to implement a change in the way the company works. That decision was supported by two pillars of organizational excellence – Organizational Cultural Shift to the Project Way of Working, and Adaptive Frameworks and Life Cycles.

2 Material in this section was provided By Dr. Leon Herszon, PhD, PMP, CSM, CSPO, DASSM, the Chief Agility Officer and Sr. VP of Global Enterprise Solutions at International Institute for Learning (IIL). Leon.herszon@iil.com - www.linkedin.com/in/leonherszon . ©2021 by The International Institute for Learning. All rights reserved. Reproduced with permission.

How It Started

The idea came after attending a Scrum training (Certified Scrum Master, Certified Scrum Product Owner, and Scrum at Scale). It became clear that the main principles of agile had great potential to improve the way IIL was doing business, covering areas like sales, marketing, media production, finances, operations, and so on. Yes, you read that right – the focus was not initially on information technology.

It was also clear that such a significant process change and cultural shift would be challenging and would have to be properly planned and implemented. Expectations were set in advance that this would not be an easy or short endeavor. Buy-in and support from the CEO was paramount.

The Reason

In a discussion with IIL's CEO in 2016, I made the argument that improving our business agility and implementing an agile approach would help IIL in several ways. A brief description of the potential benefits for the organization was then presented.

- **Improves performance:** Based on all the data collected from cases of various companies, industries and regions of the world, it proposes that agile implementations improve results. Shorter time to market, increase in sales, reduction of costs, better teamwork, just to name a few. We thought that would be worth trying.
- **Empowers people:** This means allowing the people doing the work to define the HOW while top executives define the WHAT (goals) and WHY. Many organizations still have top executives telling the team the WHAT and also HOW they should work. This mindset change is important to allow the teams to self-organize and define the best way to achieve their goals. All of that within the organization policies and ethics, of course.
- **Allows experimentation:** This is one of my preferred benefits, and is closely related to any company's survival. Without experimentation, there is no innovation, and without innovation the company sooner or later loses its competitive advantage. Allowing people and teams to experiment, to try new things and to fail is paramount for improvement. The organization should be open to allow failure and learn from it. The only way to not make mistakes is. . . do not try anything new, don't do anything different. As we've all heard before, trying to get different results by repeating the same thing is the definition of insanity.
- **Improves teamwork:** The concept of a team as a group of individuals who share a common goal, who need each other to achieve that goal, and whose work together creates something more than just the sum of the parts (www.wit.edu), was emphasized and shared with all involved. Our expectation was to create an environment with openness, courage, respect, commitment, and focus. Once you mitigate or eliminate the individual and selfish mindset (me, myself, and I) and start focusing on a team effort to reach the goal, the result is more collaboration, open conversation, better communication, and transparency.
- **Motivates people:** According to the Oxford dictionary, motivation is the general desire or willingness of someone to do something. Considering the benefits described above, one can argue that being on a team that helps each other and works toward a shared goal tends to motivate people. This might not be true for a team member who prefers to work alone and not to collaborate – these individuals usually would not stay in an agile organization for long.

Besides the benefits, we shared the usual situations an organization might face that could influence the decision to invest on such transformation journey, for instance:

- Financial issues (cash flow, reduced revenue, high costs, etc.)
- Low team morale, people demotivated
- No transparency and no clear strategy shared within the organization
- Fear of trying new things
- Losing market share
- Slow response to market
- Unhappy clients
- No predictability of results; etc.

After presenting the benefits, the reasons why organizations embark on this initiative, and our implementation plan, we got the "go ahead" to start the transformation process.

Initial Questions and Implementation Plan

During the planning process, there were several questions that had to be answered prior to the implementation. Some of them are listed below.

1. **Which agile approach to use?**
   Out of all existing agile approaches being used that could fit our organization, we decided to use Scrum as our guide. The reasons were (a) Scrum was the most used agile approach worldwide; (b) we already had the knowledge in house; and (c) we were convinced that we could make the proper adjustments to our reality.
2. **Should the implementation start in one area (pilot) or the entire organization (Big Bang)? How to do it?**
   Even though there are benefits for each strategy, we decided to start with the Sales area. Not the most common or easiest area one would think to start with, but considering the nature of IIL's business, it seemed to make sense. Supporting our decision was the paper "Scrum in Sales" by Solligen, Sutherland, and Ward, who shared cases on how it was feasible to implement Scrum within sales teams.
   To assure we had a common language and understanding within the sales team, we delivered a training for the entire team, initially using an external consulting company. After that each Scrum Master got certified as Certified Scrum Master and each Product Owner as a Certified Scrum Product Owner by ScrumAlliance.
3. **Should we use a more prescriptive model or adapt to the nature of our business and area?**
   As mentioned above, instead of following the Scrum approach on a prescriptive way, we adjusted to our own reality and way of doing business. The adaptation of the general principles of the Agile Manifesto (2001) below served as a guide:

   - Individuals and interactions over processes and tools
   - Working product over comprehensive documentation
   - Customer collaboration over contract negotiation
   - Responding to change over following a plan

   At the time we are writing this content, PMI's Disciplined Agile approach (*Choose Your Wow!,* PMI 2020) would be closer to how we implemented our agile transformation. Disciplined Agile recommendation is to consider the context of your organization/project and adapt it to your way of working.

4. **Which roles to use, and who would do what?**

   We followed a more traditional Scrum role assignment with a Scrum Master, a Scrum Product Owner, and Team Members.

   For internal engagements, the CEO was our Chief Product Owner, and she defined the goals for the entire organization (the WHAT and the WHY). Furthermore, each team had a Product Owner who would represent our client's (external or internal) interest.

   The team, facilitated by the Scrum Master, would discuss HOW to achieve each goal. The Scrum Master had the role of facilitating the work, removing impediments, and acting as a servant leader. A summary of the roles is subsequently described.

   - **Product Owner (PO):** The product owner would assure that the client needs (internal or external) would be met, share with the team the items to be delivered and WHY, and work with the team to rank the items backlog. The PO acts as a liaison between the client/Chief PO and the team.
   - **Scrum Master (SM):** The Scrum Master would set up the meetings (planning, daily, retrospective) according to the best practices, supporting the PO and team, motivating the team, facilitate daily scrums, and removing impediments directly or by escalating to decision makers.
   - **Team:** The team defines HOW the work is going to be done and WHO is going to do what, in a self-organizing way. The client (internal or external) does not interfere directly with the team and does not attend stand-up meetings. We gave chances for team members to rotate as Scrum Masters.

5. **How often will be our sprint?**

   This would certainly depend on which area we are referring to. The Scrum best practices recommend that sprints should be limited to one to four weeks' duration.

   For instance, sales teams had goals to achieve every four weeks, but they would do daily stand-up meetings and meet weekly to discuss performance, do a retrospective, and plan for the upcoming week.

   Other department could have one-week sprints, others two-week sprints, depending on what was being done.

6. **How to break down our people into teams?**

   Without entering into too many details and using the sales team again as an example, we started by breaking down the entire sales force into nine teams considering the different time zones – four teams for the Americas, three teams for EMEA, and two teams for Asia Pacific. We made sure that each team had at least one senior salesperson in it. The teams were limited to nine people each according to the recommended Scrum best practices.

7. **Would it be useful to use a supporting tool and if so, which one?**

   After analyzing several options, we decided to use Trello because it is a simple but powerful tool that was easy to learn and provided proper visual support.

   We decided that the Trello board would have a few columns as mandatory: GOALS, TO DO, DOING, DONE, and IMPEDIMENTS. Teams were free to add other columns to their Trello board like Resolved Impediments, Past Results, Goal Status, Pending, Ideas, etc.

   As previously mentioned, not all participants were co-located so an online collaborative tool would be important. For communication and remote daily stand-up meetings, we used Skype or another available platform. This is especially useful during the pandemic times.

The Journey: Challenges and How to Overcome Them

Every agile transformation is different and might face challenges during its implementation. According to Scrum Inc., the list below shows a few dysfunctions that need to be addressed when implementing Scrum (also valid for other agile approaches):

- Too many projects in the pipeline
- Everything is top priority
- Pressure to get things done delays projects and reduces quality
- New team for every project
- Lack of understanding of Scrum
- Management not focused on removing impediments; and so on

According to the *10th Annual State of Agile Report* (2016, VersionOne) and supported by further publications, the top five barriers to agile adoption are:

- 46% Company culture at odds with agile values
- 41% Lack of agile experience
- 38% Lack of management support
- 38% Lack of cultural transition support
- 38% Inconsistent agile practices and process

Furthermore, on a survey done a few years go by J. LeRoy Ward, he asked, "What was the one thing you did when implementing agile that really made the biggest impact?" This question was asked to PMO and former PMO directors, Director or Product Portfolio Management, Global Head of Project & Program Management Practice, Leader of Digital Transformation, key industry consultants, and respected industry thought leaders, covering several industries like financial services, transportation, information technology, software development, food processing, telecommunications, non-for-profit, etc. The answers can be summarized below:

- Use a common agile method.
- Engage client and empower product owner.
- Get the executives on board.
- Pilot a project to show it works.
- Establish the right metrics.
- Train everyone.
- Develop an agile strategy and roadmap.

Even though there were several answers, one element was present on all responses – the importance of executive and business support. We had a great advantage during the implementation since our CEO & Founder supported the initiative since the beginning.

We certainly had our share of challenges and had to implement a few actions to mitigate their impact during our transformation journey.

1. **Resistance to changes**

   Most people don't like changes and there are many ways to manage this process. It was helpful that I was involved and manage several different types of changes like strategic changes, operational changes, and technological changes, which included reactive changes (when you are answering to an external factor), proactive changes (when you are taking the initiative to change), incremental changes (when you are making smaller, incremental changes to the existing process), and so on. Having someone with experience managing the change is important for a successful transformation initiative.

Change management is a vast field but usually include a few steps including but not limited to (a) understanding that individuals react differently to changes and defining how to manage each case, (b) handling resistance by involving all stakeholders in the change process with constant communication, (c) having a change management process in place, (d) constantly celebrate the wins and learn from mistake, and (e) assuring executive support. Even though listed at the end, without executive support the chances of implementing significant company-wide changes are minimal.

2. **Individualism instead of teamwork**

   Some team members reacted strongly against working on teams and having to work together with others toward a common goal. What we learned is that management does not need to be concerned on how to find these individuals since they would either leave the company (not comfortable on a collaborative environment), or the team would share their concerns with management.

   We had cases like that and the ones who could not adapt, they left our organization, but most were able to accept the change and be a productive part of the team. Note that when the organization values more individual performance over collaboration, agile transformation initiatives tend to fail in the long run and things usually go back to the "old ways." This is one of the worst outcomes, considering all the effort and energy used throughout the years.

3. **Concerns about experimenting and failing**

   This concern is quite common in organizations that have the "culture of error," where mistakes are penalized and impact people's career. Looking for the guilty person instead of the cause for failure can become a roadblock for innovation.

   We were fortunate to have an environment where teams could experiment and share the successes and failures with other teams. Lessons were learned and great ideas came from experimenting.

4. **Stand-up, planning, or retrospective meetings are just status report meetings**

   Initially we had a few negative reactions related to the daily stand-up meetings. Some people thought it was a waste of time to meet frequently, even for 15 minutes. Others had a hard time to keep within the 15-minute limit. Another group found it disruptive of their existing process – must stop my day to have a meeting. The planning and retrospective meetings were better accepted, and we didn't get significant pushback on these.

   There was also a change on what to cover during the daily stand-up meetings. People were used to have status or simple report meetings, instead of focusing on what was accomplished the previous day toward the goal, what is the goal for today, and share any impediments. This was addressed by explaining how an effective stand-up meeting is held, showing that these meetings can finish earlier if needs be, rotating the Scrum Master role so each team member would receive the knowledge on how to facilitate meetings (one of the better ways to learn is to try to teach the topic, right?), and understanding that people have different timings when adopting something new.

The Results and Summary

After close to five years, including almost two years into the Covid-19 pandemic, the way of working changed. Some of the changes and improvements were planned, others came as consequence of external influences.

In our case, some of the results came sooner (around six months) and others are still to be obtained. In general, we could list the following results and benefits from going through this journey:

- **Transparency:** Every team member can see each other's results, the team's performance, and the entire organization.
- **Streamlined teams:** We have less teams but more effective and efficient ones.
- **Teamwork:** Instead of an abundance of individual performers, we now have more cohesive and integrated teamwork environment.
- **Accuracy:** When considering sales, the forecast accuracy improved a lot. Management has a better sense of upcoming sales and do proper planning.
- **Shorter time to market:** The response time to market changes got shorter as it did when launching new products and services, which allowed the company to retain more clients and business.
- **Empowerment:** The CEO is less involved with daily operations and teams are empowered to define the HOW and deliver results.
- **Tools:** The tools serve the organization, not the other way around. Some of the tools we used initially were replaced by others that would allow teams to be more effective. Good tracking of information and support for decision-making.
- **Meetings:** From daily stand-ups, weekly planning, and retrospectives, we moved to a more flexible approach where the teams decide which meetings, duration, and format, add more value to reaching the goals. Less prescriptive and more adaptive to each area.
- **Experimentation:** We started to experiment more often, gather different ideas and trying them out. Many ideas did not work, but some were instrumental in new ways to serve our clients or improve our processes. Sharing the lessons learned was also a crucial improvement.
- **Impediments:** Instead of receiving separate emails from different people asking to address the same issue, it became more effective to share the impediments on a team meeting. The Scrum Master could address the obstacles and if they could not be resolved, escalate them to the proper level.
- **Onboarding process:** The onboarding and training of new members is faster on an agile team environment.

In summary, if someone asks when the agile implementation is completed or the transformation achieved, the answer is not an exact time. It is a journey that has no specific end in which results are being obtained along the way. This is true for IIL. Adjustments were made to the ever-changing market and external economic, political, social, and technological factors. The journey continues.

8.9 The Fuzzy Front End

Selecting a framework will not be as easy as many people believe. Most frameworks do not come with structured life-cycle phases or gate reviews. The type of project will determine what the life-cycle phases should be as well as the number and type of gate reviews.

In the waterfall approach, almost all project work is linear and establishing traditional life-cycle phases and gate reviews is relatively easy. The selection process will be more difficult in the future because the more complex and strategic nature of projects will make them nonlinear.

In traditional project management, most projects start out with a well-defined scope statement, a listing of requirements, and a detailed business case. But as we get more involved in the future with strategic and complex projects, the front end of the project may be highly unstructured and based on just an idea or goal from which to start. On these types of projects, the beginning is usually referred to as the fuzzy front end (FFE) and the amount of information in the FFE may dictate what life-cycle phases and gate reviews should be selected. The FFE is the set of activities employed before the more formal and well-defined requirements and specifications can be completed.

The front end of future projects that may include innovation will be the greatest area of weakness in the new product development process. This is mainly because the FFE is often chaotic, unpredictable, and unstructured. Unfortunately, this is where many of the decisions are made for later development and commercialization. It includes all activities from the search for new opportunities through the formation of a germ of an idea to the development of a precise concept. The FFE phase ends when an organization approves and begins formal development of the concept.

Unfortunately, as history has shown us, most project managers are brought on board a project after the FFE activities take place. In the future, PMs must participate in these activities and be willing to provide information on resource skills needed for certain ideas, timing issues, and rough cost estimates. The information provided by the PMs can be invaluable during the project selection process. The FFE may be the time when a framework is selected and having the project manager and possibly team members involved in the decision is an excellent approach.

Although the FFE may not be an expensive part of the total project cost, it can consume 50 percent of development time and it is where major commitments are typically made involving time, money, and the product's nature, thus setting the course for the entire project and final product or deliverable. Consequently, this phase should be considered as an essential part of development rather than something that happens "before development," and its cycle time should be included in the total project cycle time.

There has been a great deal of research on the FFE for projects requiring innovation. Most of the research is related to incremental innovation, which is a small change to an existing product. There has been limited research regarding radical or disruptive innovation, which involves new products or services. With incremental innovation, which tends to reinforce existing core competencies, information regarding technology and the markets are known and many of the ideas that come forth are internally generated. Therefore, a business case can be developed where strategic business objectives drive the decision-making process in the FFE. This may allow for the use of traditional methodologies. With radical innovation the reverse is true and flexible methodologies may be necessary.

Technology and competition may be unknown and therefore the outcome of the FFE can drive the strategic planning process from which the business case is then prepared. Knowledge about the technology and the technology trajectory selected may come from just one individual, or perhaps a small team, and personal desires may affect his/her decisions. The decisions must still be approved by senior management, but the decision-making process can be significantly different whether we are in incremental or radical innovation. The organization may not understand the FFE process. A great many more assumptions must be made during the FFE in radical innovation and less information is normally available. FFE decision-making may require large amounts of quality information. This is one of the reasons supporting the need for a future pillar dedicated to information warehouses and business intelligence systems.

Koen et al. (2001)[3], distinguish five different front-end elements (not necessarily in any order) that must be considered during FFE activities:

1. **Opportunity identification.** In this element, large or incremental business and technological chances are identified in a structured way. Using the guidelines established here, resources will eventually be allocated to new projects.... which then lead to a structured NPPD (New Product & Process Development) strategy.
2. **Opportunity analysis.** It is done to translate the identified opportunities into implications for the business and technology specific context of the company. Here extensive efforts may be made to align ideas to target customer groups and perform market studies and/or technical trials and research.
3. **Idea genesis, which is described as an evolutionary and iterative process progressing from birth to maturation of the opportunity into a tangible idea.** The process of the idea genesis can be made internally or come from outside inputs, e.g., a supplier offering a new material/technology or from a customer with an unusual request.
4. **Idea selection.** Its purpose is to choose whether to pursue an idea by analyzing its potential business value.
5. **Idea and technology development.** During this part of the front-end, the business case is developed based on estimates of the total available market, customer needs, investment requirements, competition analysis and project uncertainty. Some organizations consider this to be the first stage of the NPPD process (i.e., Stage 0).

A universally acceptable definition or a dominant framework for the FFE of projects has not been developed so far. The reason is that the FFE for incremental innovation appears to be easier to understand than for radical innovation, and there is significantly more literature on incremental than radical innovation. This will change in the future as project management is applied to strategic projects that have new or radical expectations. However, the outcomes of the FFE in most types of projects are expected to be:

- Mission statement
- Customer needs
- Details of the selected idea
- Product definition and specifications
- Economic analysis of the product
- Development schedule
- Project staffing and the budget
- Business plan aligned with corporate strategy

Project managers in the future must be allowed to participate in the FFE activities to generate these expected outcomes rather than being brought on board at the end of the FFE process and having these outcomes forced upon the project without explanation. PMs may also have a better understanding than executives concerning staffing needs and whether the organization possesses the necessary competencies.

---

3 Koen, P. G., Ajamian, R. Burkhart, A. et al. (2001), Providing clarity and a common language in the "fuzzy front end," *Research Technology Management*, 44, 46-55.

8.10 Line-of-Sight

Strategic planning is an activity usually performed at the senior-most levels of management. Executives establish the company vision and mission statement, and then roll it out to all levels of the organization to get their support. As stated by Jack Welch, former chairman and CEO of General Electric,

Good business leaders create a vision, articulate the vision, passionately own the vision, and relentlessly drive it to completion.

The execution of the plan is performed at various lower levels through a series of projects. Not all projects require detailed knowledge about the strategic plan. For those projects where this information is critical for decision-making, such as in the FFE activities for new product development, there must be a line-of-sight between the executives, strategic planners and the project team for information sharing to make sure that project decisions are aligned with strategic business objectives. Some executives believe that information is power and refuse to share. If then sharing is done, it generally varies according to the need to know, hierarchical level, tenure, and type of project.

Without knowing the strategy, project team members can be at a loss on how to contribute effectively. They may develop conflicting goals and objectives which can interfere with management's expectations. Free thinking may then be replaced by simply following the commands given to them by others.

Strategic planning and portfolio management must be aligned with project planning and execution even in the FFE. If the alignment does not exist, we may end up cancelling potentially successful projects or wasting precious resources on projects that provide little business value. Line-of-sight is critical for this alignment to occur and assists in risk mitigation activities.

8.11 Establishing Gates

Project gates are structured decision points at the end of each life-cycle phase. Project management processes using a waterfall methodology usually have no more than six gates, whereas flexible approaches may have more. Project management is used to manage the phases between the gates. Checklists for gate reviews are critical. Without these checklists, project managers can waste hours preparing gate review reports. Good checklists focus on answering these questions:

- Where are we today (i.e., time and cost)?
- Where will we end up (i.e., time and cost)?
- What are the present and future risks?
- What assistance is needed from management?
- What assistance is needed from stakeholders?

Project managers are never allowed to function as their own gatekeepers. The gatekeepers are either individuals (i.e., sponsors) or groups of individuals designated by senior management and empowered to enforce the structured decision-making process. The gatekeepers are authorized to evaluate the performance to date against predetermined criteria and to provide the project team with additional business and technical information.

Gatekeepers must be willing to make decisions. The four most common decisions are:

- Proceed to the next gate based on the original objectives.
- Proceed to the next gate based on revised objectives.
- Delay making a gate decision until further information is obtained.
- Cancel the project.

Gatekeepers must also have the courage to terminate a project. The purpose of the gates is not only to obtain authorization to proceed, but to identify failure early enough so that resources will not be wasted but will be assigned to more promising activities.

Perhaps the most difficult phase in a project to establish gates is in the FFE. Typical FFE activities include:

- Risk management
- VUCA analysis
- Enterprise environmental factors analysis
- Competitive factors
- Determination of business benefits and value
- Business opportunity identification and analysis
- Alignment to business strategy
- Firm's competencies
- Proactive customer orientation and interfacing
- Maturity of technology
- Control of intellectual property
- Idea evaluation and screening

Based on the type of project, each of the above items can require a gate. Therefore, there can be several gates in the FFE.

8.12 The Future Fuzzy Front Gates

The definition of the FFE is changing. Historically, on traditional projects that did not require innovation, the FFE involved problem solving and decision-making, usually by the senior levels of management. Rather than having a need for brainstorming and creative problem-solving, the FFE on most of these traditional or operational projects involved selecting one of several approaches for the project based on previously used paths that were well-defined and often which included captured best practices. Therefore, gates and life-cycle phases for the FFE were deemed as unnecessary.

Companies are now realizing that doing business "the same old way" can lead to disaster. They must come up with ideas for new and innovative products and services. The need for innovative ideas may very well permeate every project especially those deemed as strategic.

The future FFE will include typical gates such as in Figure 8-6. There can also be a different set of gates and life-cycle phases in each project's FFE.

Future FFEs will therefore become structured processes with gates and life-cycle phases. There will also be forms, guidelines, templates, and checklists to provide some structure for the process, but with a great deal of flexibility for decision makers. The growth of future FFEs will be accompanied by a growth in information warehouses and business intelligence systems. Part of the information warehouses will be idea gathering and evaluation software.

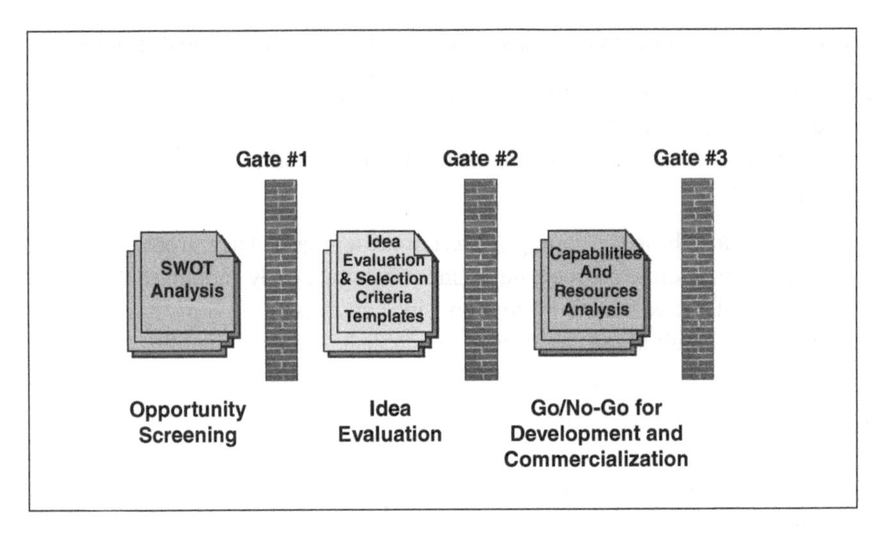

Figure 8-6 Typical FFE Gates

8.13 Excellence in Action: IdeaScale[4]

Innovation managers often cite a lack of process as a key hurdle to the progress for a good idea. In fact, in an assessment of our most successful customers (those who stated that they had met or exceeded all of their innovation goals), the leading trait that most of them had in common was a repeatable process for ideas. Compare the 81 percent that said that they did have a formal process for idea management to the fact that only 30 percent of these programs had a dedicated budget for implementing ideas. How is it possible that those programs were able to successfully meet or exceed their goals without any budget? They were experts at taking institutional energy and using it to influence leaders who already had budget.

For this reason, IdeaScale often recommends that innovators map out their process on a blank sheet of paper to get things going. We ask customers to draw out the life cycle of the idea at your organization from submission to implementation. Who are the key stakeholders, experts, and decision makers who should be involved along the way? Who will ultimately make this idea a reality? What actions do these people need to take at various moments in the progression of the idea? Maybe experts need to add information early and decision makers need to evaluate and select ideas in the middle. Maybe your decision makers are SVPs. Maybe the decision makers are the implementers.

First, draw out the life cycle of an idea at your organization assuming business as usual conditions. Then, draw out what you think would be a more effective or efficient process. But whatever the conditions of a business, we found that these five activities take place in almost every innovation program as part of the development of any new idea: understanding, ideation, refinement, assessment, and implementation. These are the building blocks of project managing from an idea to an innovation. There might be other stages or multiple rounds of certain stages or stages just for a few ideas, but you cannot deliver innovation without ideas, and you cannot move every idea forward. You must learn more and make decisions. Here is a series of stage explanations and examples.

4 This section was provided by Jessica Day of IdeaScale. ©2021 by IdeaScale. All rights reserved. Reproduced with permission.

Stage 1: Understanding

When anticipating needs and creating change, it is important to understand the need or problem that you are trying to solve. It is for this reason that most innovation philosophies start with some level of problem definition. In design thinking, they refer to it as empathize and define. In human-centered design, they reference "discovery" and "understanding." Whatever the industry, whatever the challenge, it is important that this understanding phase includes a gathering and review of existing knowledge, an empathy with those who feel the pain point and an articulation of what success looks like.

One organization that did this exceptionally well was the Department of Energy in their Sunshot challenge. Sunshot was focused on making solar power cost competitive with all other forms of electricity by 2020. The innovation challenge used multiple stages to effectively solve the costliest problems in solar installation. And in the first stage, they developed a keen understanding for the problem inviting submitters to share the costliest obstacles in the technology, manufacturing, and engineering of solar. Only after this requirements phase concluded did they begin sourcing solutions and prototyping promising ideas into new solar initiatives. This level of ground-truthing was highly effective in terms of solving the most pressing problems and engaging the public.

Stage 2: Ideation

Ideation is the term often used to describe brainstorming or coming up with new ideas. In the case of IdeaScale, people gather online and offline to discuss ideas for companies, institutions, non-profits, government entities. It is essential at this stage that you set your pride aside and accept input that may significantly change your initial vision, that you allow priorities to change, and that you gather as many ideas as possible (after all, some ideas might augment others even if they do not move forward on their own). The purpose of this stage is to get the as many ideas accepted and create momentum to get them implemented!

For example, over the years, EA SPORTS has cultivated an active community of super fans called Game Changers. EA wanted to offer their Game Changers a forum to collaborate and communicate with the product development process. IdeaScale and EA SPORTS collaborated to create an ideas forum that was fully integrated into the EA SPORTS web experience. Game Changers could share ideas on a number of different product lines and while the community was open, EA SPORTS gathered more than 7,800 ideas. Some of those ideas have been included in new versions of the game, including Madden NFL 13.

Stage 3: Refinement

Once innovation ideas begin to flow into your system, team members and leaders do additional research, requirements gathering, and begin to organize all of that data into a more robust proposal. Many people will use prebuilt templates like business model canvas or CO-STAR to further articulate complex ideas. This additional research becomes the basis of a business case that will help decision makers evaluate ideas and be the first steps toward plotting a course for implementation.

For example, The Commission for Environmental Cooperation (CEC) launched the CEC Youth Innovation Challenge which was designed to convene and support a community of young, environmentally and socially minded innovators and entrepreneurs in Canada, Mexico, and the United States and connect them with one another.

The Challenge invited youth aged 18–26 to submit science, technology, and business innovations for a chance to pitch their idea to North America's top environmental officials, receive C$5,000 in project seed funding and meet with CEC experts. The Challenge was divided into two phases: the "Idea Phase" where all community members were invited to submit ideas, and vote and comment on the ideas of others and the "Proposal Phase" whereby a short list of semifinalists built out full business proposals that would later be evaluated by a panel of judges who would select the three winners.

Stage 4: Assessment

Having a lot of amazing innovations ready to implement is great, but it can also be a bit of a problem if your organization's budget does not allow all of them to be done immediately. This is where Stage 4 comes in. You can assess the best innovation ideas and decide which ones to prioritize and support first. There are a number of ways to do this: evaluating with crowd data through votes to assess desirability and popularity vs. leaders measuring the idea on its alignment with organizational objectives, but the end result should be a prioritized list of projects that will move forward.

For example, the City of Atlanta launched the CityIdeas initiative, a city-wide employee ideas competition designed to solicit feasible solutions on how the City of Atlanta can reduce waste, cut red tape, and save money on operations.

For every submission, the idea author was required to describe their concept, estimate a potential value or savings, and select a department to champion the idea. Ideas were submitted to the head of the department for review. Eventually, the top 10 from every department were submitted to an interdepartmental evaluation committee that conducted an evaluation of each idea based on its feasibility and level of operational impact. As a team, they were able to identify the initial investment required, annual operating costs and projected savings on an annual and five-year basis.

The most promising ideas were reviewed by the COO and the Mayor who made the final award decisions. The winning ideas received award funding and the idea authors received monetary awards based on the likely cost-savings associated with their idea. The top three selected ideas alone amounted to a potential cost savings of $7.1 million annually.

Stage 5: Implementation

For ideas that reach the final stage, the most important thing is to deliver on them. If leadership has signed off on the ideas, if the community has validated them, then it is the pledge of innovation leaders to prototype and deliver on that collective vision. From here, however, the implementation project management path can differ wildly depending on the scope of the idea, the horizon for change and the team's capabilities. However, if you involve the community in the implementation process, the final product is often richer for that involvement.

For example, in 2013, brand marketing manager at TTi, Brian Stearns, was invited to Home Depot to discuss how to reinvigorate the plumbing department. He got the idea to use his company's rechargeable battery technology to put a fan and a pump on top of the ubiquitous orange Home Depot bucket. The mister became one of the fastest product launches the company ever had. Stearns had the idea in the fall, and the bucket top mister was on shelves by spring. He leveraged his own expertise to create the company's first purely social marketing campaign for the mister.

The result was recognized by Home Depot as one of the most innovative products of the year, and it was featured on HGTV and in Popular Mechanics. It generated millions of dollars in sales during its first year. The City of Anderson, South Carolina, even gave Stearns an award because the product created jobs for the city. And Home Depot asked TTi for more products in that category. What is your funnel, what are your stages, what are your decision points, and what is the criteria? These questions will help you further define the innovation management process, but at a minimum make sure that these five stages are included.

8.14 Project Selection Criteria

As shown in Figure 8-4, a critical decision that must be made in the FFE is the selection of a project. When a company utilizes a one-size-fits-all methodology, only those projects that fit the methodology are selected. Good ideas for projects may not be considered because of the risks that may occur when using the singular methodology. Project managers are traditionally assigned after the project is selected and prioritized.

In the future, as more companies accept the use of flexible methodologies, there will be significantly more choices for projects and project managers must be brought on board during the FFE for the contributions they can make to the project selection process.

As shown in Figure 8-4, project managers will be expected to participate in the WOTS analysis for project selection. Experienced project managers can be valuable assets in discussions related to the organization's strengths and weaknesses related to this potential project, such as:

Strengths:

- Competencies of the workforce
- Innovation skills needed
- Capabilities of existing technologies
- Effectiveness of sponsorship and governance

Weaknesses:

- Expected line-of-sight to senior management
- Missing competencies
- Availability of the critical resources
- Unavailable technology
- Excessive risks

Strengths and weaknesses are usually analyzed in terms of human and nonhuman resources. Strengths and weaknesses can also be discussed based on the capabilities of each functional group such as engineering, procurement, and marketing, and the support they must provide for this project.

Opportunities and threats are usually analyzed from a business perspective by the senior managers participating in the FFE. The enterprise environmental factors and VUCA analysis may impact how the opportunities and threats are looked at. Another important business factor is

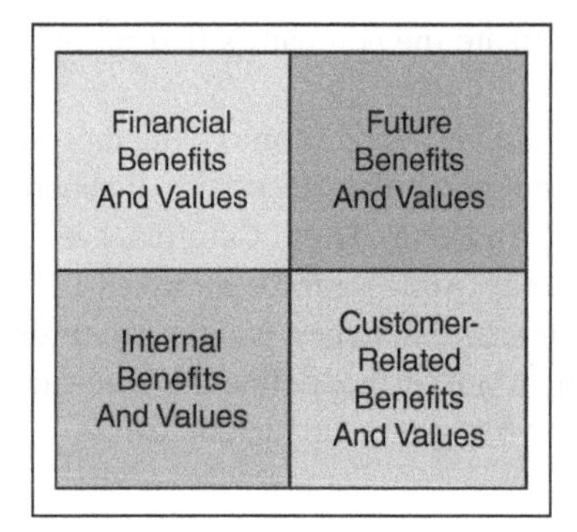

Figure 8-7 Categories of Projects

where this project fits in the portfolio of projects. Let us assume that a company has four categories of projects as shown in Figure 8-7. Internal projects focus upon continuous improvement efforts such as implementing best practices to improve the efficiency and effectiveness of operations. Financial projects are the products and services you provide to generate cash flow. Future projects could involve innovation and R&D. Customer projects are special or one-of-a-kind projects to support a customer's specific needs. Companies may require a balance of projects in each quadrant.

The projects in each of these quadrants are expected to be managed by project managers. Each quadrant may require a different type of flexible or inflexible methodology and a different set of project management processes, tools, and techniques.

Organizations can help alleviate some of the pain points in evaluating ideas and potential projects by establishing project selection criteria. The criteria may be different for each of the quadrants in Figure 8-7. The selection or suitability criteria for a new product might include:

- Using similar technology
- Using similar marketing and distribution channels
- Can be sold by current sales force
- Can be purchased by existing customer base
- Fits company philosophy, profit goals, and strategic plan
- Fits current production facilities
- Supports the organization's core values

When companies look at just one project by itself in the FFE, there is always justification for violating the selection criteria. But when a company must consider a portfolio of projects, such as Airbus discussed later in this chapter, the selection criteria become of critical importance because it may focus on multiple projects over a long horizon.

If we now look at Gate #3 in Figure 8-6, the final go or no-go decision may be influenced by the project manager's estimation of the resources needed, as shown in Figure 8-8. When project managers do not participate in the FFE, projects are often approved and prioritized without considering the competencies or availability of resources needed. When project managers participate in the FFE, better decisions on project selection are expected.

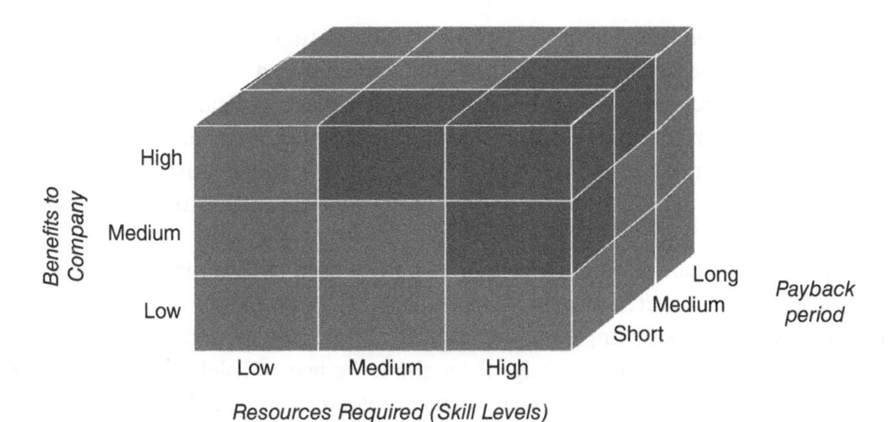

Figure 8-8 Resource Requirements Model

8.15 Excellence in Action: AstraZeneca[5]

The Use of Hybrid Project Management Methodology in Regulatory Submission Projects in Clinical Pharmacology and Quantitative Pharmacology

Introduction

Project management was introduced to the pharmaceutical industry quite recently and is still a novelty, although it has been used in other industries with great success. It can be used by pharmaceutical companies in the drug development process, from discovery to submission, launch, and life-cycle management including technical development, clinical trials, regulatory submissions of new molecular entities, technical transfers, and postmarket activities (Tripathy & Mohanty 2016) The project management approach can be used not only to lead projects but also to support programs in the portfolio. A particular example of this process is regulatory submissions, which documents specific data and analyses that must be delivered to regulatory agencies before the product is introduced to the market following approval. This process is obligatory in the pharmaceutical industry (DS Informatics 2020).

5 Material in this section was provided by:

Dorota A. Andrzejewska - Górecka, Ph.D. - Clinical Pharmacology and Quantitative Pharmacology, Clinical Pharmacology and Safety Sciences, R&D, AstraZeneca, Warsaw, Poland

Diana M. Malewicka - Clinical Pharmacology and Quantitative Pharmacology, Clinical Pharmacology and Safety Sciences, R&D, AstraZeneca, Warsaw, Poland

Alison M. Burden - Clinical Pharmacology and Quantitative Pharmacology, Clinical Pharmacology and Safety Sciences, R&D, AstraZeneca, Gaithersburg, MD, USA

Kyle Wolfe - Clinical Pharmacology and Quantitative Pharmacology, Clinical Pharmacology and Safety Sciences, R&D, AstraZeneca, Gaithersburg, MD, USA

Paramjeet Subramony, PhD - Clinical Pharmacology and Quantitative Pharmacology, Clinical Pharmacology and Safety Sciences, R&D, AstraZeneca, Gaithersburg, MD, USA

Regulatory submissions require the engagement of cross-functional teams within the organization. Information and datasets must be gathered and analyzed by highly specialized subject matter experts. Resources are often limited, and these experts typically handle multiple projects at a time. The pharmaceutical industry is highly regulated, so management of risks is key to achieve the goal of producing high-quality documents within required timelines. Project management methods and tools can help to increase the efficiency of the regulatory submission process.

Project management can be summarized as a strategy used by organizations to realize their objectives and success (Project Management Institute 2021). Project Managers assist project teams in keeping projects on track by implementing a successful management framework and facilitating the flow of information and knowledge throughout the team (Kisielnicki 2014). This flow of knowledge is especially important for research and development projects, which sit at the core of the pharmaceutical industry.

A fundamental concept of project management is the "absorptive capacity" of knowledge, which refers to a team's ability to use existing knowledge and experience to assimilate new information and apply it to present tasks. Essentially, absorptive capacity drives innovation by promoting organizational learning. Evidence in the literature demonstrates that project management contributes to the absorptive capacity of the group, thereby increasing its overall effectiveness (Vicente-Oliva et al. 2015).

Traditionally, the pharmaceutical industry focuses on the rigid and linear framework that creates lengthy timelines (Vaidyanathan et al. 2019). Projects are traditionally developed using a framework that is rigid and linear and in which most tasks cannot be completed until the completion of the previous task. Project plans are created up front, and there is no revisiting of tasks that have already been completed.

The "Agile Manifesto" is a set of principles created by a group of software engineers who came together in 2001 to define a more efficient framework for software development (Pathak & Saha 2013). The creators aimed to establish a set of values and principles that facilitate project completion. Although originally implemented for software development, so-called "agile" practices have spread to a variety of fields, including pharmaceutical research and development. The iterative approach to project management helps to promote efficiency and quality through increased collaboration and flexibility.

Scrum is a popular agile framework for project management in which an entire project is divided into smaller sections called "sprints." With these smaller iterations and frequent "Scrum meetings," teams remain flexible throughout the entire project. Typically, agile teams participate in daily Scrum meetings to discuss progress and concerns about the project. During a research-based project, a daily meeting might be redundant, but even meeting once or twice a week can be extremely valuable to a team. Scrum allows separate teams or individuals to work on different tasks asynchronously (Hidalgo 2019).

Agile processes can greatly enhance the field of pharmaceutical development by fostering innovation to bring new drugs to patients more quickly. The collaborative and iterative approach defined by agile project management promotes higher-quality work and transparency between individuals and the team. Reports in the literature have shown that a collective understanding of team goals and roles can promote success in a scientific setting (Hidalgo 2019). The implementation of agile practices can be complex, especially in the context of research. A transition process from a traditional project management framework to an agile framework requires the support of the entire organization (Pavlović et al. 2018).

In general, regulatory submission projects are led according to traditional methodology, although an agile approach can also be used. The aim of this work is to present project management methods used in regulatory submissions in the Clinical Pharmacology and Quantitative Pharmacology (CPQP) functional area in AstraZeneca. This chapter describes the process of preparing documents for regulatory submissions and presents an analysis of project management approaches used in this process. In addition, the lessons learned process and a case study are presented to show measurable benefits.

Glossary

The abbreviations used in this work and their definitions are listed below.

ADA: anti-drug antibodies
ADaM: Analysis Data Model
BLA: Biologics License Application
B&I: Biometrics and Information
CPQP: Clinical Pharmacology and Quantitative Pharmacology (functional area in AstraZeneca)
CSR: clinical study report
CST: clinical subteam
eCRT: electronic Case Report Tabulation
eCTD: electronic Common Technical Document
EMA: European Medicines Agency
ERES: Exposure-Response of Efficacy and Safety
EU: European Union
FDA: US Food and Drug Administration
ISI: Integrated Summary of Immunogenicity
MAA: Marketing Authorization Application
MAST: Marketing Application Submission Team
NDA: New Drug Application
PD: pharmacodynamic
PK: pharmacokinetics
PO: purchase order
PopPK: population pharmacokinetics
RSDT: regulatory submission delivery team
SDTM: Study Data Tabulation Model
SOW: statement of work

Regulatory Submission Projects in CPQP

In this section, the term **regulatory submission** refers to any filing, application, or submission to or with a regulatory authority. The purpose of this process is to verify whether potential new human medicines are safe and effective for their assigned use in specific groups of patients and whether they meet quality standards (Hearns-Stewart et al. 2021).

An application for a new molecule or indication is usually first submitted to US or European Union (EU) markets. The approval process is managed by the Food and Drug Administration (FDA) in the US and the European Medicines Agency (EMA) in the EU. The request for a regulatory submission to the FDA is called a "New Drug Application" (NDA) for non-biologic products or "Biologics License Application" (BLA) for biologic products, and a request to the EMA is called a "Marketing Authorization Application" (MAA).

A regulatory submission to the FDA or the EMA must be prepared as an electronic Common Technical Document (eCTD). The eCTD is a structured document whose format has been agreed upon by participating countries, including the US and the EU, for submission to regulatory authorities. The eCTD is divided into five modules. Module 1 is region specific, whereas Modules 2 through 5 are usually common to all regions:

Module 1. Regional Administrative Information and Prescribing Information
Module 2. Common Technical Document Summaries
- 2.1 Table of Contents of Modules
- 2.2 CTD Introduction
- 2.3 Quality Overall Summary
- 2.4 Nonclinical Overview
- 2.5 Clinical Overview
- 2.6 Nonclinical Written and Tabulated Summaries
  - 2.6.1 Pharmacology
  - 2.6.2 Pharmacokinetics
  - 2.6.3 Toxicology
- 2.7 Clinical Summary
  - 2.7.1 Biopharmaceutic Studies and Associated Analytical Methods
  - 2.7.2 Clinical Pharmacology Studies
  - 2.7.3 Clinical Efficacy
  - 2.7.4 Clinical Safety
  - 2.7.5 Literature References
  - 2.7.6 Synopses of Individual Studies

Module 3. Quality
Module 4. Nonclinical Study Reports
Module 5. Clinical Study Reports

In addition to the eCTD, all datasets and models used in analyses and reports must be submitted to the regulatory authority as an electronic Case Report Tabulation (eCRT) (US Food and Drug Administration 2017; European Medicines Agency 2021).

Responsibility of CPQP in Regulatory Submissions

In AstraZeneca, the CPQP functional area supports all therapeutic areas in all phases of clinical drug development, including small molecules, new molecular entities, and biologics. CPQP takes responsibility for everything connected to dosing, including:

- Designing the right trial
- Creating the right assay
- Selecting the right dose
- Preparing the right label
- Identifying an appropriate group of patients

In regulatory submission projects, CPQP is responsible for preparing Modules 2.7.1 and 2.7.2 of the eCTD. This includes summaries of the Clinical Pharmacology, Biopharmaceutics, Bioanalytical, Population pharmacokinetics (PopPK), Exposure-Response of Efficacy and Safety and Integrated

Summary of Immunogenicity (ISI) aspects of the development program. For 2.7.1 and 2.7.2, much of the content can be written ahead of phase 3 readout, especially for a first filing. Currently project management support is only provided for biologics. The process for NDA filings is similar to biologics except ISI.

Within CPQP, the Clinical Pharmacology Lead is responsible for Clinical Pharmacology deliverables in regulatory submission projects. Also part of CPQP, the project managers support Clinical Pharmacology regulatory submissions, process improvements, and business operational tasks. Project managers collaborate with CPQP team members to achieve goals in other miscellaneous areas.

Project management support is needed to ensure the last set of deliverables are met and CPQP is not the rate-limiting for the filing(s). For a Clinical Pharmacology regulatory submission, the project manager collaborates with the Clinical Pharmacology lead, who sponsors submission projects from CPQP. Together they create the project strategy. The project manager is responsible for all organizational issues, communications, and resource and timeline management, whereas the Clinical Pharmacology lead takes responsibility for scientific and technical issues and document alignment (Table 8-1).

Table 8-1 Scope and responsibilities of clinical pharmacology lead and project manager at AstraZeneca

Clinical Pharmacology Lead	Project Managers
• Leading NDA or BLA/MAA Clinical Pharmacology team • Creating the content plan for Modules 2.7.1 and 2.7.2 in collaboration with NDA-BLA/MAA team members • Collaborating with Medical Writer in preparing Modules 2.7.1 and 2.7.2 and ISI contents • Reviewing Modules 2.7.1, 2.7.2, 2.7.3, 2.7.4, and 2.5; ISI; and CSR • Responsibility for final content and delivery of Modules 2.7.1 and 2.7.2 • Collaborating with pharmacometrician and programmers to prepare datasets • Collaborating with Clinical Pharmacology team members in preparing PopPK/ERES, ISI and CSR deliverables, including creating request form, aligning budget, and reviewing deliverables • Collaborating with cross-functional teams (e.g., programming, regulatory, Medical Writers) to ensure alignment of Clinical Pharmacology deliverables • Responsible for Clinical Pharmacology input on material for meeting with FDA/MAA • Collaborating with Marketing Application Submission Team (MAST), Regulatory Submission Delivery Team (RSDT), Clinical subteam (CST); ad hoc meetings with Clinical Operations Lead, and Regulatory Lead	• Identifying all NDA or BLA/MAA team members (core team, ad hoc team) • Preparing timelines for NDA or BLA/MAA submissions related to Clinical Pharmacology and Bioanalytical deliverables (Modules 2.7.1 and 2.7.2) • Collaborating with cross-functional teams (e.g., programming, regulatory, Medical Writers) to ensure timeline alignment • Developing a plan with the Medical Writer and vendors (support with datasets) regarding monitoring the document review process • Supporting milestone tracking and risk management • Supporting NDA or BLA/MAA team with preparation for Clinical Pharmacology meeting with FDA/EMA • Organizing regular NDA or BLA/MAA meetings with team members to track milestones, review documents, conduct risk management, and assign resources • Collaborating with and attending meetings of MAST, RSDT, and CST, as well as ad hoc meetings with Clinical Pharmacology Lead, Clinical Operations Lead, and Regulatory Lead to align timelines • Providing operational support to Clinical Pharmacology Leads for contracts, vendor management for statement of work (SOW) execution, and purchase order (PO) creation • Facilitating lessons learned session with cross-functional teams in CPQP

During regulatory submission, the Project manager collaborates with other stakeholders:

- Clinical Pharmacology Team (BLA/MAA team)
- Group dedicated to preparing Module 2.7.1
- Group dedicated to preparing Module 2.7.2
- Group dedicated to preparing ISI
- Vendors
- Global Team (MAST, RSDT, CST)

The project manager collaborates with the Clinical Pharmacology team and supports the preparation of documents by setting up meetings and ways of working, creating timelines, and managing risks. The project manager is responsible for communication and aligning the documents prepared by different groups. The project manager also collaborates with global teams like the MAST, the RSDT, and the CST to collect needed information and align timelines.

In some regulatory submissions, PopPK, ERES, or CSR deliverables are outsourced and prepared by external vendors. In these cases, the project manager is responsible for negotiating timelines and budget, as well as other organizational tasks like providing accesses and scheduling meetings. The project manager is also responsible for setting up ways of working between the vendor and the NDA or BLA/MAA team.

Process for Clinical Pharmacology Regulatory Submissions

Clinical Pharmacology regulatory submission projects require input from and collaboration with many functional areas within AstraZeneca. Thus, various resources (including vendors) are needed to complete the submission process.

A Clinical Pharmacology regulatory submission project can be divided in two main steps, before and after the Go/No-Go decision. Some projects can be terminated because of a No-Go decision. It is also possible for a timeline to change significantly or for the project to be put on hold. Key milestones are presented in Figure 8-9.

The Go/No-Go decision for submission will be determined by pivotal phase III outcomes. If interim analysis (Dey & Croft, 2018) is predefined in study protocol, the first key milestone is data cutoff. The next milestone is database lock. High-level results appear one to two weeks after database lock and are presented for Go/No-Go decision. Once Go decision is made, the next key milestone is Planned Statistics Complete. Datasets will be provided by Programmers for planned popPK, ERES analysis, and ISI analysis. All of these reports, together with CSR will be used in Module 2.7.2.

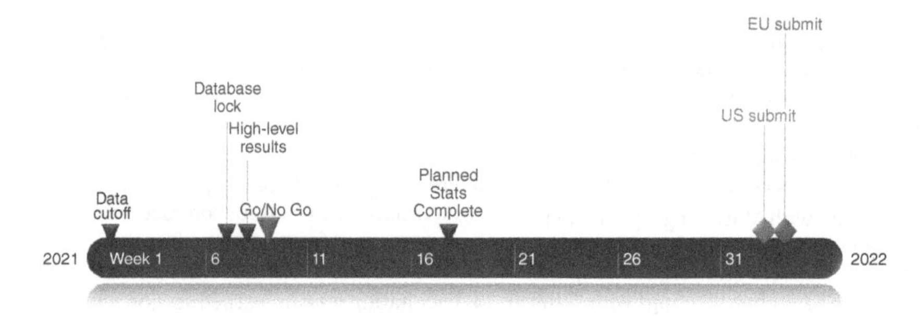

Figure 8-9 Key Milestones in the Regulatory Submission Process

The documents for Modules 2.7.1 and 2.7.2 are prepared in two stages. In the first step, the team prepares the prototype, which does not include the analysis or datasets. The second stage is dedicated to incorporation of the analysis, datasets, and conclusions. The Clinical Pharmacology cross-functional data flow is presented in Figure 8-10.

In the first step of data flow, the raw PK/ADA datasets are transferred to Clinical Data Management by the vendor or internal programing team. Programmers from Biometrics and Information (B&I) in collaboration with biostatisticians create datasets in SDTM (Study Data Tabulation Model) and ADaM (Analysis Data Model) formats. When the datasets are ready, PK Programmers will prepare datasets per data specifications for use in PK/ERES modeling. In the next step, the pharmacometrician creates the PopPK model and ERES analysis and generates the reports. The PopPK model and ERES analysis can be separate or combined reports and are approved by the Clinical Pharmacology team. The final versions of the reports are incorporated into Module 2.7.2, and at the same time, the vendor or programmers prepare the eCRT packages for publishing. Some sections from Module 2.7.2 need to be incorporated into Modules 2.7.3 and 2.7.4, which feed directly into Module 2.5. At the same time, the Clinical Pharmacology team works on Module 2.7.1, in which all reports about clinical studies supporting formulation and/or

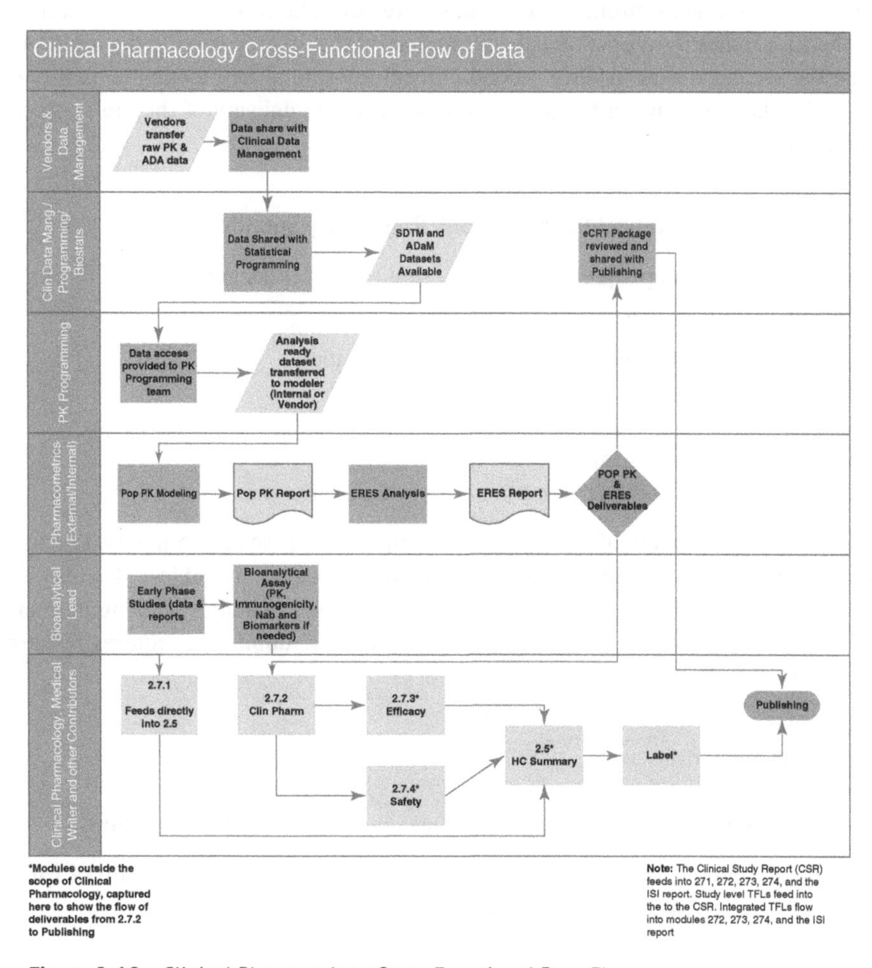

Figure 8-10 Clinical Pharmacology Cross-Functional Data Flow

device development and bioanalytical assays regarding PK, immunogenicity, neutralizing antibodies, and biomarkers are included. Module 2.7.1 feeds directly into Module 2.7.5, and CSRs are completed earlier than Modules 2.7.1 and 2.7.2. Some pieces from CSRs feed directly into Modules 2.7.1, 2.7.2, 2.7.3, and 2.7.4.

Review Process

The preparation and review of prototypes and final documents is divided into a few steps. The timelines for preparation and review of modules are discussed with the BLA/MAA team members and are aligned with the global teams (MAST, CST, RSDT). The cross-functional team begins the preparation of a prototype by creating the content plan. The Medical Writer gathers the information from the team and starts to work on the text of the document that will accompany the datasets, which is then reviewed by the team. After this review, the Medical Writer collates the team's comments and organizes the clinical interpretation meeting, after which the Medical Writer incorporates the comments and finalizes the documents. The documents are then set aside, or "parked," while the team waits for the datasets. The prototype may be reviewed once or twice, depending on the study and the team's preferences.

The final document goes through two rounds of review. The NDA-BLA/MAA team members and members from the cross-functional team take part in the first review, and the senior reviewers undertake the second review. After a second clinical interpretation meeting, a quality check is performed. The last two days of the process are dedicated to delivery of the pre-approval document and the final approval.

Project Management in Clinical Pharmacology Regulatory Submissions

In this document, the term **project** refers to any sequence of tasks and activities that include all of the following requirements:

1) Final product prepared according to a defined specification
2) Timeline
3) Budget
4) Human and nonhuman resources
5) Multifunctional teams and interdisciplinary knowledge (Kerzner 2003)

A regulatory submission qualifies as a project according to these requirements. The final product comprises prepared and published Clinical Pharmacology deliverables: Modules 2.7.1 and 2.7.2, the ISI, and the CSRs. The timeline and budget are specified and dependent on internal and external factors. Finally, a Clinical Pharmacology regulatory submission requires resources that are divided into several categories:

- Human resources
- Knowledge about studies
- Datasets and analysis
- Budget for outsourcing PopPK/ERES, CSR deliverables, or external consultation

Projects require multifunctional teams that are responsible for the preparation of various sections of the submission package and have interdisciplinary knowledge.

Traditional Approach to Clinical Pharmacology Regulatory Submission Projects

In a traditional approach to project management, a project consists of five stages:

1) Project initiation
2) Project planning
3) Project execution
4) Monitoring and controlling
5) Project closure

In project initiation, leadership obtains approval for the project, and the high-level scope, milestones, budget, key resources, and risks are identified. This stage is outside of the scope of the project manager, whose responsibility starts after the project is approved.

In the project planning stage, the project manager supports the Clinical Pharmacology lead in preparation of the project plan, high-level timelines, deliverables, and resources. The project manager confirms the resources, which come from various functions and creates the team structure. The team structure is made up of the core team and the ad hoc team. The core team collaborates on Clinical Pharmacology deliverables, and the ad hoc team provides key information on global timelines and requirements. The structure of these teams is presented in Table 8-2.

Table 8-2 Roles and responsibilities of the core and ad hoc teams

Role	Responsibilities
Core Team	
Clinical Pharmacology Lead	• Responsible for Clinical Pharmacology strategies and related deliverables for submission • Informs the NDA or BLA/MAA team on Clinical Pharmacology submission strategies and status of Clinical Pharmacology deliverables and any associated risks
Project Manager	• Organizes NDA or BLA/MAA team meetings, collects agenda items, schedules meetings, distributes meeting minutes, tracks action items • Develops and maintains Clinical Pharmacology submission timeline, maintains liaisons with clinical/regulatory submission team to ensure timeline alignment • Provides operational support to facilitate smooth execution of NDA or BLA/MAA team deliverables.
Pharmacometrics Lead	• Responsible for all pharmacometrics-related deliverables (PopPK, ERES) for submission • Raises topics related to pharmacometrics • Informs the NDA or BLA/MAA team on pharmacometrics analysis plans and status of pharmacometrics deliverables and any associated risks
Immunogenicity Lead	• Responsible for immunogenicity data interpretation and related deliverables for submission (e.g., ISI) • Provides input to immunogenicity analysis and related document writing.

(Continued)

Table 8-2 (Continued)

Role	Responsibilities
Bioanalytical Lead for PK	• Responsible for all bioanalytical assay (PK, ADA)–related deliverables for submission • Informs the NDA or BLA/MAA team on the timeline/status of assay-related deliverables (e.g., assay validation reports, bioanalytical reports) and any associated risks
Bioanalytical Lead for PD	• Responsible for all PD biomarker for submission • Informs the NDA or BLA/MAA team on biomarker strategies/plans for submission and timeline/status of biomarker-related deliverables and any associated risks
Medical Writing	• Leads development of Modules 2.7.1 and 2.7.2 and provides medical writing support to other related Clinical Pharmacology submission documents as appropriate (e.g., PopPK, ERES, ISI report) • Informs the NDA or BLA/MAA team on the timeline/status of the submission documents and any associated risks
Ad Hoc Team	
Regulatory Project Manager	• Provides input to all submission strategies and plans from regulatory perspective
Global Clinical Lead	• Provides input to all submission strategies and plans from clinical perspective
Biostatistics	• Provides input to all submission strategies and plans from biostatistical perspective • Responsible for dataset/output deliveries related to PK, ADA, PD, PopPK, and ERES
Programming	• Informs NDA or BLA/MAA about timeline and status of programming deliverables related to PK, ADA, PD, PopPK, and ERES and any associated risks
Clinical Operations Project manager	• Provides input from the overall filing plan perspective • Stays informed on Clinical Pharmacology submission plans and timelines and works closely with the project manager to ensure alignment of overall project timelines
Clinical Program Lead	• Oversees clinical efficacy/safety

In some cases, analyses need to be prepared by vendors, and in these instances the project manager coordinates with vendors on future collaborations and sets up a new contract or statement of work if needed.

The last part of project planning is a kickoff meeting, in which the project manager and the Clinical Pharmacology lead together present the team structure, roles and responsibilities, deliverables, project information, high-level timelines, and ways of working. Initially, NDA or BLA/MAA meetings take place on a monthly basis, and after a "Go" decision, the meetings occur more frequently. Project managers are responsible for setting up tools like Microsoft Teams and SharePoint sites as repositories for project documents.

The Project Execution stage is usually started before the Go decision and after the kickoff meeting. During this stage, the Project manager supports several types of stakeholders by using various tools to assign and execute project tasks (Table 8-3).

Table 8-3 Tools used by the project manager

Stakeholder	Meetings	Action items	Agenda/ meeting minutes	Risk report	Timelines	Reviewer matrix
Clinical Pharmacology Lead	Weekly	Present, update	Prepare, send	Present, update	Prepare, update, present	Prepare, present
BLA/MAA team	Weekly, biweekly, monthly based on team needs	Present, update	Prepare, send	Present, update	Prepare, update, present	Prepare, present
Group dedicated to preparing Modules 2.7.1, 2.7.2, ISI	Weekly, biweekly, ad hoc based on team needs	Present, update	Prepare, send	Gather risks and mitigate	Prepare, update, present	Use during review
Vendors (dedicated to preparing PopPK/ERES or PK/ADA)	Weekly	Present, update	Prepare, send	Gather risks and mitigate	Prepare, update, present	Use during review
Global teams	Weekly with MAST, RSDT, CST	Prepared by Global Teams	Prepared by Global Teams	Update global risks register	Align with global team	Update in global reviewer matrix

The main responsibility of the project manager is to prepare and manage the timelines for Modules 2.7.1 and 2.7.2 and ISI documents in collaboration with the Clinical Pharmacology lead and medical writers. In addition, the project manager is responsible for organizing regular or ad hoc meetings based on stakeholders' needs. For Clinical Pharmacology regulatory submissions, meetings make up an effective tool for gathering and providing information and for making timely decisions. The duration of the meetings depends on topics; status meetings can take 15–30 minutes, whereas NDA or BLA/MAA meetings can take up to 1 or 1.5 hours. Before the meeting, the Project Manager prepares the agenda by collecting topics from key members. After the meeting, the project manager is responsible for preparing meeting minutes and action items and sending them to the team.

Identification of appropriate reviewers is a key activity in Clinical Pharmacology regulatory submission projects. For this purpose, the project manager uses a reviewer matrix, which includes information regarding reviewers, senior reviewers, and approvers.

The project manager also takes responsibility for risk management, including risk identification and regular review by team members. Risks are classified based on probability and impact on the project and are included in the risk register.

The project manager monitors and controls the scope and timelines of the project for the NDA or BLA/MAA team. A request for changing the scope or timelines could come from the global team (MAST, RSDT, CST) or the NDA or BLA/MAA team members. In such instances, the project manager discusses the strategy or change request with the NDA or BLA/MAA team members and the Clinical Pharmacology Lead. A decision to increase the scope of the project is made by senior stakeholders. The Project Manager takes responsibility for identifying and managing change requests.

Table 8-4 Project phases

Project phase	Main activities and tools
Project initiation	Leadership obtains approval for the project. **Outside Project Manager's scope**
Project planning	Planning and describing main deliverables, team structure, task details timelines, resources **Tools:** Meetings with Clinical Pharmacology Lead and other stakeholders, team structure, timelines, MS Teams
Project execution	Assign and execute milestones and tasks **Tools:** Meetings, agenda/meeting minutes, progress report, details timelines, reviewer matrix, risk report
Monitoring and controlling	Project manager monitors and controls the scope and timeline of the project **Tools:** Meetings, agenda/meeting minutes, progress report, details timelines, risk report
Project closure	Support ends or continues after submission milestones **Tools:** NDA or BLA/MAA team meeting, lessons learned , project closure documents

During the Project Closure stage, the project manager is responsible for closing the documents and conducting the lessons learned session. The lessons learned session has measurable benefits regarding process improvement and knowledge database for future projects. Project phases are shown in Table 8-4.

Lessons Learned

The objective of lessons learned sessions is to collect knowledge and share best practices from one submission project to the next. Conducting lessons learned sessions and sharing the outcome provides valuable information to teams in planning submission activities. The lessons learned session is held after project completion or completion of any key milestones. For projects using agile methodology with the Scrum framework, lessons learned discussions are part of retrospective meetings.

The lessons learned process contributes significantly to the improvement of business processes and subsequent projects carried out in the organization. The lessons learned process is illustrated in Figure 8-11.

The first step is to collect completed questionnaires from the team members, using a tailored template. The facilitator, in collaboration with the project manager, organizes and conducts lessons-learned brainstorming session with team members. The participants of the lessons learned session are team members from various workstreams (clinical pharmacology, pharmacometrics, bioanalysis, statistical, translational science, medical writing, regulatory, and project management). Insights provided by project team members before the session serve as the starting point for discussions. At the end summary, recommendations and action items from the session are included in the lessons learned report, which is shared with the team and the CPQP leadership team and is stored on the SharePoint site. Some ideas for business process changes identified during lessons learned are implemented as "quick wins," whereas others may result as initiatives of business transformation projects in the future.

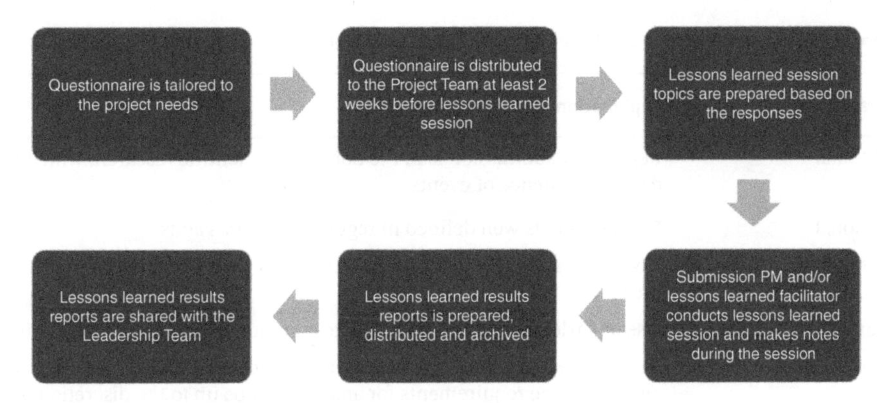

Figure 8-11 The Lessons Learned Process

Agile Approach in Regulatory Clinical Pharmacology Submission Projects

Although the implementation of agile methodologies in research could improve the quality of outputs, not all projects can fit seamlessly into an agile framework (Pavlović et al. 2018). Differences between traditional and agile approaches are shown in Table 8-5. Both traditional and agile project management approaches can be adapted for regulatory submission projects. In AstraZeneca, the NDA or BLA/MAA team is moving toward an agile approach but has not fully adopted this framework due to certain limitations imposed by the nature of work.

The different project management approaches used in Clinical Pharmacology regulatory submissions are shown in Table 8-6. The elements of project management are divided into eight categories. Categories 1, 2, and 3 use traditional methodology, and categories 4 and 6 refer to agile methodology. Categories 5, 7, and 8 combine both approaches.

In the traditional approach, a defined sequence of events is used, the end product is well defined, and documentations are structured according to regional health authority requirements. In the agile approach, many meetings are conducted, and various tools are used to help facilitate communication and collaboration.

Table 8-5 Comparison of traditional and agile methodologies

Category	Traditional methodologies	Agile methodologies
1 Sequence of events	Sequence of events is firm	Sequence of events can change
2 End product	End product is established	End product can be changed
3 Documentation	Documentation is structured	Documentation is optional
4 Technology	Technology is standard	Technology can be adaptive
5 Requirements	Requirements are defined at the beginning of the project	Requirements can change during the project
6 Meetings	Limited number of meetings	Many meetings during the project
7 Team	Structured team	Division into many self-organized subteams for various tasks
8 Decision	Decisions are made top-down	Decisions are made within the team

Adapted from Jovanovic & Berić (2018).

Table 8-6 Analysis of various project management approaches supporting regulatory submission

Agile/Traditional	Explanation
1 Traditional	Regulatory submission is in the domain of scientific research, which has a defined sequence of events
2 Traditional	End product is well defined in regulatory submissions
3 Traditional	Documentation is structured according to a highly regulated submission process
4 Agile	Use of various tools to help facilitate communication and collaboration
5 Traditional and agile	Requirements for each submission are set according to the regional health authority. The requirements for analysis can be up to the discretion of the researchers.
6 Agile	Frequent meetings between team members and stakeholders
7 Traditional and agile	Structured and subteams
8 Traditional and agile	Project management team has overall authority to make decisions within the team and top-down decisions

In some categories, the two approaches can be combined. The NDA or BLA/MAA team structure is divided into multiple subteams. Requirements for each submission are set according to the regional health authority, and analysis can be up to the discretion of the researchers. Decisions are centralized or decentralized based on project needs. For Clinical Pharmacology regulatory submissions, it is best to use a mix of traditional and agile methods for smooth execution of projects.

Case Study

Project management serves to oversee Clinical Pharmacology regulatory submissions to ensure that tasks are completed with high quality and within timelines. This approach provides many benefits, such as conducting effective meetings, providing operational support, and managing timelines, stakeholders, vendors, risks, and resources.

Table 8-7 shows the approximate number of hours spent by Project Managers in Clinical Pharmacology regulatory submissions. On average, 25 hours of Project Manager support per month allow Clinical Pharmacology Leads to focus on multiple scientific activities.

Summary

CPQP plays an important part in AstraZeneca's regulatory submissions to bring new drugs to the market for patients. CPQP takes responsibility for designing the right trial, creating the right assay, developing an appropriate dosing, preparing the right label, and identifying an appropriate group of patients. Clinical Pharmacology is responsible for the preparation and completion of Modules 2.7.1 and 2.7.2, ISI documents, and CSR deliverables. Regulatory submission projects require the input and collaboration of many functional areas especially after the readout of the Phase 3 studies when several critical data become available that need to be part of the final deliverables. Various resources are needed to complete the submission process, including document review. The Clinical Pharmacology lead from CPQP takes responsibility for all scientific and technical issues, and the project manager is responsible for overall organizational issues,

Table 8-7 Case study

Project manager activities per project	Time dedicated for activities (hours/month)
NDA or BLA/MAA team regular meeting, other meetings (organization: scheduling, preparing agendas and meeting minutes)	4
Preparing, monitoring, and aligning timelines for PopPK/ERES, CSR, and Modules 2.7.1 and 2.7.2 with key stakeholders	8
Registration and monitoring of issues and risks	4
Project action item tracking, project progress reporting	4
Operational support (SOWs, POs, invoices)	2
Vendor management and contract issues support	3
Total time savings for scientists	**25**

communications, and resource and timeline management. The Clinical Pharmacology lead and the project manager collaborate to achieve project goals with high quality and within timelines. Hybrid project management tools have been utilized in Clinical Pharmacology regulatory submissions, and this combination of traditional and agile approaches is an efficient and effective way to lead successful submission projects in CPQP.

Acknowledgments

We would like to thank David W. Boulton, Deborah J. Shuman, Stephanie Smith, Weifeng Tang and Diansong Zhou for their review and valuable comments.

8.16 Excellence in Action: Airbus[6]

Within Product Management Lifecycle, the concept phase is critical to define the key development projects for any industry, which bring more benefit to the business and most important which ones will increase the competitiveness of products.

This article describes a process called Business Driven Program Roadmap (BDPR), which is used to as a tool to prioritize and to arbiter research and development projects within Portfolio Product Management.

Within large enterprise with a complex and extend portfolio products, it is key to know the areas to research and to promote incremental developments in order to provide higher return of investments, and more competitiveness. As important as arbitration is the transparency of incremental developments projects on going, and much more how there are interlinking together between different products within Portfolio.

The mission of BDPR is to ensure that Portfolio Programmes and projects pull from a business perspective a portfolio of products and services in the mid-term (years 1 to 5) and long-term (years

6 Material in this section has been provided by Airbus.

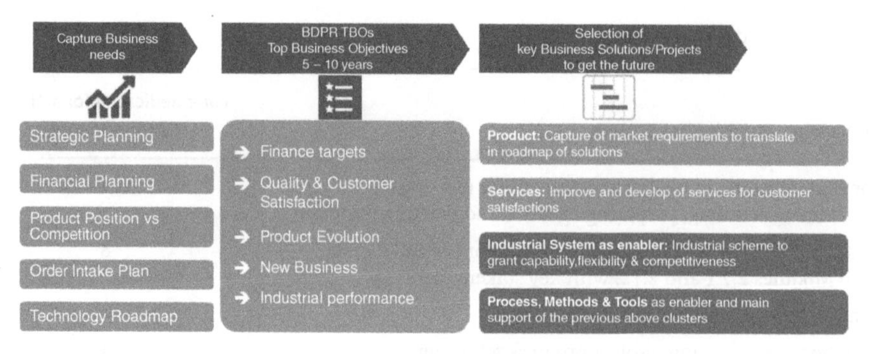

Figure 8-12 BDPR Business Driven Programme Roadmap Key Principles

6 to 10). Therefore, in the long-term BDPR shall be fully aligned with each program unit and its strategy vision, and also shall ensure alignment across all of them. Figure 8-12 illustrates typical BDPR program roadmap key principles.

The BDPR is annual cycle governance with regards to incremental developments for products and services programs and projects together with their financial implications and it forms part of the Company key governance controls. The governance is built "right to left" following Operation Plan Cycle, aiming the end result of the validation of Programmes framing by Executive Committee within Q3 of the year. The BDPR annual cycle governance is built around 1 Kick-off, 1 Steering Committee, and 1 Close-out meetings, held as part of the Executive Committees meetings as follows:

- BDPR Kick-off to Capture Business needs (Q1) with the confirmation of Top Business Objectives at within BDPR Time frame (N+5 and N+10 years). This deliverable is built in one hand from Global Marketing Analysis with the comparison of our products with the competitors and Voice of Customer to identify the potential new products features needs; on the other hand, Programme Strategy gathers inputs from Customer and Stakeholders to consolidate and to propose the potential scenarios concerning product development.
- BDPR Steering Committee with the selection of Key business projects at Q2, with the review of BDPR Deliverables to provide strategic guidance, to ensure and to confirm prioritization of incremental development projects and financial consolidation.
- BDPR Close-Out with the Prioritization and arbitration of the key projects at Q3 with the final confirmation of BDPR Data Package and their incremental development projects, which will be included within research and development budget.

In addition, to ensure full consistency (before the BDPR Kick-Off, Steering Committees and Close-Out meetings) there shall be a series of alignment sessions with Programme and enablers Functions (Engineering and Industrial).

One of the key deliverables is the BDPR roadmap, which is a collaborative time-based plan at 5–10 years that offers a look into the future by representing visually where the business is, where it wants to go, and how to get it there.

The BDPR Roadmap shown in Figure 8-13 is an effective communication and decision-making methodology which links strategic vision and market/customer needs towards a desired future state of products and services business. It represents interdependencies among business solutions/projects which anticipate alternative routes to optimize resource allocation and minimize risks. As well, it allows evaluation of technological changes and supports Programmes and Services in the prioritization and arbitration of business opportunities that matter for realizing a desired future.

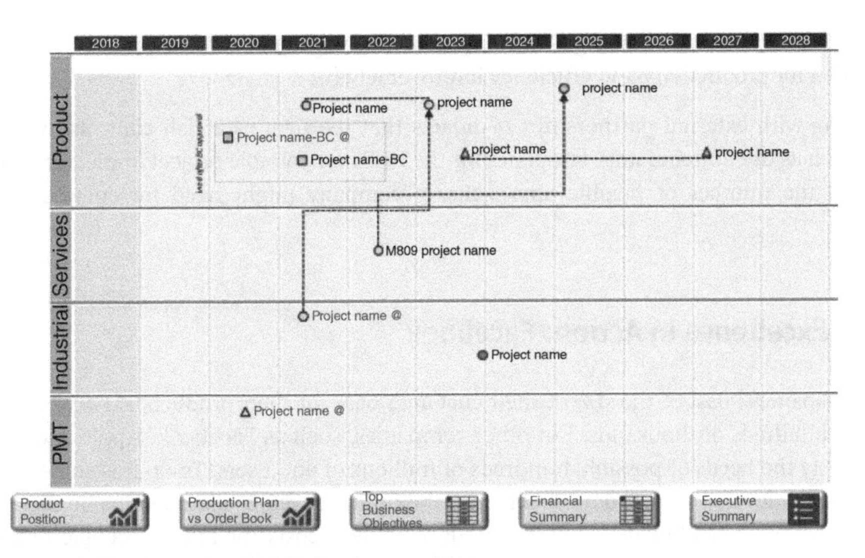

Figure 8-13 Example of BDPR Roadmap at 10 Years

Finally, BDPR as an integrated process provides the following benefits to the company:

- Provide transparency on incremental product development over portfolio and acts as a single point of truth about incremental developments within company.
- Facilitate the prioritization and arbitration of projects driven by business needs within different programs.
- Promote standardization of incremental developments in different products to increase the value added of investments in research and developments.
- Reduce bureaucracy and improve alignment among all stakeholders in a common shared governance.
- Integration between programs development needs with other layers as services, industrial and process, methods, and tools enablers.

8.17 Partnership Fuzzy Front Ends

When we look at organizational strengths and weaknesses, we tend to focus on only internal strengths and weaknesses. We also assume that the entire FFE process involves internal resources only. Companies today are recognizing that they cannot rely entirely on their own research but should solicit ideas from outside the firm including consumers, rival companies, academic institutions, licensing, and joint ventures. The boundaries between the firm and its environment are now permeable.

Companies such as Apple and Facebook tap into the knowledge of thousands of people hoping to generate ideas for new products and services. Partnering with these people during FFE activities offers several benefits to companies operating on a program of global collaboration:

- Reducing the cost for idea generation
- Reducing the cost of conducting research and development on their own
- Incorporating customers early in the development process to get buy-in and possibly partial ownership

- A more accurate understanding of the targets market segments and customers
- Potential for productivity and efficiency improvements

Working with external partnerships mandates that the firm establish controls over the processes such as use of mandatory templates for the FFE and possible project implementation. This can limit the number of flexible approaches a company might need for commercializing a good idea.

8.18 Excellence in Action: Facebook

Some companies boast of the size of their customer base for their products/services in the thousands or hundreds of thousands. But other companies, such as Facebook, Apple and Microsoft, must satisfy the needs of possibly hundreds of millions of end users. To do this successfully, and continuously develop new products/services, they must develop relationships or even partnerships with people worldwide. This lowers development costs for many strategic projects and allows companies to tap into the intellectual capital of others not on the company's payroll.

Background

To make it easy for outsiders to partner with the company on many new or strategic projects, companies like Facebook create Platforms. Facebook launched the Facebook Platform on May 24, 2007, providing a framework for software developers and other volunteers to create applications that interact with core Facebook features. A markup language called Facebook Markup Language was introduced simultaneously; it is used to customize the "look and feel" of applications that developers create. Using the platform, Facebook launched several new applications, including Gifts, allowing users to send virtual gifts to each other, Marketplace, allowing users to post free classified ads, Facebook events, giving users a method of informing their friends about upcoming events, Video, letting users share homemade videos with one another, and social network game, where users can use their connections to friends to help them advance in games they are playing. Many of the popular early social network games would eventually combine capabilities. For instance, one of the early games to reach the top application spot, Green Patch, combined virtual Gifts with Event notifications to friends and contributions to charities through Causes.

Third-party companies provide application metrics, and several blogs arose in response to the clamor for Facebook applications. On July 4, 2007, Altura Ventures announced the "Altura 1 Facebook Investment Fund," becoming the world's first Facebook-only venture capital firm.

Applications that have been created on the Platform include chess, which both allow users to play games with their friends. In such games, a user's moves are saved on the website allowing the next move to be made at any time rather than immediately after the previous move.

By November 3, 2007, seven thousand applications had been developed on the Facebook Platform, with another hundred created every day. By the second annual developers conference on July 23, 2008, the number of applications had grown to 33,000, and the number of registered developers had exceeded 400,000. Facebook was also creating applications in multiple languages.

Mark Zuckerberg said that his team from Facebook is developing a Facebook search engine. "Facebook is pretty well placed to respond to people's questions. At some point, we will. We have a team that is working on it," said Mark Zuckerberg. For him, the traditional search engines return too many results that do not necessarily respond to questions. "The search engines really need to evolve a set of answers: 'I have a specific question, answer this question for me.'"

Conclusion

Facebook appears to have successfully managed and developed strategic partnerships with application providers given the fact that they have more than 400,000 providers. This type of strategic partnership approach could be part of the future of project management.

8.19 Life-Cycle Phases

For years, academia taught that traditional project life-cycle phases begin once the project is approved and a project manager is assigned, and end after the deliverables have been created. However, when benefits realization and value management become important, as is happening now and will continue in the future, there are additional life-cycle phases that must be included as shown in Figure 8-14.

Project managers are now being brought on board earlier than before, such as in the FFE, and remaining after the deliverables have been produced to measure the business value created. Figure 8-14 is more representative of an investment life cycle than a traditional project life cycle. If value is to be created, then the benefits must be managed over the complete investment life cycle. The traditional project life cycle falls within the investment life cycle. More than six life-cycle phases could have been identified in the investment life cycle, but only these six will be considered here for simplicity.

Flexible methodologies will allow project teams to establish the life-cycle phases that best fit the project and the requirements of the stakeholders. It is possible that in the future stakeholders will have an input into the selection of the life-cycle phase and the accompanying gates.

The **idea generation phase**, which is like the FFE and often includes a feasibility study and a cost-benefit analysis, is where the idea for the project originates. The idea can originate in the client's or business owner's organization, within the senior levels or lower levels of management in the parent company or the client's firm, or within the organization funding the project. The output of the idea generation phase is usually the creation of a business case.

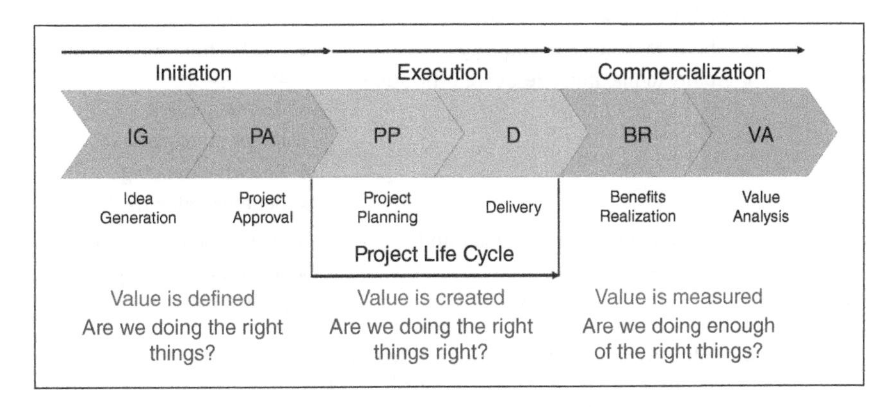

Figure 8-14 Investment Life Cycle

Although the idea originator may have a clear picture of the ultimate value of the project, the business case is defined in terms of expected benefits rather than value. Value is determined near the end of the project based on the benefits that are achieved and can be quantified. The benefits achieved may be significantly different from the expected benefits defined at project initiation because of many of the reasons stated previously that can induce changes.

Not all projects require the creation of a business case. Examples might include projects that are mandatory for regulatory agency compliance and are well understood or simply to allow the business or part of the business to continue more efficiently.

Once the business case is prepared, a request is sent to the PMO for project approval. Companies today are establishing a portfolio management PMO to control the **project approval phase** and to monitor the performance of the portfolio of projects during delivery.

The PMO must make decisions for what is in the best interest of the entire company. A project that is considered as extremely important to one business unit may be a low priority when compared to all the other corporate projects in the queue. The PMO must maximize the business value of the portfolio through proper balancing of critical resources and proper prioritization of projects. The PMO must address three critical questions as shown in Table 8-8.

The activities identified with the third question in Table 8-1 are usually part of the portfolio PMO's responsibility for determining if all the benefits were captured or if additional projects need to be added to the queue.

Table 8-8 Typical role for a portfolio PMO

Critical Questions	Areas of Consideration	Portfolio Tools and Processes
Are we doing the right things?	• Alignment to the strategic goals and objectives such as shareholder value, customer satisfaction or profitability • Evaluation of internal strengths and weaknesses • Evaluation of available and qualified resources	• Templates to evaluate rigor of the business case • Strategic fit analysis and linkage to strategic objectives • Matrix showing the relationships between projects • Resource skills matrices • Capacity planning templates • Prioritization templates
Are we doing the right things right?	• Ability to meet expectations • Ability to make progress toward benefits • Ability to manage technology • Ability to maximize resource utilization	• Benefit realization plans • Formalized, detailed project plans • Establishing tracking metrics and KPIs • Risk analysis • Issues management • Resource tracking • Benefits/value tracking
Are we doing enough of the right things?	• Comparison to strategic goals and objectives • Ability to meet all the customers' expectations • Ability to capture all business opportunities that are within the capacity and capability of the company's resources	• Overall benefits tracking • Accurate reporting using the project management information system

Most companies tend to believe that the project managers should be brought on board the project after the project has been approved and added to the queue. The argument is that project managers are not businesspeople, have limited information that could help in the approval process and are paid to make just project-based decisions. This is certainly not true today. In today's world, project managers view themselves as managing part of a business rather than just managing a project. As such, project managers are paid to make both project-based and business-related decisions on their projects.

When project managers are brought on board after project approval, they are at the mercy of the information in the business case and benefits realization plan. Unfortunately, these two documents do not always contain all the assumptions and constraints, nor do they discuss the thought process that went into creating the project.

Perhaps the most important reason for bringing the project manager on board early, such as in the FFE, is for resource management. Projects are often approved, added to the queue, and prioritized with little regard for the availability of qualified resources. Then, when the benefits are not delivered as planned, the project manager is blamed for not staffing the project correctly.

Project managers may very well be the best people qualified to critically identify the number of resources needed and the skill levels of the assigned staff. This makes it easier for the portfolio governance personnel to perform effective resource management practices according to Figure 8-15.

Even when assigning project managers early in the investment life cycle, resource management shortcoming can occur, such as:

- Not all resource demands are captured.
- Poor knowledge of the resource skill levels needed still exists.
- Resource needs can change on a project due to scope changes.
- Needed resources might not be accounted for if transformational activities are required.
- Priorities might shift due to firefighting on other critical projects.
- There could be unrealistic benefit and value estimates.

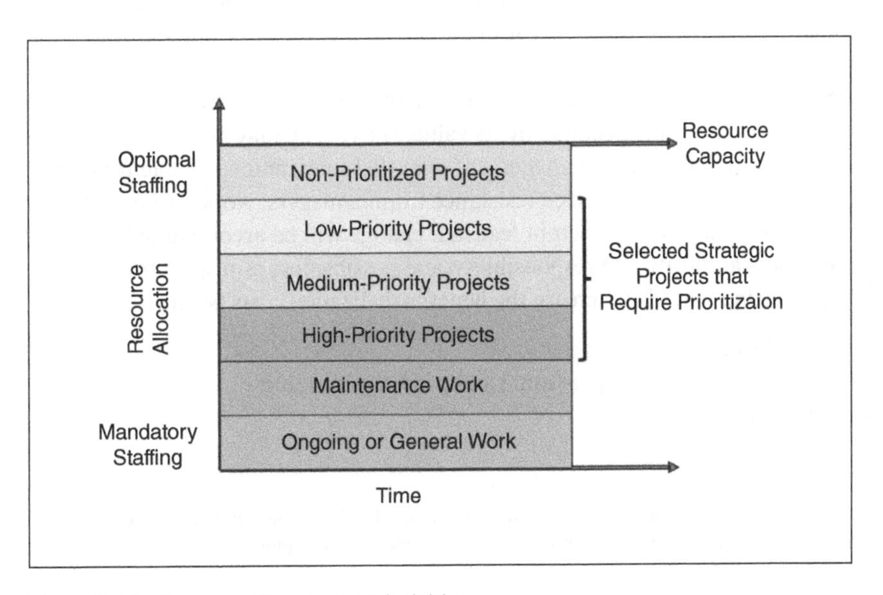

Figure 8-15 Resource Management Activities

If the shortcomings are not identified and effectively managed, the results can be:

- A failure of benefits realization planning
- No maximization of portfolio business value
- Continuous changes to the portfolio
- Continuous reprioritization
- Continuous conflicts over manpower

The third life-cycle phase is the **project planning phase**. This phase includes preliminary planning, detailed planning, and updates to benefits realization planning. Although the business case may include assumptions and constraints, there may be additional assumptions and constraints provided by the PMO related to overall business objectives and the impact that enterprise environment factors may have on the project. The benefits realization plan that may have been created as part of the business case may undergo significant changes in this phase.

The benefits realization plan is not the same as the project plan but must be integrated with the project plan. The benefits realization plan and the accompanying project plan may undergo continuous changes as the project progresses based on changing business conditions.

The fourth life-cycle phase is the **delivery phase**. This phase, as well as the project planning phase, are commonly based on the traditional domain areas that were identified in the ***PMBOK® Guide***, 6e. In this phase, the project manager works closely with the PMO, the business owner and the steering/governance committee to maximize the realization of the project's benefits.

Performance reporting must be made available to the portfolio PMO as well as to the appropriate stakeholders. If the project is no longer aligned with business objectives that may have changed during delivery, the PMO may recommend that the project be redirected or even cancelled such that the resources will then be assigned to other projects that can provide a maximization of portfolio benefits.

The last two life-cycle phases in Figure 8-14 are the **benefits realization phase** and the **value analysis phase**. The benefits realization plan, regardless in which life-cycle phase it is prepared, must identify the metrics that will be used to track the benefits and accompanying value. Benefits and value metrics identification are the weak links in benefits realization planning. Much has been written on the components of the plan, but little appears on the metrics to be used. However, companies are now creating value metrics that can be measured throughout the project rather than just at the end.[7]

The last two life-cycle phases often include benefits harvesting activities, which is the actual realization of the benefits and accompanying value. Harvesting may necessitate the implementation of an organizational change management plan that may remove people from their comfort zone. Full benefit realization may face resistance from managers, workers, customers, suppliers, and partners. There may be an inherent fear that change will be accompanied by loss of promotion prospects, less authority and responsibility, and possible loss of respect from peers.

Benefits harvesting may also increase the benefits realization costs because of:

- Hiring and training new recruits
- Changing the roles of existing personnel and providing training
- Relocating existing personnel

7 For information on creating and reporting value metrics, see Harold Kerzner, *Project Management Metrics, KPIs and Dashboards*, 3rd edition, John Wiley and IIL Co-publishers, 2017; Chapter 5.

- Providing additional or new management support
- Updating computer systems
- Purchasing new software
- Creating new policies and procedures
- Renegotiating union contracts
- Developing new relationships with suppliers, distributors, partners, and joint ventures

8.20 Project Closure

In traditional projects with well-defined scope, project closure usually takes place once the deliverables are created and accepted by the client. The project team performs both administrative and contractual closure, and then moves on to other assignments.

For the types of strategic projects expected in the future, some members of the project team may need to remain assigned to the project to validate the existence of the project's expected benefits and value. This is seen in Figure 8-16.

When strategic projects come to an end, what we have are outcomes or deliverables. Unlike traditional projects where the benefits and value may appear once the deliverable is created, strategic projects require time for the benefits and value to be extracted. Project team members may still need to be assigned to support stakeholders in identifying and extracting the benefits and value from the outcomes and deliverables. Someone must take ownership for harvesting the benefits, and it may or may not include team members that created the deliverables. The benefits harvesting time could be measured in months or years and may require readjustment activities. Therefore, the project manager in the future may still be assigned to the project long after project execution has been completed.

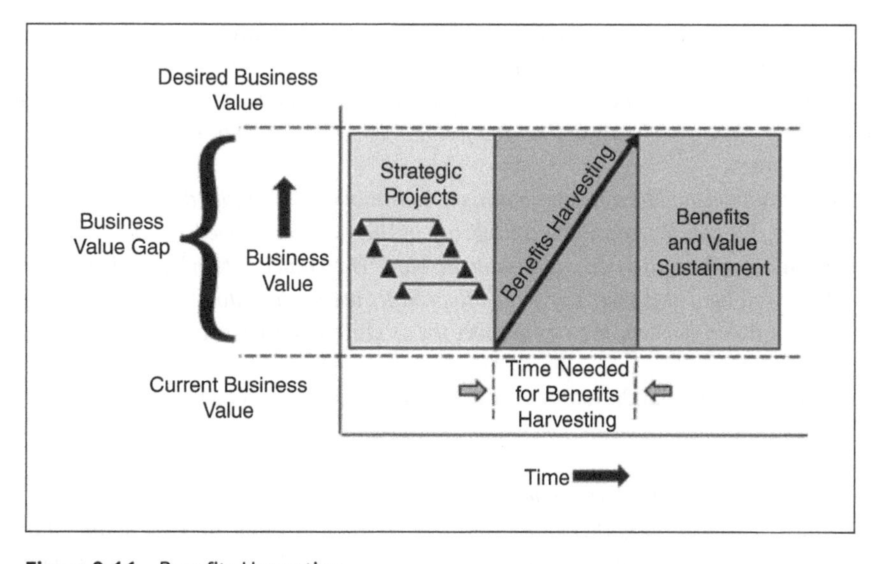

Figure 8-16 Benefits Harvesting

8.21 Excellence in Action: Motorola[8]

For more than 90 years, Motorola has been recognized as a synonym for innovation, not just in products and services, but also in leadership and project management. Motorola's strength historically has been in technical innovations in the communications and semiconductor industries but has been expanded to other industries. Motorola set the standard in some of their strategic project management practices for getting close to the end users during innovation activities and the need to understand the business model of the customers. Motorola also work closely with their customers during benefits harvesting and value sustainment.

Background

Getting close to the customers to understand their needs helps in making correct project decisions. In most companies, innovation and strategic research is simply asking customers what their needs might be now and in the future. Motorola carries their research much further. Motorola conducts "deep" customer research using their design research and innovation project teams to articulate not only how users work with Motorola's products but also how customers run their business processes and what their needs will be in the future. Motorola also observes how customers use their products. Much of this is accomplished during FFE activities.

Motorola's research goes beyond simply understanding how the product is being used. It also includes an understanding of why the product is essential to the customer's business success and how the product fits into the customers business model. This allows Motorola to perform targeted innovation. Customer knowledge is not based on just the end user's requirements. Understanding the business aspects of a product solution can give industrial design groups the opportunity to drive product development direction.

Motorola's generative research is driven by customer and partner interviews as well as observational research. As stated by Graham Marshall,

> *Based on the first round of customer visits and generative research, the team identifies specific customers to revisit during the product definition phase. At this time, the researcher will bring sample models that demonstrate form, features, and functionality. The product definition phase helps the team clearly define and test the right product fit before there is a commitment to development.*
>
> *We use model toolkits and storyboards to communicate potential design directions with our customers. We can better see the complexity of the customer's needs by posing specific questions: "What information do you need to display? How much information do you need to key in? How well lighted is your space? Is the product used in the storeroom or the storefront, or both? How far do you need to carry or move the product or the device to complete a transaction?"*

Motorola also performs validation to ensure that the focus is on a customer solution. As stated by Marshall,

> *During the course of the development program, the Innovation and Design team needs to validate the integrity of the gathered information, product direction, and development trade-offs. This ensures that, as product development progresses, the design remains targeted on the customer's needs.*

---

8 This section has been adapted from Graham Marshall, At Motorola, Design Research Becomes a Strategic Asset, *Design Management Journal.* 2009, Vol. 4 Issue 1, p61-67. 7p. DOI: 10.1111/j.1942-5074.2009.00007.x.

Conclusion

By understanding the customer's business and maintaining close customer contact, Motorola has demonstrated the benefits that can be derived from innovative customer-focused solutions. Simply stated, Motorola has evolved into a targeted business solution provider, and this begins during FFE activities.

8.22 New Causes of Complete or Partial Failure

No matter how hard we try to become good at benefits realization and value management, there are always missteps that can lead to disaster. Fourteen such causes of failure that can occur along the entire project life cycle include:

1. The business owner or stakeholders is not actively involved.
2. Decision makers are unsure about their roles and responsibilities, especially in the early life-cycle phases.
3. The project is approved without a business case or benefits realization plan.
4. A high level of uncertainly and ambiguity exists in defining the benefits and value such that they cannot be described adequately in a document such as a benefits realization plan.
5. Overly optimistic or often unrealistic estimates of benefits are made to get project approval and a high priority.
6. There is a failure to recognize the importance of effective resource management practices and the link to benefits realization management.
7. A heavy focus is maintained on the project's deliverables rather than on benefit realization and the creation of business value.
8. The wrong definition of project success is used.
9. The project is managed with traditional rather than investment life-cycle phases.
10. Having the wrong metrics, unreliable metrics, or no metrics makes it difficult to track benefits and value.
11. Benefits and value are not tracked over the complete life cycle.
12. No criteria is established for when to cancel a failing project.
13. There is no transformational process where the benefits and value can be achieved only from organizational change management.
14. There is no process to capture lessons learned and best practices, thus allowing mistakes to be repeated.

The last bullet is often the solution to correct the first 13 causes from reoccurring.

8.23 Conclusion

Because of the importance of benefits and value, today's project managers are more like business managers than the pure project managers of the past. Today's project managers are expected to make business decisions as well as project-based decisions. Project managers need to know more about the business than their predecessors.

With the growth in measurement techniques, companies have begun creating metrics to measure benefits and value. While many of these measurement techniques are still in the infancy stages, we can expect more rapid growth in years to come.

References

Dey, M. and A. Croft (2018). Implementation of data cut off in analysis of clinical trials. Pharma SUG 2018, Seattle, WA. https://www.pharmasug.org/proceedings/2018/DS/PharmaSUG-2018-DS19.pdf.

DS Informatics. (2020). Pharmaceutical regulatory submission preparation and management. https://www.news-medical.net/whitepaper/20200508/Pharmaceutical-Regulatory-Submission-Preparation-and-Management.aspx.

European Medicines Agency (2021). *ICH Guideline M4 (R4) on Common Technical Document (CTD) for the Registration of Pharmaceuticals for Human Use*. Amsterdam, The Netherlands. https://www.ema.europa.eu/en/documents/scientific-guideline/ich-guideline-m4-r4-common-technical-document-ctd-registration-pharmaceuticals-human-use_en.pdf.

Hearns-Stewart, R. M., J. Farley, K. J. Lee, S. Connelly, N. Lowy, P. Stein and K. Bugin (2021). The integrated review: FDA modernizes the review of new drug marketing applications. *Ther Innov Regul Sci* 55 (3): 467-472. https://doi.org/10.1007/s43441-020-00240-1.

Hidalgo, E. S. (2019). Adapting the scrum framework for agile project management in science: case study of a distributed research initiative. *Heliyon* 5 (34): e01447. https://doi.org/10.1016/j.heliyon.2019.e01447.

Jovanovic, P. and I. Berić (2018). Analysis of the available project management methodologies. *J Sustain Bus Manage Solute Emerg Econ* 23 (3): 1-13. https://doi.org/10.7595/management.fon.2018.0027.

Kerzner, H. (2003). *Project Management: A Systems Approach to Planning, Scheduling, and Controlling*, 8th ed. Hoboken, NJ: John Wiley & Sons.

Kisielnicki, J. (2014). Project management in research and development. *Found Manage* 6 (3): 57–70. https://doi.org/10.1515/fman-2015-0018.

Pathak, K. and A. Saha (2013). Review of agile software development methodologies. *Interact J Adv Res Comput Sci Softw Eng* 3 (2): 270–275.

Pavlović, K. B., I. Berić and L. Berezljev (2018). Agile transformation in clinical research. *Eur Proj Manage J* 8(1): 65–70.

Project Management Institute. (2021). What is project management? https://www.pmi.org/about/learn-about-pmi/what-is-project-management.

Tripathy, S. and B. B. Mohanty (2016). Project management values driven in pharma industry, case study. *Int J Drug Regul Aff* 4(2): 27–32.

US Food and Drug Administration (2017). Guidance for Industry: M4 Organization of the Common Technical Document for the Registration of Pharmaceuticals for Human Use. Rockville, MD, Center for Biologics Evaluation and Research. https://www.fda.gov/files/drugs/published/M4-Organization-of-the-Common-Technical-Document-for-the-Registration-of-Pharmaceuticals-for-Human-Use-Guidance-for-Industry.pdf.

Vaidyanathan, S., G. Freeland, M. López and D. Greber. (2019). Agile can work wonders in pharma. https://www.bcg.com/publications/2019/agile-work-wonders-pharma.

Vicente-Oliva, S., Á. Martínez-Sánchez and L. Berges-Muro (2015). Research and development project management best practices and absorptive capacity: empirical evidence from Spanish firms. *Int J Proj Manage* 33(8): 1704–1716. https://doi.org/10.1016/j.ijproman.2015.09.001.

9

Pillar 9: Evolving Nature of PMOs and Governance

9.0 Introduction

If we look back to the concept of Project Management Offices, or PMO, the only thing that did not change in the past 15 or 20 years was the acronym PMO.

Everything else evolved in a relevant way to respond to the new types of projects, to the nature of organizational transformation, and the VUCA environment, which plays a leading role in how we adapt and react to change and disruption.

Tailoring the PMO to the organizational needs is a crucial success factor. PMOs are not a "control" or "inspection" area or group. It is a partner to the project manager and the delivery teams to react to changes, improve risk responses and focus on value.

9.1 How Governance Can Be Applied in an Agile and Volatile World

Usually, the first thing that comes into mind when thinking about governance is bureaucracy, reporting information, control chain, reduction of speed, among other terms not friendly to the current business scenarios.

On the other side, lack of governance may generate chaos, reduce the corporate ability to apply internal compliance policies and procedures, and generate products and services that could become a business risk due to the lack of maturity in their development.

The solution is to develop a structure supporting your organization but not adding additional irrelevant work and bureaucracy. It is a "light" and "lean" process, framework, and structure that is focused on supporting people and teams to deliver outcomes and, at the same time, fostering a cultural change and shaping desired behaviors.

The following sessions will present several case studies. The first one is from SITA, one of the leading IT and telecommunication services providers to the air transport industry, and the challenges to manage and operate a PMO in a VUCA environment.

The second case study is the American software company ServiceNow and its PMO approach as "strategy realization offices." It also includes the four competencies for the modern PMO.

The third study was provided by the PMO Global Alliance, a global community for project management offices with members in more than a hundred countries. They offered different cases from their members and introduced the PMO Value Ring.

Finally, the last study provides mathematical modeling to calculate some "intangible" values of PMO structures, converting failure rates and reduced losses into benefits that the PMO can provide.

9.2 Excellence in Action: SITA[1] – Airport Systems Integration Projects Cry for Flexible Governance

You have likely taken a plane at least once. On most continents, just a few hours in the air takes you to a place where a different language is spoken, where the climate and culture are different – not to mention the scenery, ambience, or the food. Airports and airlines strive to make your journey as smooth as possible, creating an experience of effortlessness when whizzing from one reality to another. Behind this apparent easiness hides – of course – quite an amount of project management.

When you fly from A to B, more than 20 organizations are involved: airlines, travel agents, ground handlers, airport operations, immigration, security. . . When you cross a border, up to 19 entities and organizations are concerned with a traveller's identity, credentials, and eligibility to travel. Each of the stakeholders is constraint by strict security and health and safety requirements. Sustainability is increasingly a further must-have consideration. When you drink a coffee at the airport, waiting for your plane, you may not realize that every passenger around you has required a *different* combination of parties involved in making their travel as pleasant as possible. This makes an airport a maze of interaction – and airport system integration projects are thus probably some of the most complex and challenging to manage.

Project management is becoming more and more VUCA: volatile, uncertain, complex, and ambiguous:

- **Volatile**, meaning that there is a huge amount of ongoing change, different changes coming at different speeds. Technology changes may be to some degree predictable, as are regulatory and government changes. Other changes can be "risk managed," but one never knows when they may happen. These include natural events such as snow, volcano eruptions, typhoons, earthquakes, but also terrorist attacks and pandemics.
- **Uncertain**, meaning that they are hard to predict. This relates in the airport context mainly to the natural events, pandemics, and terrorist attacks. Whereas the airports master planning for this type of events, the suppliers seldom do, covering themselves with contractual clauses instead.
- **Complex,** due to the multiplicity of moving components, as illustrated above.
- **Ambiguous**, in the meaning that relevant information is available, but the overall meaning is still unknown. Assumed futures will prove wrong. A prime example of this was the prediction of increasing air travel (with corresponding investments) that was shattered by the Covid-19 pandemic.

This is particularly true for airport projects.

1 The material in this section has been provided by Sarianna Benain, Senior Project Portfolio Manager (SITA) ©2021. She specializes in troubled project recovery and project profitability management. Sarianna currently manages customer-facing airport projects in SITA Europe, alongside her role of a Continuous Improvement SPOC for Europe. Sarianna holds PMP, Prince2, and Lean Six Sigma Green Belt certifications. "Opinions are those of the author and not necessarily the views of her employer."

For sure, raging advances in technology bridge the complexity. Solutions support flexible business models. New platforms are developed to lower costs and to increase agility. Clever ways to integrate systems easily are invented. In practice, this is not enough.

The problem in system integration project management lies in the fact that each involved stakeholder and system is at a *different* level of project management maturity, at a *different* step on the digitalization, virtualization, Agile transformation, cloud-enablement or robotisation journey. The suppliers are based in different countries, have differing command of any shared language, have different values and business practices.

In a typical airport systems integration project where you integrate 20+ systems and suppliers (e.g., gate allocation for flights, air traffic control, prediction of number of security lanes, flight information displays, noise monitoring, airport billing. . .), each may use a different technology from 20-year-old to an unproven future technology, have varying maturity levels in project management from nothing at all via "traditional" project management to any "modern" Agile methodology. The projects navigate a multitude of governance models and methodologies – or lack of them. One thing all the 20+ suppliers around the same project have in common, is the necessity to have a legal two-party contract for the work.

The individual system providers may have managed to complete a successful Agile transformation in their own company – this is notoriously not fast nor easy and takes years to achieve. They may have noble objectives to provide the customer nimble solutions in an Agile way; then the sad reality of customer contracting hits and destroys the noble principles. Airports are often at least partially owned by the state and governed by complex anti-corruption and other regulations.

This set-up imposes a static and slow Waterfall mindset in contracting scope needs to be defined to the *N*th degree in advance, change must be avoided and strictly controlled. The contracting process may take years, which in the VUCA environment and with the speed of change means that by the time the contract is signed, the world has changed, your offering has changed, and the customer needs have changed. As a project manager, you find yourself in a contractual strait-jacket, constraint by the minimal space for movement within an outdated and slow change management processes.

Multiply the contract sclerosis with 20+ interacting companies and systems and add to it ever changing regulations. Mix this with a reality where processing a change takes longer than the time for a new change to arrive, impacting the change that you haven't yet formalized. What you get is a project where a superhero project manager is a must: s/he must have an innovative mindset, in-depth knowledge of multiple methodologies, of contract and stakeholder management, possess negotiation and problem-solving skills, focus on value, balance flexibility and control, and master risk.

One transformation can make success easier: there is urgency in modernizing and innovating governments' regulatory bodies and corporations' control and contracting practices. Contracts need to define value and benefit objectives instead of scope, provide frameworks with space to find the optimal solution and make changes fast and easy. Budgets need to be defined incrementally.

Stakeholders need to get involved in integrated many-to-many design thinking *during* the project instead of defining upfront requirements for 1-2-1 turnkey solution contracts in isolation. We need a sea-change not only in technology and project management methodology, but also in contracting practices.

Look around the next time you are sipping your coffee at an airport – the smarter the travel, the smarter the technologies, project management *and* contracting practices have been to get there.

9.3 Excellence in Action: ServiceNow[2] – From Project Management to Strategy Realization

PMOs are entering the next stage of evolution, with the potential to position themselves as "strategy realization offices" whose mission is to ensure that enterprise portfolio of projects is aligned with the company's strategic priorities and generates an expected return on investment. In a business climate marked by uncertainty, business leaders need this function to help them achieve their strategic goals and navigate change.

In the past, the PMO has focused on risk reduction and compliance, positioning itself as a center of governance. In a business climate where outcomes come first, PMOs must shift gears. An enterprise PMO's (EPMO) commitment to scale to manage more complex projects are a step forward but is not enough on its own to deliver strategic agility.

The pivot toward strategic agility requires the PMO to position itself to have visibility across all initiatives, programs, and projects running in and across business lines. The PMO should be able to analyze a portfolio of investments across strategies, and report on investment health and value.

It should be able can analyze the dependencies among different enterprise plans and help select the right approach and framework to deliver desired business outcomes. It should be able to measure the adoption of the changes introduced by programs and products to ensure the benefits of the business case are being realized.

With this insight and perspective, PMOs can be integral to digital transformation efforts.

Four Competencies for the Modern PMO

The ability to leverage this intelligence for better decision-making will be a key differentiator for successful PMOs. In a future of continuous change, PMOs will need to strengthen their competencies in four areas, including the skills to drive organizational change and user adoption of new capabilities (Figure 9-1).

The modern PMO continually improves by pursuing excellence in four different competencies

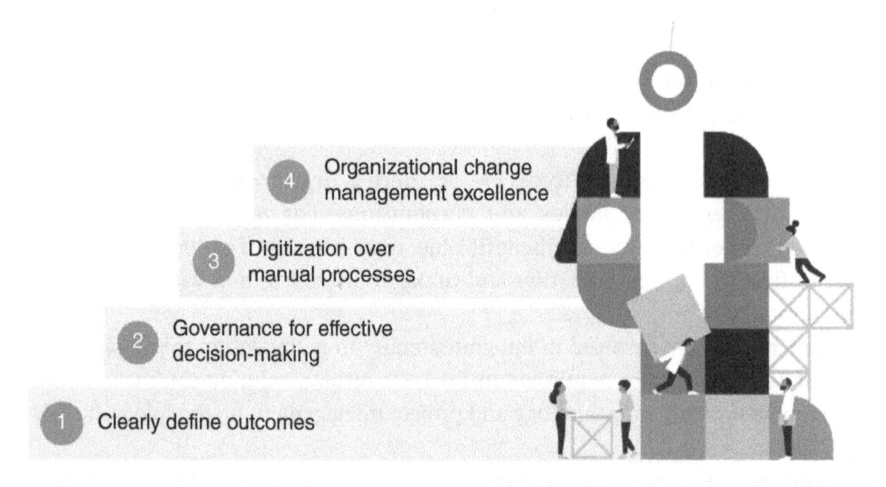

Figure 9-1 Four Competencies for the Modern PMO

2 Material in this section has been provided by Simon Grice, (Senior Director, Innovation), Doug Page (Senior Manager, Product Management), Rani Pangam (Senior Director, IT Project Management) and Tony Pantaleo (Director, Product Success).

1. Clearly Defined Outcomes

Amid the noise of change, the PMO needs play a stronger role in ensuring that business and investment cases have defined, measurable outcomes, as well as standard methods for reporting progress to senior executives. Organizing projects into programs and orienting programs around value streams can provide the architecture for this reporting, especially if the company has moved to investment block funding.

This architecture also promotes an apples-to-apples comparison of ROI across projects, as business interest shifts from traditional on-time and on-budget metrics to business outcomes and value.

Prerequisites for the PMO:

- An operating model organized around portfolio and program management
- A consulting model that defines the program's business case, including measurable outcomes
- Clear measurement analytics to track projects and outcome realization
- Project manager skills oriented around data analysis and experience

2. Governance for Effective Decision-Making

Once outcomes have been established, the PMO needs to establish decision-making frameworks that teams and project managers can use to guide resource allocation and activity to realize those outcomes. The goal of governance should not be to enforce standard ways of working, but rather to equip teams with the means to manage specific outcomes across multiple projects or portfolios, making sure the right decisions are made to maximize outcomes and mitigate hurdles, at the right time.

At the strategic level, governance involves asking if investments are adding their expected value and supporting corporate objectives (Figure 9-2):

- Are these the right projects?
- What is their expected value? Has the value been realized?
- Which goals do they align to?
- Are resources properly allocated to maximize outcomes?
- Where are the bottlenecks or hurdles to realizing value?
- How fast can we turn an idea into a product or service?

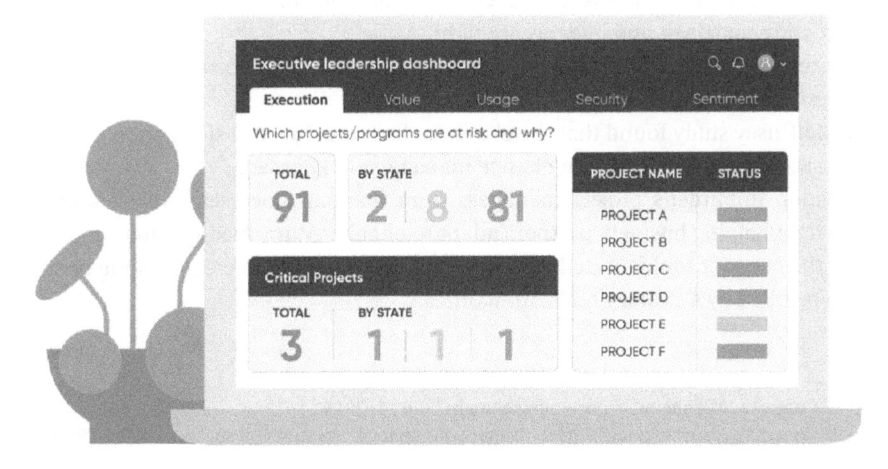

Figure 9-2 Strategic Governance via Dashboards

> Leaders require a clear, enterprise view of projects and status. PMO dashboards provide a visual reference to show how a portfolio is performing against key milestones. With a few clicks, the leaders can drill into project details.

At the project level, governance is not dictated, but is tailored, with the advice and tools offered by the PMO, to deliver the business outcomes and ensure proper resource allocation for specific projects, enabling a balance between discipline and agility:

- What methodology should we use to support delivery?
- What triggers would lead us to cancel an over-budget project in favor of a newer innovation?
- What metrics should we use to determine if a project drifting off course from its original goals, risking its outcomes?
- How can add temporary resources, when needed, to reach the market faster?

Prerequisites:

- Defined methodologies and toolsets for project management, aligned to specific project types
- Decision frameworks that align methodologies and tools to project need
- Project manager skills oriented around advisory guidance and enablement

3. Digitization over Manual Processes

To step into a more strategic role, most project managers will need to free up capacity currently locked up in overhead and administration. *Digital workflow* technology can be the enabler. Bots already have the capability to help offload project administration processes. Automation will continue to replace manual, repetitive processes. These models provide the structured discipline needed to drive rapid iteration toward predefined outcomes throughout the entire enterprise.

Prerequisites:

- A clear business architecture or operating model for project management that can be used to assess need for technology improvement
- Training for project managers in the use of advanced technologies

4. Organizational Change Management Excellence

In rapidly changing environments, the last mile for realizing value is organizational change management (OCM). Yet many businesses underestimate the value of OCM, especially when projects are under pressure to deliver and budgets are tight.

McKinsey research shows that roughly 70 percent of projects fail to achieve their goals due to "employee resistance and lack of management support."[3]

Another McKinsey study found that the chances of success in a transformation can jump from 30 percent to 80 percent with the right change management approach.[4]

As digitization unburdens project managers from manual processes, project managers can spend more time helping business partners adapt to change. Many business lines underestimate the effort or the need for sophisticated tools and methods required to create lasting change, a key gap that can be filled by a strategy realization office.

---

3 McKinsey & Company, "Changing change management," July 1, 2015

4 McKinsey & Company, "How to double the odds that your change program will succeed," April 29, 2019

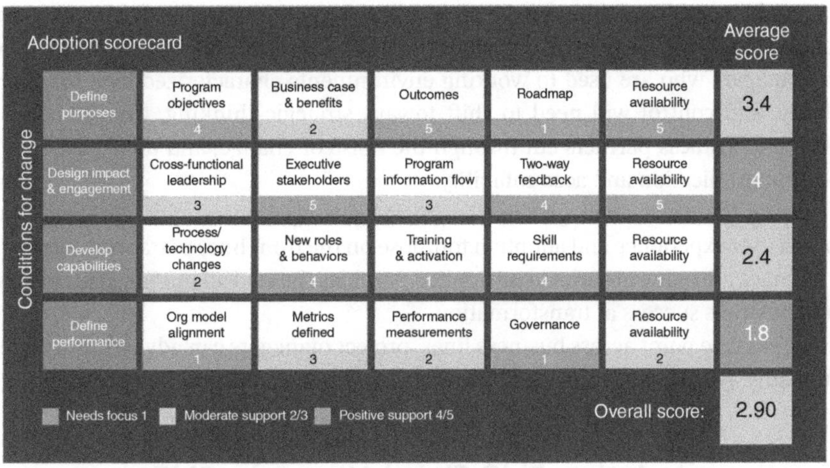

Adoption scorecard

Conditions for change						Average score
Define purposes	Program objectives 4	Business case & benefits 2	Outcomes 5	Roadmap 1	Resource availability 5	3.4
Design impact & engagement	Cross-functional leadership 3	Executive stakeholders 5	Program information flow 3	Two-way feedback 4	Resource availability 5	4
Develop capabilities	Process/ technology changes 2	New roles & behaviors 4	Training & activation 1	Skill requirements 4	Resource availability 1	2.4
Define performance	Org model alignment 1	Metrics defined 3	Performance measurements 2	Governance 1	Resource availability 2	1.8

Needs focus 1 | Moderate support 2/3 | Positive support 4/5

Overall score: 2.90

Figure 9-3 Adoption Scorecard

Scorecards oriented around adoption of new capabilities can help PMOs understand organizational change success and roadblocks, against benchmarks based on historical data. Simple scoring frameworks can enable the PMO to assess if a program is on track to reach its adoption outcomes, pinpoint areas that may be causing bottlenecks, and identify impact on related areas. Such frameworks also help establish explicit focus on organizational change and adoption as a PMO competency, where it can be directly connected to and integrated with other project management activities.

An adoption scorecard provides a holistic view of change management risks to a project's success. Project managers can identify and address root causes that are affecting value realization.

Prerequisites:

- Defined methodologies and tool sets for OCM
- Project manager skills oriented around OCM
- Defined, measurable outcomes and success criteria for OCM
- Defined approaches to assess change capacity relative to proposed project and program investments

Armed with these tools and skills, PMOs can position themselves to help businesses more effectively achieve their digital transformation objectives. They can also serve as catalysts in improving how companies develop and deliver products and services – with greater focus on outcomes instead of processes, and with the agility to help business leaders adapt to meet changing market demands.

Adapting Project Manager Skills and Mindsets

The pivot of the PMO to "strategy realization" will demand a major shift in the skills and mindsets of individual project managers, from rigorous adherence to governance frameworks to a new working paradigm oriented around effective strategic communications, management across methodologies, and the ability to help leaders and teams navigate through change to outcomes.

Project managers who are used to working environments characterized by a focus on details, predictability, and control will need to shift toward strategic thinking, the use of data-driven insights to help business partners cut through the noise of change, and the facilitation of cross-functional communication and accountability.

Most importantly, project managers need to become *change leaders*, with increased focus on the quality of the user experience and adoption for those on the front lines of transformation. Gauging whether changes are being absorbed and applied provides the critical human aspect to measuring project and business success at transformation.

With their vantage point across business lines, project managers can advise business leaders on successful pathways to change through business environments marked by uncertainty.

9.4 Excellence in Action: PMO Global Alliance[5] – PMOs in Transformation

For more than three decades, implementing PMOs has been a common solution for organizations that seek better performance and more effective project results.

PMOs have evolved significantly over the years, redesigning their purpose, amplifying their business impact, and becoming increasingly critical to the success of many organizations around the world, but all this evolution has not made the lives of PMO Leaders any easier. PMOs are being increasingly challenged about the effective value they have been generating for their organizations.

Over the past few years, a project culture has emerged in many organizations, and now they not only use more sophisticated and effective practices, techniques, and tools, but they are also more mature. Therefore, there are more challenging expectations for the benefits that a PMO should deliver.

Thus, rethinking our PMOs has become not only a necessary and an urgent task but also increasingly frequent, given the significant world changes that have generated a profound impact on both organizations and people.

It is undeniable that PMOs are undergoing a major transformation process by moving their focus from technical aspects to value generation, but this is a complex transformation, which is often not even fully understood by those who are responsible for leading it.

So, what is the real value generated by a PMO? How do we measure and demonstrate it?

The answers to these questions are the key to the success that many PMOs around the world incessantly seek as an immediate and urgent response to the increasing demands and expectations of their organizations.

For the next-generation PMO, delivering value will no longer be a simple goal, and ensuring that its value is being recognized by the organization will no longer be a mere wish. These are the real reasons that will justify its existence and allow the PMO to thrive.

For future PMOs, generating value will no longer be a simple goal and ensuring that the organization recognizes their value will no longer be a mere wish. These are the real reasons that will justify their existence and allow them to thrive.

---

5 Material in this section has been provided by Americo Pinto, Founder of the PMO Global Alliance, a global community for project management offices, with active members in more than a hundred countries.

If you believe that the current success of your PMO is enough to ensure that it will continue to exist for years to come, then understand that this is the first step to the decline and inevitable failure of your PMO.

Learning from the Global Pmo Community

The PMO

Enterprise PMO at a big retail company in North America.

The Context

The company's Enterprise PMO had been doing outstanding work providing support to the organization's portfolio by serving the senior executives in the definition and monitoring of portfolio projects, all of them adequately connected to the organizational strategy.

In addition, the company's Executives, especially the President, had been publicly acknowledging the importance of the PMO and reaffirming their support to the initiative.

The mapping and regular monitoring of the company's project portfolio in a 10-year perspective was one of the most praised services provided by the PMO.

The Issue

A few months after the beginning of the COVID-19 global epidemic, the company was plunged into a deep financial crisis. As a result, the PMO executive asked the PMO manager to redirect her focus of work from the long-term portfolio management activities to the short-term ongoing portfolio projects performance control.

The new direction frustrated the PMO manager, and she considered it as an incomprehensible setback.

For her, that change in priority did not make sense, as PMOs should provide services of high strategic value and not meet operational needs.

Lessons Learned

After initial and understandable frustration, the PMO manager understood that the company was going through an unconventional moment, which made new and urgent needs arose.

It is not difficult to understand that discussing the project portfolio for the next ten years at that moment would be completely detached from the company's crisis reality.

The new directions were intended to ensure that the company could deliver critical projects planned for that year on schedule and without cost overruns.

Of course, it doesn't mean that the portfolio and the strategy were no longer important for the executives. Still, certainly amidst a severe restriction of resources, it was necessary to take the risk of concentrating all efforts to guarantee that the projects could be delivered as expected.

The PMO should be part of this effort, understanding and adapting to the organization's new priorities.

There are no mandatory functions for the PMO, just as there are no functions that the PMO should never perform.

The value of the functions provided by PMOs depends on the current needs, which may differ from company to company and from moment to moment.

The PMO must always be opened to understand and adapt to the new demands, whatever they are.

A New Mindset for Next-Generation PMOs

The PMO Global Alliance (www.pmoga.world) is the first and largest community of PMOs and PMO professionals in the world. Over the past six years, it has conducted a series of studies aimed at generating a better understanding of the PMO phenomenon.

This work combined the participation of hundreds of seasoned PMO leaders from several countries, all of whom contributed their knowledge and experience to the development of a new and innovative mindset for PMOs, one that is practical and connected to the real world, and one which allows anyone to implement or redesign PMOs that can generate value their organizations and be effectively recognized for it.

Common sense may lead us to think that the PMO is just another support area for the organization's business, such as the information technology, accounting, or legal departments. However, the first step to understanding the newly proposed mindset is to realize that the PMO is, in fact, a "service provider" that needs to justify its existence by generating clear benefits to its "contractors." Otherwise, it will lose support and its "contract" may even be cancelled, which means that the PMO will be gradually demobilized and eventually shut down.

This simple change of names may not seem so relevant but understanding the impacts of this change is fundamental for the PMO to be recognized for the value it can generate for the organization.

There is nothing more important to a service provider than their customers. Knowing them deeply is a critical success factor because understanding their needs will enable the PMO to offer services that can effectively meet their expectations.

Learning from the Global Pmo Community

The PMO

Enterprise PMO at a healthcare company in Western Europe.

The Context

The company's PMO focuses on supporting project managers in better planning and controlling their projects.

Project managers belong to the business areas, not to the PMO. Therefore, they don't have in-depth knowledge and experience in project management practices, needing the PMO support to carry out their activities.

The Issue

The PMO was set up based on the identification of project managers' needs, collected in interviews carried out with 84 professionals from four business units.

The PMO's functions were defined to deliver solutions that meet project managers' expectations, meeting urgent support needs, and a long-term commitment to knowledge transfer to the business areas.

Project managers had high expectations of the PMO, which were not fully met because the demand for support was greater than the PMO's capacity to meet them.

Lessons Learned

The PMO functions were established based on the needs of its customers, which is an excellent practice. However, two additional practices should be implemented to ensure the complete alignment between the PMO customers' expectations and the PMO functions.

A formal process must be designed for each PMO function, which will compose the PMO services catalog. This document demonstrates how the PMO delivers its services and define roles and responsibilities for each function.

Processes should also have key indicators and goals that will allow the PMO to monitor its performance and how it is generating value for its customers.

Goals should reflect the PMO's capacity to meet demand. For example, for the "Provide Planning Support" function, the PMO assumes it will be possible to serve up to five project managers per week, given its current resource capacity.

This level of service must be communicated and agreed upon with the PMO customers. If the demand goes beyond the PMO capacity and requires a higher level of service, it will be necessary to review the PMO team headcount.

The Value of a PMO

Recent attempts to empirically demonstrate the monetary value of a PMO to the organization's business have been frequent. However, despite these academic initiatives' value for a better understanding of the PMO phenomenon, they have unfortunately presented inconclusive results when attempting to demonstrate the expected impact of a PMO on the organization's business.

Research has shown that the search for a universal model is hampered by the many variables that can influence the way a PMO generates impact on a business, such as the organization's culture, the organization's project management maturity, the executive sponsorship, and, most importantly, the specific needs of each organization, and how the PMO was set up to meet them.

"Value perception" is the expected result when we meet our customers' expectations. This concept was adapted to the reality of PMOs, based on extensive and previously consolidated knowledge in the areas of marketing, customer service, and consumer psychology.

The perception of value is a "feeling," and, as such, it is subjective, which may cause its importance to be underestimated, as compared to more objective technical indicators that can't reflect the organization's real perception about the relevance of the PMO to the business.

However, it is possible to manage and measure how the PMO generates value perception for its customers. The experience of hundreds of world-class PMOs that have been studied over the last few years in the PMO Global Alliance Community indicates that the PMO's ability to generate value perception is a critical factor for its survival and continuity.

PMOs that do not consistently generate perceived value—no matter how well they excel at the technical quality of their work—are constantly challenged and questioned, and they must live with the constant risk of being discontinued.

How long would you be willing to maintain a contract with a service provider who is not capable of generating the benefits that justify such a contract? The answer to this question should be close to the lifespan that your PMO will have if it cannot make your customers realize the value that is being generated for your organization.

The Ideal PMO Model

When implementing or redesigning PMOs, many organizations search the literature for models that can be used as reference. From time to time, new models emerge, and many become trends and/or objects of desire for many companies – at least, until a newer, trendier model appears.

If we consult the main books on PMOs published around the world in the last 15 years, then we will see that each book proposes completely different models. There are so many options that it is almost possible to use the entire alphabet to represent them:

- Agile Management Office (AMO)
- Benefits Management Office (BMO)
- Change Management Office (CMO)
- Enterprise Project Management Office (EPMO)
- Project Management Center of Excellence (PMCoE)
- Project Management Office (PMO)
- Strategic Management Office (SMO)
- Value Management Office (VMO)

These are just some of the PMO models most often cited in the literature, all of which are used as main references by many organizations when they want to set up a PMO.

But among so many options, which of these models is the right one? If one of them is the correct one, or one which is more complete, more modern, or more visionary, or one which has the greatest chances of success, then it is possible to conclude that all the others are wrong or outdated, and therefore should be disregarded or abandoned.

This is the logic of the *ideal PMO*, the perfect model that usually represents the trend of the moment and which, in general, is the "magic pill" that is supposed to solve all our problems.

In other words, just select the model that you believe to be perfect, and your problems will vanish, and if it doesn't work, then wait for the next new trend – which will certainly have an even more glamorous name – and try again.

Surprisingly, this is the exact strategy used by countless organizations around the world when setting up their PMOs. In their search for quick and easy solutions, they ignore one of the most striking features, and one of the few consensuses about PMOs – their uniqueness.

The various existing project, program, and portfolio management standards show that by bringing together professionals with the necessary qualifications and experience, it is possible to reach a consensus on the best practices for managing these domains.

We can agree that there is a "best way" to manage scope, time, costs, and risks, or to select and prioritize projects in a portfolio. There may be more-or-less-sophisticated variations, but these are still in line with the best practices that are considered valid for most cases.

On the other hand, PMOs are a different phenomenon that adds elements of complexity, which make reaching a consensus impossible—or, at least, quite questionable.

It is possible to find reliable and universal references on practices to make projects, programs, and portfolios more successful. However, when it comes to PMOs, there are not any empirically proven, universal, ideal, or more-effective models.

The complexity of the PMO phenomenon explains why there are so many different models in the literature, all of them technically defended and justified by their valuable authors, who created their proposals from their own experience and knowledge of PMOs. However, imagine that all these authors were blindfolded and standing around a large elephant. They are asked to touch it and say the name of the animal in front of them. It's likely that none of our authors would say the animal is an elephant; instead, they would suggest different animals entirely.

Upon first analysis, we could say that all authors were wrong in their answers to our question because none of them mentioned the elephant as one of the options. However, this is because they only touched parts of the animal, such as its trunk, ear, or tail. After analyzing the situation further, it is possible to understand that the answers were not exactly incorrect but appropriate for the perspective they had, depending on the part of the animal they were closest to. Although they were all dealing with the same elephant, they made their suggestions from their own points of view and experiences, which makes them all correct in some way.

The different existing PMO models proposed by several experienced and recognized authors are each one's vision of the PMO phenomenon from their different perspectives. They are certainly valid and valuable if we consider them not as a path to be chosen but as alternatives that can be combined as needed.

Common sense leads us to believe that we are putting the maxim "one size doesn't fit all" into practice when we choose to select one of the several PMO models to be implemented in our company. We are choosing the best option among the many options available. However, the truth is that when we implement one of these models, we invariably push the organization to fit a perspective that often makes no sense.

In practical terms, pre-established models may lead our PMO to offer functions that – although recommended by the model – are not capable of generating the benefits that are a "priority" for our customers, which is one of the main reasons why the value of many PMOs is not recognized – even though they offer services with an elevated level of maturity. As Peter Drucker said, "*There is nothing so useless as doing efficiently that which should not be done at all.*"

Experience shows that PMOs which stand out for generating high value-perception often exhibit characteristics from several models identified in the literature. Instead of seeking alignment with a specific model, they are genuinely committed to identifying their customers' expectations and setting up a PMO that does what it takes to deliver them.

The different PMO models are experiences and points of view that can be valuable to inspire us to create a PMO that is unique and fully tailored to both our customers' expectations and our organization's needs. So, use these models smartly and always keep in mind that – as attractive as they are – "magic pills" do not exist – especially when the subject of the discussion is a PMO.

Learning from the Global Pmo Community

The PMO

Enterprise PMO at a governmental agency in southern Asia.

The Context

Executives finally became convinced that a PMO could bring benefits to the organization. For this, the organization hired a renowned consultancy to implement the PMO. They proposed an innovative proprietary PMO model, which they said was the most modern and complete for any organization that wants to be aligned with the coming trends. The recommended model requested implementing a list of functions, which are mandatory for any advanced PMO.

(continued)

(continued)

The Issue

The PMO model suggested by the consulting firm was implemented in six months, and from that period onwards, there was growing resistance to the work of the PMO. The consulting firm argued that those problems were typical and expected, and good results would soon arise. Instead, many of the PMO's actions were ignored by executives and even sabotaged by project managers. As a result, the sponsorship initially received from top management disappeared after a few months, and the PMO's decline seemed irreversible.

Lessons Learned

The organization, for lack of maturity and experience, tried to implement one of the popular PMO models, and the implementation process completely ignored the organization's needs, the PMO's customer expectations, the organizational culture, and even the organization's capability to absorb all the functions given in the suggested model.

Pre-established models can be a great source of inspiration to set up PMOs that fit specific organizational needs. However, they never should be implemented blindly, as magic pills that can be used in any situation and any organization, without adaptation or alignment with current expectations.

How to Identify the PMO Customers' Expectations

We already know that there is nothing more relevant to a PMO than its customers, and the first step to creating a PMO that is committed to generating value-perception is to identify who its customers are. A PMO's customers are all those who are served or affected by a PMO's work in some way. Some of the most common PMO customers are directors, project managers, functional managers, and project team members because they have common interests, needs, and expectations.

The second step is to find the customers' expectations, which will allow the PMO to focus on meeting their real needs.

The following list is part of the PMO Value Ring methodology, which was developed by the PMO Global Alliance Community. It presents the 26 most-common functions in PMOs around the world:

1. Support Project Planning
2. Perform Specialized Tasks for Project Managers
3. Manage Resource Allocation Between Projects
4. Manage Interfaces with Project Clients
5. Manage Organizational Changes
6. Manage People in Projects
7. Manage Projects or Programs
8. Manage Stakeholders in Projects
9. Conduct Audits in Projects
10. Manage Project Documentation

11. Manage Lessons Learned Database
12. Perform Benchmarking
13. Manage Lessons Learned Meetings
14. Promote Project Management within the Organization
15. Provide Mentoring for Project Managers
16. Provide Training and Project Competence Development
17. Provide Project Management Tools and Information Systems
18. Provide Project Management Methodology
19. Monitor and Control Projects or Programs
20. Provide Project or Program Status Reports for Upper Management
21. Provide a Strategic Project Scoreboard
22. Support Project Portfolio Definition
23. Manage Project or Program Benefits
24. Monitor Project Portfolio Performance
25. Participate in Strategic Planning
26. Provide Advice to Upper Management for Decision-Making

At first, one can imagine that the best way to discover the PMO customers' expectations would be to ask them to indicate which functions (i.e., services) they expect the PMO to provide. In this way, the above list could be used as a "menu of functions" to collect the expectations of each PMO customer.

This idea sounds good at first, but unfortunately it doesn't work in practice. By offering a list of potential PMO functions to customers and requiring them to state their expectations, we are making a fatal mistake; this list is not written in the PMO customers' "language," and, as a result, they cannot make the decisions we are asking for. The PMO customers' "language" is not technical; it is much more benefits-oriented, which means they are concerned about effective results that will solve practical problems they have in their real lives.

Let's be honest; a director is not really interested in the reports provided by the PMO, only in the full availability of reliable and complete information whenever decision-making is required. The project manager's expectation is not that the PMO can provide project management methodology, only that their projects will be more efficient, cost-effective, and predictable because they are being managed in a structured way. In the language spoken and understood by PMO customers, the functions provided by the PMO don't really matter, only the benefits and results.

Now imagine a physician and their various patients, each with a range of illnesses that cause distinct types of pain. The patients have a clear expectation that the doctor can eliminate their pain and cure them. If the doctor can meet the patients' expectations, then the value of their work will certainly be recognized. Each type of pain has a cause and requires a different treatment. There is no single medicine that can cure all types of pain, so each patient will need a combination of medicines that meet their specific needs. Different medicines treat distinct types of pain and cure different diseases. The doctor must prescribe the most suitable medicines to eliminate pain and cure all the patients.

In our story, the physician is the PMO, and the patients are the PMO's customers. The patients' pain represents the PMO customers' benefits. For example, a director may complain about the "pain" of not clearly seeing the connection between the company's strategy and its projects. A project manager may say that their worst "pain" is a lack of reliable information for decision-making. A team member may feel the "pain" of being unmotivated to work on projects. The medicines that can be prescribed for these three patients are the potential functions that a PMO may provide. Certain functions are more likely to generate specific benefits, just as certain medicines are better suited to cure specific illnesses.

For example, to cure the "pain" of a lack of connection between strategy and projects, the PMO may provide the "Support Portfolio Management" function. To cure the "pain" of a lack of reliable information for decision-making, the best medicine would be to "Provide Integrated Systems and Project Status Reports." Finally, if a team member's "pain" is a lack of motivation to work on projects, then the most appropriate medicine may be "Provide Training" or "Provide Mentoring."

Identifying the expectations of our PMO customers by asking them to inform us of the functions that they would like the PMO to offer would be like a doctor simply offering a list of all available medicines and asking the patient to choose which ones they would like to be prescribed. Obviously, our "patient" (i.e., the PMO customer) is not an expert and can't decide which functions would be best suited to address their expectations and needs. Therefore, the answer to this dilemma is to ensure that our communication with the PMO customers will use a language they can understand, one that focuses on benefits and results.

The following list is also part of the PMO Value Ring methodology and presents 30 potential benefits commonly generated by PMOs, based on the experience of our global community:

1. Better availability of resources with skills in project management
2. Better availability of information on lessons learned from previous experiences
3. Effective transfer of knowledge in project management
4. Better communication amongst areas within the organization
5. Better communication amongst the project team
6. Better communication with upper management
7. Improved reliability of provided information
8. More reliable time and cost estimates
9. Improved availability of information for better decision-making
10. Better control over project teams
11. Better project time and cost control
12. Better third party and subcontractor control
13. Better project support from upper management
14. Increased motivation and individual commitment
15. Better definition of roles and responsibilities
16. Greater agility in project management decision-making
17. Better allocation of resources across projects
18. Reduction in project lifecycles within the organization
19. Reduction of risk exposure
20. Greater integration amongst areas within the organization
21. Increased productivity on projects
22. Better quality results for projects
23. Improved client satisfaction
24. Better project prioritization

25. Increased visibility of the relationship between projects and strategy
26. Increased visibility of the relationship amongst projects
27. Greater organization-wide commitment to results
28. Increased visibility of project progress
29. Increased visibility of resource demand
30. Increased predictability for decision-making

If you are now thinking of using the above list of benefits as a "menu" with which to collect customers' expectations, then you are certainly on the right track, as this process will allow you to use the appropriate language. By asking each customer to select and prioritize some benefits from the list above, you can understand the set of priority benefits for the PMO.

In the PMO Value Ring methodology, each customer is asked to select five priority benefits from the list above, sorted by degree of priority. After identifying the priority benefits that should be generated by the PMO, we will reach a critical moment when the relationships between the 26 potential PMO functions and the expected benefits that have been selected by its customers must be analyzed.

For example, if the expected priority benefit is to have "Better availability of resources with skills in project management," then common sense shows that the "Provide Training and Project-Competence Development" function has much more potential to generate the desired benefit than the "Monitoring and Controlling Projects or Programs" function.

By assessing how well each function can potentially generate each benefit, it becomes possible to identify the PMO's most critical functions to deliver the expected benefits, and it will further allow us to observe that certain functions recommended by preestablished models may not be the most appropriate for our PMO, given our customers' benefit expectations.

The PMO Value Ring methodology is supported by a database with the collective experience of hundreds of PMO leaders from various countries and industries; therefore, each of the 26 potential PMO functions can generate each of the 30 potential PMO benefits.

Through the web-based tool[6] – you can automatically collect your PMO customers' benefit expectations and identify which functions are most recommended for the PMO, based on the correlations between functions and benefits generated from the collective intelligence of the PMO Global Alliance Community members.

6 The PMO Global Alliance community has developed the PMO Value Ring methodology to contribute to the worldwide project management community. The methodology concepts can be optionally supported by a web-based tool, which automates the steps above and provides a database with the collective experience of hundreds of PMO Leaders from various countries and industries.

Among other valuable information, this database also provides the probability that each of the 26 potential PMO functions generates each of the 30 potential benefits of a PMO.The web-based tool allows the user to automatically collect the PMO Customers' benefit expectations and provide recommendations on the most critical functions to the PMO. These recommendations are based on correlations between functions and benefits, which are defined from regular surveys with a sample composed of the most experienced PMO Global Alliance community members.

The PMO Value Ring methodology and its web-based tool currently have more than 20,000 free users in over 90 countries and are available atwww.pmovaluering.com.

Learning from the Global PMO Community

The PMO

Business Unit PMO at a telecommunications company in Africa.

The Context

The PMO had been in place for nearly 10 years, with ups and downs. It had been offering a consistent set of functions for its customers and the overall evaluation about the PMO was quite good.

The Issue

A merger with another company had led the PMO leader to question whether the PMO was still capable of meeting the expectations of the new group of customers.

Lessons Learned

For the first time, the PMO customers' benefits expectations would be identified to verify the current alignment of the PMO's functions with the company's needs.

The PMO established three groups of customers: executives, project managers, and project team project members, with the respective relevancies: 50 percent; 35 percent 15 percent.

The PMO provided a list of potential benefits and asked each customer to select the top five benefits most critical to them.

In total, the PMO customers selected 13 benefits from the list of 30. Considering the number of times each benefit was selected and the relevance of who chosen it, a set of 11 benefits could represent the PMO customers' priority expectations, which should be the focus of attention for a review of the functions offered by the PMO.

How to Make the PMO Value Be Recognized

After identifying the PMO Customers' priority benefits that should be generated, we reach a critical moment, when the relationships between the 26 potential PMO functions and the expected benefits that have been selected by its customers must be analyzed.

For example, if the expected priority benefit is to have "Better availability of resources with skills in project management," it's possible to say that even the common sense shows that the "Provide Training and Project Competence Development" function seems to have a much higher potential to generate that wished benefit than the "Monitoring and Controlling Projects or Programs" function.

By assessing how much each function is potentially capable of generating each benefit, it becomes possible to identify the PMO's most critical functions to delivering the expected benefits. And it will also allow us to observe that certain functions, perhaps recommended by preestablished models or that we thought would always be necessary for any PMO, may not be the most appropriate for our PMO, given the benefit expectations of our customers.

It is now clear why you should never follow preestablished PMO models as if they were the turnkey solution for your organization. Those models will indicate PMO configurations with the necessary functions to implement them, which are not necessarily those that will make you be able to deliver the benefits your customers really need.

The consequence is your PMO will be wasting resources and money by offering the wrong "medicines" that cannot generate value perception but treat "pains" that are not the ones that bother your customers.

An alarm should sound every time a PMO provides functions and services with a high level of maturity, but it still fails to have its value recognized by the organization.

If your PMO is constantly being questioned and you blame your customers for not being able to see the "beauty" of what you are doing, remember that "beauty is in the eyes of the beholder."

Learning from the Global Pmo Community

The PMO

Enterprise PMO at an energy company in the Middle East.

The Context

The PMO was being implemented by an experienced professional who led a PMO in a large company for over 15 years. After completing the deployment process, the same professional would become the leader of the newly created PMO.

The Issue

Due to his extensive previous experience in another company, the future PMO leader believed that the best option was to create a PMO similar to the one he had previously led and that had been very successful.

Lessons Learned

After collecting the PMO customers' benefit expectations and understanding which priorities to consider, the PMO leader assessed and reviewed which functions would have the most significant potential to generate the priority benefits. This led him to discover that at least 40 percent of the functions that the new PMO were supposed to provide were different from what he had experienced in his previous work. In addition, some functions that he believed were essential for a PMO would not be able to deliver any of the expected benefits for his new company's PMO.

Implementing the New Mindset

The new mindset proposed by the PMO Value Ring methodology places people – the PMO's customers – as a central focus to achieve success. Everything else, including all the PMO technical aspects, must be a consequence of your organization's specific needs and customers' expectations.

The methodology proposes a cycle of six practical steps to design a "Value Perception-Driven PMO":

1. Identify the PMO's key customer groups and their relevance to achieving balanced and consistent success.
2. Assign the PMO customers to their respective groups and collect their benefit expectations using the most appropriate "language."

3. Prioritize the most critical benefits expectations based on the relevance and the number of PMO customers who mentioned them.
4. Establish from your own experience – or refer to a trusted source – the probability of each PMO function generates each benefit.
5. Identify which functions should be performed by the PMO, prioritizing those most likely to deliver its customers' expectations.
6. Define the PMO mix of functions and a set of key indicators and goals that can measure and give visibility to the PMO performance and how it is generating value for its customers.

Experience shows that an annual cycle is an ideal frequency recommended for restarting this cycle, but be aware that significant changes in the organization may require anticipating the start of a new cycle, as those changes may generate new needs and different expectations.

The Impact of the PMO Maturity

According to the PMO Value Ring methodology, the level of sophistication in which the PMO functions are performed defines the concept of PMO maturity.

Many PMOs can provide a project management methodology. However, they can perform that function at different levels of refinement.

For example, suppose we took two PMOs from different companies, both providing project management methodologies.

Analyzing them closely, we realized that the methodology adoption rate is only 40 percent in the first company, and a few project managers know how to use it.

On the other hand, we have the second company, where the methodology achieved a 100 percent adoption rate, and all project managers are trained.

Assuredly, we could expect that the benefits generated by a methodology are much more likely to be observed in the second company, where the PMO is more mature in that specific function.

The new mindset highlights the importance of selecting the functions that fit the benefits expected by the PMO customers, but those functions must be provided with the required maturity to deliver the expected benefits.

In other words, providing the right functions without proper maturity means doing the right thing in the wrong way.

Learning from the Global PMO Community

The PMO

Enterprise PMO at an industrial company in Central Europe.

The Context

The PMO had been providing services with the highest level of sophistication and maturity, to the point of being formally recognized as one of the best PMOs in their country, in a local competition.

The Issue

A few months after being recognized as one of the most mature in his country, the PMO was closed by the company's executive board.

Lessons Learned

A more in-depth study of this case showed that although the PMO was providing services with a high level of maturity, it was not generating effective value for the company in the perception of the executive level. It denotes a misalignment between the functions provided by the PMO and the benefits expected by the executive level.

The Future Is Now

There is nothing more critical to a successful PMO than making its customers consistently recognize the value of the PMO what will be achieved through the delivery of the expected benefits. Otherwise, sooner or later, it is very likely that the continuity of the PMO will be at risk.

Many PMOs around the world show that there is a growing concern with creating value for the organization. But many PMO Leaders still rely on magic pills to set up and manage their PMOs, which leads to a herd effect that makes failure seems more commonplace than it should be.

The PMOs that will thrive in the coming years must have leaders who accept and embrace their server role. They must be genuinely concerned with the customers' needs and committed to creating flexible, real-world solutions that generate practical value for their organizations.

In this way, the PMO's value will be perceived and recognized by its customers, and the PMO will become an indispensable partner for the organization's success.

9.5 Excellence in Action: Determining the Mathematical ROI of a PMO Implementation[7]

Summary

The objective of this paper is to present, discuss and apply a mathematical model based on the use of Monte Carlo simulation in conjunction with research on project success/failure rates of projects to develop a 10-step model to calculate the mathematical return on investment (ROI) for the Project Office implementation.

The paper aims to provide guidance on how intangible results resulting from the project planning and control can be linked to potential savings in time and cost comparing with projects poorly managed (Kwak & IBBS 2000). It is not the scope of this paper to demonstrate the positive impact of good project management practices. The main objective is to discuss possible ways of measuring results in order to have a clearer cost benefit analysis regarding the value of a PMO Setup (Hubbard 2010).

7 Material in this section has been provided by Ricardo Viana Vargas and was originally published at the PMI Global Congress 2013 in New Orleans, USA.

This paper also discusses the main challenges to quantify benefits considering cultural, social and value perception dimensions in order to translate benefits into clear and measurable numbers.

The Importance of Clear Benefit Measurement

Business improvement processes like the Project Management Office implementation are, most of the time, linked to indirect benefits achievement. In the past, program, project, or process success was measured by activity: number of people involved, money spent, days to complete. Little consideration was given to the benefits derived from these activities because they were considered impossible to be clearly measured (Philips & Philips 2007).

The intention of clear measurement of benefits can be based in the following arguments:

- Price/money is a proxy for value.
- Measurable outcomes contribute to a better alignment and integration with financial systems performance.
- More tangible results support the identification of critical sources of value.
- Promotes communication and makes results quantitatively tangible.

Understand the clear impact on project results of the project management processes, tools and existing support and how this structure contributes to better project results became a key driver to understand the value of project management (EIU 2009).

Model Overview

The proposed model is based on 10 (ten) processes that are organized into 6 (six) groups (Figure 9-4). Both the processes and the groups are interrelated in order to produce the needed steps to understand the real costs and benefits brought by the Project Management Implementation.

The processes are defined following the structure proposed by the PMBOK Guide (PMI 2016) with Inputs, Tools, and Techniques and Outputs.

Project Portfolio

The Project Portfolio group describes the process that should be in place to understand the scope of what should be managed by a potential Project Management Office (PMO). The intent is to make sure that the potential projects that will be supported by the project management office are identified and the cost, time frame, and benefits (value) of these projects are calculated.

The Project Portfolio group is divided in the following processes

- Create the portfolio of projects
- Calculate financial return of projects in the portfolio
- Categorize projects

Create the Portfolio of Projects

This is process is responsible for the creation of the portfolio of projects. Based in working groups and the support of experts, it aims to create a list of the projects that will be managed by the PMO including some preliminary information like the Project objectives, estimated duration and budget (Figure 9-5).

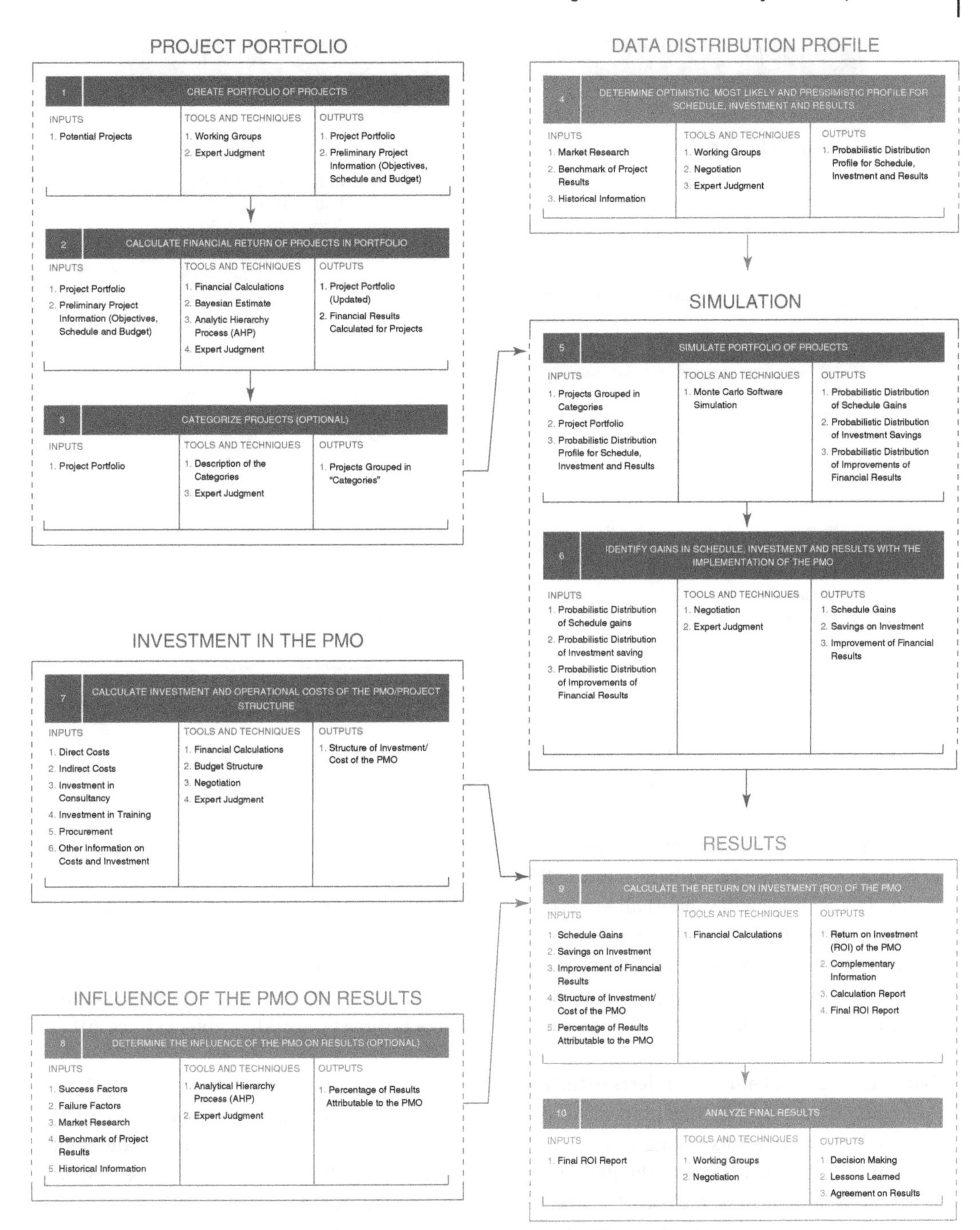

Figure 9-4 Ten Processes to Calculate the Return on Investment of a Project Management Office

1	CREATE PORTFOLIO OF PROJECTS	
INPUTS	**TOOLS AND TECHNIQUES**	**OUTPUTS**
1. Potential Projects	1. Working Groups 2. Expert Judgment	1. Project Portfolio 2. Preliminary Project Information (Ojectives, Schedule, and Budget)

Figure 9-5 Create Portfolio of Projects

ID	PROJECT	DURATION	BUDGET	FIN. RESUL. ($)	ROI	AREA	RISK	COMPLEXITY
1	Review of Product Mix	6	460,000	128,800	28%	Marketing and Sales	High	High
2	Zero Accidents	12	300,000	123,000	41%	Industrial	Low	Medium
3	Internationalization of Production Units	23	6,350,000	11,430,000	180%	Planning	Very High	High
4	Modernization of the Instrumentation System	8	2,420,000	1,573,000	65%	Industrial	Medium	Medium
5	E-commerce	4	350,000	126,000	36%	Information Technology	Medium	Medium
6	Corporate Office Projects	7	450,000	364,500	81%	Planning	Low	Low
7	New Markets	13	360,000	248,400	69%	Marketing and Sales	High	High
8	University Tiger Screws	7	350,000	258,110	74%	Human Resources	Low	Low
9	New Line for the Oil Industry	18	2,850,000	598,500	21%	Research and Development	High	High
10	New Distribution Center	19	3,600,000	2,124,000	59%	Logistics	Very High	High
11	Import Finished Products	22	2,080,000	4,430,400	213%	Marketing and Sales	Very High	High
12	Opening of Capital	24	1,200,000	660,000	55%	Financial	High	High
13	Social Media	5	225,000	41,116	18%	Marketing and Sales	Very Low	None
14	ERP System	9	1,240,000	347,200	28%	Information Technology	High	High
15	New Maintenance Policy	17	680,000	95,200	14%	Industrial	Medium	Medium
	Total	194	22,915,000	22,548,226				

Figure 9-6 Example of Basic Project List

The Project Portfolio can be presented in different ways but the most suitable to support the upcoming process is a list with the name of the project, estimated duration and budget (Figure 9-6). If the Project Office will support all kinds of projects at the corporate level, the list of projects can include a very different set of initiatives.

The preliminary project information can include all supporting information about the project, including main objectives, outputs, expected benefits and basic scope (Figure 9-7). This preliminary information can be also called Project Brief or Outline Business Case (UK Cabinet Office 2011).

Calculate Financial Return of Projects in the Portfolio

After identifying the potential portfolio of projects to be managed, it becomes important to calculate the benefits in terms of financial results of each project. This is one of the most challenging steps toward the calculation of the ROI of a Project office. Using the preliminary information, all proposed benefits will be measured to find clear outcomes (Figure 9-8).

In some cases, this is easily measurable in terms of increase in the production, marketplace, etc. In other projects, the intangible results must be converted into an estimated outcome. As an example, one main outcome of the "University Tiger Screws" project is to develop new capabilities in the current staff to reduce the investments with additional hiring of personnel.

A range of potential savings is defined and through the pairwise comparison using the Analytic Hierarchy Process (Saaty 1980; Saaty 2009; and Vargas 2010), the Expected Value of the benefit could be estimated (Figure 9-9).

The main output of this project is an updated list of projects including the estimated financial benefits.

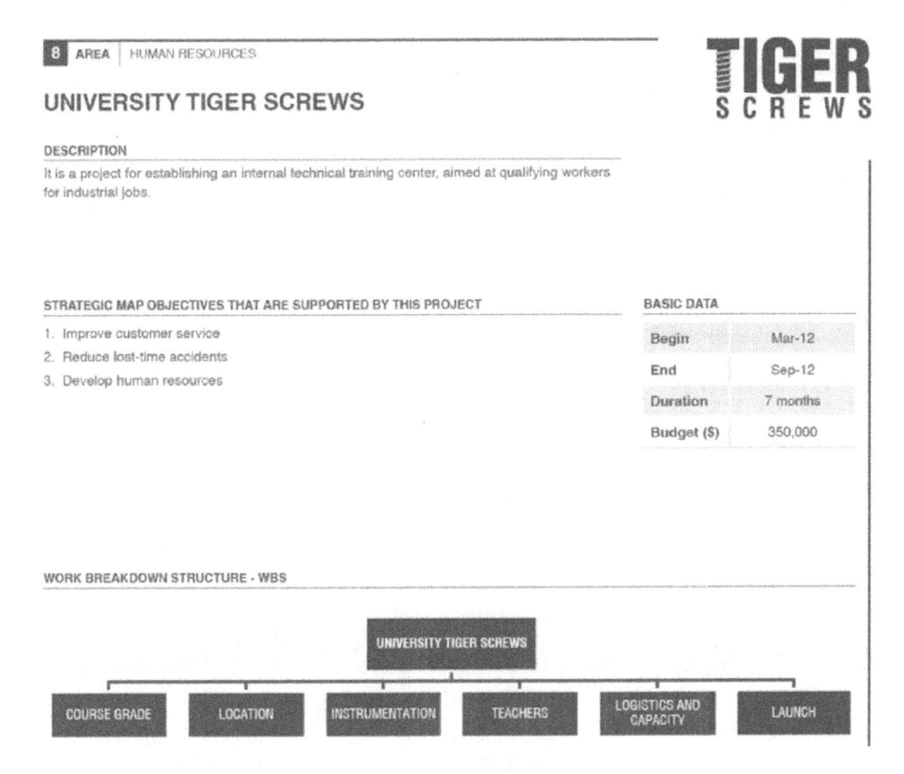

Figure 9-7 Example of Project Preliminary Information

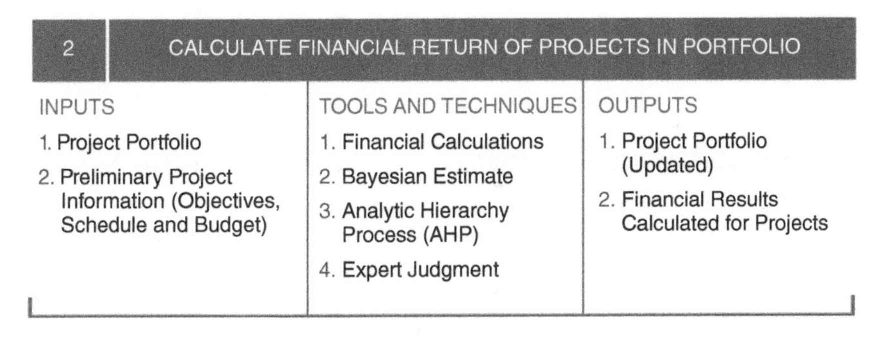

Figure 9-8 Calculate Financial Return of Projects in the Portfolio

Categorize Projects (Optional)

For organizations with a wide range of projects, the categorization of projects could add value in the analysis and stratification of efforts (Figure 9-10).

This optional process group the projects into different categories (Figure 9-11) like

- Departments
- Risk
- Value
- Sponsoring group
- Geographic location

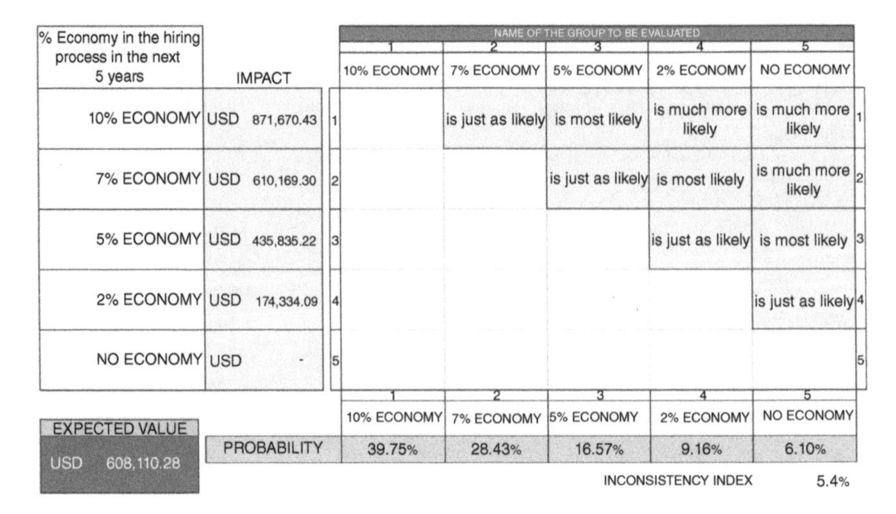

% Economy in the hiring process in the next 5 years
IMPACT
NAME OF THE GROUP TO BE EVALUATED
1 2 3 4 5
10% ECONOMY 7% ECONOMY 5% ECONOMY 2% ECONOMY NO ECONOMY
10% ECONOMY USD 871,670.43 1 is just as likely is most likely is much more likely is much more likely 1
7% ECONOMY USD 610,169.30 2 is just as likely is most likely is much more likely 2
5% ECONOMY USD 435,835.22 3 is just as likely is most likely 3
2% ECONOMY USD 174,334.09 4 is just as likely 4
NO ECONOMY USD - 5 5
EXPECTED VALUE
USD 608,110.28
PROBABILITY 39.75% 28.43% 16.57% 9.16% 6.10%
INCONSISTENCY INDEX 5.4%

Figure 9-9 Use of AHP to Estimate the Expected Value of the Project's Benefit

3	CATEGORIZE PROJECTS (OPTIONAL)	
INPUTS	TOOLS AND TECHNIQUES	OUTPUTS
1. Project Portfolio	1. Description of the Categories 2. Working Groups 3. Expert Judgment	1. Projects Grouped in "Categories"

Figure 9-10 Categorize Projects

ID	PROJECT	DURATION	BUDGET	FIN. RESUL. ($)	ROI	AREA	RISK	COMPLEXITY
1	Review of Product Mix	6	460,000	128,800	28%	Marketing and Sales	High	High
2	Zero Accidents	12	300,000	123,000	41%	Industrial	Low	Medium
3	Internationalization of Production Units	23	6,350,000	11,430,000	180%	Planning	Very High	High
4	Modernization of the Instrumentation System	8	2,420,000	1,573,000	65%	Industrial	Medium	Medium
5	E-commerce	4	350,000	126,000	36%	Information Technology	Medium	Medium
6	Corporate Office Projects	7	450,000	364,500	81%	Planning	Low	Low
7	New Markets	13	360,000	248,400	69%	Marketing and Sales	High	High
8	University Tiger Screws	7	350,000	258,110	74%	Human Resources	Low	Low
9	New Line for the Oil Industry	18	2,850,000	598,500	21%	Research and Development	High	High
11	Import Finished Products	22	2,080,000	4,430,400	213%	Marketing and Sales	Very High	High
12	Opening of Capital	24	1,200,000	660,000	55%	Financial	High	High
13	Social Media	5	225,000	41,116	18%	Marketing and Sales	Very Low	None
14	ERP System	9	1,240,000	347,200	28%	Information Technology	High	High
15	New Maintenance Policy	17	680,000	95,200	14%	Industrial	Medium	Medium
	Total	194	22,915,000	22,548,226				

Figure 9-11 Categorized List of Projects with the Calculated Benefits Highlighted

Data Distribution Profile

The Data Distribution Profile aims to determine the best risk profile of the portfolio to archive the benefits and it contains the process *Determine Optimistic, Most Likely and Pessimistic Profile for Schedule, Investments and Results.*

Using market research, historical information from previous projects and benchmarking, the objective of this process is to define the optimistic, pessimistic, and most likely scenarios for the duration, costs, and financial results of each project (Figure 9-12).

Different external sources can be used to support the decision as follows

- Standish Group Chaos Manifesto (STG 2013)
- The IPA Institute Database of Capital Projects (IPA 2013)
- PMI Pulse Report (PMI 2013)
- Reports and research from management consulting companies

This process requires a lot of negotiation to set the right thresholds for the project without being biased by individuals with over optimistic or over pessimistic behaviors.

The determination of the profiles can be done considering that the project duration, costs, and financial results follow the same distribution (Figure 9-13) or a different set of distributions for each element.

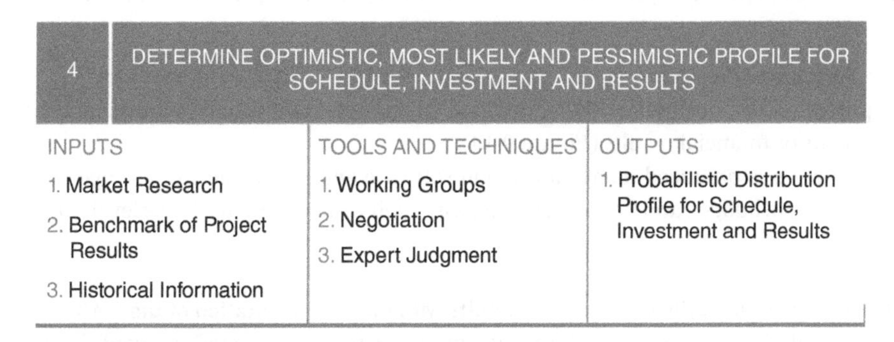

4
DETERMINE OPTIMISTIC, MOST LIKELY AND PESSIMISTIC PROFILE FOR SCHEDULE, INVESTMENT AND RESULTS
INPUTS
1. Market Research
2. Benchmark of Project Results
3. Historical Information
TOOLS AND TECHNIQUES
1. Working Groups
2. Negotiation
3. Expert Judgment
OUTPUTS
1. Probabilistic Distribution Profile for Schedule, Investment and Results

Figure 9-12 Determine Optimistic, Most Likely, and Pessimistic Profile for Schedule, Investment, and Results

	Without PMO			With PMO		
COMPLEXITY	OPTIMISTIC	MOST LIKELY	PESSIMISTIC	OPTIMISTIC	MOST LIKELY	PESSIMISTIC
High Complexity	+25%	+50%	+75%	+0%	+5%	+15%
Medium Complexity	+25%	+50%	+75%	+0%	+5%	+15%
Low Complexity	+15%	+30%	+45%	+0%	+5%	+15%
No Complexity	+10%	+20%	+30%	+0%	+5%	+15%

Figure 9-13 Probabilistic Forecasting Based on Project Complexity Level. In this case, a high complexity project with a value of $1,000,000 will cost between $1,250,000 and 1,750,000 without PM support and $1,000,00 and 1,150,000 with proper PM support.

Simulation

The Simulation group describes the process associated with the Monte Carlo simulation of the duration of the projects in the portfolio, associated costs, and financial results.

Monte Carlo was a nickname of a top-secret project related to the drawing and to the project of atomic weapons developed by the mathematician John von Neumann (Poundstone 1993). He discovered that a simple model of random samples could solve certain mathematical problems that couldn't be solved up to that moment.

The simulation refers, however, to a method by which the distribution of possible results is produced from successive recalculations of project data, allowing the development of multiple scenarios. In each one of the calculations, new random data is used to represent a repetitive and interactive process. The combination of all these results creates a probabilistic distribution of the results.

The feasibility of outcoming distribution relies on the fact that, for a high number of repetitions, the model produced reflects the characteristics of the original distribution, transforming the distribution into a plausible result for analysis. The simulation can be applied in schedules, costs, and other project indexes.

The Simulation group is divided in the following processes

- Simulate Portfolio of Projects
- Identify Gains in Schedule, Investment, and Results with the Implementation of the PMO

Simulate Portfolio of Projects

This is process is responsible for the simulation of the schedule gains, investment savings and improvements of financial results (Figure 9-14).

The simulation is produced using simulation software and the results are a range of improvements in duration, budget, and financial results with their respective confidence level (Figures 9.15, 9.16 and 9.17).

Identify Gains in Schedule, Investment, and Results with the Implementation of the PMO

After the simulation is concluded, the results are collected for a predefined confidence level to identify the measurable improvements (Figures 9.18 and 9.19).

5	SIMULATE PORTFOLIO OF PROJECTS	
INPUTS	TOOLS AND TECHNIQUES	OUTPUTS
1. Projects Grouped in Categories 2. Project Portfolio 3. Probabilistic Distribution Profile for Schedule, Investment, and Results	1. Monte Carlo Software Simulation	1. Probabilistic Distribution of Schedule Gains 2. Probabilistic Distribution of Investment Savings 3. Probabilistic Distribution of Improvements of Finacial Results

Figure 9-14 Simulate Portfolio of Projects

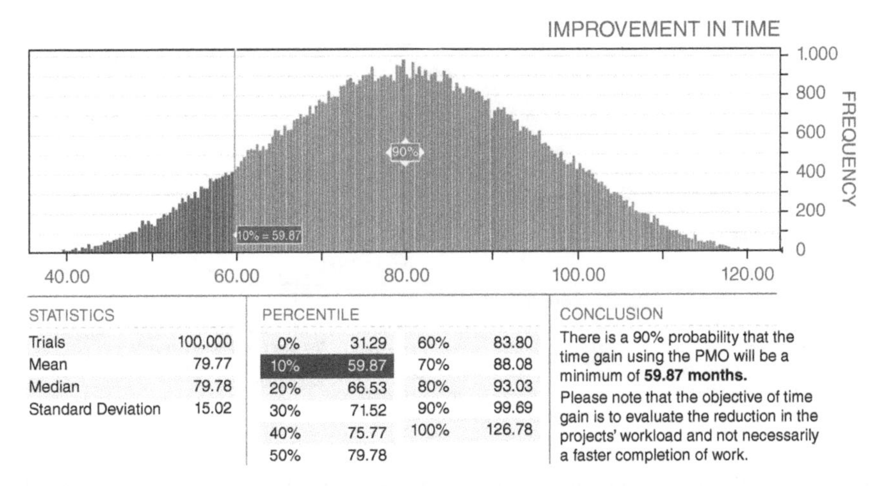

Figure 9-15 Simulation Output for Savings in the Total Time of the Projects for Using the PMO. In this case, there is a 90% confidence that the savings will be above 59,87 months.

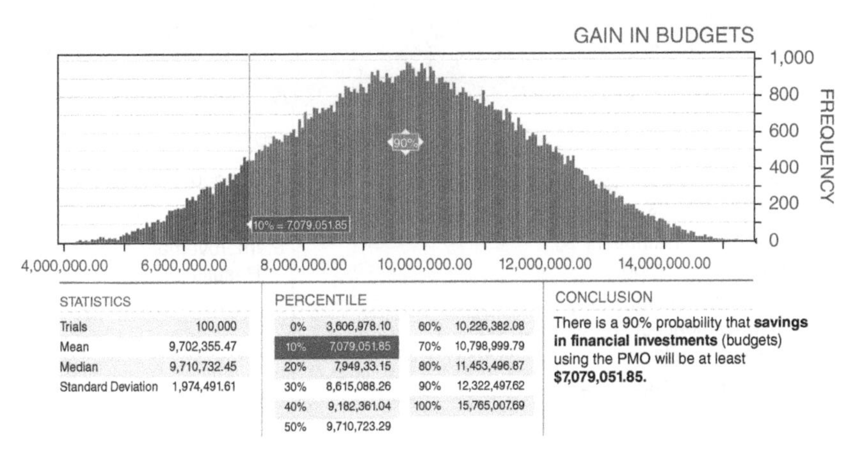

Figure 9-16 Simulation Output for Savings in the Budget of the Projects for Using the PMO. In this case, there is a 90% confidence that the savings will be at least 7,079,051.85 months.

Investments in the PMO

The other aspect that must be considered when evaluating the ROI of a project implementation is to calculate the amount of costs the organization will incur to create and maintain the PMO (Figure 9-20).

The most common costs associated with the PMO (Aubry, Hobbs, Müller & Blomquist 2010) are:

- Personal cost
- Software and hardware
- Advisory services
- Training
- Others

The main output of this process is the total cost of the PMO setup and operation for a predefined time frame (Figure 9-21).

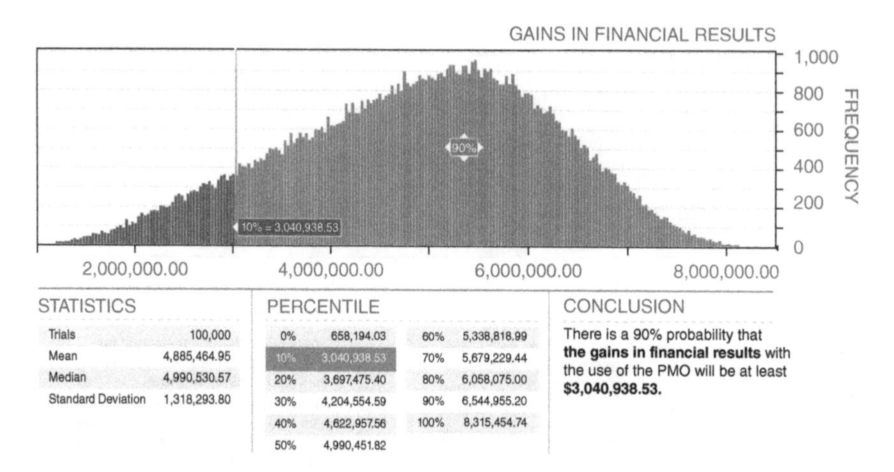

Figure 9-17 Simulation Output for Gains in Financial Results of the Projects for Using the PMO. In this case, there is a 90% confidence that the savings will be at least 3,040,938.53 months.

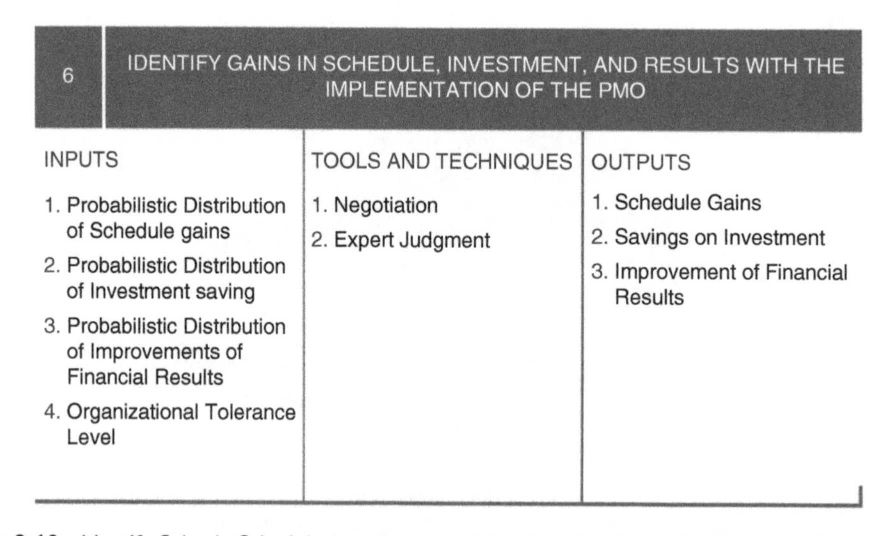

6
IDENTIFY GAINS IN SCHEDULE, INVESTMENT, AND RESULTS WITH THE IMPLEMENTATION OF THE PMO
INPUTS
1. Probabilistic Distribution of Schedule gains
2. Probabilistic Distribution of Investment saving
3. Probabilistic Distribution of Improvements of Financial Results
4. Organizational Tolerance Level
TOOLS AND TECHNIQUES
1. Negotiation
2. Expert Judgment
OUTPUTS
1. Schedule Gains
2. Savings on Investment
3. Improvement of Financial Results

Figure 9-18 Identify Gains in Schedule, Investment, and Results with the Implementation of the PMO

GAINS IN FINANCIAL RESULTS ($) Resulting from budget reduction and an improvement in the financial results.	**10,119,990.38**
FINANCIAL GAINS / PORTFOLIO VALUE (%)	**44.16%**
IMPROVEMENT IN TIME / EFFORT RELIABILITY OF 90%	**59.87 months**

Figure 9-19 Gains Based on the Simulation Results (Figures 9.16, 9.17, and 9.18)

7	CALCULATE INVESTMENT AND OPERATIONAL COSTS OF THE PMO/PROJECT STRUCTURE	
INPUTS	TOOLS AND TECHNIQUES	OUTPUTS
1. Direct Costs 2. Indirect Costs 3. Investment in Consultancy 4. Investment in Training 5. Procurement 6. Other Information on Costs and Investment	1. Financial Calculations 2. Budget Structure 3. Negotiation 4. Expert Judgment	1. Structure of Investment/ Cost of the PMO

Figure 9-20 Calculate Investment and Operational Costs of the PMO/Project Structure

	YEAR 1	YEAR 2	YEAR 3	YEAR 4	YEAR 5	TOTAL	TOTAL PV
Infrastructure	50.000	50.00	20.000	30.000	50.000	200.000	161.368,16
Consulting	800.000					800.000	800.000,00
Personal	420.000	420.000	420.000	420.000	420.000	2.100.000	1.669.277,96
Equipment	100.000					100.000	100.000,00
Other	10.000	10.000	10.000	10.000	10.000	50.000	39.744,71
Total	1.380.000	480.000	450.000	460.000	480.000	3.250.000	2.770.390,83

Figure 9-21 Example of a PMO Setup and Operation Cost in a Five-Year Time Frame. All values should be adjusted to present value.

Influence of the PMO in the Results

Based on the studies discussed in the step 4 of the process, it is important to highlight that not all benefits and positive results exclusively stem from the very existence and operation of the PMO. Many other external factors can benefit from those results and are beyond the control of the project manager and his/her team.

Some examples of benefit / disbenefit drivers not related to project management implementations are (UK CABINET OFFICE 2011)

- External economic factors like currency exchange rate, interest rates
- Market changes
- Changes in the legislation
- Changes in the senior leadership
- Others

In this process (Figure 9-22), it is proposed that the Analytical Hierarchy Process (AHP) be used to compare the likelihood of benefits coming from the PMO to other possible sources of benefits (Saaty 1980 and Vargas 2010).

The output of this process is the weight of the PMO in relationship with other sources of benefits (Figure 9-23).

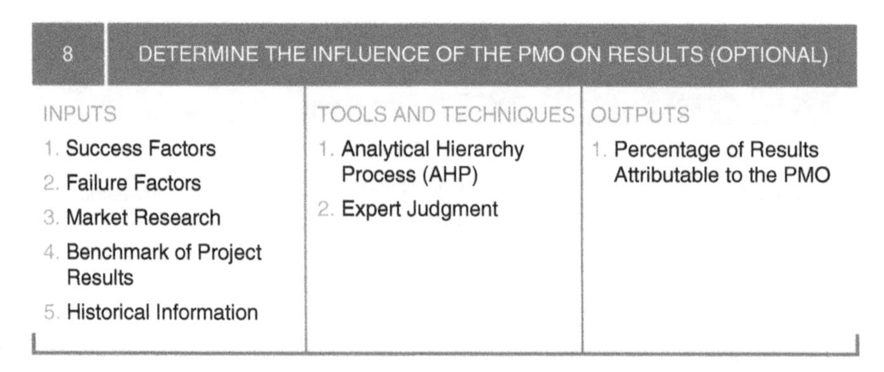

8
DETERMINE THE INFLUENCE OF THE PMO ON RESULTS (OPTIONAL)
INPUTS
1. Success Factors
2. Failure Factors
3. Market Research
4. Benchmark of Project Results
5. Historical Information
TOOLS AND TECHNIQUES
1. Analytical Hierarchy Process (AHP)
2. Expert Judgment
OUTPUTS
1. Percentage of Results Attributable to the PMO

Figure 9-22 Determine the Influence of the PMO on Results

	1 MARKET CHANGES	2 LEGISLATION	3 PROJECT MANAGEMENT (PMO)	4 LOW TECHNICAL SKILLS	5 OTHERS	
MARKET CHANGES 1		Likely	Less likely	Very likely	Likely	1
LEGISLATION 2			Very unlikely	Very likely	As likely as	2
PROJECT MANAGEMENT (PMO) 3				Highly likely	Very likely	3
LOW TECHNICAL SKILLS 4					Less likely	4
OTHERS 5						5
INCONSISTENCY INDEX: 6,8%	1 MARKET CHANGES	2 LEGISLATION	3 PROJECT MANAGEMENT (PMO)	4 LOW TECHNICAL SKILLS	5 OTHERS	
PROBABILITY	23,36%	11,61%	52,20%	3,63%	9,21%	

Figure 9-23 Example of AHP Comparing Different Sources of Benefits with Project Management Implementation. In this case, it is suggested that 52.2% of the benefits could be justified by the setup and operations of a PMO.

Results

The final group of process intends to calculate the return on investment and analyze and discuss the results.

The Results group is divided in the following processes

- Calculate the return on investment (ROI) of the PMO
- Analyze final results

Calculate the Return on Investment (ROI) of the PMO

This process compares the results obtained in the simulation and compare them with the investments related to the PMO and the percentage of results attributable to the PMO (Figure 9-24).

The output of this process is the calculated return on investment with complementary information (Figure 9-25).

9	CALCULATE THE RETURN ON INVESTMENT (ROI) OF THE PMO	
INPUTS	TOOLS AND TECHNIQUES	OUTPUTS
1. Schedule Gains 2. Savings on Investment 3. Improvement of Financial Results 4. Structure of Investment/ Cost of the PMO 5. Percentage of Results Attributable to the PMO	1. Financial Calculations	1. Return on Investment (ROI) of the PMO 2. Complementary Information 3. Calculation Report 4. Final ROI Report

Figure 9-24 Calculate the Return on Investment (ROI) of the PMO

GAINS IN FINANCIAL RESULTS ($) Resulting from budget reduction and an improvement in the financial results.	**10,119,990.38**
FINANCIAL GAINS / PORTFOLIO VALUE (%)	**44.16%**
IMPROVEMENT IN TIME / EFFORT RELIABILITY OF 90%	**59.87 months**

IMPORTANCE OF THE PMO ON RESULTS (%)	**52.20%**
FINANCIAL GAINS ADJUSTED FOR IMPORTANCE OF THE PMO ON RESULTS ($)	**5,282,634.98**
PMO INVESTMENT ($)	**2,770,390.83**

PMO RETURN ON INVESTMENT ($)	2,512,244.15
PMO RETURN ON INVESTMENT (%)	90.68%

Figure 9-25 Financial Calculation Based on the Simulation Results and the Cost / Relevance of the PMO in the Results

10	ANALYZE FINAL RESULTS	
INPUTS	TOOLS AND TECHNIQUES	OUTPUTS
1. Final ROI Report	1. Working Groups	1. Decision Making
	2. Negotiation	2. Lessons Learned
		3. Agreement on Results

Figure 9-26 Analyze Final Results

Analyze Final Results

After receiving the final ROI Report, working groups and the PMO sponsoring group need to meet to analyze and discuss the results to make the final decision (Figure 9-26).

Final Remarks

The proposed model is a *masterline* of the value calculation and can thus be customized and adapted to different scenarios. It is important to highlight that this model is a simplification of very complex environment, where different perceived values can provide different directions to different stakeholders.

To avoid resistance and criticism during the simulation of intangible results, it is important to do this work in *teams* to avoid personal biases in the process.

Finally, it important to understand the challenge of determining ROI without knowing which projects are selected and the strategy that supports them. A project office that takes care of several different multimillion projects is a very different effort from a project office that has simple work packages to be controlled.

9.6 Conclusions

Project management offices can be created and managed in different shapes and forms according to the organization's business, its project management maturity level, the type of products and services they deliver.

However, one aspect is the same for each of these efforts: to provide support, guidance, and relevant information to improve results and deliver better outcomes.

References

Aubry, M., Hobbs, B., Müller, R. & Blomquist, T. (2010). *Identifying Forces Driving PMO Changes. Project Management Journal* 41 (4): 30–45. Newtown Square: Project Management Institute.

EIU (2009). *Closing the Gap – The Link Between Project Management Excellence & Long-Term Success.* London: Economist Intelligence Unit.

Hubbard, D. (2010). *How to Measure Anything.* 2nd Edition. Hoboken: John Wiley & Sons.

IPA (2013). *Research as a Foundation.* Ashburn: Independent Project Analysts. http://ipainstitute.com/Home/About/Research

Kwak, Y.H., and Ibbs, C.W. (2000). *Calculating Project Management's Return on Investment.* Newtown Square: *Project Management Journal* 31 (2). http://www.lamarheller.com/projectmgmt/calculatingpmroi.pdf

Philips, J. J., and Philips, P. P. (2007). *Show me the Money: How to Determine the ROI in People, Projects and Programs.* San Francisco: Berrett-Koehler Publishers.

PMI (2019). *Pulse Report of the Profession.* Newtown Square: Project Management Institute. http://www.pmi.org/Knowledge-Center/Pulse.aspx

PMI (2016). *The Project Management Body of Knowledge.* 6th Edition. Newtown Square: Project Management Institute.

Poundstone, W. (1993). *Prisoner's Dilemma.* Flushing: Anchor Publishing Group.

Pritchard, C. L. (2001). *Risk Management: Concepts and Guidance.* 2nd Edition. Arlington: ESI International.

Saaty, T. L. (1980). *The Analytic Hierarchy Process.* New York: McGraw-Hill International.

Saaty, T. L. (2009). *Extending the Measurement of Tangibles to Intangibles. International Journal of Information Technology & Decision Making* 8 (1): 7–27, http://ssrn.com/abstract=1483438.

STG (2013). *The Chaos Manifesto.* Boston: The Standish Group. https://secure.standishgroup.com/reports/reports.php

UK Cabinet Office (2011). *Managing Successful Programmes: Forth Edition.* London: The Stationery Office.

Vargas, R (2004). *Earned Value Probabilistic Forecasting Using Monte Carlo Simulation.* Washington: AACE – Association for Advancement of Cost Engineering 48th Annual Meeting.

Vargas, R. V. (2010). *Using Analytical Hierarchy Process to Select and Prioritize Projects in Portfolio Management.* Dallas: PMI Global Congress North America. Available at http://www.ricardo-vargas.com/articles/analytic-hierarchy-process/

10

Pillar #10: Significant Growth in Value-Driven and Business-Related Metrics

10.0 The Growth of Project Metrics

As stated in previous pillars, project managers are realizing that they are no longer just managing a traditional project to create an outcome or deliverable. Instead, project managers are managing part of the business and are being treated as business managers. The traditional metrics that we use, such as time, cost, and scope, may be insufficient for assisting in making some project decisions and may not capture the true business value of the deliverable of a project.

Today, we live in a world where digitalization is increasingly important. Also, significant advances are being made in measurement techniques that are being applied to projects. As such, there has been a significant growth in types of metrics as seen in Figure 10-1.

The five levels identified in Figure 10-1 can be described as follows:

1. **Basic Project Management (PM 1.0).** This is traditional or operational project management, which focuses on well-defined requirements and primarily time, cost, and scope metrics only. Once the deliverables are created, the project manager hands the project over to someone else and moves on to another assignment.
2. **Business-Driven Project Management (PM 2.0).** The project manager is viewed as a business manager and participates in some business-related decisions that impact the projects. The PM must capture and report certain business metrics so that management can make timely decisions based on facts and evidence rather than guesses.
   The business metrics referred to here are usually metrics related to the customers of the project's deliverables and may include the number of new customers, number of customers lost, and maintenance and support information for a targeted customer base. The business metrics are not necessarily strategic metrics but are used to assist the project sponsors in the decisions they must make for the projects where they are providing governance.
3. **Value-Driven Project Management (PM 3.0).** As mentioned in previous pillars, projects are now aligned to strategic business objectives where success is measured by the business benefits and value created from the deliverables. The PM must identify, capture, and report benefits and value metrics, and provide this information to executives and managers of strategic business portfolios of projects.

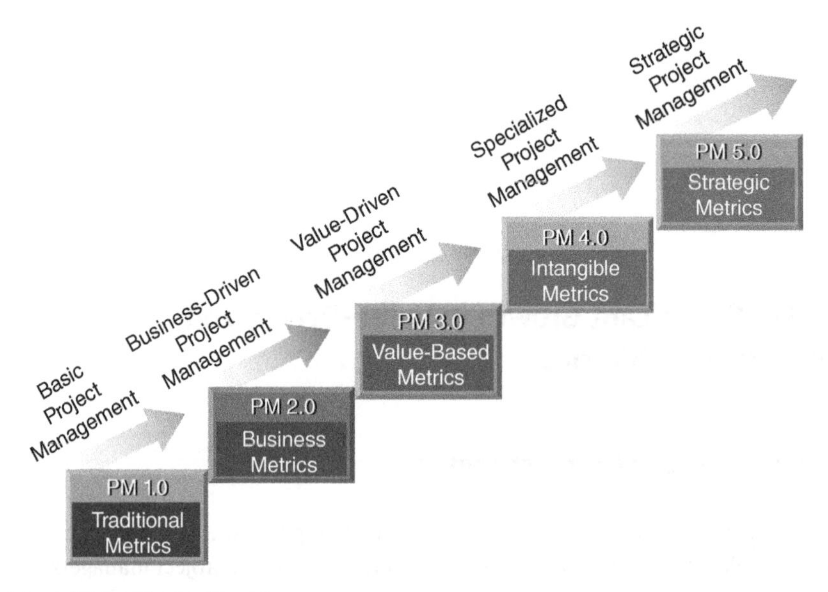

Figure 10-1 Metrics Growth

4. **Specialized Project Management (PM 4.0).** The outcome of some projects requires the measurement of intangibles rather than just tangible outcomes. Intangible metrics will be discussed later in this chapter. Intangible metrics may focus on continuous improvement efforts for better management or governance of projects and programs.
5. **Strategic Project Management (PM 5.0).** As business managers, some project managers may be asked to manage strategic projects, such as those involving innovation and new business models, and must therefore monitor and report strategic business metrics in addition to other metrics.

10.1 The Growth of Metric Measurement Techniques

Selecting metrics and KPIs from the five levels in Figure 10-1 is not that difficult, provided they can be measured. This is the major obstacle with metric selection. On the surface, metrics seem easy to measure, but there are complexities. With traditional project management, metrics are established by the enterprise project management methodology and most often fixed for the duration of the project's life cycle. In the future, where project teams will have a choice in which flexible methodology to use, metrics can change from project to project, during a life-cycle phase and over time because of:

- The way the customer and contractor jointly define success at project initiation
- The way the customer and contractor come to an agreement at project initiation as to what metrics should be used on a given project, and when
- New or updated versions of tracking software and measurement techniques
- Improvements to the enterprise project management methodology and the accompanying project management information system
- Changes in the enterprise environmental factors

Even with the best possible metrics, measuring some necessities such as business value created can be difficult. During project selection, benefits less costs often indicate the project's business value and determine if the project should be done. The challenge is that not all costs are quantifiable. Moreover, some costs are easy to measure, while others are more difficult. The easy metrics to measure are often called soft or are established by the enterprise project management methodology and fixed for the duration of the project's life cycle. The hard metrics to measure are often considered as intangible metrics. Table 10-1 illustrates some of the easy and hard metrics to measure. Table 10-2 shows some of the problems associated with measuring both hard and soft metrics.

Measurement techniques in the future will most likely be a compromise between a quantitative and qualitative measurement. The timing of the measurement is also critical. During the life cycle of a project, it may be necessary to switch back and forth from qualitative to quantitative assessment of the metric, and the actual metrics or KPIs may then be subject to change.

On some strategic projects, using metrics to assess business benefits and value at project closure may be difficult. We may need to establish a time frame for how long we are willing to wait after the deliverables have been created to measure the final or real value or benefits from a project. This is particularly important if the actual value cannot be identified until sometime after the project has been completed. Therefore, it may not be possible to appraise the success of a project at closure if the true economic value cannot be realized until sometime in the future.

Table 10-1 Typical Financial Value Metrics

Easy (Soft/Tangible) Metrics	Hard (Intangible) Metrics
	Stockholder satisfaction
	Stakeholder satisfaction
ROI calculators	Customer satisfaction
Net present value (NPV)	Employee retention
Internal rate of return (IRR)	Brand loyalty
Cash flow	Time to market
Payback period	Business relationships
Profitability	Safety
Market share	Reliability
	Goodwill
	Image

Table 10-2 Problems with Measuring Value Metrics

Easy (Soft/Tangible) Metrics	Hard (Intangible) Metrics
Assumptions are often not fully disclosed and can affect decision-making.	It is almost always based on subjective attributes of the person doing the measurement.
Measurement is very generic.	It is more of an art than a science.
Measurement never meaningfully captures the correct data.	Limited models are available to perform measurement.

10.2 Selecting the Right Metrics

Because of the growth in measurement techniques, companies are now tracking a dozen or more metrics on projects. While this sounds good, it has created the additional problem of potential information overload. Having too many performance metrics may provide the viewers with more information than they need, and they may not be able to discern the true status or what information is critical. It may be hard to ascertain what is important and what is not, especially if decisions must be made. Providing too few metrics can make it difficult for the viewers to make informed decisions. There is also a cost associated with metric measurement, and we must determine if the benefits of using this many metrics outweigh the costs of measurement. Cost is important because we tend to select more metrics than we need.

There are three generic categories of metrics:

- **Traditional metrics.** These metrics are used for measuring the performance of the applied project management discipline more so than the results of the project and how well we are managing according to the predetermined baselines (e.g., cost variance and schedule variance). These metrics are common to all types of projects, including innovation.
- **Key performance indicators (KPIs).** These are the few selected metrics that can be used to track and predict whether the project will be a success. These KPIs are used to validate that the critical success factors (CSFs) defined at the initiation of the project are being met (e.g., time-at-completion, cost-at-completion, and customer satisfaction surveys). These KPIs can be unique to a particular project.
- **Value (or value reflective) metrics.** These are special metrics that are used to indicate whether the stakeholders' expectations of project value are being or will be met. Value metrics can be a combination of traditional metrics and KPIs (value-at-completion and time to achieve full value).

Each type of metric has a primary audience, as shown in Table 10-3.

There can be three information systems on a project:

- One for the project manager
- One for the project manager's superior or parent company
- One for the each of the stakeholders, customers, or partners

Project manager must be careful not to micromanage their project and establish 40 to 50 metrics, many of which do not provide useful information. Typical metrics on traditional projects may include:

- Number of assigned versus planned resources
- Quality of assigned versus planned resources

Table 10-3 Audiences for Various Metrics

Type of Metric	Audience
Traditional metrics	Primarily the project manager and the team, but may include the internal sponsor(s) as well
Key performance indicators	Some internal usage but mainly used for status reporting for the client and the stakeholders
Value metrics	Can be useful for everyone but primarily for the client as well as members of the co-creation team

- Project complexity factor
- Customer satisfaction rating
- Number of critical constraints
- Number of cost revisions
- Number of critical assumptions
- Number of unstaffed hours
- Percent of total labor hours on overtime
- Cost variance
- Schedule performance index
- Cost performance index

This is obviously not an all-inclusive list. These metrics may have some importance for the project manager but not necessarily the same degree of importance for the client and the stakeholders.

Clients and stakeholders are interested in critical metrics or KPIs. These chosen few metrics are reported to the client and stakeholders and provide an indication of whether success is possible; however, they do not necessarily identify if the desired business value will be achieved. The number of KPIs is usually determined by the amount of real estate on a computer screen. Most dashboards can display 6–10 icons or images where the information can be readily seen with reasonable ease.

To understand what a KPI means requires a dissection of each of the terms:

- *Key:* A major contributor to success or failure
- *Performance:* Measurable, quantifiable, adjustable, and controllable elements
- *Indicator:* Reasonable representation of present and future performance

Obviously, not all metrics are KPIs. There are six attributes of a KPI, and these attributes are important when identifying and selecting the KPIs.

- *Predictive:* Able to predict the future of this trend
- *Measurable:* Can be expressed quantitatively
- *Actionable:* Triggers changes that may be necessary
- *Relevant:* The KPI is related to the success or failure of the project
- *Automated:* Reporting minimizes the chance of human error
- *Few in number:* Only what is necessary

Applying these six attributes to traditional metrics is highly subjective and will be based on the agreed-on definition of success, the critical success factors (CSFs) that were selected, and possibly the whims of the stakeholders. There can be a different set of KPIs for each stakeholder based on each stakeholder's definition of project success and final project value. This could significantly increase the costs of measurement and reporting, especially if each stakeholder requires a different dashboard with different metrics.

10.3 Benefits Realization and Value Management

Introduction

The future of project management will be value-driven project management. As shown in Figure 10-1, project managers will be required to identify, track and report metrics related to business benefits and value.

Organizations in both the public and private sectors have been struggling with the creation of a portfolio of projects that would provide sustainable business value. All too often, companies would add all project requests to the queue for delivery without proper evaluation and with little regard if the projects were aligned with business objectives or provided benefits and value upon successful completion. Projects were often submitted without any accompanying business case. Many projects had accompanying business cases that were based on highly exaggerated expectations and unrealistic benefits just to get authorization to procced. Other projects were created because of the whims of management, and the order in which the projects were completed was based on the rank or title of the requestor. Simply because an executive says "Get It Done" does not mean it will happen. The result was often project failure or a waste of precious resources, and in some cases, business value was eroded or destroyed rather than created.

Understanding the Terminology

It is important to begin with an understanding of the basic terminology related to benefits and value that were discussed at the beginning of this chapter. Without a clear understanding of the terminology, project teams may lose insight on the meaning of value-driven project management.

A **benefit** is an outcome from actions, behaviors, products, or services that is considered important or advantageous to specific individuals, such as business owners, or a group of individuals, such as stakeholders. Generic benefits might include:

- Improvements in quality or productivity
- Cost avoidance or cost reduction
- Efficiency of operations
- Increase in revenue generation
- Improvements in customer service

Benefits, whether they are strategic or nonstrategic, are normally aligned to the organizational business objectives of the sponsoring organization that will eventually receive the benefits. The benefits appear through the **deliverables** or **outputs** that are created by the project. It is the responsibility of the project manager to create the deliverables.

Benefits are identified in the project's business case. Some benefits are tangible and can be quantified. Other benefits, such as an improvement in employee morale, may be difficult to measure and therefore be treated as intangible benefits. However, they might still be measurable.

There can also be dependencies between the benefits where one benefit is dependent on the outcome of another. As an example, a desired improvement in revenue generation may be dependent on an improvement in quality.

Benefits realization management is a collection of processes, principles, and deliverables to effectively manage the organization's investments.[1] Project management focuses on maintaining the established baselines whereas benefits realization management analyzes the relationship that the project has to the business objectives by monitoring for potential waste, acceptable levels of resources, risk, cost, quality, and time as it relates to the desired benefits.

---

1 For additional information on benefits realization management see Craig Letavec, ***Strategic Benefits Realization***, J. Ross Publishers, 2004, and Trish Melton, Peter Iles-Smith and Jim Yates, ***Project Benefits Management; Linking Projects to the Business***, Butterworth-Heinmann Publishers, an imprint of Elsevier Publishers, 2008

Decision-makers must understand that, over the life cycle of a project, circumstances can change, requiring modification of the requirements, shifting of priorities, and redefinition of the desired outcomes. It is entirely possible that the benefits can change to a point where the outcome of the project provides detrimental results, and the project should be cancelled or backlogged for consideration later. Some of the factors that can induce changes in the benefits and resulting value include:

- **Changes in business owner or executive leadership.** Over the life of a project, there can be a change in leadership. Executives that originally crafted the project may have passed it along to others that either have a tough time understanding the benefits, are unwilling to provide the same level of commitment, or see other projects as providing more important benefits.
- **Changes in assumptions.** Based on the length of the project, the assumptions can and most likely will change, especially those related to enterprise environmental factors. Tracking metrics must be established to make sure that the original or changing assumptions are still aligned with the expected benefits. This is critical when considering a VUCA environment.
- **Changes in constraints.** Changes in market conditions (i.e., markets served and consumer behavior) or risks can induce changes in the constraints. Companies may approve scope changes to take advantage of additional opportunities or reduce funding based on cash flow restrictions. Metrics must also track for changes in the constraints.
- **Changes in resource availability.** The availability or loss of resources with the necessary critical skills is always an issue and can impact benefits if a breakthrough in technology is needed to achieve the benefits or to find a better technical approach with less risk.

Project **value** is what the benefits are worth to someone. Project or business value can be quantified, whereas benefits are usually explained qualitatively. When we say that the ROI should improve, we are discussing benefits. But when we say that the ROI should improve by 20 percent, we are discussing value. Progress toward value generation is easier to measure than benefits realization, especially during project execution. Benefits and value are generally inseparable; it is difficult to discuss one without the other.

Redefining Project Success

For more than five decades, we have erroneously tried to define project success in terms of only the triple constraints of time, cost, and scope. We knew decades ago that other metrics should be included in the definition such as value, safety, risk, and customer satisfaction, and that these were attributes of success. Unfortunately, our knowledge of metrics measurement techniques was just in the infancy stage at that time, and we selected only those metrics that were the easiest to measure and report, namely time, cost, and scope.

Today, metric measurement techniques are maturing to the point where we believe that we can measure just about anything.[2] Perhaps the greatest level of research has been in measuring and reporting business value. Value could very well become the most important word in the project manager's vocabulary, especially in the way that we define project success. A project can be defined as:

- A collection of sustainable business value scheduled for realization

2 For additional information, see Douglas W. Hubbard, *How to Measure Anything: Finding the Value of Intangibles in Business*, 3rd edition (Hoboken, NJ: John Wiley, 2014).

The definition of project success has almost always been the completion of a project within the triple constraints of time, cost, and scope. Therefore, the definition of success might be:

- Achieving the desired business value within the competing constraints

This definition of project success that includes reference to value becomes extremely important when reporting the success of benefit realization and value management activities. With traditional project management, we create forecast reports that include the time at completion and cost at completion. Using the new definition for success, we can now include in the forecast report benefits at completion and value at completion. This now elevates project performance reporting to the corporate boardroom.

There is another inherent advantage to using value as part of the project's success criteria. We can now establish a termination, or "pull the plug," criteria defined in terms of values or benefits that tell us when we should consider cancelling a project before additions funds and resources are squandered. All too often, projects linger on and continue wasting valuable resources because nobody has the heart to abandon the failing project. Establishing cancelation criteria in the business case or benefits realization plan, accompanied by tracking metrics, may solve this issue.

The Business Case

Benefits realization and value management begin with the preparation of the business case. There are four major players in benefits realization and value management projects:

1. A governance committee composed of members that possess at least a cursory level of knowledge of project management
2. The benefits or business owner
3. The change management owner, if organizational change management is necessary to harvest the benefits at project completion
4. Project and/or program managers

The business owner is responsible for the preparation of the business case as well as contributing to the benefits realization plan. Typically, business case development includes these steps:

- Identification of opportunities such as improved efficiencies, effectiveness, waste reduction, cost savings, new business, etc.
- Benefits defined in both business and financial terms
- A benefits realization plan
- Estimated project costs
- Recommended metrics for tracking benefits and value
- Risk management
- Resource requirements
- High-level schedules and milestones
- Degree of project complexity
- Assumptions and constraints
- Technology requirements – new or existing
- Exit strategies if the project must be terminated

Templates can be established for most of the items in the business case. What is new for most companies is the creation of a benefits realization plan as part of a business case. A template for a benefits realization plan might include the following:

- A description of the benefits
- Identification of each benefit as tangible or intangible
- Identification of the recipient of each benefit
- How the benefits will be realized
- How the benefits will be measured
- The realization date for each benefit
- The handover activities to another group that may be responsible for converting the project's deliverables into benefits realization

10.4 Measuring Benefits and Value

The growth in metric measurement techniques has made it possible to measure just about anything. This includes both benefits and value. But at present, since many of the measurement techniques for newer metrics are in the infancy stage, there is still difficulty in obtaining accurate results. Performance results will be reported both quantitatively and qualitatively. There is also difficulty in deciding when to perform the measurements – incrementally as the project progresses or at completion. Measurements of benefits and value are more difficult to determine incrementally as the project progresses than at the end.

Value is generally quantifiable and easier to measure than benefits. On some projects, the value of the benefits of the project cannot be quantified until several months after the project has been completed. As an example, a government agency enlarges a road to hopefully reduce traffic congestion. The value of the project may not be known until several months after the construction project has been completed and traffic flow measurements have been made. Value measurements at the end of the project, or shortly thereafter, are generally more accurate than ongoing value measurements as the project progresses.

Benefits realization and business value do not come from simply having talented resources or superior capabilities. Rather, they come from how the organization uses the resources. Sometimes, even projects with well-thought-out plans and superior talent do not end up creating business value and can even destroy existing value. An example might be a technical prima donna that views this project as his/her chance for glory and tries to exceed the requirements to a point where the schedule slips and business opportunities are missed. This occurs when team members believe that personal objectives are more important than business objectives.

Converting Benefits to Value

Value is what the benefits are worth either at the end of the project or sometime in the future. Even though the benefits may be on track for achievement, the final value may be different from the planned value based on the deliverables produced, the financial assumptions made, and stakeholder or consumer acceptance. Here are two examples of converting benefits to value:

- A company approved the development of a customized software package with the expected benefit of reducing order entry processing time, which would be a savings of approximately

Table 10-4 Metrics for Specific Types of PMOs

Project Management	Traditional PMO	Portfolio PMO
• Adherence to schedule baselines • Adherence to cost baselines • Adherence to scope baselines • Adherence to quality requirements • Effective utilization of resources • Customer satisfaction levels • Project performance • Total number of deliverables produced	• Growth in customer satisfaction • Number of projects at risk • Conformance to the methodology • Ways to reduce the number of scope changes • Growth in the yearly throughput of work • Validation of timing and funding • Ability to reduce project closure rates	• Business portfolio profitability or ROI • Portfolio health • Percentage of successful portfolio projects • Portfolio benefits realization • Portfolio value achieved • Portfolio selection and mix of projects • Resource availability • Capacity and capability available for the portfolio • Utilization of people for portfolio projects • Hours per portfolio project • Staff shortage • Strategic alignment • Business performance enhancements • Portfolio budget versus actual • Portfolio deadline versus actual

$1.5 million annually. The cost of developing the package was estimated at $750,000. The value calculation was as follows:

$$\text{Value} = (60 \text{ workers}) \times \frac{(5 \text{ hours})}{\text{week}} \times \frac{(\$100)}{\text{hour}} = \$1.5 \text{ million in yearly savings}$$

- A company decided to create a dashboard project performance reporting system to reduce paperwork and eliminate many nonproductive meetings. The value calculation was made as follows:
  - Eliminate 100 pages or reports and handouts each month at a fully burdened cost of $1000/page, or a savings of $1.2 million.
  - Eliminate 10 hours of meetings per week for 50 weeks, with 5 people per meeting and at $100 per hour, or a savings of $250,000

$$\text{Value} = \$1,200,000 + \$250,000 = \$1.45 \text{ million in yearly savings}$$

In both cases, the company received multiyear benefits and value from the projects.

Portfolio Benefits and Value

The project tracking metrics of time, cost, and scope are design to track individual projects. However, there are specific metrics that can be used to measure the effectiveness of a portfolio of projects. Table 10-4 shows the metrics that can be used to measure the overall value created by

Table 10-5 Interpretation of the Metrics

Benefit Metric	Project Manager's Interpretation	Customer's Interpretation	Consumer's Interpretation
Time	Project duration	Time-to-market	Delivery date
Cost	Project cost	Selling price	Purchasing price
Quality	Performance	Functionality	Usability
Technology and scope	Meeting specifications	Strategic alignment	Safe buy and reliable
Satisfaction	Customer satisfaction	Consumer satisfaction	Esteem in ownership
Risks	No future business from this client	Loss of profits and market share	Need for support and risk of obsolescence

project management on individual projects, a traditional PMO, and a portfolio PMO. The metrics listed under project management and many of the metrics under the traditional PMO are considered as micro metrics focusing on tactical objectives. The metrics listed under the portfolio PMO are macro-level metrics that represent the benefits and value of the entire portfolio. These metrics can be created by grouping together metrics from several projects. Benefits and value metrics are also used to help create the portfolio metrics.

Both the traditional and portfolio PMOs are generally considered as overhead and subject to possible downsizing unless the PMOs can show through metrics how the organization benefits by their existence. Therefore, metrics must also be established to measure the value that the PMO brings to the parent organization.

It is important to understand that some of the micro metrics we use for tracking benefits may have a different meaning for the customer or ultimate consumer. As an example, let us assume that you are managing a project for an external client. The deliverable is a component that your customer will use in a product they are selling to their customers (i.e., your customer's customers or consumers). Table 10-5 shows how each of the metrics may be interpreted differently. It is important to realize that benefits and value are like beauty; they are in the eyes of the beholder. Customers and contractors can have a different perception of the meaning of benefits and value as well as the associated metrics.

10.5 Excellence in Action: Philips Business Group Hospital Patient Monitoring[3]

Background

Michael Bauer, Capability Leader and Head of the Global SSMO (Solutions and Services Management Office) at Philips Business Group HPM Services & Solution Deliverability (Hospital Patient Monitoring), and Mary Ellen Skeens, Director of Solution Services Capabilities at Philips

3 Material in this section has been provided by Michael Bauer, Capability Leader and Head of the Global SSMO (Solutions and Services Management Office) at Philips Business Group HPM Services & Solution Deliverability (Hospital Patient Monitoring) and Mary Ellen Skeens, Director of Solution Services Capabilities at Philips Business Group HPM Services & Solution Deliverability.

Business Group HPM Services & Solution Deliverability, describe how Customer Success Management and Outcomes realization are successfully integrated with a Scalable Solution Design & Delivery Services framework.

In the 4th edition of *Project Management Best Practices Achieving Global Excellence*, Michael Bauer provided an overview of the Philips SOLiD Framework, along with key takeaways on how a scalable approach enables organizations to achieve Solution Implementation & Services Excellence.[4]

In *Innovation Project Management: Methods, Case Studies, and Tools for Managing Innovation Projects*, Michael Bauer and Mary Ellen Skeens described healthcare driving Solution Innovation, understanding customer needs, and considering solution complexity, as well as enablers for achieving Solution Design & Delivery Service Excellence.[5]

In this section, we review the dynamics of the shift to healthcare customer value-driven solution projects, including:

- Major Healthcare trends and need for Solutions & Outcomes
- Solution Innovation, Development and Commercialization
- Solution Design & Delivery Services and specific Capabilities
- Integrating frameworks for Solution Services, Customer Success, and Outcomes
- Holistic and fully integrated approach for the Customer Lifecycle
- Customer Success in Solution Projects and how to measure its Value to Customers
- Solution Excellence and Continuous Improvement for Customer Success

Enabling Customer Success in Healthcare Solutions Business: Royal Philips

Royal Philips (NYSE: PHG, AEX: PHIA) is a leading health technology company focused on improving people's health and well-being and enabling better outcomes across the health continuum – from healthy living and prevention, to diagnosis, treatment, and home care. Philips leverages advanced technology and deep clinical and consumer insights to deliver integrated solutions. Headquartered in the Netherlands, the company is a leader in diagnostic imaging, image-guided therapy, patient monitoring, and health informatics, as well as in consumer health and home care. Philips generated 2020 sales of EUR 17.3 billion and employs approximately 77,000 employees with sales and services in more than 100 countries. News about Philips can be found at www.philips.com/newscenter.

The **Hospital Patient Monitoring (HPM) Business Group** is a software and solutions business encompassing patient monitoring and its capabilities. Reaching over 500 million people every year, HPM solutions are advanced intelligence platforms, providing key insights and information to clinicians when and where they need it. The ultimate priority for the HPM Business Group is to enable smart decision-making for caregivers, administrators, and patients such that workflows are improved, costs are controlled, efficiency is increased, and, importantly, better health outcomes are supported.

---

4 Kerzner, H. (2018). *Project Management Best Practices Achieving Global Excellence 4th edition*. Hoboken, NJ: John Wiley, pp. 448–457.

5 Kerzner, H. (2019). *Innovation Project Management: Methods, Case Studies, and Tools for Managing Innovation Projects*. Hoboken, NJ: John Wiley, pp. 190–202.

Mega Trends in Healthcare Toward Innovative Solutions

The healthcare industry is quickly evolving. Digital technology and innovative solutions are shaping the industry to support individuals taking charge of their own health.

There are four key trends driving disruptive change in healthcare technology. They include:

1. The shift from volume to value-based care, due to global resource constraints. The World Health Organization (WHO) estimates that 18 million more healthcare workers are needed to close the gap to meet demands of the system in 2030.[6]
2. The growing population of older patients and increase in chronic conditions such as cardiovascular disease, cancer, and diabetes. The world's older population is forecasted to outpace the younger population over the next three decades.[7]
3. Patients are exerting more control over healthcare decisions and choosing which healthcare organizations they utilize as consumers. With access to digital healthcare tools and the incentive of reducing out of pocket expenses, patients are making more carefully informed decisions regarding care.[8]
4. COVID has accelerated healthcare digitalization,[9] triggering growing demand for integrated care coordination solutions to enable quick decision-making. Physicians can now leverage digital and artificial intelligence solutions to automate data collection and translate it into useful information to make evidence based medical decisions.[10]

These trends have resulted in healthcare organizations striving to find solutions to reach the goals of improving clinical, patient, and financial outcomes while also addressing the well-being and engagement of healthcare employees.[11] A concrete example of these solutions is early warning systems that use algorithms to alert clinicians that a patient is deteriorating so they can intervene to prevent the acute deterioration and therefore prevent ICU (intensive care unit) admissions. These solutions support appropriate patient management throughout the different care areas and decrease in-hospital, unexpected mortality rates, therefore positively influencing financial outcomes.

Philips has adopted a solution-oriented approach in providing value to customers via integrated solution offerings. Within this approach, Philips defines a **Solution** as a combination of Philips (and third-party) systems, devices, software, consumables, and services, configured and delivered in a way that solves customer-(segment) specific needs.

---

6 See Health Workforce Trends from WHO (World Health Organization): www.who.int/health-topics/health-workforce#tab=tab_1

7 He, W., Goodkind, D., Kowal, P., (2016). *An Aging World: 2015 (pp. 6), essay, International Population Reports.*

8 Cordina, J., Kumar, R., Martin, C.P., and Jones, E.P. (2020, March 1). Healthcare consumerism 2018: An update on the journey. McKinsey & Company. https://www.mckinsey.com/industries/healthcare-systems-and-services/our-insights/healthcare-consumerism-2018.

9 Find more information here: hbr.org/2020/12/what-the-pandemic-means-for-health-cares-digital-transformation

10 World Economic Forum and The Boston Consulting Group (2017). (rep.). *Value in Healthcare Laying the Foundation for Health System Transformation.* Retrieved from http://www3.weforum.org/docs/WEF_Insight_Report_Value_Healthcare_Laying_Foundation.pdf

11 Bodenheimer, T., Sinsky, C. (2014). From Triple to Quadruple Aim: Care of the Patient Requires Care of the Provider, *Annals of Family Medicine* 12 (6): 573–576.

Varying Customer Needs and Different Solution Complexities

Solutions address the customer need to effectively maximize speed and consistency of clinical decisions, actions, and usage of patient information for reduced clinical variation and improved clinical performance within their IT ecosystem.

Designing and delivering **Solution Projects** is a local activity performed at hospital organizations in every country, often in the local language. Philips operates with both local and centralized resources to support this. This global/local organizational design often leads to virtual working environments with specific requirements to efficiently drive solution value creation. The requirements and maturity levels in each country, market and hospital customer greatly vary. Each project in a hospital is unique and varies in duration (from weeks to years), in size (up to multimillion euros/dollars) and in complexity (from stand-alone solution for one clinician to regional distributed solution for thousands of users). The range of size and complexity for Solution Projects in healthcare is broad, it includes simple products, highly configurable informatics systems, as well as software and services including technical and clinical consulting. It is influenced by different customer situations, demand, and existing and new technologies. Typical dimensions showing varying customer needs and requirements are as follows:

- From single department to multi-hospital deployment across country borders
- From standalone solutions in group practice or small departments to complex solutions with different systems, software, services fully integrated in the hospital infrastructure across multiple departments
- From simple clinical processes to highly designed workflows intended to optimize patient outcomes
- From "Greenfield" implementations across all modalities and applications to customized solutions into an existing hospital environment

Healthcare organizations around the globe have varying levels of maturity in terms of digital transformation. This can range from rural health clinics with limited infrastructure to integrated delivery networks with electronic health records, robust LAN/WAN network infrastructure, and Lean organizational culture to rapidly integrate technology innovations into clinical practice. This maturity level plays a large role in how successful healthcare customers are in achieving solution value creation, The Healthcare Information and Management Systems Society (HIMSS) has developed a Digital Health Indicator to guide healthcare organizations in assessing their level of maturity in terms of progress toward digital health ecosystem.[12]

Figure 10-2 gives an overview of the Complexity Drivers in healthcare projects.

The variability in customer needs drives the Solution Commercialization process. Important elements considered include scalable solution requirements in product and service design, solution delivery readiness, and quality of execution in markets.

When designing and delivering low complexity, single solution projects in one hospital department on a simple, stand-alone network, the sales account manager leads the definition of the solution deliverables, the project manager and the project team will implement basic delivery tasks. They include stakeholder identification, plan development, performing installation, controlling scope, and obtaining customer acceptance. When a high complexity solution is delivered within a health system, with many stakeholders and a variety of solutions, the Solution Design & Delivery model becomes much more detailed including standard work for the various Solution Project team members. The clinical specialist performs Discovery to uncover the clinical solution requirements. Based on these requirements, the Solution architect develops the Reference Architecture Specification and Solution Design. During the Solution Delivery phase, the project

12 Find more information on the HIMSS Digital Health Indicator here: www.himssanalytics.org/dhi.

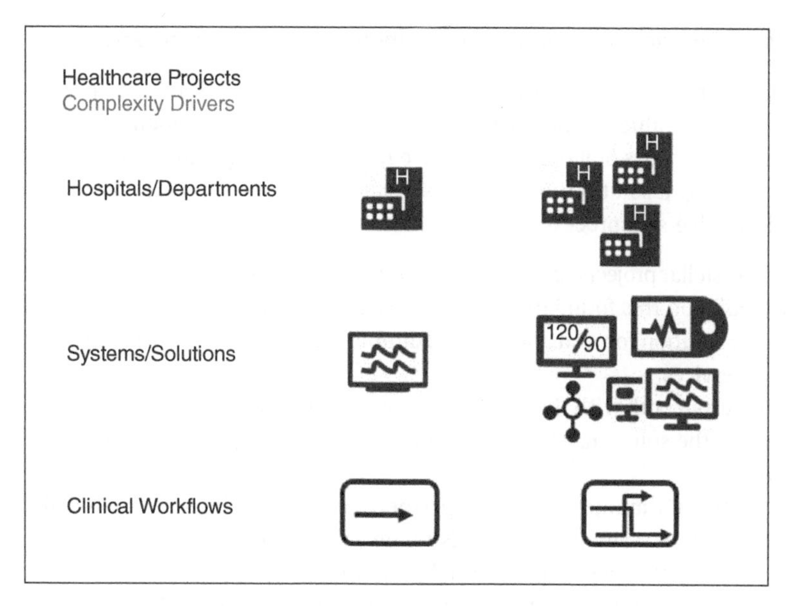

Figure 10-2 Healthcare Projects: Different Drivers Influence Complexity © Koninklijke Philips N.V., 2018. All rights reserved.

manager and multidisciplinary solution project team will then execute additional tasks from the five PMI process groups. These include performing a customer expectation analysis, developing a stakeholder RACI matrix, performing a workflow analysis, performing solution integration testing, controlling risk, cost and labor budgets, and conducting Lessons Learned reviews.

The different Complexity Drivers lead to different complexity levels (see Figure 10-3), which are broken into three different levels (low, medium, high).

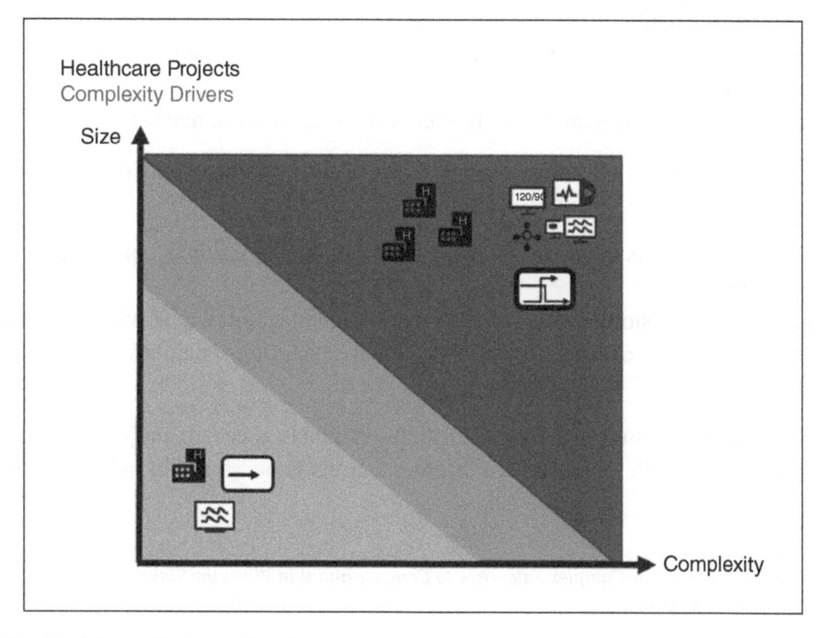

Figure 10-3 Healthcare Projects: Complexity Levels © Koninklijke Philips N.V., 2018. All rights reserved.

Implications for Solution Innovation Development and Commercialization

Solution Innovation Development and Commercialization processes are key enablers for effective and efficient Solution value creation. These processes support the development and launch of new solutions, define new ways of working and drive necessary changes to supporting infrastructure.

In order to develop a solution that includes systems, devices, software, consumables, and services, we follow a five-step process:

- First we rely on stellar project management for product and services realization and market launch.
- Second, we build on this foundation to add new capabilities (process, tool, content) as needed for a complete solution for a customer. An example is new ways of transacting business such as subscription services.
- Third, we underpin both with the human resource capabilities (skill) required to design, deploy, and implement the solution. This may include new roles, or new ways of working across the organization.
- Fourth, as we transform into a solutions partner, we are assessing our organizational structure and ensuring an E2E systems approach is followed. Customer experience is driving all of our decisions regarding changes to our internal processes and organization.
- Last but very important, the ability to measure customer success and the solution specific outcome needs to be fully embedded from the beginning.

All these points highlight the need for business transformation to evolve current ways of working, specifically:

1. A new process for Solution Commercialization
   - Drives customer solution requirements for consideration in product and service designs and the value it brings to the customer
   - Feeds solution specific content to the solution standard work framework
   - Considers diagnosing the customer problem, including the identification of specific KPIs, from pre-sales engagement with the customer[13]
   - Ensures continued alignment and execution on accepted requirements
   - Commercialized through a gated process that is scalable, repeatable, and leverages enabling realization processes
   - Ensures delivery readiness and quality execution of solution in markets
2. Underpinning Solution Projects with analytic capabilities to measure Customer Success and Outcomes
   - Defines adoption measures to understand customer consumption of solution, including volumes and utilization
   - Includes tracking customer interactions including management of service-level agreements
   - Facilitates translation of data into insights, enabling solution capabilities to be linked to customer outcomes

Due to the nature of Solution Innovation, the traditional aspects of project management are needed but are not enough. Focus on the human aspects is necessary to help impacted parties

13 Thull, J. (2010). *Mastering the Complex Sale: How to Compete and Win When the Stakes Are High, 2nd Edition*, Hoboken, NJ: John Wiley, pp. 87.

(internal, partners, customers) to transition through change. It helps ensure adoption and to sustain the change. Key to this is strong, aligned messaging, delivered by all levels of leadership, articulating business and customer benefits and the contribution to that benefit.

Another consideration is the trend toward recurring business models (e.g. subscriptions) as well as long-term partnerships with customers. This requires a different approach how to engage with the customer and how to operate on the supplier side. TSIA published the Land, Adopt, Expend, Renew (LAER) customer engagement framework, which could be seen as the industry reference for technology-as-a-service (XaaS) companies.[14] It describes how to shape holistic customer engagement models, enable desired customer outcomes and as a consequence, optimize financial results for the supplier too. While TSIA highlights the importance of the LAER model to recurring business models, we believe the focus on customer success and outcome realization is important in traditional business models too. This starts from the beginning with engagement with the customer. We need to consider early on how to analyze the value a Solution would bring to a customer and how to resolve their underlying complex problem.[15]

Solution Services and Capabilities

Solution projects need a specific combination of services, scoped and executed by a multidisciplinary team consisting of sales account managers, project managers, solution architects, technical consultants, clinical consultants, and field service engineers. Some of the solution-related services are specific to the Solution, some are more generic and independent of the Solution. For any Solution, it is a combination of both (see Figure 10-4 as overview).

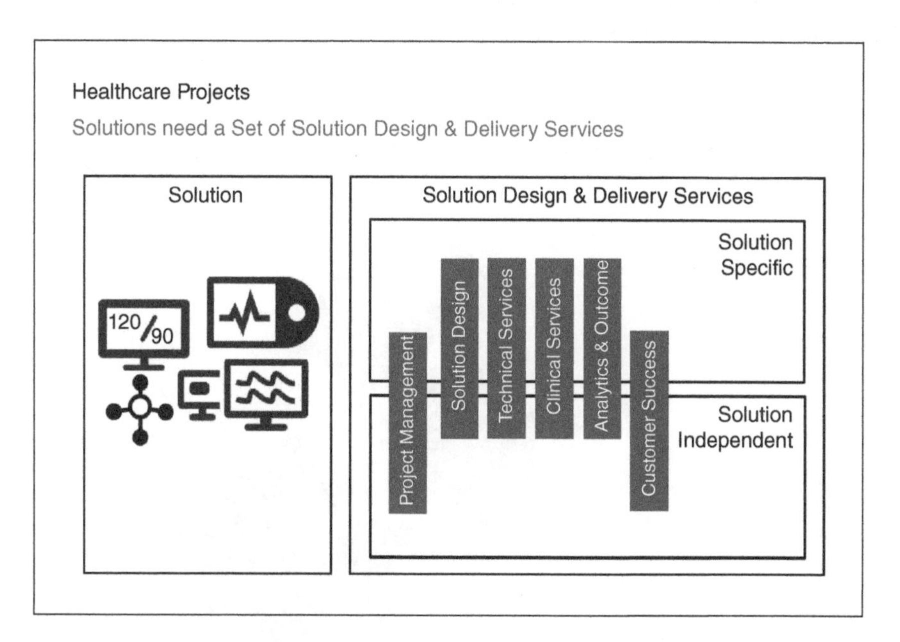

Figure 10-4 Solution Projects Need Set of Services and Capabilities © Koninklijke Philips N.V., 2021. All rights reserved.

14 TSIA, Frost, S. (2021). *The 7 Levels of LAER Transformation*, p. 3.
15 Thull, p. 117.

A set of capabilities are required to provide Solution Design & Delivery Services. Philips considers the following capabilities important for Solution Design & Delivery Services:

- **Skills:** Well-educated, certified, skilled (hard, soft), continuously trained, cross-functional project team out of sales account managers, solution architects, clinical transformation manager, projects managers, and professional services consultants with a professional mindset, appearance, and behavior. This also includes recruiting the best talent[16] and providing a common culture, career path and business environment that facilitates consistency across teams.
- **Processes/Methodology:** Highly efficient, standardized, lean, repeatable, and well-documented processes that are continuously improved.
- **Tools:** Highly integrated, automated, and efficient tools, templates, and applications from the project acquisition until the end of the project.
- **Content:** Role-specific content (templates, training material, diagrams).

Some of the Solution-related capabilities are specific to the Solution, some are more generic and independent of the Solution. For any Solution it is a combination of both. The Solution-specific capabilities are directly linked to the Solution Innovation. To be fully successful with Selling, Designing, & Delivering Solution Projects these capabilities need to be prepared, designed, and deployed to executing organizations in the countries.

With the objective for Solution Services Excellence, Customer Success, and Outcomes, we defined three types of integrating frameworks (see Figure 10-5 as overview):

- Solution Design & Delivery Services Framework
- Customer Success Management Framework
- Solution Analytics and Outcomes Framework

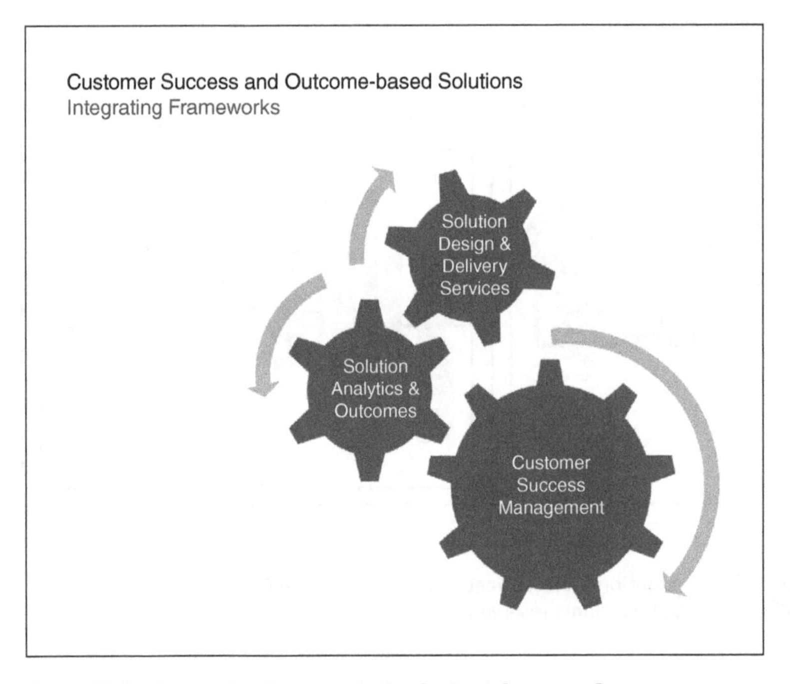

Figure 10-5 Integrating Frameworks for Optimal Customer Success

16 See as well for importance of Project Management talent management: PMI's Pulse of the Profession In Depth Study: Talent Management, March 2013.

Solution Design and Delivery Services Framework

The SOLiD Design & Delivery Framework has an integrator function to combine a Solution-specific and Solution-independent capabilities for any Solution, which enables every role to contribute successfully to the project (see Figure 10-6 as overview).

In close collaboration with the Philips Solutions & Services Community around the globe, the SOLiD Framework was developed. The SOLiD Framework is now the Philips solution approach for designing, managing, executing, and servicing customer facing Solution Implementation Projects and Services. SOLiD is an abbreviation and stands for:

- **S**calable, which allows flexibility to meet the demands of our low-, medium-, and high-complexity projects.
- **O**perationally agile, meaning a rapid, customer-focused development approach was utilized.
- **L**ean, only including the tasks that would add value to the Project & Services Team and even more important to hospital customers.
- **i**T focused, including the structure, tools and processes needed to successfully manage Projects & Services in an IT Solutions environment; and lastly, SOLiD will help to
- **D**eliver consistent results and bring business value by providing a standard & lean way of working.

The underpinnings of this framework are the process groups of Initiating, Planning, Executing, Monitoring/Controlling and Closing as defined in the PMBOK (Project Management Body of Knowledge) by the Project Management Institute (PMI).[17] Each process group is then further broken down into more specific processes and procedures detailing the implementation of Solution Projects & Services. All project team member roles are included in the Framework, along with the related activities each is responsible for during the Solution Design and Delivery phases. This definition enables the organization to deliver high quality implementations and includes a holistic approach to Solution Design and Testing. An important element of the approach is the definition of

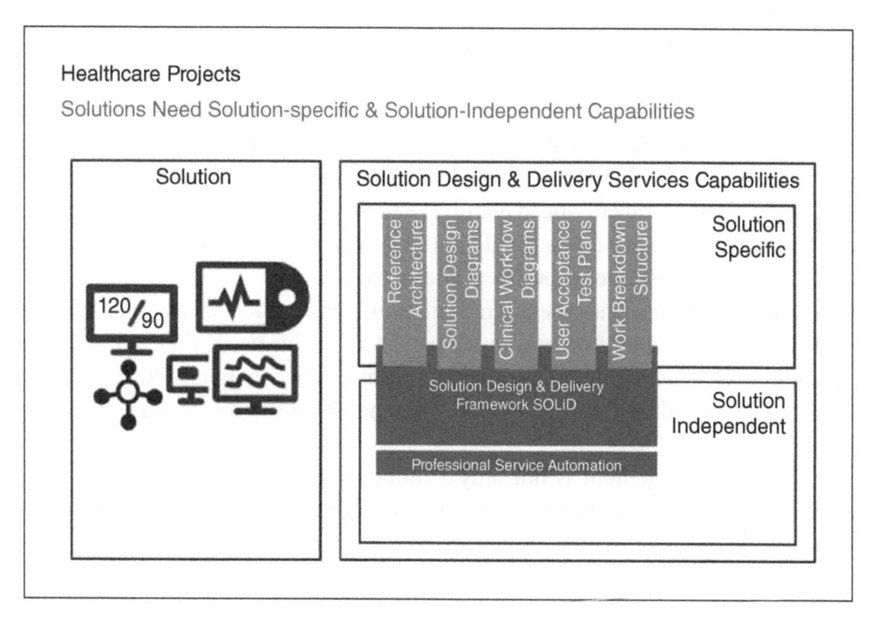

Figure 10-6 Solution Design and Delivery Services Need Set of Capabilities

17 See www.pmi.org/pmbok-guide-standards/foundational/pmbok for more information.

a customer Reference Architecture for the Solution. This serves as the vision for the solution where use models, application/configuration, and infrastructure elements are defined. The Reference Architecture Specification is utilized to develop the system design and define the solution test plan.

Scalability in Project Implementations is key to allow the right, flexible, agile, and efficient approach per project but to leverage from a rich tool set. Solution Projects are defined by their level of complexity. Typical factors when defining complexity are total cost of the project, number of team members involved, number and size of deliverables, complexity of deliverables and complexity of the customer environment and timeframes involved.

Customer Success Management Framework

Customer Success Management (CSM) starts in pre-sales and goes hand-in-hand with the Solution Design and Delivery but continues throughout the full Solution adoption along the entire lifecycle of the engagement with the customers. It could incorporate different engagements with the customer, e.g., multiple realized solutions like patient monitoring in an ICU (intensive care unit) or an X-ray system in a radiology department.

CSM focuses specifically on full Customer Success, adoption of the solution, and realization and optimization of clinical, technical, operational, financial value. TSIA provides a holistic Customer Success Capabilities Framework.[18] It starts with defining a Customer Success Strategy and then explains the realization from opportunity management, customer success delivery and operations, and analytics. TSIA incorporates Customer Experience (CXP) and Customer Journey Mapping into the CSM framework.

Management toward full Customer Success requires the creation and execution of Customer Success plan and regular follow-up and reviews with the customer. This is typically done from Customer Success Manager located in the country of the customer, here hospitals or healthcare organizations.

The Customer Success Manager requires a toolset that provides a holistic perspective on the plan, the related actions, their realization, and customer perspective. This might include different solutions in different hospital units or departments. Full visibility of all customer transactions is one aspect, this spans from opportunities, existing contracts, solution design and delivery programs, entitlements, to service incidents or change requests. More and more customers opt for software maintenance contracts which entitle the customer to install and use the latest software revision, usually with new and enhanced features, which helps to enable enhanced customer success. The smooth, fast, and easy operational handling of service incidents and software upgrades is very important for the customer experience and success.

Another key aspect is to measure the performance with various KPIs, providing insights into all engagements with the customer from a supplier perspective. This needs then to be combined with solution adoption and specific outcome realization from a customer perspective. We will see later how Philips focuses specifically on Customer Experience KPIs relative to Customer Success KPI and Outcome KPIs.

Customer Success Management is not only a framework, tool, set of KPIs, or a specific role (Customer Success Manager), rather it is a mindset and the adoption of Customer Success thinking and proactive behavior across the organization.

---

18 TSIA, Nanus, P. (2019). Is it time to rethink your customer success strategy? Technology & Services World San Diego, p. 29.

Solution Analytics & Outcomes Framework

A Solution Analytics & Outcomes Framework is foundational for effective Customer Success Management. Solution Analytics facilitate the improvement of customer clinical, operational, and financial outcomes by deriving insights from data. In healthcare this requires use of data science, analytics skills, and deep clinical expertise. Various hospital departments such as intensive care unit (ICU), operating room (OR), cardiac catherization lab, echocardiography lab, etc. measure operational and clinical key performance indicators (KPIs). Examples of operational KPIs include OR room utilization and volume of procedures per month.

The healthcare solutions supporting these departments aggregate data from medical devices and can integrate to other systems such as Laboratory Information System or Electronic Medical Record. These solutions can provide a wealth of data to support the management of the departmental KPIs. For example, Patient Management Solutions can provide insights into how well a nursing unit is managing patient alarms by analyzing the patient alarm data (i.e. alarms per patient bed by nursing unit).

The generic TSIA LAER model describes three levels of analytics in the Adopt phase:[19]

- Descriptive Analytics summarize current and past data sets.
- Predictive Analytics leverage statistical data analysis models and algorithms to identify data patterns and correlations to predict the future.
- Outcome Analytics support translating operational improvements into value for the customer.

Each of these Analytics elements supports expanding and eventual renewal of the solution.

Healthcare data analytics is very important to improve clinical, financial, operational outcomes. Few healthcare organizations today execute on the huge potential of data. It is a key element for the future transformation of healthcare systems including concepts like artificial intelligence (AI). Data is foundational for Outcomes KPIs and Value realization. In addition to Healthcare Solutions, Data Professional Services may be needed to translate the data into meaningful insights for the customer. Examples of these services include:

- Data Quality Assurance includes testing, validation, and cleansing of data sets.
- Data Reporting includes generation of reports based on specification.
- Data Visualization includes graphical representation of data in a way that can be easily consumed by end users.
- Privacy Impact Assessments help to identify the different legal requirements related to data privacy. Related services include data anonymization, data export to specific formats, and data migrations.
- Cybersecurity Assessment helps prevent potential data breaches, malware intrusions, and identifies system vulnerabilities.
- Data Governance includes design of a data strategy that is effective and useful. It will define where and how data is gathered, as well as the main purposes and research areas the institution would like to focus on.

Solution Services along the Customer Lifecycle

Philips strategized a fully integrated approach on how to offer and implement Solutions and Services from a process and methodology perspective. This is getting more important as the Philips

19 TSIA, Conolly, J. (2021). Client Management and Customer Success in Managed Services, *tsia interact*, p. 17.

portfolio transitions more and more into a Solutions and Services business. A more holistic approach is key for scoping and designing, delivering, and servicing Solutions at the customer throughout the Customer Lifecycle (see Figure 10-7 for graphical overview).[20] Philips considers here the technique of Customer Journey Mapping to define all the touchpoints the customer has with a business throughout the entire lifecycle. These touchpoints start from first contacts with the company, spans over pre-sales, solution discovery, solution design, the full lifecycle until the engagement with the company ends.[21] Those touchpoints happens via different channels; they could happen digital and personnel, with sales and service, or other representatives. Philips strives for long-term strategic partnerships with its customers; this means these engagements go deep and have the ambition to enable full Customer Success. Especially these long and deep engagements require the optimization of the Customer Journey Map toward full Customer Success and Outcomes.

The Customer Lifecycle begins with **Solution Discovery**. This involves intensive dialogue with the customer to fully understand the customer needs. It is followed by the Solution Design phase during pre-sales, where reference architectures and design guidelines help to shape a strong customer solution. This phase is essential for following Solution phases, it builds the real foundation. "Having a solid foundation is an essential element for delivering project excellence."[22] The work performed in the Solution Design phase is captured and documented into a SOW (statement of work), which is referenced throughout the rest of the project. McKinsey emphasizes the importance of technical and commercial capabilities: "Companies that invest in this capability are able to achieve win rates of 40 to 50 percent in new business and 80 to 90 percent in renewal business."[23] Following the **Solution Design** phase, a multiyear Solution Lifecycle & Success Plan is aligned with the customer. Then the **Solution Delivery** phase is executed to

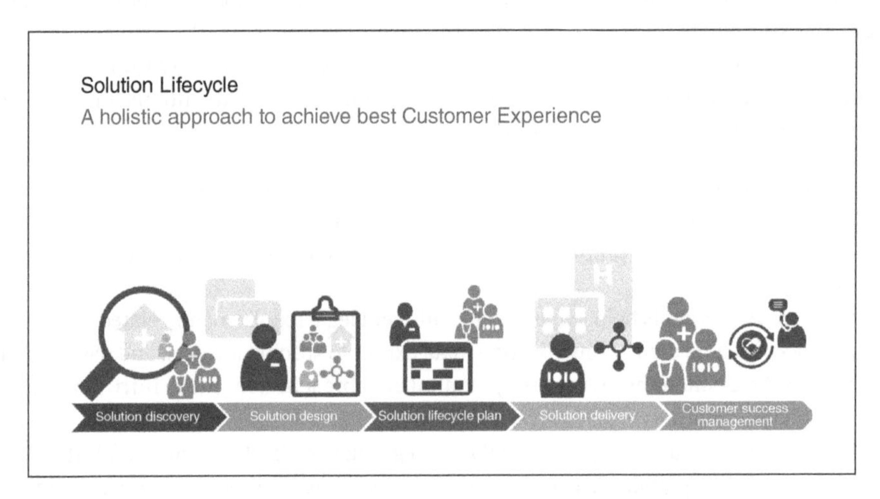

Figure 10-7 Solution Services along Customer Lifecycle © Koninklijke Philips N.V., 2021. All rights reserved.

20 Find more information about the Health Care Technology Life Cycle from the University of Vermont: its.uvm.edu/tsp

21 Find more information on Customer Journey Mapping here: www.visual-paradigm.com/guide/customer-experience/what-is-customer-journey-mapping

22 Martin, M. G. (2010). *Delivering project excellence with the statement of work*. Management Concepts Incorporated.

23 McKinsey & Company (2016 September). *Let's talk about sales growth* [Audio podcast].

implement the solution initially and additional services are provided over the lifecycle to fully create the customer value.

In the following **Customer Success Management** phase, the focus is to ensure that "achieve their desired outcomes while using your product or service."[24] This requires a deep and ongoing engagement and visibility into the customer processes, related to the product, system, solution, service provided or implemented. Continuous Customer Engagement is key for full success and enablement of the desired customer outcome (including continuous partnership and collaboration going forward).

Out of the entire Customer Lifecycle several key aspects need to be highlighted (see Figure 10-8 for the continuous customer engagement):

- **Solution Discovery**
  - Understand customer's clinical, technical, and operational requirements
  - Drive consensus with customer stakeholders on solution vision
  - Identify and define expected outcomes
  - Leverage diagnostics and analytics
- **Solution Design**
  - Technically feasible and implementable
  - Supportable by the provider and the customer
  - Financially transparent and profitable
  - Align with customer expectations

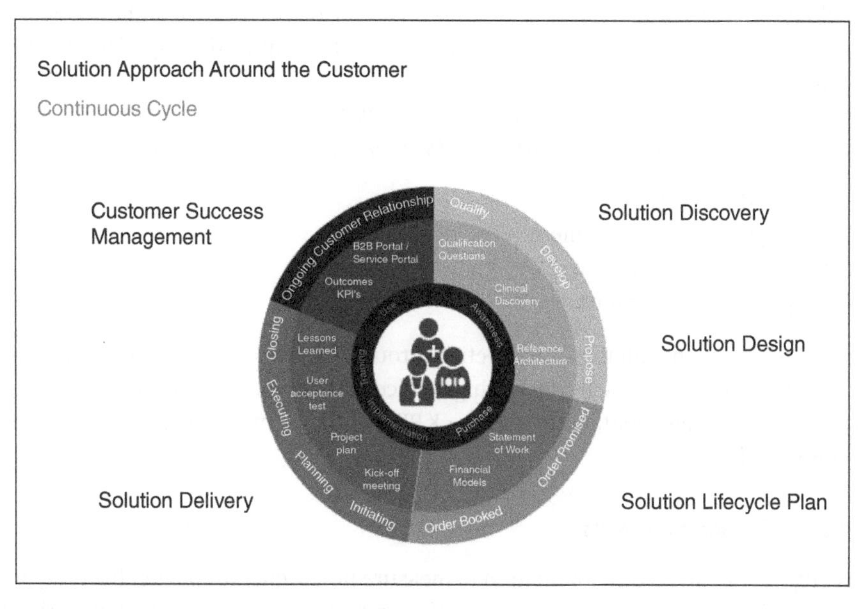

Figure 10-8 Continuous Engagement around the Customer © Koninklijke Philips N.V., 2021. All rights reserved.

24 Find overview for Customer Success Management here: www.gainsight.com/guides/the-essential-guide-to-customer-success

- **Solution Delivery**
  - Successful implementation, in line with what was scoped
  - Enable a lean and scalable project management approach
  - Provide the right tools to deliver an exceptional Customer Experience
  - Align way of working across all markets
  - Be service oriented internally and externally
- **Customer Success Management**
  - Set up customer for success and execute success plan
  - Drive solution adoption
  - Engage via well-structured customer journey toward their desired business outcomes

Customer Success in Solution Projects and How to Measure Its Value to Customers

Philips strives for long-term strategic partnerships with its customers, and this goes beyond "just" delivering traditional project deliverables within the constraints of time, cost, and scope. Designing and delivering true solutions for healthcare customers requires a deep engagement with them, including the analytic identification of the benefits, outcomes, and values – from a customer perspective.

Having that in mind, we see a need for stronger differentiation of measurements and KPIs beyond traditional perspectives and to develop toward full customer focused KPIs to show the full value creation and benefits in partnership with the customer. The time horizon is naturally longer than the solution design and delivery project itself.

Some of these outcomes based or Customer Success KPIs are directly linked into the Solution Innovation cycle and are very specific to the clinical domain; others are more generic. It is important to highlight that all of them are required to get a holistic perspective on how well a supplier performs and how much benefit the customer could realize out of the supplier's products, systems, solutions, or services.

In the following part we want to describe the evolution in customer-related KPIs we see from provider-centric KPIs to customer-centric KPIs (see Figure 10-9 as overview):

1. Service Level Management
2. Customer Experience Measurement
3. Customer Success Management
4. Customer Outcome Realization

It is important to say that all these perspectives around the customer are needed; they are complementary. This holistic approach considering Service Level Management, Customer Experience, Customer Success Management, and Outcome KPIs requires a much deeper understanding and analytic framework.

Service-Level Management KPIs

This is usually the "contractual" foundation to measure how well a supplier delivers against commitments. The ITIL (Information Technology Infrastructure Library) provides here a rich toolbox of standards and frameworks.[25] Depending on the setup, a variety of KPIs are possible.[26] Their focus is on successful execution of defined processes, and they are used for regular reviews with internal or external customers.

25 Find more information about the ITIL framework: www.axelos.com/best-practice-solutions/itil/what-is-itil
26 Find overview of ITIL related KPIs here: wiki.en.it-processmaps.com/index.php/ITIL_Key_Performance_Indicators

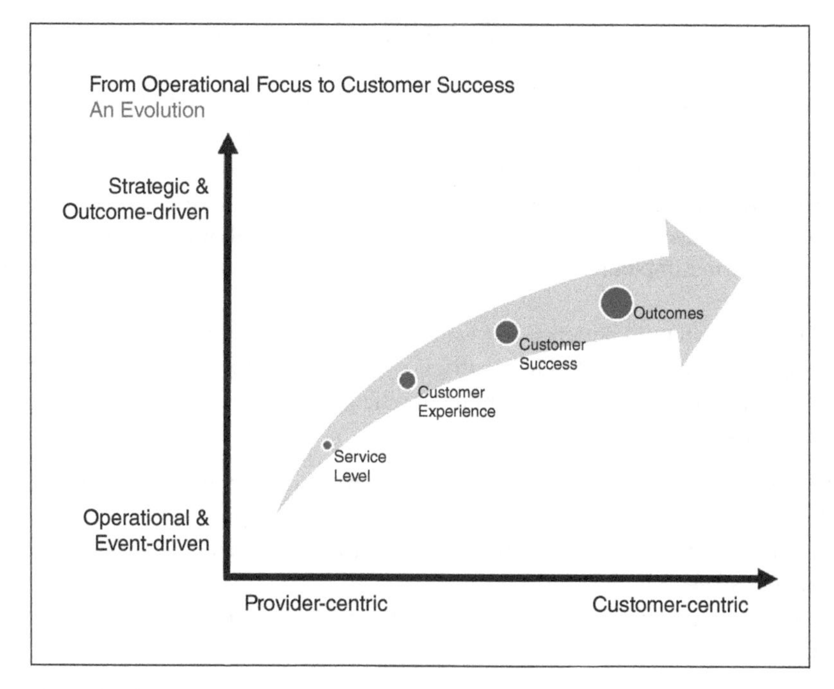

Figure 10-9 Evolution from Provider Centric to Customer Centric KPIs

Typical focus areas are availability of the contracted platform or services, operational performance of incident management (due to capacity or security incidents) or change processes of the provider. Typically, these KPIs are provider centric. They are important to the customer, but they don't show the benefits or outcome the customer has for their own processes.

Examples for service-level KPIs:

- System uptime %: Percentage of time that solution is available for use
- Preventative maintenance completed on time %: Includes system updates and patches applied within a specific timeframe defined in service level agreement
- Abandon rates in call centers

Customer Experience KPIs

Philips is conscious that each organization leaves an imprint with the customer, an experience made up of rational and emotional aspects that determines what healthcare customers associate with the Philips brand, and what Philips means to them. This is especially pronounced in a services business. Customer experience is at the heart of a relationship that translates into whether customers repeatedly rely on the organization's capabilities and embrace them as a trusted advisor.[27]

Therefore, another important aspect is how well the organization actively and holistically "designs" the customer experience end to end (E2E) in terms of all capabilities (e.g. tools,

27 Find more about customer experience concepts:, www.beyondphilosophy.com, www.qualtrics.com/experience-management/customer/customer-experience

processes, and skills) considering a full customer journey.[28] Philips strives to apply this customer experience–focused approach across the entire Customer Lifecycle from the point in time the customers share their vision through Solution Design, Delivery, and Continuous Engagement & Improvement.

In this context, Solutions Design & Delivery, Services Excellence is key to ensure Philips reliably and repeatedly delivers the desired customer experience. Hence, building and sustaining Project Solution Design & Delivery Services Excellence and reaching a high level of project management maturity with solution implementation projects is an ambition of vital importance for both, the customer and Philips.

Example for a Customer Experience KPI is the NPS (net promoter score).[29]

Philips requests customers provide feedback on their experiences with Installations (IQ), Services (SQ), and Relationship. All in conjunction help to get a good reflection from customers. Example questions from the survey include:

- Based on your Philips Implementation, how likely are you to recommend Philips? What is the reason for your score?
- Based on your most recent service event, how likely are you to recommend Philips? What is the reason for your score?
- How strongly do you prefer Philips as a strategic partner? How do you feel about your relationship with Philips?

Although CXP KPIs are very important to get insights on how a customer perceives its suppliers, it does not show the exact benefits or outcome the customer has for their own processes.

Customer Success Management KPIs

The concept of Customer Success Management shifts the perspective from the supplier to the customer. "Customer Success is the business methodology of ensuring customers achieve their desired outcomes while using your product or service."[30]

Customer Success Management requires actual visibility into the customer processes, related to the product, system, solution, and service provided or implemented. This requires a much deeper relationship between the supplier/provider and customer.

"Customer Success is when your customer achieves their Desired Outcome through their Interactions with your company."[31]It is all about that the customer realizes the desired outcome while using the supplier's offering of solutions. It links back to the pre-sales, the expectations and promises made, based on an analytic or diagnostic approach.[32]

Customer Success Management consists of a holistic framework, tools, customer facing roles, adoption plans, regular follow-up, and engagement with customers.

---

28 Find more information on Customer Journey Mapping here: www.visual-paradigm.com/guide/customer-experience/what-is-customer-journey-mapping

29 Find more information on NPS here: www.netpromoter.com/know

30 Find overview for Customer Success Management here: www.gainsight.com/guides/the-essential-guide-to-customer-success

31 Murphy, L. (2020): sixteenventures.com/customer-success-definition

32 Thull, p. 87.

TSIA highlights here in particular three KPIs with regards Customer Success Management:[33]

- Retention Rate
- Churn Rate
- Expansion Rate

Customer Outcomes Realization KPIs

Healthcare Solution Projects intend to address specific customer pain points and needs. The ultimate measure of success of a Healthcare Solution is achieving the expected Outcomes KPI targets. These may be clinical, operational, or financial related and are generally linked to the healthcare organization's mission and aligned with the quadruple aim[34]: patient satisfaction, staff satisfaction, better clinical outcomes, and financial sustainability.

Clinical outcomes may be difficult to corelate to a Healthcare Solution as there may be multiple factors affecting the Outcome measure, including patient population and acuity. It is important to measure KPIs that may be translated into one or several outcomes. For example, patient flow time can trigger actions that correlate to financial savings, additional surgical volumes and decreased patient wait time.

Examples of Outcome KPIs:

- Readmission Rate: patient readmissions that occur within 30 days of discharge
- Medication Errors per 1,000 Patients: inappropriate medication given to patient while under the care of a healthcare professional
- Healthcare Associated Infection Rate: infections contracted by patients while in a hospital being treated with invasive medical devices and/or medical procedures
- Healthcare staff satisfaction and turnover rates
- Nursing hours per patient day
- Patient NPS

Solution Excellence and Continuous Improvement for Customer Success

Philips strives for Solution Excellence, which requires continuous improvement around all services and capabilities. Even though it is not an absolute objective per se, it is considered a proactive way to anticipate and fulfill the needs of customers with regards solution project management, customer success and outcomes realization. The following aspects are critical to build and improve Solution-related Capabilities:

- **Solution Design & Delivery Services Excellence matters** – key aspect to value and improve capabilities (skills, processes, and tools). Some of the capabilities are solution specific, some capabilities are generic for any solution.
- **Process Harmonization and Standardization**[35] is highly important for the success of an organization operating globally and reducing complexity. Tight integration in the upstream processes (e.g., sales, bid management, analytics for outcome measurements) and downstream

---

33 TSIA, Nanus, P. (2020). *The Building Blocks of Customer Success*, p. 23.

34 Bodenheimer and Sinsky, pp. 573–576.

35 "High-performing organizations are almost three times more likely than low-performing organizations (36 percent vs. 13 percent) to use standardized practices throughout the organization and have better project outcomes as a result." Source: PMI's Pulse of the Profession (2013), The High Cost of Low Performance 2013 p. 10.

processes (e.g., entire lifecycle) are very important too. This includes Framework Integration as well.
- **Change management.** Identify, drive, and implement improvements and changes in the organization.
- **Continuous learning.** Train, review, and mentor all project team members as required.
- Facilitation of **Community of Practice** for all the different professions – key aspect to enable to Share, Learn, Leverage, Network, and Communicate.

The key takeaways for achieving Customer Success and Outcome-based Solutions could be summarized as follows:

- **Organizations striving** to find solutions to reach the goals of improving operational and financial **outcomes** for their customers (in our case healthcare).
- Holistic and fully integrated approach for the **Customer Lifecycle** is key for scoping, designing, delivering, and servicing Solutions for healthcare customers.
- A **Customer Journey Map** is a structured approach to review, understand, and design all touchpoints the customer has for best quality and outcome.
- With the ambition of **Solution Services Excellence, Customer Success and Outcomes** we see three integrated frameworks:
  - **Scalable and role-specific Solution Design & Delivery Framework** enables success for different project complexities and all Project team members.
  - **Customer Success Management Framework** includes a clear Customer Success Strategy with well managed execution along the multiyear engagement with the customer based on a structured tool set. It is key to create a customer centric culture and mindset within organization.
  - **Solution Analytics and Outcomes Framework** is important to translate data into analytic insights and to quantify solution value for customers.
- **Measuring Customer Success** requires evolving KPIs beyond traditional perspectives, which are usually supplier centric, and to develop toward full customer-focused KPIs to show the full sustainable value and benefits in partnership with the customer.

10.6 Metrics for Measuring Intangibles

Effective project management education, especially at the executive level, is now addressing long-term benefits and value resulting from not only the outcome of projects, but also from the way that projects were managed. Many of the benefits and value were the result of intangible asset creation that was difficult to measure. Fortunately, measurement techniques have advanced to the point where we believe that we can measure anything. Projects now have both financial and non-financial metrics, and many of the nonfinancial metrics are regarded as intangible metrics. An example of an intangible metric might be the effectiveness of governance, as shown in Figure 10-10.

The value of intangibles can have a greater impact on long-term considerations rather than short-term factors. Management support for the value measurements of intangibles can also prevent short-term financial considerations from dominating project decision-making. Measuring intangibles is dependent on management's commitment to the measurement techniques used. Measuring intangibles does improve performance, provided we have valid measurements free of manipulation.

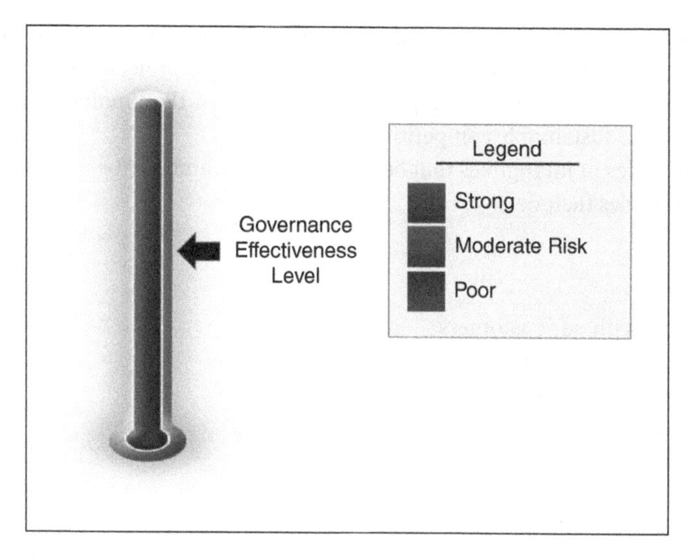

Figure 10-10 Governance Effectiveness

The following are often regarded as intangible project management assets and can be measured:

- Project management governance (Did we have proper governance, was it effective and did the governance personnel understand their roles and responsibilities?)
- Project management leadership (Did the project manager provide effective leadership?)
- Commitment (Is top management committed to continuous improvements in project management?)
- Lessons learned and best practices (Did we capture lessons learned and best practices?)
- Knowledge management (Are the best practices and lessons learned part of our knowledge management system?)
- Intellectual property rights (Is project management creating patents and other forms of intellectual property rights?)
- Working conditions (Are the people on the project teams satisfied with the working conditions?)
- Teamwork and trust (Are the people on the project teams working together as a team and do they trust one another's decisions?)

Businesses need to understand and create new sets of metrics and key performance indicators (KPIs) that can measure the intangibles and how they impact future decision-making and performance.

Even though most executives seem to understand the benefits of measuring intangibles, there is still resistance, such as:

- Intangibles are long-term measurements, and most companies focus on the short-term results.
- Companies argue that intangibles do not impact the bottom line.
- Companies are fearful of what the results will show.
- Companies argue that they lack the capability to measure intangibles.

The traditional metrics that we use, such as time, cost, and scope, may not capture the true business value of the deliverable of a project if the result is an intangible asset. Today, we live in a world where digitalization is becoming increasingly important. Also, significant advances are being made in measurement techniques that can be applied to projects. As such, the outcomes resulting from intangibles are becoming more important than the tangibles for the business to succeed and develop a sustainable competitive advantage.

There are many types of intangibles that benefit businesses and will be used as companies plan for the future. Examples include:

- Goodwill
- Customer satisfaction
- Our relationships with our customers
- Our relationship with our suppliers and distributors
- Brand image and reputation
- Patents, trademarks, and other intellectual property
- Our business processes
- Executive governance
- The company's culture and mindset
- Human capital, including retained knowledge and the ability to work together
- Strategic decision-making
- Strategy execution

Businesses need to understand and create new sets of metrics and KPIs that can measure the intangibles and how they impact future decision-making and performance.

For years, we had difficulty answering questions about intangibles:

- Are all corporate assets tangible, intangible, or both?
- Can we define intangible assets?
- Can the intangible assets be expressed in financial terms and their impact on the corporate balance sheet?
- Can intangible assets be measured?
- Are they value-added, and can we establish intangible project management metrics?
- Do they impact the performance of the organization in the future?

Intangible assets today are more than just goodwill or intellectual property. They also include maximizing human performance. Intangible assets include such items as corporate culture, intellectual capital, and the accompanying knowledge management system, executive and project leadership and governance, employee talent, employee satisfaction, trust and credibility, and workforce innovation capability. Understanding and measuring intangible asset value improves performance. While many of these may be hard to measure, they are not immeasurable.

10.7 The Need for Strategic Metrics

Because of advances in metric measurement techniques, models have been developed by which we can show the alignment of projects to strategic business objectives. One such model appears in Figure 10-11. Years ago, the only metrics we would use were time, cost, and scope. Today, we can include metrics related to both strategic value and business value. This allows us to evaluate the health of the entire portfolio of projects, as well as individual projects.

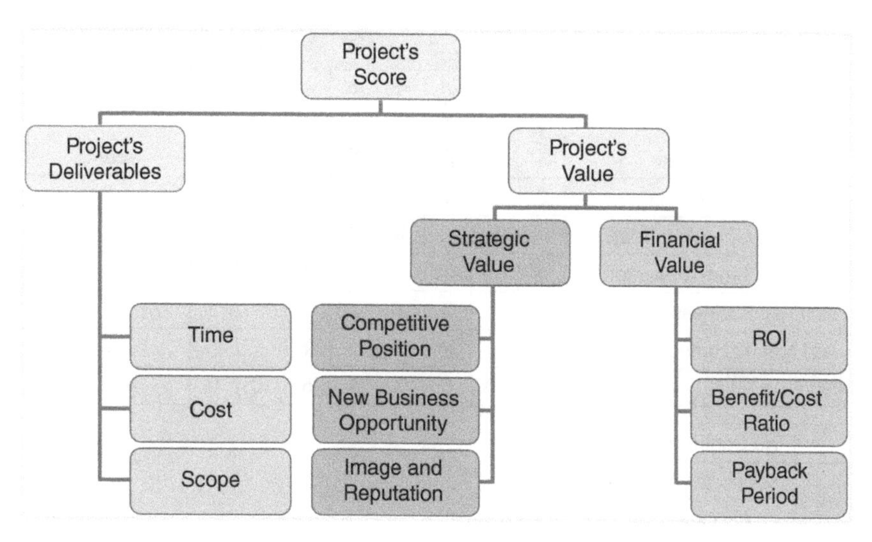

Figure 10-11 Project Scoring Model

Since all metrics have established targets, we can award points for each metric based on how close we come to the targets. The metrics selected help support strategic decisions that may be necessary. Figure 10-12 shows that the project identified in Figure 10-11 has received thus far 80 points out of a possible 100 points. Figure 10-13 shows the alignment of projects in the portfolio to strategic objectives. If the total score in Figure 10-12 is 0–50 points, we would assume that the project is not contributing to strategic objectives at this time, and this would be shown as a zero or blank cell in Figure 10-13. Scores of 51–75 points would indicate a "partial" contribution to the objectives and shown as a 1 in Figure 10-13. Scores of 76–100 points would indicate fulfilling the objectives and shown as a 2 in Figure 10-13. Periodically, we can summarize the results in Figure 10-13 to show management Figure 10-14, which illustrates our ability to create the desired benefits and final value.

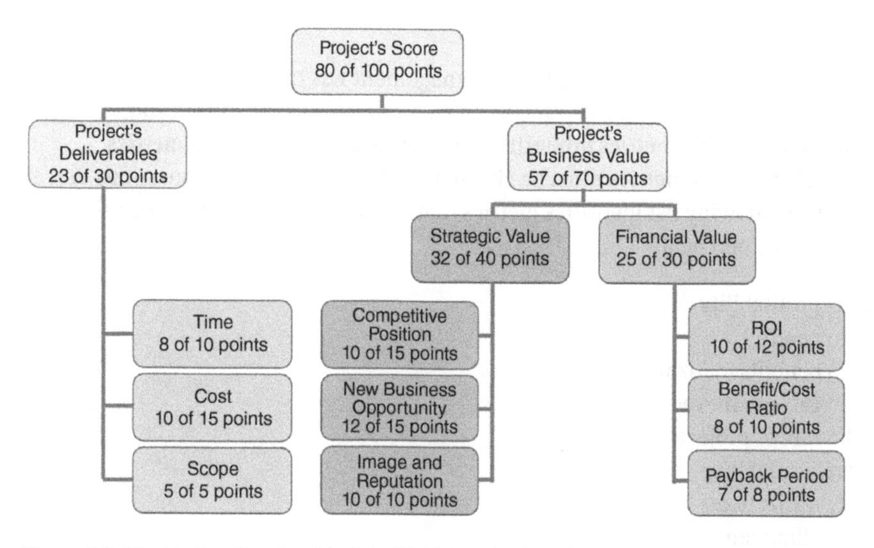

Figure 10-12 Project Scoring Model with Points Assigned

	Projects								
Strategic Objectives:	Project 1	Project 2	Project 3	Project 4	Project 5	Project 6	Project 7	Project 8	Scores
Technical Superiority	2		1			2		1	6
Reduced Operating Costs				2	2				4
Reduced Time-To-Market	1		1	2	1	1		2	8
Increase Business Profits			2	1	1	1		2	7
Add Manufacturing Capacity	1		2	2		1		1	7
Column Scores	4	0	6	7	4	5	0	6	

	No Contribution
1	Supports Objective
2	Fulfills Objective

Figure 10-13 Matching Projects to Strategic Business Objectives

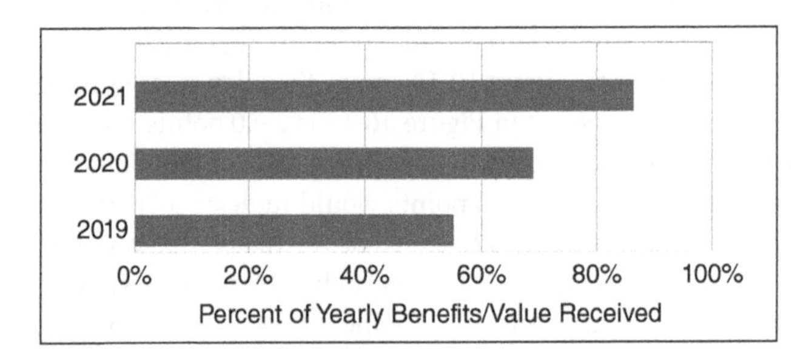

Figure 10-14 Periodic Benefits and Value Achieved

In previous pillars, we stated that project management has become a strategic competency and that most projects are aligned to strategic business objectives. To select and evaluate projects, there must exist strategic metrics other than value-driven and intangible metrics.

The strategic business metrics must be able to be combined to answer questions that executives might ask. The following list identifies metrics that executives need to make decisions concerning business and portfolio health.

- Business profitability
- Portfolio health
- Portfolio benefits realization
- Portfolio value achieved
- Portfolio mix of projects
- Resource availability
- Capacity utilization
- Strategic alignment of projects
- Overall business performance

Project managers will be required to provide project performance information that supports the use of these metrics. The project-based strategic business metrics must be able to be combined to create this list of metrics that executives need for business decision-making and strategic planning. Typical questions that senior management must address, which requires use of these metrics, include:

- Do we have any weak investments that need to be canceled or replaced?
- Must any of the programs and/or projects be consolidated?
- Must any of the projects be accelerated or decelerated?
- How well are the projects aligned to strategic objectives?
- Does the portfolio have to be rebalanced?
- Is value being created?
- Do we know the risks and how risks are being mitigated?
- Can we predict future corporate performance?
- Will we need to perform portfolio resource re-optimization?

10.8 Project Health Checks

Projects seem to progress quickly until they are about 60–70 percent complete. During that time, everyone applauds that work is progressing as planned. Then, perhaps without warning, the truth comes out and we discover that the project is in trouble. This occurs because of:

- Our disbelief in the value of using project metrics
- Selecting the wrong metrics
- Our fear of what project health checks may reveal

Some project managers have an incredible fixation with project metrics and numbers, believing that metrics are the Holy Grail in determining status. Most projects seem to focus heavily on only two metrics: time and cost. These are the primary metrics in all earned value measurement systems (EVMS). While these two metrics "may" give you a reasonable representation of where you are today, using these two metrics to provide forecasts into the future are gray areas and may not indicate future problem areas that could prevent a successful and timely completion of the project. The metrics used for health checks may be different from the traditional metrics being used for performance reporting. At the other end of the spectrum, we have managers who have no faith in the metrics and therefore focus on vision, strategy, leadership, and prayers.

Rather than relying on metrics alone, the simplest solution might be to perform periodic health checks on the project rather than waiting for serious issues to surface. In doing this, three critical questions must be addressed:

- Who will perform the health check?
- Will the interviewees be honest in their responses and not influenced by internal politics?
- Will management and stakeholders overreact to the truth?

The surfacing of previously unknown or hidden issues could lead to loss of employment, demotions, or project cancellation. Yet project health checks offer the greatest opportunity for early corrective action to save a potentially failing project by mitigating risks early. Health checks can also discover future opportunities as well as validating that the project is still aligned to strategic corporate objectives. It is essential to use the right metrics.

Understanding Project Health Checks

People tend to use audits and health checks synonymously. Both are designed to ensure successful repeatable project outcomes, and both must be performed on projects that appear to be heading for a successful outcome as well as those that seem destined to fail. There are lessons learned and best practices that can be discovered from both successes and failures. Also, detailed analysis of a project that appears to be successful might bring to the surface issues that show that the project is really in trouble.

Table 10-6 shows some of the differences between audits and health checks. Although some of the differences may be subtle, we will focus our attention on health checks.

Situation: During a team meeting, the project manager asks the team, "How's the work progressing?" The response is: "We're doing reasonably well. We are just a little bit over budget and a little behind schedule, but we think we have solved both issues by using lower salaried resources for the next month and having them work overtime. According to our enterprise project management methodology, our unfavorable cost and schedule variances are still within the threshold limits and the generation of an exception report for management is not necessary. The customer should be happy with our results thus far."

These comments are representative of a project team that has failed to acknowledge the true status of the project because they are too involved in the daily activities of the project. Likewise, we have project managers, sponsors, and executives who are caught up in their own daily activities and readily accept these comments with blind faith, thus failing to see the big picture. If an audit had been conducted, the conclusion might have been the same, namely that the project is successfully following the enterprise project management methodology and that the time and cost metrics are within the acceptable limits. A forensic project health check, on the other hand, may disclose the seriousness of the issues.

Just because a project is on time and/or within the allotted budget does not guarantee success. The end result could be that the deliverable has poor quality such that it is unacceptable to the

Table 10-6 Audits vs. Health Checks

Variable	Audit	Health Checks
Focus	On the present	On the future
Intent	Compliance	Execution effectiveness and deliverables
Timing	Generally scheduled and infrequent	Generally unscheduled and when needed
Items to be searched	Best practices	Hidden, possible destructive issues and possible cures
Interviewer	Usually someone internal	External consultant
How interview is led	With entire team	One-on-one sessions
Time frame	Short term	Long term
Depth of analysis	Summary	Forensic review
Metrics	Use of existing or standard project metrics	Special health check metrics may be necessary

customer. In addition to time and cost, project health checks focus on quality, resources, benefits, and requirements, just to name a few. The need for more metrics than we now use should be apparent. The true measure of the project's future success is the value that the customers see at the completion of the project. Health checks must therefore be value focused. Audits, on the other hand, usually do not focus on value.

Health checks can function as an ongoing tool by being performed randomly when needed or periodically throughout various life-cycle stages. However, there are specific circumstances that indicate that a health check should be accomplished quickly. These include:

- Significant scope creep
- Escalating costs accompanied by a deterioration in value and benefits
- Schedule slippages that cannot be corrected
- Missed deadlines
- Poor morale accompanied by changes in key project personnel
- Metric measurements that fall below the threshold levels

Periodic health checks, if done correctly and using good metrics, eliminate ambiguity such that true status can be determined. The benefits of health checks include:

- Determining the current status of the project
- Identifying problems early enough such that sufficient time exists for corrective action to be taken
- Identifying the critical success factors that will support a successful outcome or the critical issues that can prevent successful delivery
- Identifying lessons learned, best practices, and critical success factors that can be used on future projects
- Evaluating compliance to and improvements for the enterprise project management methodology
- Validating that the project's metrics are correct and provide meaningful data
- Identifying which activities may require or benefit from additional resources
- Identifying present and future risks as well as possible risk mitigation strategies
- Determining if the benefits and value will be there at completion
- Determining if euthanasia is required to put the project out of its misery
- The development of or recommendations for a fix-it plan

There are misconceptions about project health checks. Some of these are:

- The person doing the health check does not understand the project or the corporate culture, thus wasting time.
- The health check is too costly for the value we will get by performing it.
- The health check ties up critical resources in interviews.
- By the time we get the results from the health check, it is either too late to make changes or the nature of the project may have changed.

Who Performs the Health Check?

One of the challenges facing companies is whether the health check should be conducted by internal personnel or by external consultants. The risk with using internal personnel is that they may have loyalties or relationships with people on the project team and therefore may not be totally honest in determining the true status of the project or in deciding who was at fault.

Using external consultants or facilitators is often the better choice. External facilitators can bring to the table:

- A multitude of forms, guidelines, templates, and checklists used in other companies and similar projects
- A promise of impartiality and confidentiality
- A focus on only the facts and hopefully free of politics
- An environment where people can speak freely and vent their personal feelings
- An environment that is relatively free from other day-to-day issues
- New ideas for project metrics

Life-Cycle Phases

There are three life-cycle phases for project health checks:

1. Review of the business case and the project's history
2. Research and discovery of the facts
3. Preparation of the health check report

Reviewing the business case and project's history may require the health check leader to have access to proprietary knowledge and financial information. The leader may have to sign nondisclosure agreements and noncompete clauses before being allowed to perform the health check.

In the research and discovery phase, the leader prepares a list of questions that need to be answered. The list can be prepared from standard practices discussed in the *PMBOK® Guide*. The questions can also come from the knowledge repository in the consultant's company and may as templates, guidelines, checklists, or forms. The questions can change from project to project and industry to industry.

Some of the critical areas that must be investigated include:

- Performance against baselines
- Ability to meet forecasts
- Benefits and value analyses
- Governance
- Stakeholder involvement
- Risk mitigation
- Contingency planning

If the health check requires one-on-one interviews, the health check leader must be able to extract the truth from interviewees who have different interpretations or conclusions about the status of the project. Some people will be truthful, whereas others will either say what they believe the interviewer wants to hear or distort the truth as a means of self-protection.

The final phase is the preparation of the report. This should include:

- A listing of the issues
- Root cause analyses, possibly including identification of individuals that created the problems
- Gap analysis
- Opportunities for corrective action
- A get-well or fix-it plan
- New metrics that should be used for tracking

Project health checks are not "Big Brother Is Watching You" activities. Rather, they are part of project oversight. Without these health checks, the chances for project failure significantly increase. Project health checks also provide us with insight on how to keep risks under control. Performing health checks and taking corrective action early is certainly better than having to manage a distressed project.

10.9 Action Items

Too many companies are plagued with meeting mania where project managers go from meeting to meeting each day, often with little being accomplished. Meeting mania is a costly curse. Assume you are managing a one-year project that requires that the project manager attend 10 meetings a month. If each meeting has 15 participants in attendance, lasts for two hours, and has a fully loaded hourly labor rate of $200, then the yearly meeting costs for this project will exceed $700,000. If the company is working on 50 projects concurrently, the yearly meeting cost for all projects will exceed $35 million!! Effective preparation for meetings can significantly reduce this amount.

Meetings are conducted to share information, report performance, and make decisions. Unfortunately, many meetings end up with action items rather than with decisions. According to Wikipedia, the Free Encyclopedia, **action items** are usually created during a discussion by a group of people who are meeting about one or more topics and during the discussion it is discovered that some kind of action is needed. The action required is then documented as an action item and usually assigned to someone, usually a member of the group. The person to whom the action is assigned is then obligated to perform the action and report back to the group on the results.

Action items are usually documented in the meeting minutes and are recorded in the task list of the group. As people complete action items, the items are documented as being completed and the item is removed from the list of outstanding action items.

Many attributes can be associated with an action item, such as:

- **Identifier:** Unique mark to reference event or item
- **Description:** Brief explanation of activity to be performed
- **Work stream:** Business requirements, technical design, user interface, commit checklist, commit gate materials, etc. (optional)
- **Issue or Risk:** Associated with a project issue or risk
- **Status:** Open, In Progress, Resolved, Canceled
- **Urgency/Priority:** What is the impact to your project's critical path?
- **Comments:** Description of what is now being done to solve the issue
- **Owner:** Who is responsible for actively working the issue?
- **Created Date:** Date issue was opened
- **Planned completion date:** When will this issue be solved?
- **Actual completion date:** Date issue was closed

Action items often occur because the people in attendance at the meeting may not have the authority to resolve the issues. One way to partially resolve this decision-making problem is to find out which people can and cannot make decisions for their respective functional group. This should be done at the onset of every project. If some people are not authorized to make decisions, then their functional managers would be invited to attend the meetings. Some functional managers might ask when during the meeting their involvement would be essential because they do not want

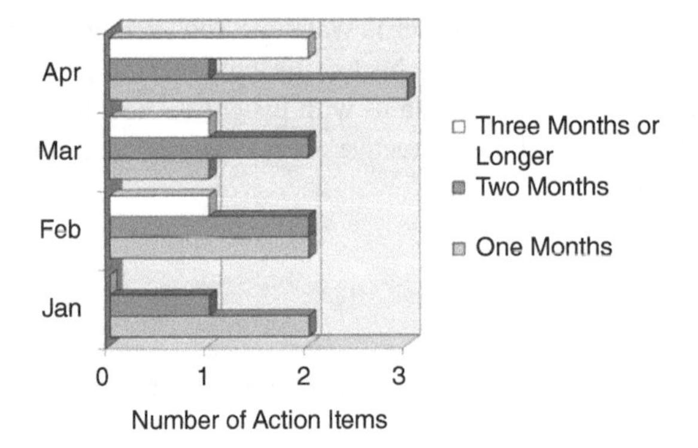

Figure 10-15 Open Action Items

to attend a two-hour meeting where only 15 minutes would involve their functional area. A detailed meeting agenda can make it easier for functional managers to know exact what time to attend.

While meeting agendas and knowing which team members possessed decision-making authority for their functional areas can reduce some meeting mania, there is still the issue of action items that require customer involvement and customer decision-making. Meetings that involve customers are significantly costlier than just internal meetings because of travel time, airfare, meals, and lodgings. Most of the meetings with the clients involve handouts as discussion points. Giving the client a copy of the handouts at the beginning of the meeting and then expecting them to understand everything rapidly and be prepared to make an on-the-spot decision is unrealistic. This creates more action items and quite often more travel costs than budgeted for. The solution might be to send the client a copy of the handouts at least a week before the meeting. This allows them time to digest the material and be prepared to make decisions.

Corporate governance is now asking project managers to prepare a metric that states how long certain action items have been in the system and have not been resolved. This is shown in Figure 10-15. In some cases, unresolved action items may reflect poorly on the project manager's leadership ability.

10.10 Failure of Traditional Metrics and KPIs

While some people swear by metric and KPIs, there are probably more failures than success stories. Typical causes of metric failure include:

- Performance being expressed in traditional or financial terms only
- The use of measurement inversion; using the wrong metrics
- No link of performance metrics to requirements, objectives, and success criteria
- No link to whether the customer was satisfied
- Lack of understanding as to which metrics indicate project value
- No feedback from customers on value-in-use

Metrics used for business purposes tend to express all information in financial terms. Project management metrics cannot always be expressed in financial terms. Also, in project management we often identify metrics that cannot effectively predict project success and/or failure and are not linked to the customer's requirements.

10.11 Establishing a Metrics Management Program

The future of project management must include metrics management. What many people neglect to consider is that the selection of metrics is often based on the constraints imposed on the innovation team:

- Having too few constraints gives the innovation team too much freedom and management may not be able to track performance.
- Having too many constraints may lead to severe limitations.
- The imposed constraints must be supported by metrics that are relatively easy to measure.
- Constraints (and the accompanying priority) can change over the life cycle of a project, thus causing changes in the metrics selected.

We can now identify certain facts about metrics management:

- You cannot effectively promise deliverables to a stakeholder unless you can also identify measurable metrics.
- Good metrics allow you to catch mistakes before they lead to other mistakes.
- Unless you identify a metrics program that can be understood and used, you are destined to fail.
- Metrics programs may require change, and people tend to dislike change.
- Good metrics are rallying points for the project management team and the stakeholders.
- There are also significant challenges facing organizations in the establishment of some business metrics, such as value-based metrics.
- Project risks and uncertainties may make it difficult for the project team to identify the right metrics and perform effective measurement.
- The more complex the project, the greater the difficulty is in establishing meaningful metrics.
- Competition and conflicting priorities among projects can lead to havoc in creating a metrics management program.
- Added pressure by management and the stakeholders to reduce the budget and compress the schedule may have a serious impact on metric selection.

Metric management programs must also consider relationships with suppliers and contractors as well:

- Suppliers, contractors, and stakeholders must understand the metrics being used.
- The metrics must be in a form that all parties can understand to take effective corrective action if needed.
- Some decision-making metrics must be sufficiently detailed rather than high level.
- If possible, the metrics should be real-time metrics such that governance personnel can react rapidly to changing market conditions.
- Enough metrics must be selected such that changes in the marketplace can be determined.
- A combination of metrics may be necessary to understand the marketplace and competitive positioning. One metric by itself may not suffice.
- For partnerships and joint ventures, knowledge transfer metrics must be established and tracked throughout the project life cycle to validate that the alliance is working well.
- Metrics allow us to validate that we have consensus building and systematic planning cooperation, rather than complexity and rigidity.
- Terminating alliances can be costly.

Metric management programs must be cultivated. Some facts to consider in establishing such a program include:

- There must be an institutional belief in the value of a metrics management program.
- The belief must be visibly supported by senior management.
- The metrics must be used for informed decision-making.
- The metrics must be aligned with corporate objectives as well as project objectives.
- People must be open and receptive to change.
- The organization must be open to using metrics to identify areas of and for performance improvement.
- The organization must be willing to support the identification, collection, measurement, and reporting of metrics.

Best practices and benefits can be identified because of using metrics management correctly and effectively. Some of the best practices include:

- Confidence in metrics management can be built using success stories.
- Displaying a "wall" of metrics for employees to see is a motivational force.
- Senior management support is essential.
- People must not overreact if the wrong metrics are occasionally chosen.
- Specialized metrics generally provide more meaningful results than generic or core metrics.
- The minimization of the bias in metrics measurement is essential.
- Companies must be able to differentiate between long-term, short-term, and lifetime value.

10.12 Conclusion

Major changes to project management will continue to occur in the coming decade due to new material and thought leadership that PMI and other professional organizations have been advocating in the Standards for Project Management and the strategic role that project management will play in the accompanying worldwide growth.

Most of the critical future of work changes that are expected to happen through this next decade can be clustered into themes covered in this book that will encompass the proposed 10 pillars of the project management future. Themes emerge around project management as a strategic vehicle, changes in the role of project managers, the criticality of the project economy, and the correlation between transparency and autonomy in delivery cultures and business excellence.

The connected future of business is human, projectized, and digitized. It is the authors' wish that this book provides a foundational starting point enabling executives and professionals to jointly transform the future of work and to discover the unlimited strategic value that project management provides as the vehicle for driving sustainable change and organizational excellence.

About the Authors

Harold Kerzner, PhD, MS, MBA

Dr. Harold Kerzner is Senior Executive Director for Project Management for the International Institute for Learning (IIL). He has taught engineering and business administration at three universities. Dr. Kerzner has traveled around the world conducting project management lectures and delivering keynote speeches. A globally recognized expert on management practices, he is the author of several books on project management, including *Project Management: A Systems Approach to Planning, Scheduling and Controlling* and *Innovation Project Management*.

Dr. Al Zeitoun, PhD, PgMP

Dr. Al Zeitoun is a Future of Work, business optimization, and operational performance excellence thought leader with global experiences in strategy execution. He holds five advanced PMI Certifications and his experiences encompass leading organizations; delivering their Enterprise Transformation as with Booz Allen Hamilton; and executing complex missions such as in the $24B UAE Nuclear program. Dr. Zeitoun champions the program and project management profession and authored influential publications on future business topics, including for Siemens and for Duke CE's Dialogue. He served on the PMI Global Board of Directors for five years, delivered 100+ global keynotes, leads key panels for IIL and DIPMF, and teaches the edX/University of Maryland Professional Certificate in *Program Management: Enabling Value Driven Change*.

Dr. Ricardo Viana Vargas, PhD, PMP

Dr. Ricardo Vargas is a chief advocate of the project economy. He has specialized in implementing global initiatives, capital projects, and product development, managing more than $20 billion in international projects in the past 25 years. He has written 15 books in the field. Ricardo also created and led the Brightline Initiative from 2016 to 2020 and was the director of project management and infrastructure at the United Nations, leading more than 1,000 projects in humanitarian and development projects. He has shared his expertise and passion with millions of professionals around the globe through "5 Minutes Podcast" with more than 11 million views since 2007, and his books have been translated into six languages. Dr. Vargas was the first Latin American elected as chair of the Project Management Institute. He has also directed the Brightline Initiative, a Project Management Institute think tank, dedicated to bridging the gap between strategic design and delivery.

Index

B

C

D

G

H

J

K

L

R

S

T

W